Pion Production and Absorption in Nuclei—1981
(Indiana University Cyclotron Facility)

Site of the workshop:
Canyon Inn, McCormick's Creek State Park, Spencer, Indiana
Photo by Kent Berglund

AIP Conference Proceedings
Series Editor: Hugh C. Wolfe
Number 79

Pion Production and Absorption in Nuclei—1981
(Indiana University Cyclotron Facility)

Edited by
Robert D. Bent
Indiana University

American Institute of Physics
New York   1982

Copying fees: The code at the bottom of the first page of each article in this volume gives the fee for each copy of the article made beyond the free copying permitted under the 1978 US Copyright Law. (See also the statement following "Copyright" below). This fee can be paid to the American Institute of Physics through the Copyright Clearance Center, Inc., Box 765, Schenectady, N.Y. 12301.

Copyright © 1982 American Institute of Physics

Individual readers of this volume and non-profit libraries, acting for them, are permitted to make fair use of the material in it, such as copying an article for use in teaching or research. Permission is granted to quote from this volume in scientific work with the customary acknowledgment of the source. To reprint a figure, table or other excerpt requires the consent of one of the original authors and notification to AIP. Republication or systematic or multiple reproduction of any material in this volume is permitted only under license from AIP. Address inquiries to Series Editor, AIP Conference Proceedings, AIP.

L.C. Catalog Card No. 82-70678
ISBN 0-88318-178-9
DOE CONF- 8110152

Proceedings of the

Workshop on Pion Production

and Absorption in Nuclei

Indiana University Cyclotron Facility

Bloomington, Indiana

October 22-24, 1981

PREFACE

Pion production and absorption in nuclei, a high momentum transfer process involving short range interactions, bears on many fundamental questions of current interest in intermediate-energy nuclear physics. During the past several years, substantial experimental advances have been made, particularly in studies of the (p,π) and (π,p) reactions. Theoretical developments during this period, however, have not kept pace with the experiments. With the recent commissioning of a new pion spectrograph at IUCF, plans for possible new spectrographs at other laboratories and the consequent impending experimental programs, the time seemed right for a workshop on (p,π) reactions and related topics to take stock of the present experimental and theoretical situation. It was hoped that such a workshop would stimulate theoretical analyses of existing data and provide guidance for future experimental programs.

The above considerations led to the workshop on Pion Production and Absorption in Nuclei which was held at the Canyon Inn in McCormick's Creek State Park near Bloomington, Indiana from 22-24 October 1981. The purposes of this workshop were: to examine the fundamental questions that underlie current theories of nuclear pion production; to review recent experimental developments; to explore current models of pion production, the connections among different models and the successes and failures of the models in explaining existing data; and to stimulate new theoretical developments and ideas for crucial experiments.

A site was chosen for the workshop which would be both pleasant and conducive to the free and informal exhange of ideas among persons actively interested in the subject of the workshop. Ample time was provided in the program for discussion both during and outside the formal scientific sessions, and speakers and participants were encouraged to make use of posters, which were displayed in a room next to the lounge where people congregated for refreshments and discussion after the evening sessions.

The manuscripts are grouped here as they were delivered at the workshop. The discussion that occurred during the scientific sessions was recorded on magnetic tape and most of this is included in the proceedings, except for that which occurred during the two panel discussions; the latter is not reproduced verbatim but is included in the proceedings as summarized by D. Koltun and H.W. Fearing.

I would like to thank the members of the Scientific Program Committee (A.D. Bacher, M. Dillig, W. Gibbs, B. Hoistad, H. Meyer, P.P. Singh and G.E. Walker) for their advice and help in planning the program of the workshop, the members of the Local Organizing Committee (R. Claussen, C. Foster, D. McGovern, J. Thornburg, P. Thompson and R. Westerfield) for their cheerful and proficient handling of all of the local arrangements, and the session chairmen (R.R. Silbar, H. Toki, E.G. Auld, J.V. Noble, and M.M. Sternheim) for their tact and good humor in keeping physicists from talking too long and conducting the lively discussions in an orderly way so that they could be recorded.

My special thanks go to the IUCF secretaries Diana McGovern, Penny Lepley and Becky Westerfield for their dedication in accomplishing the arduous task of transcribing the discussions from magnetic tapes. The cooperation of the participants in editing their comments promptly was appreciated.

Most of all, I am grateful to the speakers for their excellent contributions to the workshop and for providing written versions of their talks, and to the participants for their active role in making the workshop stimulating and worthwhile.

The financial support of the workshop provided by the National Science Foundation, the Department of Energy, and Indiana University, and the staff needed to run the workshop and produce the proceedings provided by the Indiana University Cyclotron Facility are gratefully acknowledged.

Bloomington, Indiana   Robert D. Bent
January, 1982

SPONSORS

U.S. National Science Foundation

U.S. Department of Energy

Indiana University Cyclotron Facility

Department of Physics, Indiana University

Indiana University Foundation

SCIENTIFIC PROGRAM COMMITTEE

A.D. Bacher	IUCF	M.H. Macfarlane	Indiana
R.D. Bent (Chairman)	IUCF	D.F. Measday	UBC
M. Dillig	Erlangen	H.O. Meyer	IUCF
W.R. Gibbs	LASL	P.P. Singh	IUCF
B. Höistad	Texas	G.E. Walker	Indiana

LOCAL ORGANIZING COMMITTEE

R. Claussen	J. Thornburg
C. Foster	P. Thompson
D. McGovern	R. Westerfield

SESSION CHAIRMEN

E.G. Auld	R.R. Silbar
R.D. Bent	M.M. Sternheim
J.V. Noble	H. Toki

CONFERENCE PARTICIPANTS

L. ANTONUK, University of Maryland, College Park, MD, USA
E. ASLANIDES, Universite Louis Pasteur, Strasbourg, Cedex, France
E. AULD, University of British Columbia, Vancouver, B.C., Canada
A. BACHER Indiana University, Bloomington, IN, USA
M. BANERJEE, University of Maryland, College Park, MD, USA
P. BARNES, Carnegie Mellon University, Pittsburgh, PA, USA
W. BENENSON, Michigan State University, East Lansing, MI, USA
R. BENT, Indiana University, Bloomington, IN, USA
M. BETZ, TRIUMF, Vancouver, B.C., Canada
C.-Y. CHEUNG, Carnegie-Mellon University, Pittsburgh, PA, USA
B. CLARK, Ohio State University, Columbus, OH, USA
J. CLARK, Los Alamos National Laboratory, Los Alamos, NM, USA
M. CLOVER, Los Alamos National Laboratory, Los Alamos, NM, USA
F. COESTER, Argonne National Laboratory, Argonne, IL, USA
J. CONTE, Indiana University, Bloomington, IN, USA
T. COOPER, Indiana University, Bloomington, IN, USA
P. COUVERT, Centre D'Etudes Nucléaires, Saclay, France
V. CUPPS, Indiana University, Bloomington, IN, USA
M. DILLIG, Universität Erlangen-Nürnberg, West Germany
J. EISENBERG, Tel Aviv University, Tel Aviv, Israel
G. EMERY, Indiana University, Bloomington, IN, USA
W. FALK, University of Manitoba, Winnipeg, Canada
H. FEARING, TRIUMF, Vancouver, B.C., Canada
C. FOSTER, Indiana University, Bloomington, IN, USA
M. GALLIO, I.N.F.N. Sezione Di Torino, Torino, Italy
J.-F. GERMOND, Institut de Physique, Neuchatel, Switzerland
W. GIBBS, Los Alamos National Laboratory, Los Alamos, NM, USA
G. GILES, University of British Columbia, Vancouver, B.C., Canada
C. GLOVER, Indiana University, Bloomington, IN, USA
C. GOODMAN, Indiana University, Bloomington, IN, USA
M. GREEN, Indiana University, Bloomington, IN, USA
N. GRION, Instituto di Fisica, Univ. of Trieste, Trieste, Italy
R. HACKMAN, Case Western Reserve University, Cleveland, OH, USA
J. HALL, Indiana University, Bloomington, IN, USA
B. HÖISTAD, University of Texas at Austin, Austin, TX, USA
M. HUBER, Universität Erlangen-Nürnberg, Erlangen, West Germany
P. HWANG, Indiana University, Bloomington, IN, USA
M. HYNES, Los Alamos National Laboratory, Los Alamos, NM, USA
J. IQBAL, Indiana University, Bloomington, IN, USA
W. JACOBS, Indiana University, Bloomington, IN, USA
D. JENKINS, Virginia Poltechnic University, Blacksburg, VA, USA
J. JOHNSTONE, Univ. of British Columbia, Vancouver, B.C., Canada
G. JONES, University of British Columbia, Vancouver, B.C. Canada
W. JONES, Indiana University, Bloomington, IN, USA
J. KEHAYIAS, Indiana University, Bloomington, IN, USA
B. KEISTER, Carnegie-Mellon University, Pittsburgh, PA, USA
L. KISSLINGER, Carnegie-Mellon University, Pittsburgh, PA, USA

D. KOLTUN, University of Rochester, Rochester, NY, USA
Y. LE BORNEC, Institute de Physique Nucléaire, Orsay, France
D. LICHTENBERG, Indiana University, Bloomington, IN, USA
K.-F. LIU, University of Kentucky, Lexington, KY, USA
L.-C. LIU, Los Alamos National Laboratory, Los Alamos, NM, USA
G. LOLOS, University of British Columbia, Vancouver, B.C., Canada
T. LONDERGAN, Indiana University, Bloomington, IN, USA
L. LUDEKING, Indiana University, Bloomington, IN, USA
H. MCMANUS, Michigan State University, East Lansing, MI
H.-O. MEYER, Indiana University, Bloomington, IN, USA
G. MILLER, University of Washington, Seattle, WA, USA
D. MILLER, Indiana University, Bloomington, IN, USA
H. NANN, Indiana University, Bloomington, IN, USA
J. NISKANEN, TRIUMF, Vancouver, B.C., Canada
J. NOBLE, University of Virginia, Charlottesville, VA, USA
C. OLMER, Indiana University, Bloomington, IN, USA
J. O'NEILL, Carnegie-Mellon University, Pittsburgh, PA, USA
D. OTTEWELL, TRIUMF, Vancouver, B.C., Canada
M. PICKAR, Indiana University, Bloomington, IN, USA
K. PITTS, Indiana University, Bloomington, IN, USA
R. POLLOCK, Indiana University, Bloomington, IN, USA
R. RIEDER, Carnegie-Mellon University, Pittsburgh, PA, USA
E. RÖSSLE, Universität Freiburg, Freiburg, Fed. Rep. Germany
M. SABER, Indiana University, Bloomington, IN, USA
A. SAHARIA, University of Rochester, Rochester, NY, USA
H. SARAFIAN, Michigan State University, East Lansing, MI, USA
G. SCHMIDT, Universität Karlsruhe, Karlsruhe, West Germany
W. SCHOTT, Technische Universität München, Garching, Germany
P. SCHWANDT, Indiana University, Bloomington, IN, USA
R. SILBAR, Department of Energy, Washington, DC, USA
P. SINGH, Indiana University, Bloomington, IN, USA
F. SOGA, Inst.for Nucl. Study, University of Tokyo, Tokyo, Japan
E. STEPHENSON, Indiana University, Bloomington, IN, USA
M. STERNHEIM, University of Massachusetts/Amherst, Amherst, MA, USA
D. STORM, University of Washington, Seattle, WA, USA
K. STRICKER-BAUER, University of Maryland, College Park, MD, USA
A. THOMAS, TRIUMF, Vancouver, B.C., Canada
T. THROWE, Indiana University, Bloomington, IN, USA
G. TOKER, University of Pittsburgh, Pittsburgh, PA, USA
H. TOKI, Michigan State University, East Lansing, MI, USA
Y. TZENG, University of Texas, Austin, TX, USA
R. VAN DANTZIG, NIKHEF-K, Amsterdam, The Netherlands
S. VAN DER WERF, Univ. of Groningen, Groningen, The Netherlands
J. VAN ORDEN, University of Maryland, College Park, MD, USA
J. VARY, Iowa State University, Ames,, IA, USA
S. VIGDOR, Indiana University, Bloomington, IN, USA
V. VIOLA, Indiana University, Bloomington, IN, USA
P. WALDEN, TRIUMF, Vancouver, B.C., Canada

G. WALKER, Indiana University, Bloomington, IN, USA
T. WARD, Indiana University, Bloomington, IN, USA
H. WEBER, University of Virginia, Charlottesville, VA, USA
W. WHARTON, Carnegie-Mellon University, Pittsburgh, PA, USA
N. WILLIS, Institut de Physique Nucléaire, Orsay, France
J. WILLS, Indiana University, Bloomington, IN, USA
W. ZIEGLER, Univ. of British Columbia, Vancouver, B.C., Canada

TABLE OF CONTENTS

I

INTRODUCTION
FUNDAMENTAL INTERACTIONS AND PROCESSES

Pion Production and Absorption in Nuclei: Introductory Remarks
on the (p,π) and (π,p) Reactions - J.M. Eisenberg.............. 3

NN→πd and NN→NN; A Review of Experimental Results - G. Jones... 15

Questions in a Microscopic Theory of Pion Production/
Absorption - M.K. Banerjee...................................... 37

Quarks in Nuclei - G.A. Miller.................................. 47

Theories of Pion Production in Nucleon-Nucleon Collisions -
M. Betz, B. Blankleider, J.A. Niskanen and A.W. Thomas......... 65

Summary: Fundamental Interactions and Processes - D.S. Koltun.. 93

Panel Discussion Summary - D.S. Koltun.......................... 101

II

PION PRODUCTION AND ABSORPTION IN NUCLEI:
EXPERIMENTAL RESULTS

An Overview of Experimental Trends and Systematics in Pion
Production and Absorption in Nuclei - B. Höistad............... 105

Recent Developments and Results in (p,π) at IUCF -
M.C. Green.. 131

Near Threshold Proton Induced Neutral Pion Production from
Deuterium - M.A. Pickar... 143

Orsay and Saturne New Results on (p,π) and (Ion,π)
Experiments - Y. Le Bornec and N. Willis........................ 155

Neutron Induced Pion Production Processes - E. Rössle,
W. Dutty, J. Franz, L. Lehmann, G. Nicklas and H. Schmitt...... 171

New Experimental Results on Pion Production at LNS (Saclay) -
P. Couvert.. 187

The (p,π) Program at TRIUMF: Past, Present and Future -
G.J. Lolos.. 201

The (p,π) Reaction at 800 MeV - H. Nann........................ 219

III

CURRENT MODELS: CALCULATIONS AND COMPARISONS WITH DATA

The Relativistic One Nucleon Model – E.D. Cooper and
H.S. Sherif 231

Models of Pion Production and a Quark Model of the Physics
at Short Distances – L.S. Kisslinger 243

Results of Microscopic Calculations – B.D. Keister 265

Proton-Induced Pion Production in the Rescattering Model –
M. Dillig, F. Soga and J. Conte 275

Study of the (p,π) Reaction in the Two Nucleon Model –
F. Soga and M. Dillig 289

Pion Production by Target Emission – W.R. Gibbs 297

Models for (p,π) Reactions – H.W. Fearing 319

Summary of the Discussion on Connections Among Models of
Pion Production – H.W. Fearing 333

IV

RELATED REACTIONS
PION INDUCED REACTIONS
COMPLEX PROJECTILES
SUMMARY

One-Nucleon Knockout Reactions – J.T. Londergan 339

High Resolution Study of (π^+,pp), (π^+,pd) and other (π^+,xx)
Reactions on ^{6,7}Li, ^{14}N, and ^{16}O – W.R. Wharton ... 371

Charged Pion Production by ^{20}Ne on NaF at E/A = 140 MeV –
W. Benenson 381

Coherent Pion Production in Nucleus-Nucleus Collisions –
M.G. Huber 389

Pion Induced Fission – G.-F. Germond 411

Summary of the Workshop – G.E. Walker 419

Introduction
Fundamental Interactions and Processes

Top: G. Jones, M.K. Banerjee, J.M. Eisenberg
Middle: G.A. Miller, R.R. Silbar, H. Toki, A.W. Thomas
Bottom: F. Coester, J.A. Niskanen, H.J. Weber, D.S. Koltun
Photos by Kent Berglund

PION PRODUCTION AND ABSORPTION IN NUCLEI: INTRODUCTORY REMARKS ON THE (p,π) AND (π,p) REACTIONS

J. M. Eisenberg*
Department of Physics and Astronomy,
Tel Aviv University, Tel Aviv 69978, Israel

ABSTRACT

A brief review is presented of highlights in the history of the exclusive (p,π) and (π,p) reactions in nuclei with A>3, placing special emphasis upon the basic theoretical questions that have not yet been fully answered in their analysis. Some remarks are appended concerning eikonal insights into the nondiffractive nature of these reactions at high energy.

INTRODUCTION

These remarks are intended to set the stage for a workshop on pion production and absorption, with special emphasis on the exclusive (p,π) and (π,p) reactions. In an effort to avoid being more platitudinous than is absolutely necessary in fulfilling such a task, I shall confine myself exclusively to the (p,π) and (π,p) topics, which already offer a quite sufficient number of puzzles and a richness of physical ingredients that can serve well as an introduction to many concerns in intermediate-energy nuclear physics. I shall open with a review of some of the historical highlights in the experimental study of (p,π) and (π,p) for A>3, both ancient history and modern history. However, it is no part of my intention to add to the excellent reviews[1-4] that have been written on these processes and to which the reader is directed for systematic discussions and exhaustive references. (See also ref. 5 for general background material.) After the brief effort at history, I shall list a number of theoretical questions that have troubled and complicated the analysis of (p,π) and (π,p) from the beginning until the present. It seems important to have both the history of experiments and ongoing theoretical concerns clearly before us as we set out upon the presentation and evaluation of recent results, ongoing programs, and plans for the future. Last, I shall discuss briefly a quite recent synthesis of some theoretical results based on an eikonal approach to (p,π) or (π,p) which seem to me to illustrate well the nondiffractive nature of these processes and the consequences that ensue therefrom.

---

*Supported in part by the U.S.-Israel Binational Science Foundation and by the Israel Academy of Sciences and Humanities -- Basic Research Foundation.

HISTORICAL HIGHLIGHTS OF EXPERIMENTS

The earliest experimental work[6] on the (p,π^+) reaction for A>3, from 1958, carried within it the seeds of much of the more modern work. It was carried out at 209 MeV, that is, quite near threshold, thus allowing for reasonable energy resolution on the outgoing pion and giving at least a modicum of information on the final nuclear state reached for the light nuclei (carbon and aluminum targets) involved. Furthermore, this very early work already provided some pion asymmetry measurements for polarized incident protons - measurements that have been taken up anew only rather recently and that hold out promise of providing considerable constraint on the possible theoretical reaction mechanisms that may enter in (p,π) or (π,p) processes. Very shortly after these earliest studies, a (p,π^0) experiment was performed[7] on ^{7}Li as a background measurement in the search for a neutral, J=1, T=0, meson (based on its assumed decay to three photons). These two early experiments yielded data that we would consider a bit crude by modern standards and very sparse, and, so far as is known to me, did not at that time spur any theoretical work that attempted to account for them.

Nearly ten years elapsed before the next experimental study was reported[8], this time in the (π^+,p) mode and in response to the early, primitive theoretical interest and predictions[9] concerning the reaction. Shortly thereafter there were reported a series of (p,π^+) - and eventually even (p,π^-) - measurements[10] carried out at CERN at 600 MeV on light nuclei.

The modern era of (p,π) work was inaugurated in 1971 at Uppsala with the first high-resolution study[11] of $^{12}C(p,\pi^+)^{13}C$ at 185 MeV, thereby elegantly stealing a march on the high-intensity medium-energy accelerators that were just about to become available for use. Once these accelerators began to operate, a complete and very beautiful collection of data soon became available covering a variety of nuclei, energies ranging from exceedingly close to threshold on up to 800 MeV, and asymmetry data for the $(\vec{p},\pi^+)$ reaction. These data included - and I here merely try to illustrate the scope of the experimental activity[1-3] - measurements from Orsay and Saclay on $^{16}O(\pi^+,p)^{15}O$ at 66 MeV, as well as other cases, and on $^3He(p,\pi^+)^4He$ at 415 and 716 MeV, along with companion studies. The high-energy regime was further explored at LAMPF with (p,π^+) studies on a number of nuclei at 800 MeV.

Significant new information has been provided within the last few years by pion asymmetry measurements[12] at TRIUMF in $(\vec{p},\pi^+)$ reactions near 200 MeV on several light nuclei including the fundamental $\vec{p}p \to d\pi^+$ case. At about the same time a series of very striking experiments[13] has been carried out at Indiana in which pions of exceedingly low energy - within a few MeV or less of threshold - have been studied systematically, eventually with the inclusion of asymmetry measurements[14].

Much of this modern work will be reviewed subsequently in this workshop. My only purpose in noting it here is to underscore the truly dramatic strides forward that have been made on the experimental side of (p,π) and (π,p) reactions over the last ten years. As a matter of shear numbers, the interest aroused in these reactions is reflected in Fearing's bibliography[1] by the list of over 250 papers relating to this topic in one way or another.

THEORETICAL QUESTIONS TO WHICH YOU ARE ENTITLED TO DEMAND ANSWERS

The theoretical analysis of (p,π) and (π,p) reactions has not met with the same unambiguous and unqualified success as has the experimental work. This is because of a number of very deep-reaching questions that are raised in the course of theoretical investigation. These questions are still very much on the scene, with rather less than definitive answers, and I wish to list some of them here in an effort to set the framework for categorizing and evaluating the theoretical attempts.

1. <u>Single-nucleon vs. two-nucleon mechanisms (SNM vs. TNM)</u>. The earliest theoretical work[9,15] on the exclusive (π^+,p) reaction for A>2 dealt with a model in which only a single nucleon was active, the positive pion being essentially deposited on the initially bound neutron as it became the exiting proton. At the very high momentum transfers $q \gtrsim \sqrt{(2mM)} \sim 500$ MeV/c involved in (π,p) or (p,π), however, it is natural also to consider mechanisms involving two or more active nucleons in order to share this large q amongst them. (Indeed, the very early paper by Henley[16], treating inclusive pion production by proton bombardment of ^{12}C, was already based on such a view.) These two-nucleon mechanisms were studied[17,18] in models based on delta-isobar intermediate states, and the ability of the TNM to compete with the SNM in accounting for (p,π) and its inverse was established. The TNM can also address quite naturally the intrinsically two- or more-nucleon process (p,π^-), but must then offer an explanation for its suppression relative to (p,π^+) as seen in experiment[1-3].

Of course, the introduction of a two-nucleon mechanism raises a problem of double-counting: To what degree has the effect of the second participating nucleon already been taken into account in the distortion[15] of the waves in the SNM approach? This issue has not yet been probed systematically. Moreover, only very recently have efforts been made[19,20] to sum the SNM and TNM terms coherently with full treatment of interference effects. Other varieties of TNMs in the sense of two-step dynamic processes[21,22] have also been considered, though somewhat desultorily, and a proposal has also been put forth[23] that the ejected pion may originate from a collective target effect, the role of the incident proton being to place the pion on shell. In sum, modern theoretical work on (p,π) or (π,p) should be expected to deal with the competing and interfering effects of production on one or more nucleons.

2. **Pionic wave catastrophe.** As soon as pionic wave distortion was included in the calculation of (π,p), it was noted[15] that the use of the well-known Kisslinger potential in this far-off-shell situation can generate quite spectacular and unphysical enhancements in the cross sections. This comes about because the Kisslinger potential yields a Schroedinger equation of the form

$$(\nabla^2+k^2)\phi = (-b_0 k^2 \rho + b_1 \vec{\nabla}\cdot\rho\vec{\nabla})\phi \qquad (1)$$

for a pion field ϕ of asymptotic momentum k and with nuclear density ρ and s- and p-wave coefficients in the πN amplitude b_0 and b_1; this equation in turn gives an effective momentum k_{eff} in the medium satisfying

$$k^2_{eff} = \frac{1+b_0\rho}{1-b_1\rho} k^2 . \qquad (2)$$

For low-energy scattering $b_1 \sim 6$ fm^3, while $\rho(0) \sim 0.17$ fm^{-3}, so that such a potential generates a wealth of high-momentum components in the pion wave function that are readily picked up in the (p,π) process. The proper optical treatment should thus be done in momentum space with suitable vertex cutoff functions - not an easy numerical task[24]. Theorists who have not gone this full route have occasionally preferred to use eikonal distortion for the pion (or even no distortion at all) in regions where this is of doubtful legitimacy simply in order to avoid the ambiguities or nonsense that result from simplistic treatments of the off-shell behavior of ϕ. Once again, modern theoretical work should be expected to deal adequately with this off-shell behavior.

3. **Nonrelativistic reduction of the πNN vertex.** Almost since the beginning of theoretical work on (p,π) it has been known[25,26] that SNM calculations of the process will be especially sensitive to ambiguities in the nonrelativistic reduction of the relevant πNN vertex. One way to illustrate the problem is to note that if the nuclear central field arises as the result of the interplay between relativistic potentials of scalar U and (time-like component of) four-vector V character, that is, from a Dirac equation

$$[\vec{\alpha}\cdot\vec{p} + \beta(M+U)]u = (E-V)u, \quad u = \binom{f}{g} , \qquad (3)$$

then the γ_5-vertex will assume the form

$$\bar{u}_n \gamma_5 u_o = -(f_n^\dagger g_o - g_n^\dagger f_o) \stackrel{\sim}{=} \frac{1}{2M+U-V} f_n^\dagger \vec{\sigma}\cdot\vec{k} f_o. \qquad (4)$$

Thus we shall need to know not only the summed field $U+V$ that binds the nucleon in a nonrelativistic limit but also the difference field U-V, about which we may not know all we wish, and which produces a large (U-V $\sim$ - 700 MeV) change in the effective nucleon mass.

In other words we shall need to deal with more of the relativistic aspects of nucleon motion than is under good control at present, as is clear directly from the intrinsic dependence in eq. (4) on the Dirac "small" components g. First efforts in dealing with the relativistic aspects of (p,π) have been made[27], and, again, it is incumbent on modern theoretical work to define a stand with regard to this issue.

4. <u>Orthogonality</u>. It has been pointed out[28-30] that the trivial requirement of orthogonality between the incident nucleon in the continuum and the final bound nucleon in (p,π) is problematic in practicable SNM calculations where distortion is achieved by a complex, energy-dependent potential and binding by a different potential. That is, the transition amplitude we must calculate,

$$A_p(\underline{k}) = \int \phi^*(\underline{x}) \, \Omega(\underline{k}) \, e^{i\underline{k}\cdot\underline{x}} \, \psi_p^{(+)}(\underline{x}) \, d\underline{x} \, , \quad (5)$$

will not generally satisfy the basic orthogonality requirement

$$A_p(\underline{k}) \xrightarrow[k \to 0]{\Omega = 1} 0 \quad (6)$$

in practice. Moreover, there would appear to be no very simple prescription for correcting this situation[29]; rather, a consistent coupled-channel approach should be used. Mercifully, this problem probably plagues only the very-near-threshold (k<0.5 fm^{-1}, T_π<30 MeV) analysis seriously[30].

5. <u>High-momentum components in nuclei</u>. Assuming that theoretical work has been convincing in its treatment of the various uncertainties in the (p,π) mechanism, the payoff should come in the form of information on the high-momentum transfer behavior of nuclei (or even of nuclear single-particle wave functions should one be able to prove TNM << SNM). But we know so little about nuclear behavior at q ≳ 500 MeV/c - even elastic proton scattering fits break down miserably there - that it will not be easy to obtain corroboration of other assumptions. Perhaps the point is that the very large longitudinal momentum transfer restricts us to the deep nuclear interior over the whole kinematic range of (p,π), so that distinctions between SNM and TNM, pion off-shell models, nonrelativistic limits, and the like may not even be meaningful notions.

6. <u>Quarks</u>. The reader may be quite legitimately shocked by the listing of so many hadronic difficulties without any mention having been made thus far of quarks and gluons, QCD, and, perhaps, some phenomenology of bags. This is quite simply because I have no idea as to where the underlying structure of the hadrons may fit into (p,π) reactions. It seems to me to be necessary to know more about the solution of the confinement problem than we appear to at the moment, and about the special features of the pion as a Goldstone

boson in these theories, before QCD can add significantly to our understanding of (p,π); for (p,π) to help QCD, on the other hand, we must first feel some confidence in the treatment of the many-body features touched upon above.

All of this is not intended to lead to the conclusion that the theory of (p,π) and (π,p) is hopeless, but rather to stress that our level of theoretical understanding has risen to a maturity that now calls for especially careful and critical work. There are areas of broad success: Magnitudes and shapes of differential (p,π) cross sections emerge from calculation with some degree of success, in this very high q region, when viewed over a range of energies, nuclei, and transitions. Asymmetries are more problematic, but the harder TNM calculations have not yet been tried. Insights have been emerging from comparisons[31] of (π,p) and (p,π) to companion (γ,p) and (p,γ), or to (p,d). The study of pion production or absorption induced by clusters[32-35] as in $(^3He,\pi)$ or (π,d) is in its early stages and may offer further elucidation of (p,π) and (π,p). Last, one must note that many of the theoretical issues sketched above may legitimately be resolved by phenomenological rather than fundamental approaches, and that these in turn may give rise to an accretion of experience and corresponding accrual of self-confidence regarding our ability to handle much of the physics involved in (p,π) and (π,p) reactions.

EIKONAL INSIGHTS

As a sort of existential statement of faith concerning the possibilities of further theoretical understanding on (p,π), I should like to close these introductory remarks with a very brief outline of some recent results - spurred, in fact, partly by preparations for this workshop - on (p,π) at the higher energy range, say $500 \lesssim T_p \lesssim 800$ MeV. The outline given here will touch only upon the main results; a more systematic treatment, with careful attention to corrections, can be found in ref.[36]. (Some of the points noted here are hinted at in some of Noble's work[4], though with a quite different thrust.) We treat forward directions, say $\theta < 40°$, and use an eikonal approximation in which the (p,π) amplitude involves, in distorted wave approximation (DW) and after azimuthal integration,

$$T^{DW} \sim \int_0^\infty b\, db\, J_{|M_L|}(q_\perp b)\, \exp\{-\bar{\gamma} \int_{-\infty}^\infty \rho[(b^2+\zeta^2)^{\frac{1}{2}}]d\zeta\}\, F(b), \quad (7)$$

where M_L is the projection of the bound-state orbital angular momentum L; J is a cylindrical Bessel function, its argument containing the (relatively small) perpendicular momentum transfer $q_\perp$ and the impact parameter variable b. The exponential expresses the effects of distortion and absorption, involving the attenuation factor

$\gamma = \frac{1}{2} \sigma (\text{tot})(1-i\alpha)$, where $\sigma(\text{tot})$ is the total projectile-nucleon cross section and α is the ratio of imaginary to real amplitude, averaged over pion and proton.

The quantity

$$F(b) = \int_{-\infty}^{\infty} dz \, \exp(iq_{\|} z) \, g[z,(b^2+z^2)^{\frac{1}{2}}] \, (b^2+z^2)^{-\frac{1}{2}} \exp[-\kappa(b^2+z^2)^{\frac{1}{2}}], \quad (8)$$

is the crux of the issue, containing the very large longitudinal momentum transfer $q_{\|}$ and an integration over the longitudinal variable z, as well as the nuclear bound-state wave function, taken quite generally as a polynomial in z and $r = (b^2+z^2)^{1/2}$ times the decay factor $r^{-1}\exp(-\kappa r)$, where $\kappa = (2M|B|)^{1/2}$ for separation energy $|B|$. Expressing the polynomial g as a differential operator, and recalling the known result for the integral of Eq.(8) when g=1, we have

$$F(b) = 2g(-i\partial_{q_{\|}}, -\partial_{\kappa}) \, K_0[(q_{\|}^2+\kappa^2)^{\frac{1}{2}} b] \cong g(\frac{2}{q_{\|}b})^{\frac{1}{2}} e^{-q_{\|}b}, \quad (9)$$

where K_0 is a modified Bessel function. The last form in Eq.(9) follows from the asymptotic behavior of K_0, which is all we require since $q_{\|}$ is so large (>2.5 fm^{-1}); it implies that $b \lesssim q^{-1} \lesssim 0.4$ fm, that is, the impact parameter is restricted to very small values. This situation is quite different from that of scattering, say, where relevant impact parameters are near the nuclear radius; in other words, (p,π) is <u>nondiffractive</u> because of the very large longitudinal momentum transfer entering there.

As a further consequence of this nondiffractive feature, the distortion factor in Eq.(7) can be well approximated by its value along the nuclear diameter (i.e. at b=0; corrections to this arising from spin-orbit effects, among others, are discussed in ref.[36]),

$$\exp[-\bar{\gamma} \int_{-\infty}^{\infty} \rho \, d\zeta] \cong \exp[-2\bar{\gamma}\rho_0 c], \quad (10)$$

where c is the radius parameter in the nuclear Fermi distribution of central density ρ_0. This can now be taken out from under the integral, yielding the remarkably simple result that the DW amplitude is almost trivially proportional to the plane-wave (PW) one,

$$T^{DW} \cong e^{-2\bar{\gamma}\rho_0 c} \cdot T^{PW}. \quad (11)$$

I have not noted so far, as perhaps I should have, that all this is correct only for the SNM because it has an obvious extension to the TNM provided the dynamic interaction region is of much smaller range than the other dimensions in the problem.

The physical reason for this result rests on conservation of angular momentum: The conversion of proton to pion at impact parameter b, in a semiclassical approach, deposits $q_\parallel$ b units of angular momentum into the nuclear system. But if the nucleon is to go into a bound state of angular momentum L we must have $q_\parallel b \sim L$ or $b \sim L/q_\parallel$. [In Eq.(9) the effects of L > 1 are hidden in g.] Thus for such large longitudinal momentum transfer, and small final L, the impact parameter must be small in order to prevent a large mismatch of angular momentum. A similar result will apply for other such reactions, for example (p,γ) or (p,d), and, indeed, Eq.(11) yields an immediate explanation for the finding[31] that pion distortion does not play a big role when (p,π) results are compared directly to (p,γ). It should also be noted that Eq.(11) effortlessly gives correct magnitudes for (p,π⁺) cross sections, especially bearing in mind the large uncertainties in wave functions in the nuclear interior.

One can then go further with this theme, though perhaps with somewhat less conviction, and examine the high-longitudinal-momentum-transfer consequences for the plane-wave result. Thus

$$T^{PW} \sim \int_0^\infty j_L(qr) e^{-\kappa r} r^\mu dr, \quad \mu \geq L + 2, \quad (12)$$

where now j_L is the spherical Bessel function and the full momentum transfer q appears. We have again taken the wave function to go as $r^L e^{-\kappa r}$ near the origin (and suppressed the details of a Taylor series expansion for the wave function near the origin, whence the various values of μ). The integral in Eq.(12) is easily evaluated and leads to

$$T \sim q^{-L-4} \quad \text{or} \quad d\sigma/d\Omega \sim q^{-2L-8}, \quad (13)$$

up to leading orders in q^{-1}, with corrections involving $\kappa/q \ll 1$ and the like. Again this is generalizable for a zero-range TNM, but then L is some summed value of the orbital angular momenta of the participating particles. The result of Eq.(13) can be compared in a rough way with experiments in this kinematic region. For the 1s shell one seems to have $d\sigma/d\Omega \sim q^{-6\pm 2}$ rather than q^{-8}, while for the 1p shell experiment gives, crudely, $q^{-12\pm 2}$ in place of q^{-10}. Although this agreement seems fair, the more detailed examination[36] suggests that Eq.(13) is not the whole story and perhaps one is seeing here - albeit not convincingly - the effects of interference with non-SNM features.

Last, note that Eq.(11), as applied to the SNM, implies zero asymmetry ε for $(\vec{p},\pi^+)$ measurements, since it is well known that ε = 0 for the PWBA. The next order corrections can be evaluated[36] and suggest, for the spin-orbit contributions, a positive ε on the order of a couple of tenths, and for optical path length corrections a variable sign with roughly the same magnitude. Thus the asymmetry would appear to be smaller and less regular than seen at 200 MeV, an inference that would be interesting to check experimentally.

ACKNOWLEDGEMENTS

It is truly a pleasure to thank the many colleagues with whom I have had intense (p,π) interaction: some years ago J. LeTourneux and W. B. Jones, later J. V. Noble and H. J. Weber, and still more recently D. S. Koltun and A. Saharia.

REFERENCES

1. H. W. Fearing, Prog. Part. Nucl. Phys. 7, 113 (1981); a companion, categorized bibliography is H. W. Fearing, TRIUMF report TRI-80-3, November, 1980.
2. B. Holstad, Adv. Nucl. Phys. 11, 135 (1979).
3. D. F. Measday and G. A. Miller, Ann. Rev. Nucl. Part. Sci. 29, 121 (1979).
4. J. V. Noble, Univ. Virginia report, 1976.
5. J. M. Eisenberg and D. S. Koltun, Theory of Meson Interactions in Nuclei (Wiley-Interscience, New York, 1980).
6. E. Heer, A. Roberts and J. Tinlot, Phys. Rev. 111, 640 (1958).
7. R. P. Ely and D. H. Frisch, Phys. Rev. Lett. 3, 565 (1959).
8. T. R. Witten, M. Blecher and K. Gotow, Phys. Rev. 174, 1166 (1968).
9. J. LeTourneux and J. M. Eisenberg, Nucl. Phys. 87, 331 (1966).
10. J. J. Domingo et al., Phys. Lett. 32B, 309 (1970).
11. S. Dahlgren, B. Holstad and P. Grafström, Phys. Lett. 35B, 219 (1971).
12. E. G. Auld et al., Phys. Rev. Lett. 41, 462 (1978).
13. P. H. Pile et al., Phys. Rev. Lett. 42, 1461 (1979).
14. T. P. Sjoreen et al., Phys. Rev. C24, 1135 (1981).
15. W. B. Jones and J. M. Eisenberg, Nucl. Phys. A154, 49 (1970).
16. E. M. Henley, Phys. Rev. 85, 204 (1952).
17. A. Reitan, Nucl. Phys. B29, 525 (1971); B50, 166 (1972).
18. Z. Grossman, F. Lenz and M. P. Locher, Ann. Phys. (N.Y.) 84, 348 (1974).
19. P. Couvert and M. Dillig, contribution to the Ninth Int. Conf. on High Energy Physics and Nuclear Structure, Versailles, July, 1981.
20. B. D. Keister and L. S. Kisslinger, ibid.
21. J. M. Eisenberg, J. V. Noble and H. J. Weber, in High-Energy Physics and Nuclear Structure, G. Tibell, ed. (North-Holland, Amsterdam, 1974) p.270, and unpublished work.
22. G. A. Miller, Nucl. Phys. A224, 269 (1974).
23. W. R. Gibbs, in Workshop on Nuclear Structure with Intermediate-Energy Probes; H. Baer et al., eds. (Los Alamos Scientific Laboratory report LA-8303-C) p.233, and unpublished work.

24. M. Tsangarides, Ph.D. thesis, Indiana University, October, 1979 (IUCF internal report 79-4).
25. M. V. Barnhill, Nucl. Phys. A131, 106 (1969).
26. M. Bolsterli, W. R. Gibbs, B. F. Gibson and G. J. Stephenson, Jr. Phys. Rev. C10, 1225 (1974).
27. L. D. Miller and H. J. Weber, Phys. Rev. C17, 219 (1978).
28. J. V. Noble, Phys. Rev. C17, 2151 (1978).
29. J. M. Eisenberg, J. V. Noble and H. J. Weber, Phys. Rev. C19, 276 (1979).
30. H. J. Weber and J. M. Eisenberg, Nucl. Phys. A312, 201 (1978).
31. B. M. K. Nefkens, contribution to the Ninth Int. Conf. on High Energy Physics and Nuclear Structure, Versailles, July, 1981.
32. N. S. Wall, J. N. Craig and D. Ezrow, Nucl. Phys. A268, 459 (1976).
33. L. P. Fulcher and M. K. Banerjee, Univ. Maryland report 76-020, 1976.
34. M. Betz and A. K. Kerman, Nucl. Phys. A345, 493 (1980).
35. K. G. R. Doss et al., Carnegie-Mellon Univ. preprint, 1981.
36. J. M. Eisenberg and D. S. Koltun, Univ. Rochester/Tel Aviv Univ. preprint, October, 1981.

DISCUSSION

Julian Noble (U. of Virginia): The first remark I want to make has to do with the spin-asymmetry thing, since in fact the error in the paper is obviously due to me. I'm the first one who wrote down what the formula was and I think everybody copied the wrong formula. What I think actually happened is that the formula in the paper is wrong but everybody has probably done the calculation correctly, because you end up calculating the cross-section for one spin orientation of the initial nucleon and then for the other spin orientation of the initial nucleon, and one uses the Madison convention for one's directions and then one takes the asymmetries. So I think that, in fact, the actual calculations are not generally done with projection operators. They're done properly, and so I believe that the results with negative asymmetries may be right. But of course I do want to say that the final result in my paper was a spurious one because it had to do with the effects of high Fourier components coming from a square well that I assumed for the bound state wave functions, and therefore had nothing whatsoever to do with reality. So I don't claim anything for my old paper, except that I was interested in looking at some of the old data.

I want to make a remark also about the nonrelativistic reduction that you showed. You showed something horrible happening for the γ_5 interaction but, of course, as I demonstrated in two papers a couple of years ago, if one insists on having a γ_5 interaction which has enough π-σ coupling terms, so that the end result is chirally invariant, then you have additional contributions to the pion scattering from the U potential, and the end result is that the corrections make γ_5 equivalant to the $\gamma_\mu \gamma_5$ coupling, at least to the leading orders of perturbation theory. So that's not a major problem if one is consistent.

Then on this business of the one-nucleon versus the two-nucleon model, I just want to say that there are a lot of misleading statements that are being circulating in this regard. You can't possibly make pions off a free nucleon, so they're all two-nucleon models in some sense. The question is whether what makes the pions is some kind of average property of the nucleus or whether you need to worry about the detailed behavior of every other nucleon. If it's the latter, then you are stuck, I'm afraid. But I think it's the former, that in fact the pion production mechanism can be well understood in terms of average properties of the nucleus; for example, these average U and V potentials.

L.S. Kisslinger (Carnegie-Mellon Univ.): I have a brief comment, which is that the alternative to keeping track of the coordinate of one nucleon is not necessarily to keep track of the coordinates of all the nucleons in the nucleus. If one keeps track of the coordinates of two nucleons, that's a very big step past one nucleon.

I also have a question. My understanding is that we tend to have surface reactions with pions. Doesn't that throw some doubt on the formulation which you presented?

J.M. Eisenberg (Tel Aviv Univ.): If you do something elastic it is diffractive in the sense that it's surface peaked, and if you remember you saw all those decades going down, say, in ^{208}Pb. But if you do (p,π) it is nondiffractive. What you have is the central region, and the reason you have the central region is that you have a battle between this tremendous decay induced by the large longitudinal momentum transfer and the factor which wants you to stay at the surface because it insists that it will damp anything that requires too much penetration into the medium. The longitudinal momentum transfer is so large that it wins that battle. In fact, when we started out to look at this, we simply tried to generate the same kinds of considerations that went into the work of Amado, Dedonder and Lenz, looking at diffractive features with a method of steepest descent. What we found was that it was nonsense. It wasn't what was involved in this problem. What was involved here was very small impact parameters.

Kisslinger: I agree with you, except for the following thing. The battle that you mentioned can also be between the high momentum components coming from two nucleon cluster properties rather than the whole nucleus, and I think that's the battle. It is a surface correction.

Eisenberg: The data don't show it to be a surface reaction.

NN → πd AND NN → NNπ; A REVIEW OF EXPERIMENTAL RESULTS

G. Jones
University of British Columbia, Department of Physics,
Vancouver, B.C., Canada V6T 2A6

INTRODUCTION

Pion production in the nucleon-nucleon interaction has, because of its fundamental interest, been a subject of experimental investigation for a whole generation of physicists. However, it is just in the last few years, largely as the result of development of high quality beam lines at the meson factories together with exploitation of recent developments in particle detection technology, that extensive high quality data has begun to emerge. It is this data, and its impact on the experimental definition of the pion production reaction that I wish to address in this paper. Ample references have been made to the early work in previous articles and reviews.[1-4] In this paper, I do not intend to refer explicitly to any of the prehistoric work published in the 1950's, and indeed will only mention in passing some results of the primitive 1960's. Essentially all work referred to, then, will be that published in the decade since 1970.

The first part of the paper will be concerned with the pion production reaction in which the bound nucleon state, the deuteron, is also formed. Because of the two-body nature of the final state, it is the only nucleon-nucleon pion production reaction that is amenable to investigation by studying the inverse pion absorption reaction. It is the study of this absorption reaction, in fact, that accounts for much of the quality data that has appeared during this past year. The final part of the paper will deal then with pion production reactions leading to a three-body final state. In both sections, I will describe, within the constraints of my own limited knowledge, plans and proposed experiments at the various intermediate energy facilities. In all cases, I will restrict my attention to that of single pion production only. Multiple pion production will be ignored.

Pion production at intermediate energies is dominated by the generation of a Δ-isobar in the intermediate state. Models[5,6] based on such a mechanism have been generally successful in explaining the gross features of the experimental data in this energy region. At the lower energies (up to about 700 MeV), the singlet nucleon-nucleon spin state dominates[5] while at the higher energies the contribution of the triplet state becomes a major factor as well. Thus, a full understanding of the reaction mechanism requires detailed measurements of the spin dependence of the reaction. In recent years, interest in the spin dependence of the nucleon-nucleon reaction has received additional stimulus by the observation of significant energy dependent structure in both the $\Delta\sigma_L$ (Ref. 7) and $\Delta\sigma_T$ (Ref. 8) total nucleon-nucleon cross-section differences. The interpretation of this structure, whether associated with exotic dibaryon resonances, or whether it merely reflects the opening of additional angular momentum channels in the intermediate state has been the subject of much recent speculation.

0094-243X/82/79015-22 $3.00 Copyright 1982 American Institute of Physics

1. NN $\rightleftarrows$ dπ

1.1 PP $\rightleftarrows$ dπ^+

1.1.1 General. Until quite recently, the quality of the data available for this elementary pion production reaction was of surprisingly poor quality. In Fig. 1 is depicted the information available in 1978[1] concerning the shape of the differential cross section. The ratio of the coefficients of an expansion in $\cos^2\theta$ is plotted as a function of the centre-of-mass pion momentum (in units of μc), η. Most of the reason for this poor state of affairs was the low quality of the data obtained by the initial exploratory measurements. Some of the reason, however, is the result of the particular way the data was analyzed. Because the original theoretical discussions of the reaction[9] involved an expansion in even powers of $\cos\theta$:

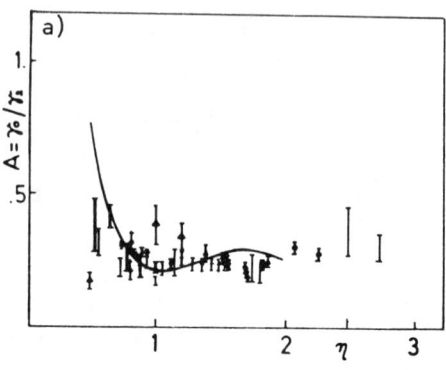

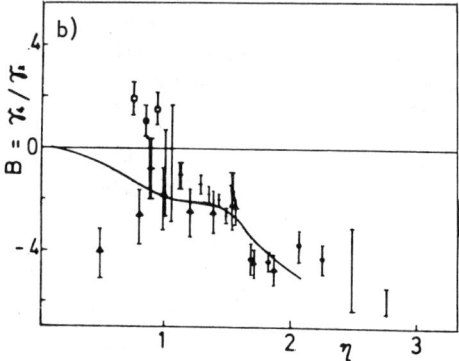

Fig. 1. a) γ_0/γ_2 as a function of pion momentum (c.m.). b) γ_4/γ_2 as a function of pion momentum (c.m.).

$$\frac{d\sigma}{d\Omega} \propto \sum_i \gamma_{2i} \cos^{2i}\theta , \quad (1)$$

the experimental analyses tended to follow suit and extract the appropriate coefficients from their experimental data. A number of authors have pointed out, however, that such expansions are dangerous, being prone to yield values for the coefficients which depend on the order to which the expansion is truncated. Such problems occur to a much smaller extent when expansions utilizing orthogonal polynomials are employed. Thus, there has been a growing tendency in recent years for the use of Legendre polynomials for the differential cross section and associated Legendre polynomials for the spin-dependent cross sections which characterize pion production arising from the collision of polarized and unpolarized protons. In fact, Niskanen has recently suggested[10] a convention employing such expansions for representing the various spin-dependent cross sections possible using both polarized proton beams and targets. For a polarized beam on an unpolarized target, he suggests the expansion:

$$4\pi \frac{d\sigma}{d\Omega} = \sum_{k(\text{even})} a_k^{00} P_k(\cos\theta) + \vec{P}\cdot\vec{n} \sum_k b_k^{y0} P_k^1(\cos\theta) \quad (2)$$

where the P_k and P_k^1 are the Legendre and associated Legendre functions, $\vec{P}$ is the polarization of the proton beam and $\vec{n}$ is a unit vector in the $\vec{k}_i \times \vec{k}_\pi$ direction ($\vec{k}_i$ and $\vec{k}_\pi$ are the incident proton and outgoing pion momenta, respectively). In the following, I shall disregard the superscripts and refer simply to the (even) a_k, and b_k values. Because of the normalization of the Legendre polynomials, the total cross section is then simply given by:

$$\sigma_t = a_0 . \qquad (3)$$

[For the inverse reaction, $\pi d \to 2p$, the total cross section is (assuming detailed balance) $a_0/2$ because of the identity of the two outgoing protons.]

The advantage in employing such an expansion is very nicely demonstrated by the results of Hollas et al.[11] for the $np \to d\pi^0$ reaction at 795 MeV. In this reaction, where the detected particle (deuteron) is the only charged particle involved in the reaction, the relative differential cross section was determined over a very wide range of angles, $2° \leq \theta \leq 178°$ (c.m.s.)! At this energy, terms up to order $\cos^6\theta$, [or $P_6(\cos \theta)$] were required. For an expansion in terms of a simple power series in $\cos \theta$:

$$\frac{d\sigma}{d\Omega} \propto A + \cos^2\theta + B \cos^4\theta + C \cos^6\theta + ... \qquad (4)$$

the authors examined the dependence of the values of the fitted parameters on the order assumed for the expansion. Thus, for Eq. (4) the following results were obtained:

Table I. Cos θ expansion coefficients: $A = \gamma_0/\gamma_2$, $B = \gamma_4/\gamma_2$, $C = \gamma_6/\gamma_2$.

A	B	C
0.232 ± 0.003	−0.510 ± 0.009	—
0.308 ± 0.004	+0.239 ± 0.060	−0.650 ± 0.066

The addition of the sixth order term affects all coefficients significantly, leading even to a change of sign of the coefficient of the $\cos^4\theta$ term (representing changes in the values of the coefficients by up to 40 standard deviations)! The situation is much more stable for an expansion in Legendre polynomials [the 'unpolarized' first half of the right-hand side of Eq. (2)].

Table II. Legendre polynomial expansion coefficients.

a_2/a_0	a_4/a_0	a_6/a_0
0.809 ± 0.011	−0.250 ± 0.013	—
0.827 ± 0.012	−0.247 ± 0.013	−0.071 ± 0.017

In this case, the changes in the values of the coefficients are within the limits associated with their errors. For these reasons, it seems timely to repeat again the request that, henceforth, authors cease presenting data in terms of expansions in $\cos \theta$.

1.1.2 **Differential Cross Section.** During this past year, the results of several different high precision experiments have become available. Ritchie et al.[12] measured the $\pi^+ \to 2p$ reaction at seven energies between 20 and 65 MeV incident (lab) energy. Boswell et al.[13] carried out similar measurements over a different range of pion energies, yielding data for seven energies between 80 and 417 MeV. Meanwhile, measurements on the $pp \to d\pi^+$ reaction were carried out by Hoftiezer et al.[14] at seven energies between 518 and 583 MeV. Measurements using the HRS spectrometer at LAMPF are being carried out for a range of proton energies by Nann et al., with data already published[15] at 800 MeV. Legendre polynomial expansions of all this data have been performed, and the results are displayed in Figs. 2(a) and 2(b) and in Table III. In some cases, where the authors analyzed their data

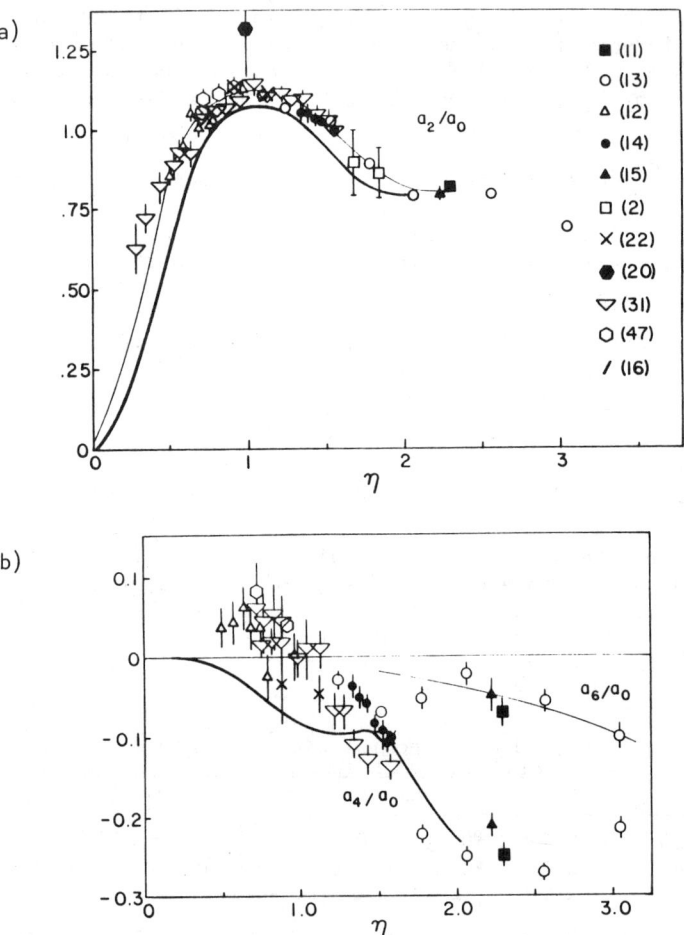

Fig. 2. a) a_2/a_0 as a function of pion momentum (c.m.). The solid line is the theoretical prediction.[16] b) a_4/a_0 and a_6/a_0 as a function of pion momentum (c.m.). The thin lines serve only to guide the eye.

Table III. Legendre polynomial coefficient ratios for the data shown in Fig. 2.

Reaction	Ref.	η	a_2/a_0	a_4/a_0	a_6/a_0	$\sigma_t(pp \to d\pi)$ mb	$\sigma_t(\pi d \to 2p)$ mb
$\pi d \to 2p$	12	0.500	0.860(0.022)	0.038(0.017)	—	0.206(0.002)	4.31(0.035)
		0.582	0.964(0.022)	0.043(0.028)	—	0.291(0.002)	4.67(0.040)
		0.64	1.065(0.020)	0.062(0.027)	—	0.311(0.002)	4.25(0.033)
		0.693	1.010(0.022)	0.038(0.031)	—	0.434(0.003)	5.20(0.040)
		0.744	1.015(0.021)	0.038(0.032)	—	0.499(0.004)	5.33(0.040)
		0.793	1.030(0.020)	-0.026(0.026)	—	0.537(0.004)	5.19(0.040)
		0.971	1.131(0.019)	0.002(0.029)	—	1.040(0.010)	7.40(0.07)
pp → dπ	22	0.884	1.137(0.040)	-0.035(0.048)	—	0.79(0.01)	6.44(0.07)
		1.120	1.107(0.018)	-0.046(0.024)	—	1.68(0.01)	9.80(0.08)
pp → dπ	20	1.00	1.33(0.065)	0.030(0.090)	—	1.31(0.04)	8.94(0.27)
pp → dπ	14	1.338	1.062(0.026)	-0.038(0.015)	—	2.53(0.04)	11.67(0.18)
		1.383	1.065(0.022)	-0.049(0.014)	—	2.74(0.03)	12.11(0.15)
		1.428	1.038(0.018)	-0.058(0.011)	—	2.95(0.03)	12.52(0.13)
		1.476	1.030(0.017)	-0.088(0.012)	—	2.96(0.03)	12.03(0.10)
	"	1.525	1.026(0.016)	-0.078(0.011)	—	2.93(0.03)	12.23(0.12)
		1.549	1.017(0.017)	-0.093(0.011)	—	3.12(0.03)	12.20(0.15)
	"	1.572	1.001(0.020)	-0.109(0.012)	—	3.16(0.04)	11.87(0.07)
$\pi d \to 2p$	13	0.986(0.015)	1.009(0.016)	-0.110(0.011)	—	3.13(0.03)	
		1.016(0.016)	0.119(0.009)	—	3.11(0.03)		
		1.091	1.115(0.013)	0.100(0.009)	—	3.16(0.03)	
		1.242	1.071(0.007)	-0.030(0.009)	—	3.20(0.48)	12.6(1.9)
		1.520	1.026(0.007)	-0.069(0.009)	—	2.93(0.29)	9.60(0.96)
		1.788	0.892(0.006)	-0.222(0.009)	—	1.65(0.16)	4.60(0.46)
		2.073	0.785(0.007)	-0.250(0.010)	-0.053(0.010)	1.6(0.2)	1.6(0.2)
		2.581	0.791(0.007)	-0.270(0.010)	-0.021(0.012)	0.711(0.089)	0.64(0.06)
		3.053	0.684(0.010)	-0.217(0.013)	-0.057(0.012)	0.331(0.031)	
		1.533	1.020(0.011)	-0.101(0.048)	-0.101(0.017)	3.02(0.13)	11.8(0.5)
$\pi d \to 2p$		1.682	0.90(0.10)	-0.206(0.046)	—	3.22(0.14)	11.3(0.5)
(Richard Serre)		1.818	0.86(0.08)	-0.210(0.03)	—	2.74(0.12)	8.8(0.4)
		2.073	0.80(0.09)	-0.16(0.03)	—	1.68(0.07)	4.7(0.2)
		2.253	0.81(0.09)	-0.19(0.03)	—	1.16(0.05)	2.98(0.13)

cont'd.

Reaction	N					
pp → dπ	15	2.22	0.811(0.016)	−0.214(0.017)	—	1.22(0.01)
np → dπ⁰	11	2.30 or	0.795(0.018)	−0.213(0.017)	−0.047(0.020)	1.23(0.01)
np → dπ⁰	31			−0.247(0.013)	−0.071(0.017)	—
		0.292	0.827(0.012)			
		0.364	0.632(0.078)			
		0.449	0.739(0.051)			
		0.521	0.835(0.049)			
		0.587	0.889(0.045)			
		0.647	0.937(0.037)			
		0.703	0.934(0.034)			
		0.756	1.042(0.028)	0.062(0.033)		
		0.806	1.070(0.032)	0.046(0.037)		
		0.878	1.070(0.032)	0.056(0.038)	0.030(0.039)	
		0.969	1.071(0.029)	0.047(0.031)	0.010(0.029)	
		1.054	1.092(0.021)	−0.001(0.024)	0.030(0.046)	
		1.136	1.146(0.033)	0.012(0.036)	0.000(0.026)	
		1.215	1.109(0.019)	0.010(0.022)	0.044(0.026)	
		1.291	1.122(0.020)	−0.066(0.022)	0.016(0.027)	
		1.365	1.098(0.020)	−0.070(0.022)	−0.022(0.020)	
		1.436	1.108(0.015)	−0.107(0.017)	−0.053(0.021)	
		1.507	1.067(0.015)	−0.128(0.017)	−0.072(0.021)	
		1.575	1.054(0.015)	−0.098(0.017)		3.18(0.02)
			0.992(0.015)	−0.135(0.017)	−0.156(0.020)	—

only in terms of powers of cos θ, I have re-analyzed the data in terms of Legendre polynomials. In so doing, the errors quoted for σ_t resulting from my re-analyses are generally somewhat smaller than those quoted in the original paper by the authors reflecting, I suspect, neglect by the original authors of the correlations that exist in the errors associated with the coefficients of the $\cos^{2n}\theta$ terms which appear in the σ_t integral. In the paper of Ritchie et al.,[12] one of the quoted σ_t values is 10% different from mine (the value at 25 MeV incident energy), indicating either an error in the cross section data listed in the paper, or a numerical error made when extracting the σ_t value. Because some of the data reflected measurements of relative cross sections only (not absolute), I have presented the results in terms of ratios of coefficients: a_j/a_0. That is, the values plotted are the relative values of the coefficient of interest to the total cross section, σ_t. In the figures the solid lines are the predictions of Niskanen,[16] given to indicate the quality of current theoretical understanding of the reaction. It is clear that the experimental situation is now becoming quite well-defined, with excellent consistency between the pp → dπ results of Hoftiezer et al.,[14] and Nann et al.[15] together with the np → dπ⁰ results of Hollas et al.[11] and the Rössle group[31] with those of the inverse reaction by Boswell et al.[13] and Ritchie et al.[12] Such consistency provides excellent verification of the applicability of detailed balance to this reaction. It is clear, also, that there is now unequivocal evidence for the observation of f-wave pions (as reflected by the non-zero values obtained for a_6). The overall theoretical understanding of the energy dependence of the coefficients (as exemplified by the curves of Niskanen) is also quite good, the principal exception being in the energy dependence of the a_4 term, particularly at low energies. The current experimental situation below 500 MeV is still somewhat unclear. However, the results of both Ritchie et al.[12] and the Rössle group (np → dπ⁰)[31] indicate finite, positive values for a_4/a_0 for values of η below about 1.05. The results of further measurements currently underway at TRIUMF[17] should help clarify this situation.

1.1.3 Total Cross Section, σ_t Some of the precision differential cross section results referred to in 1.1.2 pertained to relative cross sections, only. Most, however, were absolute cross sections which yielded values of a_0 (and thus σ_t). The results, when transformed[18] to the πd → 2p total cross sections are displayed in Fig. 3 with the numerical values listed in Table III. On Fig. 3 are also shown the old (1967) low-energy cross sections of Rose.[19] Generally, the various sets of data from the different groups are reasonably internally consistent and, more importantly, are consistent with each other. The datum of Dolnick[20] appears somewhat high, by several standard deviations, probably reflecting the limitations inherent in the rather primitive instrumentation characterizing his two-arm particle detection system. The two data points of Aebischer et al.[22] were obtained by normalizing to "well-known" pp elastic scattering cross sections, a cross section that was measured at a given angle (about 13° cms) with the same two-arm apparatus used for the pp → dπ⁺ measurements. One of them, the 455 MeV value (η = 1.12) was

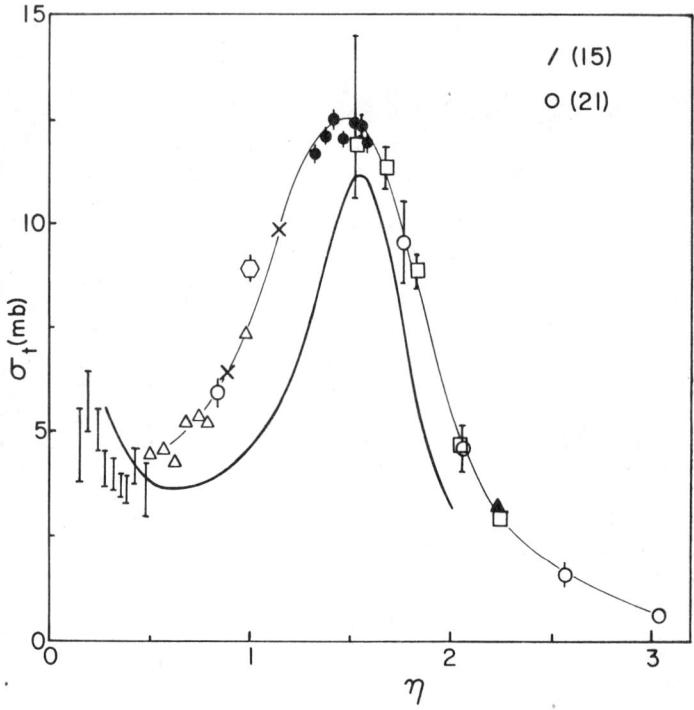

Fig. 3. Total cross section of the $\pi^+ d \to 2p$ reaction as a function of pion momentum (c.m.). Detailed balance has been assumed. The solid curve is the prediction of Niskanen.[16]

normalized to a pp cross section of 4.68 mb/sr at 13.35° (cms). As a result of subsequent pp measurements the best current estimate of the pp elastic cross section at 13.35° is 5.135 ± 0.059 mb/sr.[23] The Aebischer cross section at η = 1.12 plotted on Fig. 3 has thus been renormalized appropriately. Also shown in Fig. 3 is Niskanen's prediction[16] of the total cross section. Interestingly, this quantity is much less satisfactorily handled by his model than is the angular structure of either the unpolarized or polarized cross sections.

1.1.4 <u>Polarization Analyzing Power</u> The simplest spin-dependent parameter that can be experimentally determined is the analyzing power, $A(\theta)$, of the reaction, a quantity obtained by measuring the left-right production asymmetry induced when a transversely polarized beam impinges on an unpolarized target (or, conversely, when an unpolarized beam impinges on a polarized target). The asymmetry is given by the ratio of the two terms on the right-hand side of Eq. (2). Thus,

$$\frac{d\sigma}{d\Omega_{y_o}} = A(\theta) \frac{d\sigma}{d\Omega_{oo}} (\theta) ,$$

or
$$\sum_k b_k Y_k^0 P_k^1(\cos\theta) = 4\pi A(\theta) \frac{d\sigma}{d\Omega_{oo}}(\theta) = A(\theta) \sum_{k(\text{even})} a_k^{oo} P_k(\cos\theta). \quad (5)$$

As shown in Fig. 4, the structure of $A(\theta)$ is strongly energy-dependent in the 500 MeV region, changing from a simple single negative peak below 450 MeV to a double-humped positive peak above 450 MeV. The 350 MeV results[24] displayed in Fig. 4 were obtained using a "single-arm" configuration, whereby a particle detector displaying sensitivity to particle type (in this case a magnetic spectrograph) detects the outgoing pions, with the deuterons left undetected. The higher energy results[14] were, on the other hand, obtained using a "double-arm" experimental configuration. In this case, both the pion and the deuteron were simultaneously detected in counters of rather low particle selectivity. In this case, one relies instead, on kinematic correlations to provide the sensitivity to the reaction of interest.

For some of the measurements quoted, the precision with which the differential cross sections were measured was rather less than that of the analyzing power.[24] In this case, the accuracy with which the b_k of Eq. (5) could be determined was limited, not by the errors associated with the analyzing power measurements themselves but by the uncertainty in the basic unpolarized differential cross section (the a_k values).

In order to provide the best possible estimates of b_k, I have where necessary taken the published values for

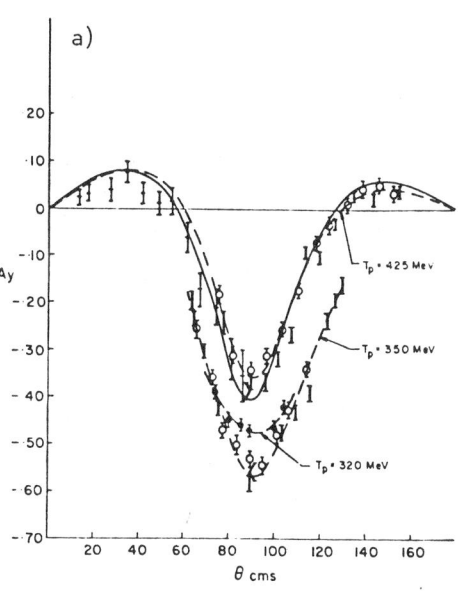

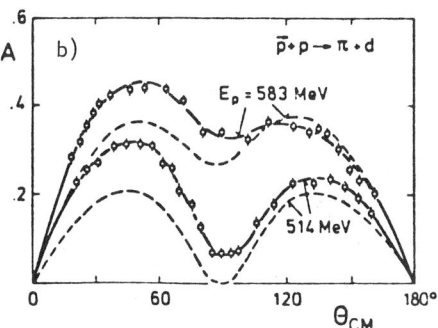

Fig. 4. a) Analyzing power for $\vec{p}p \to d\pi^+$ as a function of pion angle (c.m.) for proton energies (lab) of 320, 350 and 425 MeV. The dashed curves are fits to the data. The solid curve is a theoretical prediction.[16] b) Angular distribution of the analyzing power for 514 and 583 MeV protons. The dashed lines are theoretical predictions.[16]

analyzing powers and combined them with estimates of the unpolarized differential cross sections derived from the experimental data of Fig. 2, and fitted the results using associated Legendre polynomials as indicated by Eq. (5) and the following:

$$\sum_k \frac{b_k}{a_0} P_k^1(\cos\theta) = A(\theta) \sum_{k(\text{even})} \frac{a_k}{a_0} P_k(\cos\theta) . \qquad (6)$$

The analyzing power data of Nann et al.[15] only covered the angular range 23-92° (cms). In order to constrain the fit more effectively, their data set was expanded by including somewhat less precise data from Cverna et al.[25] at 108, 122 and 135°. The results, ratios of the b_k coefficients to the total cross section are displayed in Figs. 5(a)-5(b), and summarized in Table IV. A consistent set of data for b_1/a_0 and b_3/a_0 appears to have emerged, with the errors on the higher-order coefficients generally smaller for the double-arm experiments where the results were obtained over a larger angular range than than for some of the single-arm magnetic spectrograph results. The situation for b_2/a_0, particularly at lower energies is rather uncertain. A rather strong correlation was found between the value for this coefficient and that for some of the higher-order coefficients (especially b_4/a_0). At $\eta = 1$ (425 MeV) two b_2/a_0 data points are illustrated, separated by several standard deviations. The largest value is the result of analysis of a set of analyzing power data obtained using a magnetic spectrograph at TRIUMF.[24] As the measurements only spanned the angular range from 75° to 155° (c.m.), the data were inadequate for extracting four parameters, so only b_1-b_3 were obtained. The angular range of the data set was then enlarged by including the analyzing power measurements of Dolnick[20] which spanned the angular range: 14-87° (c.m.). For the extended set, extraction of four b_k/a_0 coefficients was possible. The lower b_2/a_0 value characterizes the result obtained by this means. It is likely that the large values obtained for $\eta = 0.52$ and 0.63 are also artificially large due to limitations in the angular range of the data sets employed.

The theoretical situation is exemplified again by the predictions of Niskanen[16] shown in the figures. Up to about 600 MeV, his prediction for b_1/a_0 is impressively good. However, above 600 MeV, this calculation fails to reproduce the observed fall-off in b_1/a_0. Again, as with the differential cross section, the theoretical higher-order coefficients are in rather poor shape. Both the b_2/a_0 and b_3/a_0 are predicted to change sign and become negative at about 550 MeV, a feature completely at variance with the experimental results. Except in the range of 500-600 MeV, however, there is as yet very little experimental data of sufficient quality to enable extraction of these higher-order coefficients. Current experiments by Nann et al.[15] at LAMPF are endeavouring to provide such information in the energy region between 550 and 800 MeV, and further experiments are proposed at TRIUMF[17] for improving the quality of the data below 500 MeV.

It is interesting to note that both "single-arm" and double-arm" experiments are providing data of comparable levels of precision. In order to obtain such precision with a single-arm system, good particle selectivity is essential not only to eliminate the normal background

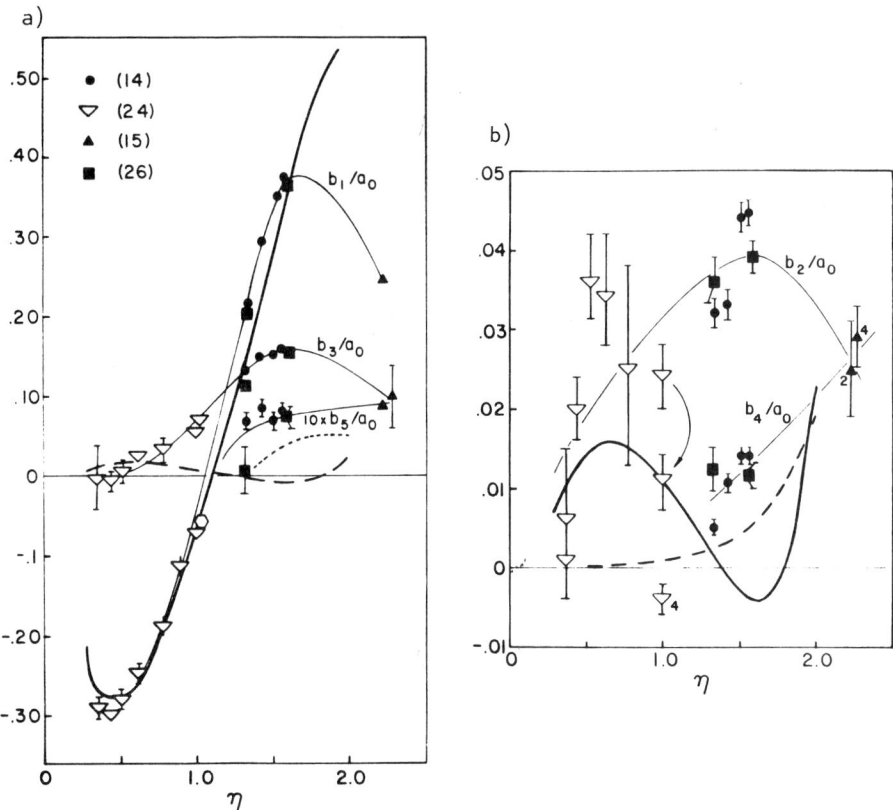

Fig. 5. a) Momentum dependence of the coefficients of the odd associated Legendre functions. The coefficients are shown relative to the total cross section (a_0). The solid curve is the theoretical expectation[16] for b_1/a_0, the dashed curve for b_3/a_0, and the dotted for b_5/a_0. The thin lines are only to guide the eye. b) Momentum dependence of the coefficients of the even associated Legendre functions. The solid curve is the theoretical expectation[16] for b_2/a_0, while the dashed is for b_4/a_0.

contributions, but also, at least in the case of the pp → dπ reaction, to distinguish the events from those of pp → pnπ where the pn pair has a low level of relative kinetic energy. To date, single-arm systems have universally employed magnetic spectrometric devices for this purpose.

1.1.5 <u>Spin-Spin Correlation Parameters</u>   As the spin of the initial pp state may be either singlet (s = 0) or triplet (s = 1) and the spin of the outgoing state is characterized by the triplet (s = 1) deuteron, the spin matrix characterizing the pp → dπ reaction has 3 × 4 matrix elements containing six independent complex amplitudes. Thus, eleven independent measurements are required to provide a complete experimental determination of the pion production amplitude. That a start has been made on such an experimental program is indicated by the measurements of the Geneva group at SIN. To date, this group has measured[26]

Table IV. Associate Legendre polynomial coefficients.

Ref.	Energy	n	b_1/a_0	b_2/a_0	b_3/a_0	b_4/a_0	b_5/a_0	Assumed a_2/a_0	a_4/a_0	a_6/a_0	χ^2/ν
17	310	0.367	-0.285(0.004)	0.006(0.010)	—	—	—	0.60	—	—	3.0
	320	0.444	-0.297(0.004)	0.019(0.005)	—	—	—	0.76	—	—	1.3
	330	0.513	-0.278(0.013)	0.036(0.006)	0.005(0.012)	—	—	0.90	—	—	2.5
	350	0.641	-0.243(0.003)	0.028(0.010)	0.022(0.003)	—	—	1.00	—	—	5.0
	375	0.774	-0.191(0.008)	0.014(0.012)	0.033(0.007)	-0.016(0.008)	—	1.05	—	—	2.4
	400	0.893	-0.113(0.010)	0.039(0.010)	0.066(0.010)	—	—	1.09	—	—	0.1
	425	1.000	-0.060(0.004)	0.024(0.004)	0.066(0.003)	—	—	1.11	—	—	3.2
	425	1.00	-0.080(0.055)	0.008(0.048)	0.055(0.022)	—	—	1.11	—	—	0.6
	425	1.00	-0.073(0.004)	0.011(0.004)	0.057(0.003)	-0.004(0.002)	—	1.11	—	—	2.4
14	514	1.34	0.214(0.004)	0.032(0.002)	0.133(0.002)	0.005(0.001)	0.007(0.001)	—	—	—	—
	540	1.42	0.295(0.004)	0.033(0.002)	0.149(0.002)	0.0105(0.0010)	0.0085(0.0010)	—	—	—	—
	569	1.52	0.350(0.004)	0.044(0.002)	0.151(0.002)	0.015(0.001)	0.0067(0.0010)	—	—	—	—
	583	1.56	0.375(0.004)	0.044(0.002)	0.160(0.002)	0.014(0.001)	0.008(0.001)	—	—	—	—
(final)	568	1.52	0.352(0.002)	0.042(0.002)	0.153(0.001)	0.0136(0.0011)	0.0082(0.001)	—	—	—	—
26	515	1.341	0.204(0.003)	0.036(0.003)	0.115(0.002)	0.0124(0.003)	—	1.06	-0.045	—	2.2
	578	1.555	0.366(0.002)	0.040(0.002)	0.158(0.002)	0.013(0.002)	0.007(0.002)	1.00	-0.11	—	1.3
15	800	2.213	0.247(0.006)	0.025(0.006)	0.090(0.006)	0.029(0.004)	0.001(0.004)	0.80	-0.24	-0.05	2.5

A_{LL} at five energies (440, 490, 515, 540 and 580 MeV) and A_{NN} at two (515 and 578 MeV). Representative samples of their results are shown in Fig. 6.

Measurements of A_{LL} and A_{SL} are also planned[27] at LAMPF, thus extending the range of energies for such measurements up to 800 MeV.

1.1.6 Deuteron Polarization

Experimental measurements involving the spin of the deuteron will obviously be required if a complete determination of the production amplitude is desired. A group at LAMPF is planning[28] a measurement of the vector polarization of the deuteron produced when polarized protons impinge on an unpolarized target.

Dependence on deuteron spin can also be ascertained using a polarized deuteron target in the inverse reaction: $\pi + \vec{d} \rightarrow 2p$. Measurements of this type have already been started at SIN with more extensive measurements involving improved apparatus currently underway. Preliminary values[29] for the vector analyzing power it_{11} obtained from the first measurements are illustrated in Fig. 7.

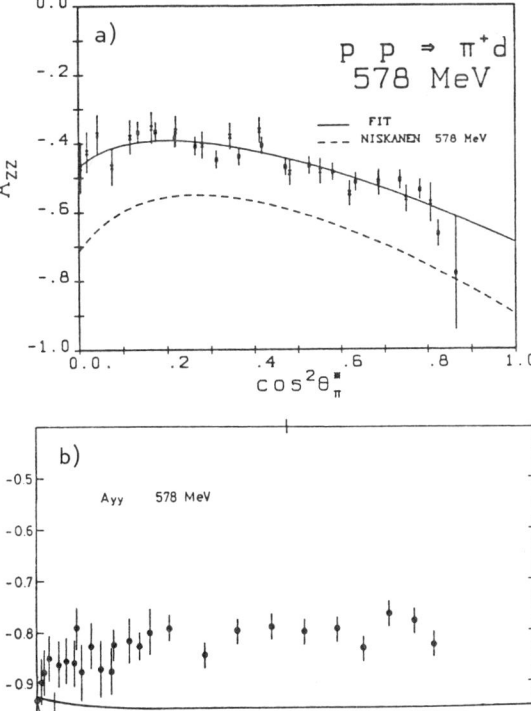

Fig. 6. Angular dependence of the spin-spin correlation parameters for the $\vec{p}\vec{p} \rightarrow d\pi^+$ reaction at 578 MeV. a) shows A_{ZZ} (or A_{LL}) vs. $\cos^2\theta$, while b) shows A_{YY} (or A_{NN}).

1.2 $pn \rightarrow d\pi^0$

This reaction has long been of interest as a test of isospin invariance in pion production reactions. In this reaction, isospin invariance precludes any contribution from the isospin singlet pn initial state. Application of the usual vector coupling algebra then implies that the cross section is just one half of the corresponding value for the $pp \rightarrow d\pi^+$ reaction (with the angular dependence being the same and thus symmetric about 90°). The very complete angular distribution[11] ($2 \leq \theta \leq 178°$) referred to in Sect. 1.1.1 for 800 MeV incident neutrons was obtained in order to provide a test of the

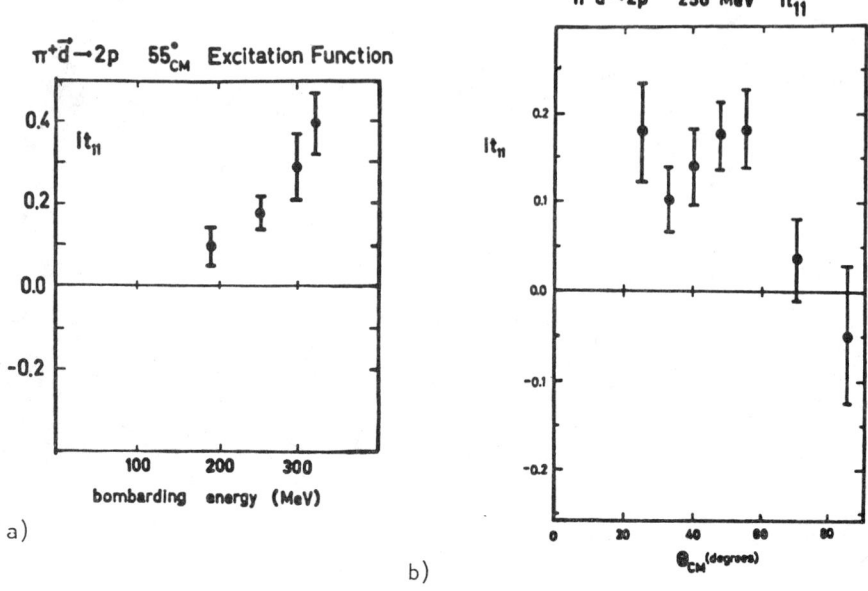

Fig. 7. a) Energy dependence of the vector analyzing power it_{11} for $\vec{\pi d} \to 2p$ for 55° (c.m.) production. b) Angular dependence of the vector analyzing power at 256 MeV incident pion energy.

magnitude of charge-symmetry breaking effects, effects that would lead to some asymmetry of the angular distribution about 90°. Hollas et al.[11] find no evidence for such an asymmetry to an accuracy of ±0.5%. For this reason the results of the analysis of the Hollas data were included in the pp → dπ data set shown in Fig. 2. Some uncertainty in deciding which value of the abscissa is most appropriate arises because of the different masses of the particles involved. Hollas et al. suggest that the reactions should be compared at the same value of the invariant mass. However, because of the dependence of the cross-section on the momentum of the outgoing particles (through phase space and centrifugal barrier considerations) I have compared the results at the same values of pion momenta (η). For most of the η range shown in Fig. 2 however, these different criteria produce very little positional shift of the data points.

Some "prehistoric" np → dπ⁰ data which disagree markedly in angular dependence with the pp → dπ⁺ data displayed in Fig. 2 have been ignored for the reasons stated in the introduction.

Some recent published data[30] from the Freiburg group at SIN yielded values of a_2/a_0 which were in substantial disagreement (several standard deviations) with the trend of the pp → dπ⁺ results shown in Fig. 2. Recent results,[31] however, which were obtained over an extended angular range are in good agreement with the pp → dπ⁺ results. This data, even though preliminary in nature, is shown in Fig. 2, and is included in Table III.

Some slight discrepancies are noted with the a_4/a_0 values for $\eta > 1$. In addition, the a_6/a_0 values (see Table III) appear to be substantially more negative than the corresponding $pp \to d\pi^+$ results in this energy range. The data is still preliminary, however, so not too much should be concluded regarding these higher order terms until analysis is complete. Also, one should be wary of inferring too much regarding the low-energy behaviour of the $pp \to d\pi^+$ reaction from the raw $np \to d\pi^0$ data, as Coulomb effects should lead to differences at the very low energies.

2. $NN \to NN\pi$

As shown by the data[32] in Fig. 8, the $pp \to pn\pi^+$ reaction dominates the pion production process for proton energies above 500 MeV, whereas for energies less than 500 MeV, the $pp \to d\pi^+$ reaction dominates. Considering the importance of the $pp \to pn\pi$ reaction in our understanding of pion production in general, surprisingly little work has been done on it in recent years. This situation is changing now, however, a change promoted at least in part by the interest generated by the spin dependent total reaction cross-section measurements ($\Delta\sigma_L$ and $\Delta\sigma_T$)[7,8] referred to earlier.

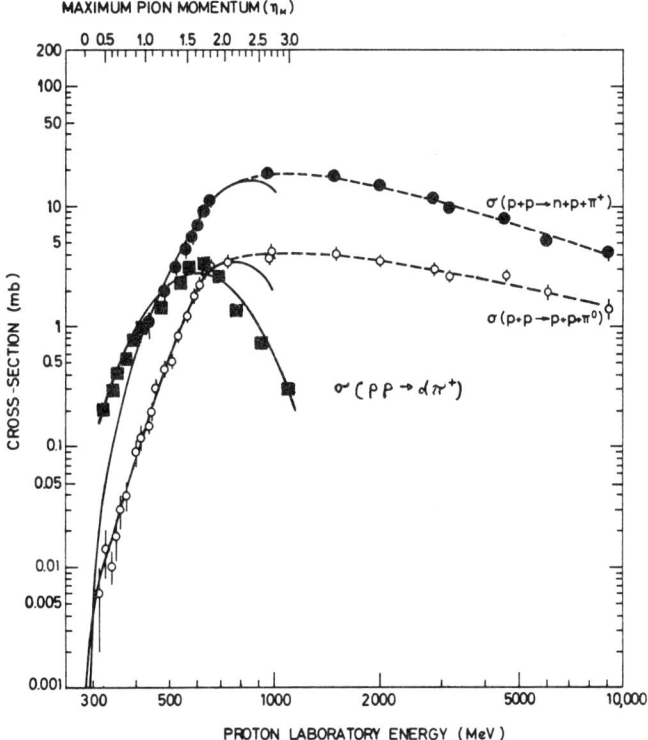

Fig. 8. Energy dependence of $pp \to d\pi$, $pp \to pn\pi^+$ and $pp \to pp\pi^0$ total cross sections.

The np → nnπ⁺ and np → ppπ⁻ reactions are interesting because they permit the investigation of pion production from the isospin singlet component of the initial nucleon-nucleon state. Because of isospin invariance, this reaction channel cannot occur via an N-Δ intermediate state. A LASL, Texas A and M and University of Texas collaboration performed some experimental measurements of these reactions at 800 MeV a few years ago.[43,46] As a result of such measurements they concluded that the singlet isospin contribution to pion production was less than 0.3 of the isospin triplet. Further experiments of this kind at LAMPF have been awaiting the re-installation of the neutron beam facility previously in use there.

A recent measurement of the total cross section for production of π⁻ by the reaction: np → ppπ⁻ at nine energies between 500 and 1000 MeV was presented to the 9th ICOHEPANS in Versailles (1981) by Dakhno et al.[33] Their data, together with older data for the reaction are shown in Fig. 9. The discrepancy with the older data points to the need for further experimental work in this area.

2.1 Differential Cross Section

2.1.1 Kinematically Complete (Exclusive) Cross Section

2.1.1.1 pp → pnπ⁺ Only two groups have reported results obtained from experiments utilizing a kinematically-complete description of the reaction products. Zulkarneev et al.[34] described such an experiment at 650 MeV (but without determination of the absolute normalization). A more extensive measurement of the reaction was carried out at 800 MeV by a Rice/Houston collaboration.[35] In this experiment, four p-π⁺ angle pairs were selected, with the proton angle varied from 15° to 30° (lab). The kinematics was chosen to provide maximal sensitivity to Δ⁺⁺ involvement in the intermediate state (with less sensitivity to Δ⁺ and non-isobar type intermediate states). Typical results for the proton momentum spectra (for protons in coincidence with a pion) are shown in Fig. 10. The strong peaking clearly shows the effects of a final state enhancement (FSI) between the pion and proton indicative of a 3/2-3/2 isobar. The experimental data was reasonably well fit by a Δ⁺⁺ production model based on π- and ρ-exchange. Another prominent feature of the data is the NN final state interaction (FSI) peak (d). Both singlet and triplet FSI enhancement factors were required in order to fit the data. Features that were difficult to reproduce were the large off-resonance cross section and the cross section at large angles.

2.1.1.2 pp → ppπ⁰ A kinematically-complete measurement of the pp → ppπ⁰ reaction at 600 MeV incident energy by Gugelot et al.[36] was obtained by detecting both protons emerging from the reaction and measuring their momenta with magnetic spectrometers. The data are discussed by Borie, Drechsel and Weber[37] who fitted the angular distributions with a moderate degree of success with a model containing both a (3,3) resonance and nucleon pole.

2.1.2 Inclusive Differential Cross Sections

2.1.2.1 pp → π⁺x Many experimental measurements of the momentum distribution of pions inclusively produced in nucleon-nucleon collisions have been made. References to the older experiments are given in

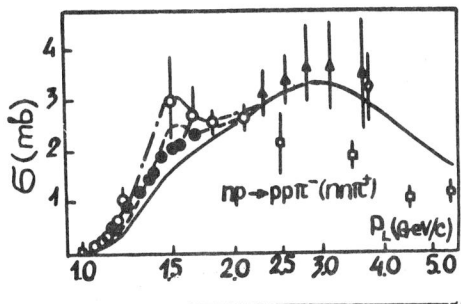

Fig. 9. Energy dependence of the total cross section for the np → ppπ⁻ reaction.

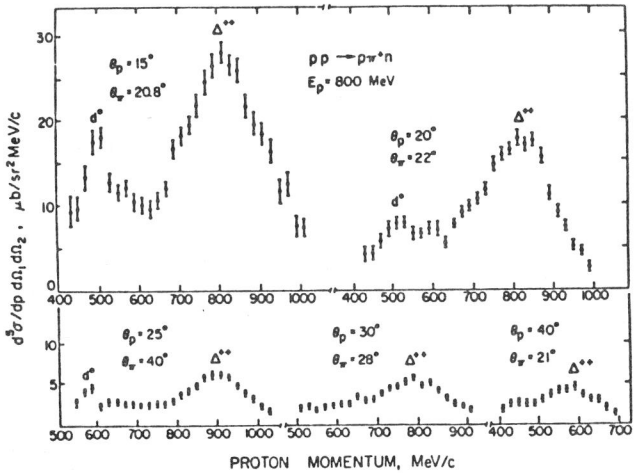

Fig. 10. Momentum spectra of protons in coincidence with π^+ for the pp → pnπ⁺ reaction at 800 MeV.

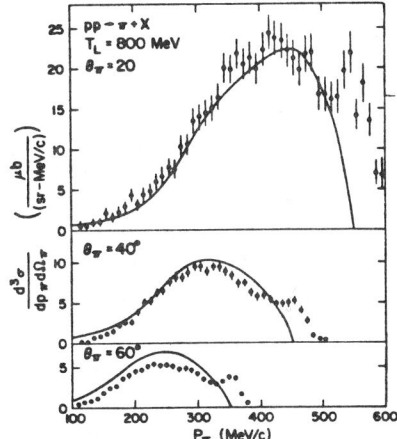

Fig. 11. Pion momentum spectra characterizing inclusive production in the pp reaction at 800 MeV.

Cochran et al.[4] Somewhat more recent results were presented by Bevington[38] at the Vancouver nucleon-nucleon conference, in which a single-arm configuration employing a magnetic spectrometer was used. The data were taken at 800 MeV at LAMPF. Typical spectra with fits by VerWest[39] are shown in Fig. 11. Here the peak at the high momentum end corresponds to those pions associated with the pp → dπ⁺ reaction. Similar inclusive pion spectra taken with a single-arm magnetic spectrograph are shown in Fig. 12. This preliminary data,[40] taken at TRIUMF using 450 MeV protons, involved the use of a

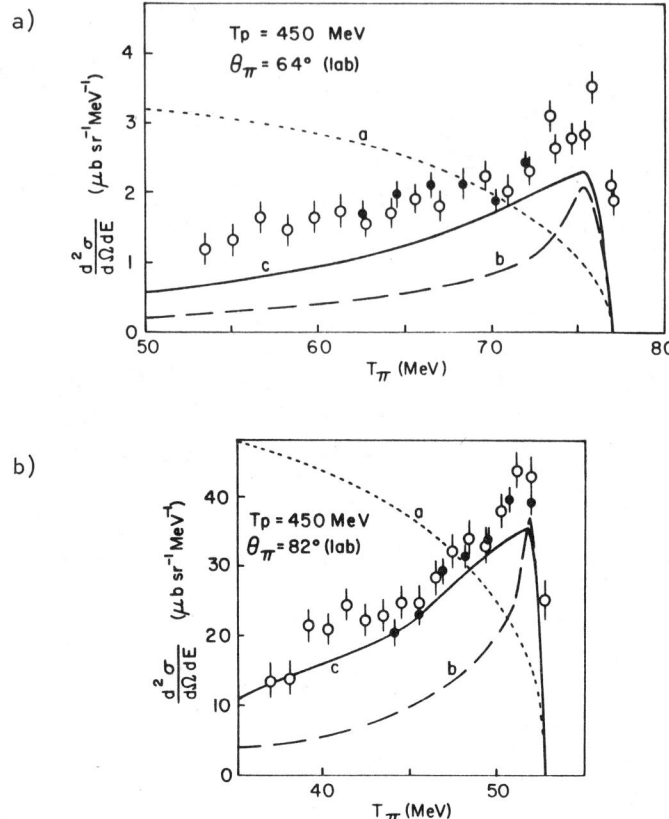

Fig. 12. Pion momentum spectra characterizing inclusive production in the pp reaction at 450 MeV. The dotted line (a) shows the phase space shape without FSI. The dashed line (b) shows the effect of a singlet FSI and the solid line (c) shows the effect of a statistical mix of singlet and triplet FSI.

spectrographic system of sufficient resolution to resolve the pp → dπ^+ peak from the pp → pnπ^+ continuum. The data in Fig. 12 shows the results after the pp → dπ^+ line has been subtracted. The pn FSI is clearly evident in this spectrum. The solid lines show the shape of the phase space dependence.[41]

2.1.2.2 np → px 0° proton spectra from the 800 MeV neutron bombardment of hydrogen were presented by Glass et al.[42] at the nucleon-nucleon meeting in Vancouver. These spectra complement similar measurements performed by the same group using an incident beam of protons instead of neutrons.[43] An example of the np → px spectrum together with a theoretical fit by VerWest[39] is shown in Fig. 13. The shoulder on the low energy side of the peak is due to the zero-energy (singlet and triplet) nucleon-nucleon FSI.

2.2 Analyzing Power

2.2.1 $\vec{p}p \to pn\pi^+$ Analyzing power measurements were also made by the Rice/Houston group at LAMPF for 800 MeV protons using the same kinematically-complete experimental arrangement as used in the differential cross section measurements described in 2.2.1.1. Their results[44] together with fits obtained using a peripheral model incorporating both π- and ρ-exchange are illustrated in Fig. 14. This group plans to obtain additional data for proton energies of 500 and 650 MeV in the near future.

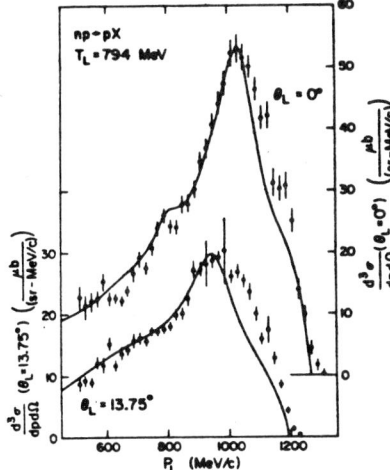

Fig. 13. Momentum dependence of the protons produced at 0° in the np → px reaction at 800 MeV.

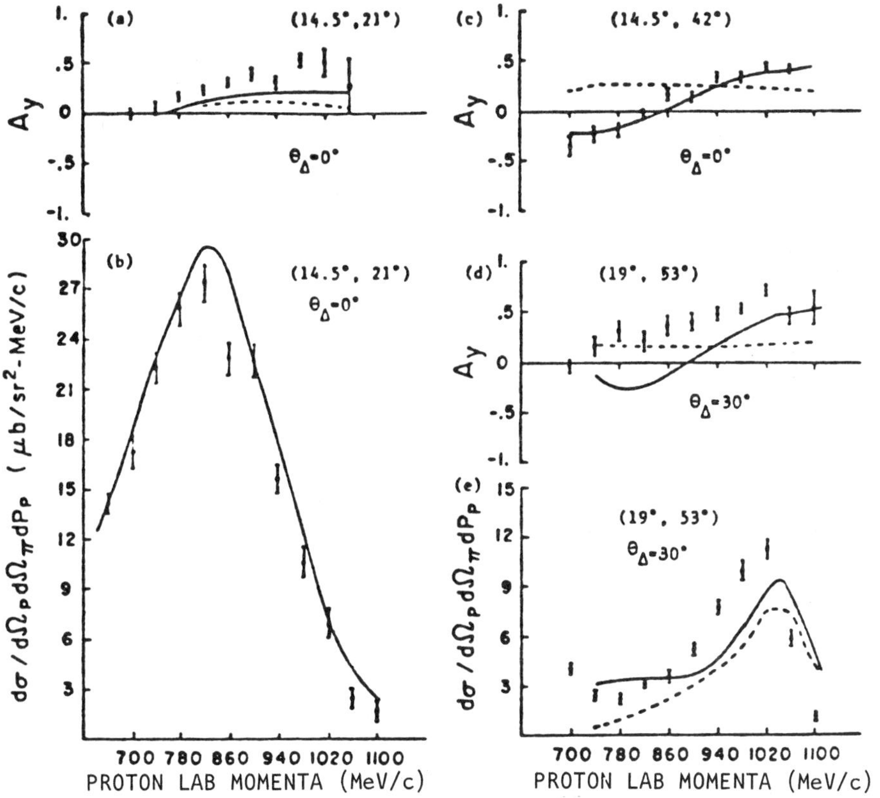

Fig. 14. Analyzing power and differential cross sections for the protons in coincidence with pions from the $\vec{p}p \to pn\pi^+$ reaction at 800 MeV.

2.3 Spin-Spin Correlation Parameter

2.3.1 $\vec{pp} \to pn\pi^+$ At TRIUMF, experiments to measure the spin-spin correlation parameter, A_{LL} are currently underway[45] with data currently being obtained at 380, 425, 480 and 516 MeV incident energy. As with the Rice/Houston experiment, this is a kinematically-complete experiment, with the neutron energy measured by time of flight and the angular coordinates of all three outgoing particles also measured (the charged particles by MWP chambers, and the neutrons by a scintillation counter hodoscope). This group plans to follow this experiment with measurements of A_{LS}, A_{SS} and A_{NN}. Kinematically-complete differential cross sections in this energy range are also being planned.

3. CONCLUSIONS

As a result of recent precision measurements, detailed experimental characteristics of the $pp \to d\pi$ reaction are now beginning to emerge. Most experiments have involved measurements of either the (unpolarized) differential cross section or the polarization analyzing power. The lowest partial wave components are now quite well determined, but more work is required in order to elucidate the energy dependence of the higher components. More complete spin-dependent measurements have also been initiated. However, the task of completely defining such spin dependence—at any energy—is still very far from complete. Another major achievement during this past year has been the demonstration of a 0.5% upper limit for charge symmetry breaking effects in the $np \to d\pi^0$ reaction.

Measurements of the pion production reaction: $pp \to pn\pi^+$ (leading to a three-body final state) have progressed to the stage where kinematically-complete experiments are now being performed using polarized beams. Such experiments are much more complex than those involving few-body final states and so it will be a long time before a comparable body of data is acquired for this type of reaction. Indeed, because of the complexity (and range of kinematic variables from which to select) such experiments tend to be designed to answer specific physical questions (role of the Δ^{++}, FSI effects, etc.). In this regard, it is clear that more effort is required using neutron beams in order to evaluate the role of the isospin singlet initial state in the pion production process. Finally, considering the complexity of such experiments, more active involvement of our theoretical colleagues during the planning and design stage of the experiments might help to extend the range of physical questions which such experiments could address.

ACKNOWLEDGEMENTS

The author gratefully acknowledges the many useful discussions he has had with his collaborators regarding a number of aspects of this presentation. In addition, the generosity of the experimental spokesmen at the various laboratories in providing data, both final and preliminary, is also much appreciated.

REFERENCES

1. G. Jones, Nucleon-Nucleon Interactions — 1977 (Vancouver), AIP Conference Proceedings 41, 292 (1978); Few Body Systems and Nuclear Forces II (Graz), Lecture Notes in Physics 87, 142 (1978).
2. C. Richard-Serre et al., Nucl. Phys. B20, 413 (1970).
3. D.F. Measday and G.A. Miller, Ann. Rev. Nucl. Sci. 29, 121 (1979).
4. D. Cochran et al., Phys. Rev. D6, 3085 (1972).
5. S. Mandelstam, Proc. Roy. Soc. (London) A244, 491 (1958).
6. E. Ferrari and F. Selleri, Nuovo Cimento 27, 1450 (1963).
7. I.P. Auer et al., Phys. Rev. Lett. 41, 354 (1978).
8. E.K. Biegert et al., Phys. Lett. 73B, 235 (1978).
9. F. Mandl and T. Regge, Phys. Rev. 99, 1478 (1955) for example.
10. J.A. Niskanen, Proc. 5th Int. Symp. on Polarization Phenoma in Nuclear Physics (Santa Fe) 1980.
11. C.L. Hollas et al., LAMPF preprint LA-UR-81-693 (1981).
12. B.G. Ritchie et al., Phys. Rev. C24, 552 (1981).
13. J. Boswell et al., private communication (1981).
14. J. Hoftiezer et al., Phys. Lett. 100B, 462 (1981), and private communication (1981).
15. H. Nann et al., Phys. Lett. 88B, 257 (1979).
16. J.A. Niskanen, Nucl. Phys. A298, 417 (1978); Phys. Lett. 79B, 190 (1978).
17. P. Walden, private communication, 1981.
18. $\sigma(\pi d \to 2p) = \frac{2}{3} \frac{(P/\mu)^2}{\eta^2} \sigma(pp \to d\pi) = \frac{1}{6} \frac{\{[1+\eta^2]^{\frac{1}{2}} + [(M_d/\mu)^2 + \eta^2]^{\frac{1}{2}}\}^2 - (2m_p/\mu)^2}{\eta^2}$

    $\times \sigma(pp \to d\pi)$,

    where m_p, M_d, μ are the proton, deuteron and pion masses, respectively.
19. C.M. Rose, Phys. Rev. 154, 1305 (1967).
20. C.L. Dolnick, Nucl. Phys. B22, 461 (1970).
21. D. Axen et al., Nucl. Phys. A256, 387 (1976).
22. D. Aebischer et al., Nucl. Phys. B108, 214 (1976).
23. R. Arndt, private communication via SAID (1981).
24. P. Walden, et al., Phys. Lett. 81B, 156 (1979) for example.
25. M. McNaughton, private communication (1980) (exp. 27/33/336).
26. E. Aprile et al., private communcation (1981).
27. J. Simmons et al., private communication (1981).
28. B.E. Bonner et al., private communication (1981).
29. E. Boschitz et al., private communication (1981).
30. W. Hürster et al., Phys. Lett. 91B, 214 (1980).
31. E. Rössle, private communication (1981).
32. W.O. Lock and D.F. Measday, Intermediate Energy Nuclear Physics, (Methuen, London, 1970).
33. L.G. Dakhno et al., 9th ICOHEPANS (Versailles) 1981, submission A32.
34. R. Ya Zulkarneev et al., Yad. Fiz. 14, 989 (1971); [Sov. J. Nucl. Phys. 14, 555 (1972)].
35. J. Hudomalj-Gabitzsch et al., Phys. Rev. C18, 2666 (1978).

36. K. Gugelot et al., Nucl. Phys. B37, 93 (1972).
37. E. Borie, D. Drechsel and H.J. Weber, Z. Phys. 267, 393 (1974).
38. P.R. Bevington, Nucleon-Nucleon Interactions — 1977 (Vancouver), AIP Conference Proceedings 41, 305 (1978).
39. B. VerWest, Phys. Lett. 83B, 161 (1979).
40. W.R. Falk et al., private communication (1981).
41. D. Beder, private communication (1981).
42. G. Glass et al., Nucleon-Nucleon Interactions — 1977 (Vancouver), AIP Conference Proceedings 41, 544 (1978).
43. G. Glass et al., Phys. Rev. D15, 36 (1977).
44. E.A. Umland, T.M. Duck and G.S. Mutchler, 5th Int. Symp. on Polarization Phenomena in Nuclear Physics (Santa Fe), 1980.
45. D. Axen, private communication (1981).
46. W.R. Thomas, LASL report LA-6962-T (1977).
47. B.M. Preedom et al., Phys. Rev. C17, 1402 (1978).

DISCUSSION

R.R. Silbar (Los Alamos Scientific Lab.): I was told by somebody, I don't remember who, that phase shift analysis people have made the statement that the singlet to triplet ratio is much smaller than the 0.3 that you indicated, and that they find very little inelastic I=0 cross section. What would Dr. Rössle say to that?
E. Rössle (Univ. of Freiburg): I will cover that point tomorrow in my talk.

M. Dillig (Univ. of Erlangen-Nürnberg): There have been some discussions concerning the comparison between the rescattering model and the Niskanen model. The groups employing the rescattering model question whether Niskanen has included enough partial waves for the outgoing pion. My question: do you think that the discrepancies at high energies reflect deficiencies in the rescattering model or are they, perhaps, just reflecting the lack of higher partial waves in the coupled channel model? Are the discrepancies of a kinematical or dynamical origin?
J. Niskanen (Univ. of Helsinki): The reason is kinematic. In my earliest results I used for the center of mass energy just half of the lab energy – not the relativistic one. I thought that I would leave 10% effects out at that stage. The 10% effect was, however, in the abscissa in a very steep function, and of course it's 30%-50% in the height of the function. Once I used the relativistic center of mass energy, I got a perfectly good fit at high energy to the total cross section. Recently I have included high partial waves also. I had hoped I could omit those. They destroy the fit to the differential cross section, but the effect on the total cross-section is small.

QUESTIONS IN A MICROSCOPIC THEORY OF PION PRODUCTION/ABSORPTION[†]

Manoj K. Banerjee
University of Maryland, College Park, Md. 20742

ABSTRACT

In this talk we show that in the treatment of nuclear processes involving pions, such as pion production/absorption or exchange currents, one should use for the πNN vertex function $g_\pi(q^2)\gamma_5\tau_\alpha$, where q is the pion four-momentum and $g_\pi(m_\pi^2) \simeq 13.4$ is the pion-nucleon coupling constant. The result is independent of the form and details of any interaction Lagrangian. We show that the difference between $g_\pi(q^2)\gamma_5\tau_\alpha$ and the fully dressed πNN vertex function contributes to amplitudes involving more bosons, e.g., scattering amplitudes (πN→πN), production amplitudes (πN→ππN), etc. Such contributions are mediated through either seagull terms or P11 intermediate states. We point out that it is neither necessary nor practical to calculate these partial contributions to scattering and production amplitudes. One can always exploit other approaches, theoretical and phenomenological, to make reasonably good models for the full amplitudes.

Almost all calculations of processes involving pions and nucleons are critically dependent on the choice of the πNN vertex function. The popular choices are pseudoscalar (PS) coupling $g_\pi\gamma_5$ and pseudovector (PV) or derivative coupling, $(g_\pi/2M)\gamma_5 \slashed{q}$ where q is the pion four-momentum. Some authors have considered linear combinations of the two. In many calculations a form factor, usually a function of the pion space momentum, is used.

To illustrate the importance of the choice of the vertex function I show in Fig. 1 the recent results of Cooper and Sherif[1] for the reaction ^{40}Ca(p,π$^+$)^{41}Ca(g.s.). These authors conclude that one should use the PV vertex.

In this talk I will present the results of work done by George Walker and myself,[2] done while we were both spending o sabbatical at the University of Washington, Seattle. Our conclusion is diametrically opposite to that of Cooper and Sherif. We conclude that one must use the PS form $g_\pi(q^2)\gamma_5$ where q is the four-momentum. Furthermore, the result is completely independent of the form of the interaction Lagrangian.

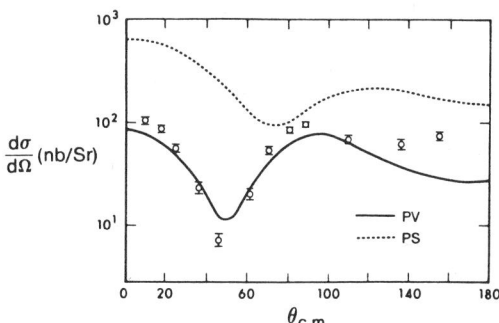

Fig. 1: Results of Cooper and Sherif for ^{40}Ca(p,π$^+$)^{41}Ca with 160 MeV protons.

Of course, a complete knowledge of the correct πNN vertex function does not ensure that we can calculate all pionic processes correctly. For each process there may be other major questions and uncertainties which must be resolved before a useful calculation can be done. For example, for the $^{40}Ca(p,\pi^+)^{41}Ca$ we may have to add to the one-nucleon mechanism, Fig. 2a, the two-nucleon mechanism, Fig. 2b. At least, the two are necessary to reflect the orthonormality of the initial and final states. Cooper and Sherif themselves have introduced a feature absent in most previous calculations, namely, relativistic nucleon wave functions.

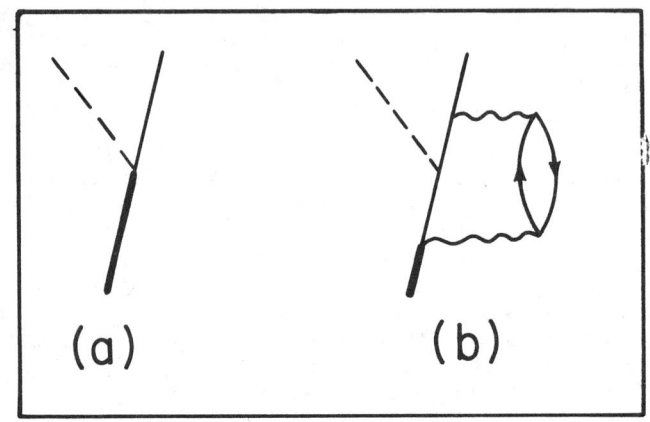

Fig. 2: (a) and (b) are Goldstone diagrams for one- and two-nucleon mechanisms. Thick line represents propagation in optical potential appropriate for the initial proton.

It will become clear that our result is obtained by paying careful attention to the event that one wishes to describe with a πNN vertex. Obviously the event is the following. Initially one has a nucleon, not an N* (a resonance with the quantum numbers of a nucleon), which propagates some distance in space-time. It then absorbs or emits a pion, but still remains a nucleon, does not get internally excited to N* or Δ. Then it propagates again as a nucleon some distance in space-time. The ability to propagate both before and after the absorption/emission is important. If this is not possible then we do not have a simple pion absorption/emission event, because then other events become coincident in space-time with the pion emission/absorption event. Of course, both propagations must take place without internal excitation, or else we will have a πNN* or a πN*N* vertex.

Let us now examine the relevant parts of a Feynman graph and see how we can isolate and describe the event just discussed. It is sufficient for our purpose to discuss only two quantities; the fully dressed nucleon propagator, $S_F'(p)$, and the fully dressed NN vertex function $\Gamma(p,q,p')$.

$$S_F'(p) = -i \int d^4x \, e^{ip \cdot x} <0|T(\psi(x),\bar{\psi}(0))|0>$$
$$= \frac{1}{\not{p} - M_0 - \Sigma(p)}$$

where M_0 is the bare mass and $\Sigma(p)$, the proper self-energy of the nucleon. It is an experimental fact that there is only one stable physical particle with the quantum numbers $B = 1$, $I = J = 1/2$, $P = +$, namely, the nucleon and that all other physical states with these quantum numbers are systems of two or more particles, the lightest being the πN system. Assuming that our field theory conforms to this experimental fact we write

$$S_F'(p) = \frac{Z_2}{\not{p} - M} + \int_{(m+m_\pi)^2}^{\infty} d\sigma \frac{\not{p}\, \rho_1(\sigma) + \rho_2(\sigma)}{p^2 - \sigma + i\eta}$$

where M is the physical mass of the nucleon and Z_2 the wave function renormalization constant. The pole term describes the propagation of a nucleon, while the cut term describes the propagation of πN or more complicated systems, which, for brevity, we call N*. Diagrammatically we represent the decomposition of $S_F'(p)$ as shown in Fig. 3.

Fig. 3: Decomposition of $S_F'(p)$.

For our purpose we attach only the thin lines to a πNN vertex. Thus Fig. 4 describes πN scattering with P11 intermediate state cut. Therefore the two vertices are πNN* and not πNN vertices. Hence, from now on we will exhibit only the pole part of $S_F'(p)$ in nucleon propagators attached to a πNN vertex function, whenever this is appropriate.

Next we examine the fully dressed πNN vertex, shown in Fig. 5, which also shows some typical vertex renormalization diagrams. For convenience of labelling external lines are attached to

Fig. 4: Two pion vertices with N* intermediate state.

the vertex diagrams. But these diagrams do not contain external propagators, neither can they be reduced into a propagator part and

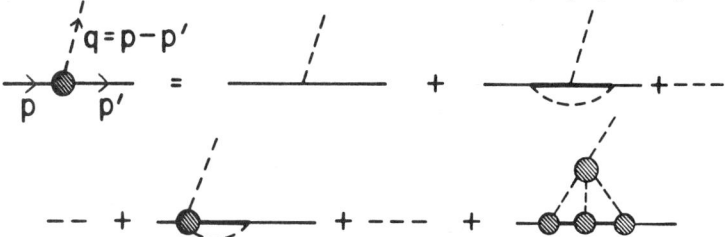

Fig. 5: Fully dressed πNN vertex and some examples of its diagrammatic content.

a vertex part by cutting a single line, pion or nucleon. A completely general form of the πNN vertex function is

$$\Gamma_{\pi NN} = \frac{1}{Z_2 \sqrt{Z_\pi}} [G_\pi(p^2,q^2,p'^2)\gamma_5 + g_1(p^2,q^2,p'^2)\gamma_5 \frac{(\not{p}-M)}{2M}$$
$$+ g_2(p^2,q^2,p'^2) \frac{(\not{p}'-M)}{2M} \gamma_5 + g_3(p^2,q^2,p'^2) \frac{(\not{p}'-M)}{2M} \gamma_5 \frac{(\not{p}-M)}{2M}]\tau_\alpha$$

With γ_5 and two independent four-momenta, say, p and p', we can construct only four linearly independent pseudoscalars and the form, shown here, is a particular choice. From our earlier remarks it is also clear that the scalar functions $G_\pi(p^2,q^2,p'^2)$, $g_n(p^2,q^2,p'^2)$, n = 1,2,3, considered as analytic functions of p^2 with the other two variables, q^2 and p'^2 fixed has no pole nor any other singularity at $p^2 = M^2$. It does have a cut starting at $p^2 = (M+m_\pi)^2$. Identical remarks are applicable in the complex p'^2 plane. (In the complex q^2 plane there is no pole at $q^2 = m_\pi^2$, but a cut starting at $q^2 = 9 m_\pi^2$). From symmetry under charge conjugation one has $G_\pi(p^2,q^2,p'^2) = G_\pi(p'^2,q^2,p^2)$, $g_1(p^2,q^2,p'^2) = g_2(p'^2,q^2,p^2)$ and $g_3(p^2,q^2,p'^2) = g_3(p'^2,q^2,p^2)$.

In the first-order pseudoscalar coupling theory $G_\pi = g_0$, the bare coupling constant, and $g_1 = g_2 = g_3 = 0$. The last three appear in third- and higher-order perturbation calculations. In the first-order pseudovector coupling theory $G_\pi = g_1 = g_2 = g_0$, while $g_3 = 0$. Once again the g_3 term appears in third and higher orders. Thus there is a simple and transparent connection between the functions $G_\pi(p^2,q^2,p'^2)$ and $g_n(p^2,q^2,p'^2)$, n = 1,2,3 and the interaction Lagrangian in first-order perturbation theory. Whether the connection survives when the vertex is fully dressed, in all orders, can be known only upon doing a complete calculation. Among other things the result depends on the choice of not only the πNN interaction but also the ππ interaction, as illustrated in the last diagram of Fig. 5.

We suggest that it is not practical to establish the connection between the fully dressed vertex and the chosen Lagrangian. We also suggest that it is not necessary to know what the connection is in order to develop a practical and theoretically sound approach to the treatment of pionic processes.

Let us examine what happens to $\Gamma_{\pi NN}$ in a few typical situations.

<u>Case I</u>: $\Gamma_{\pi NN}$ is the first event on a nucleon line of external momentum p. So p is on the nucleon mass shell and there is a spinor $u(\vec{p})$ on the right. Notice the use of the pole part (thin line) on the right of the vertex. The algebraic expression for the full graph will contain the factors

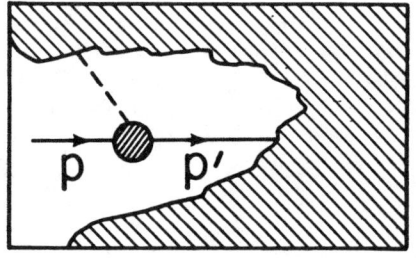

Fig. 6: $\Gamma_{\pi NN}$ as the first event on a nucleon line.

$$\ldots \times \frac{Z_2}{\not{p}'-M} \Gamma_{\pi NN} u(\vec{p}) = \ldots \times \frac{1}{\not{p}'-M} \frac{1}{\sqrt{Z_\pi}}$$
$$\times [G_\pi(M^2, q^2, p'^2) + g_2(M^2, q^2, p'^2) \frac{\not{p}'-M}{2M}] \gamma_5 \tau_\alpha u(\vec{p})$$

The g_1 and g_3 terms have disappeared because $(\not{p}-M)u(\vec{p}) = 0$ in this case. Note also that the $(\not{p}'-M)$ factor with g_2 will cancel the nucleon propagator. Thus by our previous arguments the g_2 term will not describe a pion absorption/emission event. Only the G_π term can do so.

Case II: $\Gamma_{\pi NN}$ is the last event on a nucleon line of external momentum p'. Here we have

$$\bar{u}(\vec{p}') \Gamma_{\pi NN} \frac{Z_2}{\not{p}-M} \times \ldots = \bar{u}(\vec{p}') \gamma_5 \tau_\alpha [G_\pi(p^2, q^2, M^2) + g_1(p^2, q^2, M^2) \frac{\not{p}-M}{2M}] \frac{1}{\sqrt{Z_\pi}} \frac{1}{\not{p}-M}$$

Again only the G_π term is relevant.

Case III: $\Gamma_{\pi NN}$ is the only event on a nucleon line. Now we have

$$\ldots \times \frac{1}{\sqrt{Z_\pi}} G_\pi(M^2, q^2, M^2) u(\vec{p}') \gamma_5 \tau_\alpha u(\vec{p}).$$

Of course, we are most familiar with the function

$$g_\pi(q^2) = G_\pi(M^2, q^2, M^2).$$

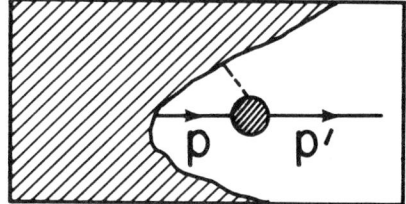

Fig. 7: $\Gamma_{\pi NN}$ is the last event on a nucleon line.

In this particular example, q^2 is space-like. But its analytic continuation to time-like q^2 is possible and is the subject of dispersion theoretic studies of the πNN vertex. We know that

$$g_\pi(m_\pi^2) \simeq 13.4$$

is the pion-nucleon coupling constant.

Finally we go to the general event where $\Gamma_{\pi NN}$ is an intermediate event.

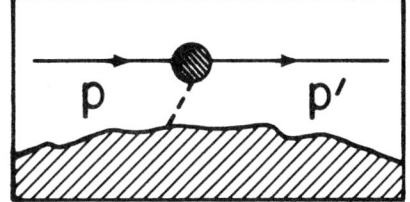

Fig. 8: $\Gamma_{\pi NN}$ is the only event on a nucleon line.

Case IV: $\Gamma_{\pi NN}$ is an intermediate event. To be specific, yet simple, we sandwich $\Gamma_{\pi NN}$ between a BNN and B'NN vertices, represented by Γ_{BNN} and $\Gamma_{B'NN}$, respectively. B and B' are two arbitrary bosons, π, ρ, ω, etc. The algebraic expression for the Feynman graph contains, among other things,

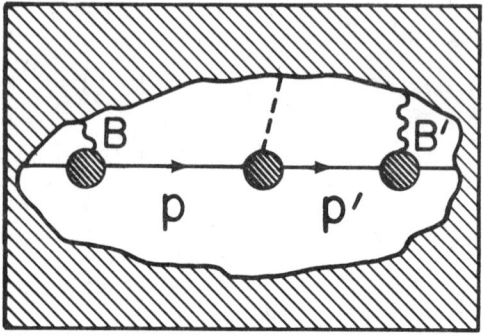

Fig. 9: $\Gamma_{\pi NN}$ is an intermediate event.

$$\ldots \times \Gamma_{B'NN} \frac{Z_2}{\not{p}'-M} \Gamma_{\pi NN} \frac{Z_2}{\not{p}-M} \Gamma_{BNN} \times \ldots$$

$$= \ldots \times \sqrt{Z_2} [\Gamma_{B'NN} \frac{1}{\not{p}'-M} G_\pi(p^2,q^2,p'^2) \gamma_5 \tau_\alpha \frac{1}{\not{p}-M} \Gamma_{BNN}$$

$$+ \Gamma_{B'NN} \frac{1}{\not{p}'-M} g_1(p^2,q^2,p'^2) \gamma_5 \tau_\alpha \frac{1}{2M} \Gamma_{BNN}$$

$$+ \Gamma_{B'NN} \frac{1}{2M} g_2(p^2,q^2,p'^2) \gamma_5 \tau_\alpha \frac{1}{\not{p}-M} \Gamma_{BNN}$$

$$+ \Gamma_{B'NN} \frac{1}{2M} g_3(p^2,q^2,p'^2) \gamma_5 \tau_\alpha \frac{1}{2M} \Gamma_{BNN}] \sqrt{Z_2} \times \ldots$$

The factors ($\not{p}$-M) and ($\not{p}'$-M) of the g_1, g_2 and g_3 terms cancel one or both of the nucleon pole terms adjacent to $\Gamma_{\pi NN}$. As a result, in the last three terms the π vertex and the B vertex and/or B' vertex either occur at the same space-time point or they are connected by N* propagation. The latter is a possibility because $g_n(p^2,q^2,p'^2)$ do have cuts corresponding to N* propagation. These three do not contribute to what we will call pion absorption/emission by a nucleon. Only the G_π term has a chance to do so.

Next we write

$$G_\pi(p^2,q^2,p'^2) = g_\pi(q^2) \quad (\equiv G_\pi(M^2,q^2,M^2))$$

$$+ [M^2 \frac{G_\pi(p^2,q^2,M^2)-G_\pi(M^2,q^2,M^2)}{p^2 - M^2}] \frac{p^2-M^2}{M^2}$$

$$+ \frac{p'^2-M^2}{M^2}[M^2 \frac{G_\pi(M^2,q^2,p'^2)-G_\pi(M^2,q^2,M^2)}{p'^2 - M^2}]$$

$$+ \frac{p'^2-M^2}{M^2}[M^4 \frac{G_\pi(p^2,q^2,p'^2)-G_\pi(M^2,q^2,p'^2)-G_\pi(p^2,q^2,M^2)+G_\pi(M^2,q^2,M^2)}{(p^2-M^2)(p'^2-M^2)}]\frac{p^2-M^2}{M^2}$$

Because of the analytic properties of $G_\pi(p^2,q^2,p'^2)$, noted earlier, the combinations in the three square brackets are all analytic in the neighborhood of $p^2 = M^2$ and $p'^2 = M^2$. Thus the last three terms have zeros at $p^2 = M^2$ and/or $p'^2 = M^2$ and, therefore have the same policidal tendency as the g_1, g_2, and g_3 terms. So, we conclude that pion absorption/emission events can be described by only the $(1/Z_2\sqrt{Z_\pi})g_\pi(q^2)\gamma_5\tau_\alpha$ term. The difference between it and the full $\Gamma_{\pi NN}$ contributes to either the seagull term or the P11 cut terms of processes such as $\pi N \to BN$, $\pi N \to B'N$ or $\pi N \to BB'N$, etc.

To ensure that we, indeed, have a pion emission/absorption event it is clear that Γ_{BNN} and $\Gamma_{B'NN}$ must also be subjected to the type of treatment just described. The general instruction is this:

For any field theoretical multipoint function involving two nucleon lines and one or more boson lines, retain only that part which can survive on the mass shell of both nucleons. The remainder must be counted as contributions to amplitudes involving more bosons.

Note carefully that in $g_\pi(q^2)$ the pion four-momentum q retains whatever its value is in the original diagram, not necessarily spacelike as would be the case when p and p' are actually on the nucleon mass shell.

There should be no misunderstanding that we are suggesting that the difference between $\Gamma_{\pi NN}$ and $(1/Z_2\sqrt{Z_\pi})g_\pi(q^2)\gamma_5\tau_\alpha$ be ignored or approximated away. Thus consider the simple examples of Feynman diagrams occuring in the NN→NNπ process. The four-point blob in Fig. 10(b) is the sum of all Feynman diagrams occurring in πN scattering amplitude except the ones which can be reduced into two Feynman graphs by cutting a single nucleon line. Indeed this last group is completely represented by the top part

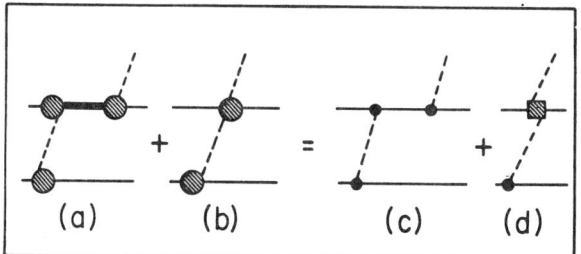

Fig. 10: Analysis of certain diagrams for NN→NNπ process.

of Fig. 10(a). Remember that it contains the nucleon pole as well as some P11 cut parts. The hatched circular box of Fig. 10(b) has only cuts, but not confined to P11 channel only. Our scheme leads to the arrangement shown in Figs. 10(c) and (d). Here the little solid circle stands for $(1/Z_2\sqrt{Z_\pi})g_\pi(q^2)\gamma_5\tau_\alpha$ and the square hatched box the full πN scattering amplitude minus the positive and the negative energy (Z graph) pole terms. The square box includes everything that Fig. 10(a) produces due to the killing of the intermediate nucleon pole. It is not practical to calculate reliably either these contributions or the circular hatched box by starting out with a field theoretic Lagrangian. On the other hand, there are reasonably good models for the square box. These models are easier to develop if we choose to lump the Z-graph part of Fig. 10(c) with it. At the very least, one can make models for the square box based on our experimental knowledge of πN scattering. This is not true for the separate parts of Figs. 10(a) and (b).

I hope that I have communicated our basic result and the logic behind it. I end by restating the main result. For the calculation of pionic processes, one should use $(1/Z_2\sqrt{Z_\pi})g_\pi(q^2)\gamma_5\tau_\alpha$ for the πNN vertex. This inference is independent of the choice of the interaction Lagrangian. Of course, the value and the q^2 dependence of $g_\pi(q^2)$ are determined by the Lagrangian, just as it determines the masses of nucleon and pion and every other relevant physical quantity.

REFERENCES

†Work supported by the U. S. Department of Energy and the National Science Foundation.
1. E. D. Cooper and H. S. Sherif, Phys. Rev. Lett. <u>47</u>, 818 (1981).
2. A detailed paper is being prepared.

DISCUSSION

<u>L.S. Kisslinger</u> (Carnegie Mellon Univ.): That's a very beautiful idea. However, I have one question about the starting point. I didn't understand why g_1, for example, cannot have a logarithmic singularity in k^2-m^2, so that the product $g_1 \times G(M^2)$, which is the physical point that one looks at, is a branch point.

<u>M. Banerjee</u> (Univ. of Maryland): This is because you can have a logarithmic singularity only because of the electromagnetic corrections. Otherwise you must start with the next available hadron mass with the quantum numbers of the nucleon.

<u>Kisslinger</u>: I thought you would say that. What I didn't understand is that if one starts with a field theory rather than physical objects, whether there can't be massless quark properties which might give developing singularities at the origin.

<u>Banerjee</u>: I would rule those things out. Cut singularities must reflect physical particles with a physical spectrum. I would consider that it is a very bad theory which has singularties that have nothing to do with physical states.

<u>Kisslinger</u>: I would agree to that. I thought your starting point was bare theory.

<u>Banerjee</u>: We start with the original theory, but then we discuss the fully dressed Lagrangian, that is, the true 3-point function. Otherwise you can have all kinds of trouble.

<u>J.M. Eisenberg</u> (Tel Aviv Univ.): Most calculations that are done so quickly go over to a phenomenological plane that the field theory questions may or may not be of great pertinence, depending on what you've summed and what you can't sum. Let me sharpen my question a little further. Sometimes comparative shopping helps. I'd appreciate it if you could tell us why one should go this route rather than, say, the route that is suggested by Weinberg's Lagrangian.

<u>Banerjee</u>: If you read the paper carefully you will see that Weinberg wrote down the Lagrangian with the following purpose. It was a mnemonic device. If you use the Lagrangian to calculate tree graphs, you've got the soft pion limit for processes involving pions. But that is not our object. And we don't even need a mnemonic device. We need a Lagrangian which will, in principle, give the mass of the nucleon, the mass of the pion, the value of g_π and everything else.

<u>Eisenberg</u>: Will I get the soft limit correctly?

<u>Banerjee</u>: If you want to do a soft limit calculation, it won't help you. There are other ways of getting the soft limit. I live in that world all the time, and I never use a Lagrangian. Let me assure you, there are other ways of doing it. Weinberg's Lagrangian is a mnemonic device. It is not the only way of getting soft pion limits. Still it does not even matter. I repeat again that our theory does not even discuss what the original Lagrangian is. You don't need a Langrangian. In fact, you don't even need

fields of nucleons and pions. It could be quark based also. Our conclusions will hold even then.

A. Thomas (TRIUMF): I agree. I think the logical development is beautiful. What I disagree with is the interpretation. You've demonstrated very well the conventional wisdom, which is that pseudo-vector and pseudo-scalar coupling are always related by a contact term. What you haven't said is that the contact term is crucial. You've shown that one must lump together the two processes which give pi nucleon scattering, we know there that γ_5 needs some compensating term. What you didn't point to is the other vertex, and I assume that your recipe means for that I should also use γ_5 coupling, and I would say that is wrong.

Banerjee: It doesn't matter what the Lagrangian is. You must use the Lagrangian if you want to calculate the nucleon mass, the pion mass, the value of g_π, etc. You should use other methods of calculating the full pion scattering amplitude. Once you have the sum of all possible graphs, including the seagull like or contact terms, you have other ways of making a model for the full sum, such as a quark model, or a low expansion, whatever. You don't have to go the Lagrangian route to calculate the πN amplitude. That is what we have in mind. But if you must use a Lagrangian, you have to select the correct one which may or may not be the pseudovector Lagrangian.

Thomas: The paper was advertised as how to calculate pion production; what vertex do I use? All you're telling us is exactly what everybody does; i.e., you lump the scattering process into some phenomenological scattering amplitude. That makes sense. I think everybody agrees with that. What you've demonstrated next is that at the creation vertex you get γ_5 plus a contact term. The point I would like to make is that the contact term is absolutely crucial, **strongly** modifying the pseudoscalar result. The sensible first approximation would be to use pseudo-vector coupling, as suggested by chiral symmetry.

Banerjee: I have already explained the role of the contact term in the talk and in response to your previous question. I did not discuss the bottom vertex. You should lump together every other diagram which contributes to the NB $\rightarrow$ πN amplitude, add the whole thing up and make a good model for that amplitude. If you think something else is important, put it in.

Kisslinger: From what you showed, it sounds like you are working up to third order in g_0.

Banerjee: No, completely, in all orders. It is a matter of organization. We just took a glimpse of one element of it; namely through the πN vertex function. Naturally the same series of arguments can be done to the πN amplitude and so on and so forth. The whole point is to maintain the nucleon without internal excitation. In fact, some of you may have recognized the simularity between what we have said and what is known as the Beg theorem. They are both of the same type of logic.

QUARKS IN NUCLEI

Gerald A. Miller
University of Washington, Seattle, WA 98195

ABSTRACT

A brief discussion of the relevance of quark degrees of freedom in Nuclear Physics is presented. Pertinent aspects of bag models of baryons are reviewed, and a six-quark bag model study of the reaction $pp \to d\pi^+$ is made.

WHY WORRY ABOUT QUARKS?

It is important to realize that there is now a theory, quantum chromodynamics (QCD), which many physicists believe is the fundamental theory of strong interactions.[1] Within QCD hadrons are described as composite particles with fundamental constituents called quarks. QCD goes beyond the original quark model of Gell-mann[2] and Zweig[3] in that it is a dynamical theory of the interactions between quarks. It says that as distances between quarks get smaller the interaction between quarks gets weaker. This property is called asymptotic freedom, as short distances correspond to large momentum transfer. Conversely, at large separations it is conjectured that the interaction between quarks is so strong and attractive that quarks can not exist as free particles. This aspect of QCD is called confinement. QCD is a very successful theory -- no firm prediction has ever been shown by experiment to be wrong.

A nuclear physicists's response to this description of QCD might be: "QCD is a nice theory but, so what?" After all, isn't nuclear physics the physics of elementary nucleons, and aren't nuclear dynamics obtained by solving the Schroedinger equation with the Hamiltonian obtained by summing kinetic energies of individual nucleons and potential energies of pairs of nucleons?

Despite the successes of this standard "nucleon model" (it leads to a shell model) there are actually a substantial number of significant failures. For example nuclear matter saturation properties such as binding energy per particle and density can not be explained quantitatively using two body forces.[4] Calculations which yield e.g. the correct binding energy are about fifty percent off in the density. Another process which presents problems for the standard theory is the reaction $\gamma d \to np$, at forward angles, for photon energies between about 20 and 100 MeV. Significant disagreements between theory and experiment occur, even if meson exchange corrections are included. Hadjimichael and Saylor[5] can fit the data, but only by significantly modifying the deuteron wave function at distances less than 1.5 fm. (There is some controversy over Hadjimichael and Saylor's selection of the data base.[6])

In both of these problems the crucial element in the calculation is the short distance part of the nucleon-nucleon wave function.

For both problems the conventional nucleon description is found wanting.

Another example is the measured hole in the charge densities of ^{3}He and ^{4}He nuclei.[7] Conventional theory finds it very difficult to reproduce the data, although inclusion of mesonic exchange currents may yield a depression in the charge density of qualitatively the right character.[8] However quantitative understanding of these data is still to be obtained.

Actually there are many other examples of troubles for the conventional theory. Any time a two nucleon wave function is involved, there is bound to be trouble. The (π^+,π^-) double charge exchange reaction, and the pion production reaction that is the subject of this conference are two examples of chronic problem areas.

Perhaps some combination of three and four body forces, or meson exchange currents may lead to the solutions of these problems. However, in my opinion, a new view of the nucleus -- namely that quark degrees of freedom play an important role -- is needed.

Many nuclear theorists have turned to QCD and quark models in their attempts to understand fundamental aspects of nuclear physics. A partial list of nuclear physicists working on quark ideas is included here.[9-19] The areas of application of quark models are very broad, so that we limit ourselves to a selected review of those aspects most relevant for their applications toward pion-nucleus interactions.

BAG MODELS FOR NUCLEAR PHYSICS

We start with a discussion of the single baryon problem, for: How can one understand the many-body problem if one doesn't have the one-body problem under firm control? Furthermore it is important to get some idea of nucleon sizes from fundamental considerations. We need to know whether nucleons are to be regarded as big blobs that continually overlap in nuclei, or instead as little dots that never touch.

To answer such questions semi-phenomenological theories are needed. This is because confinement mechanisms of QCD, which obviously play an important role in determining low energy baryonic properties, are still not fully understood. The first model we consider is the MIT bag model.[20] A bag is a finite region of space within which quarks and gluons may be confined. In the bag interior there is a constant positive energy density. Because nature always seeks the lowest energy state, the vacuum squeezes on the bag and attempts to make it smaller. This vacuum pressure is balanced by the outward force produced by quarks bouncing off the interior boundary. There is also a condition which imposes the requirement that no color flux leaves the bag. The pressure balance and flux conservation condition provide the phenomenological means for confinement in the MIT bag model. Inside the bag quarks obey the free Dirac equation. (The one gluon exchange interaction can be included as a perturbation.) This is how asymptotic freedom is implemented.

The radius of the bag, R, is about 1.0 fm, and is determined by minimizing the energy, E, as a function of R, and setting the minimum value of E equal to the average of nucleon and delta masses. This radius is larger than half the average distance (d = 1.8 fm) between the centers of nucleons in nuclear matter, so that these bags do indeed overlap.

Although the MIT bag model provides a successful description of many areas of hadron spectroscopy, from the point of view of a nuclear physicist, it does have significant problems. Namely, it is very well known from studies of nucleon-nucleon scattering, that at long range the strong interaction is obtained from the exchange of a single pion. The MIT bag model makes no allowance for this feature. Furthermore if two nucleons can exchange a pion, a single nucleon can emit and absorb a pion. Hence a nucleon must have a cloud of pions. However an MIT bag model nucleon has no cloud. A second problem is the failure of the MIT bag model to have a partially conserved axial vector current (PCAC). From studies of the muonic decay of the pion, and from the Goldberger-Treiman relationship one expects that the axial current, A_μ, of a reasonable Lagrangian density, $\mathcal{L}$, obeys the PCAC relation

$$\partial_\mu A^\mu = f m_\pi^2 \phi_\pi, \qquad (1)$$

where f is the pion decay constant, m_π the pion mass and ϕ_π the pion field operator. However the axial vector current of the MIT bag model does not obey (1), and in particular its divergence does not vanish as m_π approaches zero. (This is because the axial flux, in contrast with the color flux, does not vanish at the bag surface.)

Brown and Rho[9] (see also Chodos and Thorn[21]) fixed both problems by generalizing the baryonic model to consist of quarks inside an MIT bag surrounded by pions (treated as point particles) outside. The pions interact with quarks at the surface so that $\mathcal{L}$ can be constructed such that $\partial_\mu A^\mu = 0$ if $m_\pi = 0$. Thus chiral symmetry is respected. The pions carry axial current outside the bag, but none leaves the volume of the baryon because the pion's p-wave function vanishes if the pion is far from the bag center. One expectation of such a model is that R be smaller than in the MIT bag model. This is because the pions exert an inward pressure on the bag surface. Brown and Rho expect the bag radius to be small, R $\sim$ 0.35 fm. For such a radius the bags do not overlap in nuclei. However there are problems with this little bag picture. For example, the root-mean-square charge radius of a proton is about 0.8 fm, so that the bag volume would only be about one-tenth the effective volume at the nucleon. Hence pion effects and (since m_π is much smaller than 1/R) multi-pion effects must be very important. These terms cause considerable difficulty in computing the mass and wave function of a nucleon, so that at present there are no reliable little-bag calculations. (There are classical solutions,[22] but these badly violate isospin invariance and so must be regarded as illustrative examples only.) Of course, it is possible that the model is correct.

So far we have discussed the great big MIT bag and the little-bitty Stony Brook bag. However there is another possibility, the

Cloudy Bag.[15] In this model the Lagrangian density is the same as in the Brown-Rho bag, but the field equations are solved by treating the pion quantum field operator in perturbation theory. This is a valid approach for fairly large bag radii, because there is a correspondence between a large value at R and small pionic effect. This may be understood from a classical argument of Jaffe.[23] If the pion mass is neglected, the field equation for the pion is a Laplace equation with a source density at the bag surface. Since the pion is a pseudoscalar particle, it is in a p-wave and ϕ_π is given by

$$\phi_\pi(\vec{r}) \propto \frac{\vec{\sigma}\cdot\hat{r}}{R^2} \qquad (2)$$

for $\vec{r}$ just outside the bag surface, where $\vec{\sigma}$ is the nucleon Pauli spin operator. Hence for large R, $\phi_\pi(\vec{r})$ is small. Another argument comes from the momentum dependence of the pion-nucleon (πN) coupling constant, $g(k)$. This is given by

$$g(k) = g\frac{3j_1(kR)}{kR} \approx g\, e^{-k^2R^2/10}, \qquad (3)$$

where k is the momentum of the pion and g the dimensionless πN coupling constant at k = 0. The spherical Bessel function of order 1 enters because the pion is in a p-wave, and the argument is kR because the πN interaction occurs only at the bag surface. The factor 3/kR is for normalization. The Gaussian approximation is valid for all diagrams we have calculated. From (3) we see that for a given value at k, the bigger R is, the smaller the πN coupling. Thus for reasonably large R, pionic effects are small and can be treated in perturbation theory.

From (3) we derive a quantum field theory of pion-baryonic interactions. The bag radius is determined by computing πN scattering at energies in the (3,3) resonance region. The diagrams shown in Fig. 1 are included. By adjusting R to fit the (3,3) phase shift as a function of energy we obtain R $\approx$ 0.8 fm. Such a value could also be obtained by minimizing the pionically amended bag energy as a function of R, but this procedure is complicated by uncertainties in the zero point energies.[24]

With this value of R the nucleon wave function is completely defined and one must employ further tests. Thus we calculate the electromagnetic

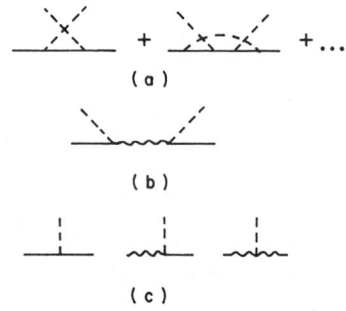

Fig. 1. Diagrams of the (3,3) resonance, where the solid lines represent nucleons, dashed ones pions and the curved line the delta: (a) Chew-Low series; (b) delta model; (c) pion-baryon vertex functions.

properties of the nucleon. The
charge radius of the neutron is
of particular interest. Part of
the time the neutron consists of
a proton bag and a π^- particle.
Since the π^- is mainly outside
the bag, the mean square charge
radius of the neutron is expected to be negative in our
model. (In the original cloudy
bag calculations pions could
enter the bag interior. However later calculations,[25] done
in a version for which pions are
excluded, show that the effects
of pions inside the bag are
negligible except for computations of the axial vector charge,
g_A). Pion effects are also expected to be significant in computations of magnetic moments because the currents of a pion circulating
around the bag contribute to the magnetic moments of the nucleons.
Computed values of proton, μ_p, and neutron, μ_n, magnetic moments in
the MIT bag model[20] are significantly too small in magnitude, although the ratio μ_p/μ_n is $-3/2$ which is essentially the experimental
ratio.

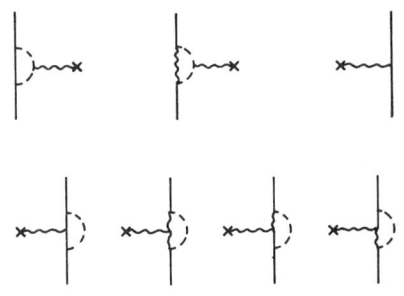

Fig. 2. Photon-nucleon interactions. The wiggly line with an x
at the end is used to describe the
photon.

The electromagnetic properties of the nucleon are computed by
evaluating the diagrams of Fig. 2, and the results are displayed in
the Table. (Center-of-mass effects[20] are included in the values
shown therein.) The calculated values of the static electromagnetic
properties of the nucleon are in excellent agreement with the experimental values.

Table I

Comparison of cloudy-bag-model results with experiment
and MIT results

	$<r^1>_p (fm^2)$ [a]	$<r^2>_n (fm^2)$ [b]	μ_p [c]	μ_n [c]
CMB (R=0.8)	0.53	-0.13	2.60	-2.01
Experiment	0.69	-0.12	2.79	-1.91
MIT (R=1.0)	0.53	0.0	2.14	-1.42

[a] The mean-square charge radius of the proton.
[b] The mean-square charge radius of the neutron.
[c] The nucleon magnetic moments, μ_p and μ_n, are given in units of
nuclear magnetons.

A variety of other applications of the model have been made.[26-29] Low energy, s-wave πN scattering parameters have been computed and are in good agreement with experiment.[26] The magnetic moments of strange baryons are sensitive to pionic effects,[27] and their inclusion leads to agreement between theory and experiment. Pion-nucleon scattering in the P_{11} channel may also be understood using the ideas of the cloudy bag model.[28,29]

Thus the cloudy bag model is a calculable semi-phenomenological theory of baryon structure.

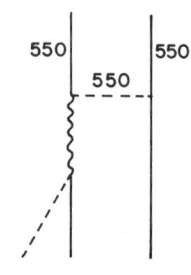

Fig. 3. Kinematics for the $pp \rightarrow d\pi^+$ process. The numbers refer to the momenta of the protons or virtual bosons.

It has PCAC and allows the OPE interaction between nucleons. It is consistent with a broad variety of data. The bag radius is 0.8 fm. This is significant in nuclear physics. With this radius, bags in nuclei often overlap at full nuclear matter densities, and the six-quark aspects of nucleon-nucleon wave functions can be expected to be important.

QUARK MODEL TREATMENT OF $pp \rightarrow d\pi^+$

Another set of processes for which quark effects enter are those that occur at high momentum transfer. Short distances are relevant for such reactions and one would naturally expect nucleons to overlap and quark effects to enter. Thus we turn to a consideration of the $pp \rightarrow d\pi^+$ process. This is a high momentum transfer process. If the reaction occurs in a combination of one boson exchange processes shown (a π or a ρ may be exchanged[30,31]) in Fig. 3, the virtual boson carries a momentum of about 550 MeV/c between the two nucleons. Hence the relevant distance scale, as computed from the uncertainty principle, is about 0.4 fm which is about half the bag radius.

Of course in standard computations of the diagram of Fig. 3, pion absorption cannot occur when nucleons are separated by 0.4 fm (r = 0.4 fm) since the deuteron wave function is essentially zero for such a separation. Still one might ask, within the framework of such calculations, where does the absorption occur? To do this recall that the transition matrix for a given angular momentum channel can be written, crudely, as the radial matrix element of an operator, $O(r)$, between the deuteron wave function $u(r)$ and a spherical Bessel function which describes the relative motion of the nucleons. To find out where pions are produced integrate not from r = 0 to infinity but from r = 0 to a parameter R_{max}. Thus define a quantity $T(R_{max})$ with

$$T(R_{max}) = \int_0^{R_{max}} rdr\, j_2(pr)O(r)u_D(r). \qquad (4)$$

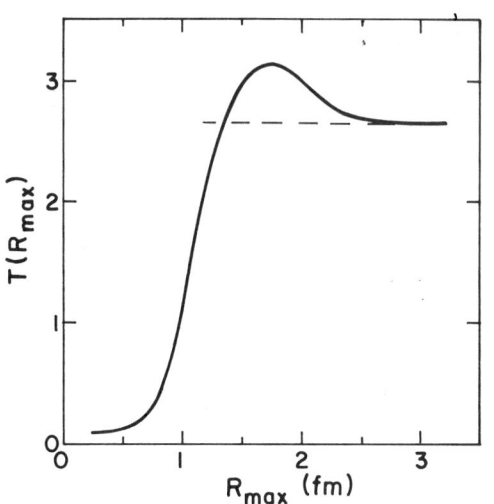

Fig. 4. Radial contributions to pion production.

The order of the spherical Bessel function is chosen as two, because the $^1D_2$ final state is the most significant one. By varying R_{max} as shown in Fig. 4 one can determine the significant region for pion production. Used in that figure is a one-pion exchange mechanism with a form factor that leads to good agreement between theory and experiment for the total cross section for the $\pi d \to pp$[32] process. The convergence for R_{max} greater than about 3.0 fm results from the slow fall off of the deuteron wave function combined with the rapid oscillation of the spherical Bessel function. As discussed above, there is no contribution at very short distances. The interesting thing to observe is that essentially all of pion production occurs for distances less than 1.5 fm. Thus, if the bag radius is about 0.8 fm, the nucleons overlap during the entire pion absorption process! This suggests that one should explicitly treat the quark degrees of freedom.

In what follows we use a six-quark shell model idea, introduced by L.S. Kisslinger at the Orsay Conference, which is discussed by him here.[11] The idea is simple, whenever two nucleons overlap treat them as six quarks in a bag. Thus we may write the NN physical wave function, $|\psi_{NN}\rangle$, as a sum of two terms, one of which is a (more or less) ordinary two-nucleon piece, $|\phi_{NN}\rangle$, which describes the long-range physics, and the other, $|6q\rangle$, is a six-quark part. In a schematic notation one has

$$|\psi_{NN}\rangle = \theta(r-r_0)|\phi_{NN}\rangle + \theta(r_0-r)|6q\rangle, \qquad (5)$$

where r_0 is a parameter of the model. Electromagnetic interactions of the deuteron are used by Kisslinger[11] to determine the value of

r_0 as about 0.8 - 0.9 fm.

The states $|6q\rangle$ are described as having one or two **active** quarks which carry the quantum numbers of the states plus an inert core. For example the $^1S_0$ nucleon state can be described as having a component with two quarks in a 1s state (plus a core) with an amplitude C^{ss}, or as two quarks in a 1p state (plus a core) with an amplitude C^{pp}. At present the spectroscopic amplitudes are to be determined from experiment, but we are thinking about how to derive them theoretically.

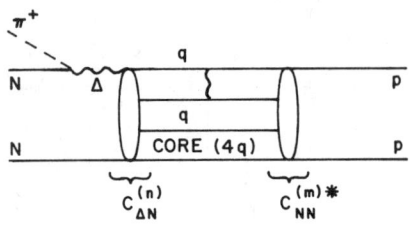

Fig. 5. Pion production in the six-quark bag model.

In this picture the $pp \rightleftarrows d\pi^+$ process proceeds as in Fig. 5. A π^+ incident on a nucleon (N) transforms it into a delta (Δ). The ΔN system coalesces to form a six-quark bag. The quarks interact and momentum is transferred. Finally the six quarks separate into two nucleons. The amplitude for a ΔN system to be transformed into a six-quark system of component n is $C_{\Delta N}^{(n)}$, and the corresponding amplitude for the NN system is $C_{NN}^{(m)}$. The T-matrix element is given schematically as

$$T = \sum_{\Delta N} \frac{\langle d\pi^+|V_{\Delta N}|\Delta N\rangle C_{\Delta N}^{(n)} \langle n|V|m\rangle C_{NN}^{(m)*}}{(E - E_{\Delta N})} \quad (6)$$

where $V_{\Delta N}$ is the interaction that converts a πN system into a Δ, and $|n\rangle$ and $|m\rangle$ represent coefficients of the six-quark wave function. The operator V is the quark-quark effective interaction which is taken from Isgur and Karl.[33] A harmonic oscillator representation is used (as in Ref. 33) to compute the necessary matrix elements.

At first glance the set of amplitudes $C_{\Delta N}^{(n)}$ and $C_{NN}^{(m)}$ seems to embody a tremendous number of free parameters. However, only one number representing an overall probability really matters. That is we take

$$\sum_n |C_{\Delta N}^{(n)}|^2 = \sum_m |C_{NN}^{(m)}|^2 \approx 3\%. \quad (7)$$

This value is consistent with Kisslinger's determinations.[11] It is important to note that all spectroscopic amplitudes are <u>independent</u> of pion energy.

At this stage it is worthwhile to observe that the more conventional one-boson exchange treatments of the pion deuteron absorption process are now at a very sophisticated level and enjoy a good deal of agreement with experiment.[34] Thus, for our calculation (this is done in collaboration with L.S. Kisslinger) to have any meaning, it

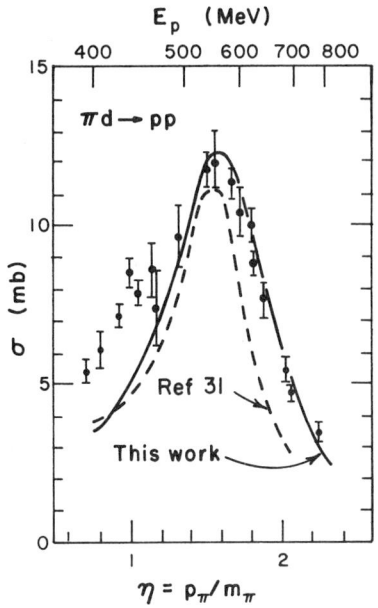
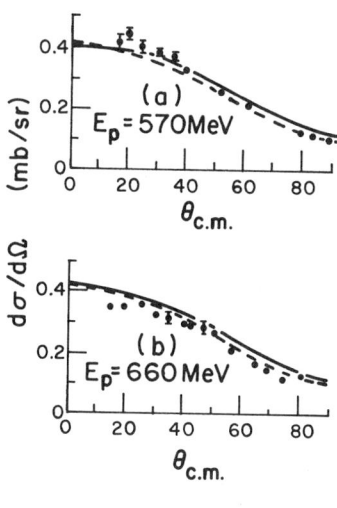

Fig. 6. Total cross section for $\pi^+ d \to pp$.

Fig. 7. Angular distributions. Solid line — our work. Dashed line -- Ref. 31.

should have to do as well in comparison with experiment as the more conventional treatments. Thus we compare our results with those of Niskanen[31] as well as with data.[35]

Shown in Fig. 6 is our calculation of the total cross section for pion absorption. For energies near the (3,3) resonance the agreement with experiment is very good. The energy dependence shown arises from the Greens function for the propagation of the ΔN system. The overall magnitude is essentially fixed from (7) and the Isgur-Karl force.

Shown in Fig. 7 are angular distributions for the pp → $d\pi^+$ reaction at proton energies (570 MeV and 660 MeV) near the (3,3) resonance peak. Again good agreement with the data is obtained. The shape of these angular distributions comes almost entirely from the restrictions imposed by angular momentum conservation.

SUMMARY

There are many unsolved problems in Nuclear Physics, and those which involve the relative two-nucleon wave function are among the toughest. The lack of success of conventional approaches encourages many to examine quark and QCD ideas, and to include these in nuclear physics. To do this one must understand the structure of the nucleon and delta. The Cloudy Bag Model, (CBM) described above, presents a calculable approach to study baryonic properties. It incorporates

PCAC and an OPE force between nucleons. Many properties of baryons have been successfully described using this approach. The bag radius in the CBM is about 0.8 fm, so that the reaction $pp \to d\pi^+$ takes place in the region where the two nucleon bags overlap. This leads us to apply a model[11] in which the short distance part of the nucleon wave function is treated as six quarks in a bag to the $pp \to d\pi^+$ reaction. Good agreement between the theoretical and experimental total cross sections and angular distributions is obtained with a quark probability of between three and five percent.

I thank my collaborators A.W. Thomas and S. Théberge and L.S. Kisslinger for many valuable discussions.

This work was supported in part by the U.S. Department of Energy.

REFERENCES

1. An excellent elementary description of QCD is given in D.B. Lichtenberg, Contemporary Physics 22, 311 (1981).
2. M. Gell-mann, Phys. Lett. 8, 214 (1964).
3. G. Zweig, CERN Report 8419/TH 412 (1964).
4. B.D. Day, Phys. Rev. Lett. 47, 226 (1981).
5. E.H. Hadjimicheal and D.P. Saylor, Phys. Rev. Lett. 45, 1776 (1980).
6. H. Arenhövel, Phys. Rev. Lett. 47, 749 (1981); E. Hadjimicheal and D.P. Saylor, Phys. Rev. Lett. 47, 750 (1981).
7. J.S. McCarthy, I. Sick and R.H. Whitney, Phys. Rev. C15, 1396 (1977).
8. E. Hadjimicheal, Nucl. Phys. A294, 513 (1978).
9. G.E. Brown and M. Rho, Phys. Lett. 82B, 177 (1979); G.E. Brown, M. Rho and V. Vento, Phys. Lett. 84B, 383 (1979).
10. M. Harvey, Nucl. Phys. A352, 326 (1981).
11. L.S. Kisslinger, Orsay Meeting on Pion Production, and this Conference.
12. F. Mhyrer, G.E. Brown and Z. Xu, Nucl. Phys. A362, 317 (1981); I. Hulthage, F. Mhyrer and Z. Xu, Nucl. Phys. A364, 322 (1981).
13. D.P. Stanley and D. Robson, Phys. Rev. Lett. 45, 235 (1980); D. Robson in "Proceedings of Nuclear Structure with Intermediate Energy Probes" (Los Alamos, New Mexico, 1980), and in International School of Nuclear Physics, April 21-30, 1981, Erice, Italy, "5th Course: Quarks and the Nucleus" (Pergamon Press) to be published; Nucl. Phys. A308, 381 (1978).
14. L.S. Celenza, W.S. Pong and C.M. Shakin, B.C.I.N.T. 81/061/109, June 1981.
15. G.A. Miller, A.W. Thomas and S. Théberge, Phys. Lett. 91B, 192 (1980); S. Théberge, A.W. Thomas and G.A. Miller, Phys. Rev. D22, 2383 (1980); A.W. Thomas, S. Théberge and G.A. Miller, Phys. Rev. D24, 216 (1981); S. Théberge, G.A. Miller and A.W. Thomas, Can. J. Phys., to be published (1981); G.A. Miller,

S. Théberge and A.W. Thomas, Comments Nucl. Part. Phys. $\underline{10}$, 101 (1981).
16. H.J. Weber, Z. Phys. $\underline{A297}$, 261 (1980); $\underline{A301}$, 141 (1981); B.L.G. Bakker, M. Bozodian, J.N. Maslow and H.J. Weber, Univ. of Virginia preprint, October 1981.
17. R.A. Freedman, W.Y.P. Hwang and L. Wilets, Phys. Rev. $\underline{D23}$, 1103 (1981).
18. H.J. Pirner and J.P. Vary, Phys. Rev. Lett. $\underline{46}$, 1376 (1981).
19. C.W. Wong, Int. School Nucl. Phys. April 21-30, 1981, Erice, Italy "5th Course: Quarks and the Nucleus" (Pergamon Press) to be published; K.F. Liu and C.W. Wong, Phys. Rev. $\underline{D21}$, 1350 (1980).
20. A. Chodos, R.L. Jaffe, K. Johnson, C.B. Thorn and V.F. Weisskopf, Phys. Rev. $\underline{D9}$, 3471 (1974); A. Chodos, R.L. Jaffe, K. Johnson and C.B. Thorn, Phys. Rev. $\underline{D10}$, 2594 (1979); T. DeGrand, R.L. Jaffe, K. Johnson and J. Kiskis, Phys. Rev. $\underline{D12}$, 2060 (1979); J.F. Donoghue and K. Johnson, Phys. Rev. $\underline{D21}$, 1975 (1980).
21. A. Chodos and C.B. Thorn, Phys. Rev. $\underline{D12}$, 2733 (1975).
22. V. Vento, M. Rho, E.M. Nyman, J.H. Jun and G.E. Brown, Nucl. Phys. $\underline{A345}$, 413 (1980).
23. R.L. Jaffe, Lectures at the 1979 Erice Summer School "Ettore Majorana (1979).
24. C.E. DeTar, Phys. Rev. $\underline{D24}$, 752, 762 (1981).
25. G.A. Miller, to be published.
26. A.W. Thomas, TRIUMF preprint (1981).
27. S. Théberge and A.W. Thomas, TRIUMF preprint TRI-PP-81-34 (1981); see also G.E. Brown, M. Rho and V. Vento, Phys. Lett. $\underline{97B}$, 423 (1980).
28. A.S. Rinat, Weizmann Inst. preprint (1981).
29. D.J. Ernst and R. MacCleod, Texas A&M University preprint (1981).
30. M. Brack, D.O. Riska and W. Weise, Nucl. Phys. $\underline{A287}$, 425 (1977).
31. J.A. Niskanen, Phys. Lett. $\underline{79B}$, 190 (1978).
32. M. Alberg, E.M. Henley and G.A. Miller, Nucl. Phys. $\underline{A306}$, 447 (1978); G.A. Miller in "Meson-Nuclear Physics -- 1979 (Houston)" p. 561, Ed. by E.V. Hungerford III, AIP, New York (1979).
33. N. Isgur and G. Karl, Phys. Rev. $\underline{D18}$, 4187 (1978).
34. A.W. Thomas, this Conference.
35. G. Jones, this Conference.

DISCUSSION

D.B. Lichtenberg (Indiana Univ.): If the nucleon bag is as large as 0.8 fm, wouldn't you expect the pion bag also to be maybe 0.6-0.7 fermi, and why can you treat that as a point?
G.A. Miller (Univ. of Washington): That's a good question. Calculations have been done by a student of Cutkosky, who has considered the effect in such calculations of the pionic decays of the nucleon and the delta. He uses a harmonic oscillator or quark model for each nucleon and pion system, and he finds there that pionic size effects in those configurations are very small. In fact, the relevant number is not the ratio of the radius of the nucleons to the radius of the pions, but actually the square of that number, and then there is some geometrical coefficient that comes in that makes that ratio even smaller. So the effect of the finite pion size is small, at least in the momentum transfer region that we are interested in, which is under 1 GeV/c.

K-F. Liu (Univ. of Kentucky): I have two questions: the first one is regarding the bag radius you obtain. It was shown by Chin that if you use the MIT bag model considering the nucleon self energy addressed by the pions and include all the intermediate states of the quarks inside a bag, s-and p-waves and so on, the series is divergent. Do you have any comments on that?
Miller: Tony Thomas and L. Dodd have studied the perturbation theories in detail and have proved some rigorous theorems which show not only is the series convergent but very rapidly.
Liu: Chin is considering the quark orbitals, not just the physical nucleon.
Miller: The quark what?
Liu: The orbitals of the quarks inside a bag, not considering the intermediate states as the physical nucleon or delta or N* and so on.
G.A. Miller: Jaffe (who did not sum over intermediate nucleons or deltas) has shown that the expansion is basically in powers of $1/R^2$, where R is the bag radius, so while it's large you are O.K.. If you make the bags small, then it is true in perturbation theory you can't do any calculations. That's the problem with a small bag. [I received Chin's paper after returning from Indiana. I do not believe his divergent result, which arises not from s or p wave quarks but from d,g,f, ... quarks. These are forbidden in second order for a spherical bag (which Chin is using)].

Liu: Another question I have is about the charge radius of the neutron. I understand that the chiral bag model has a problem of generating a time averaged charge density outside a bag. Do you have a similar problem?
Miller: No, we treat the pion field as a quantum field operator, so j_0 (or the density operator) is not zero in our calculation.

Mike Hynes (Los Alamos Scientific Lab.): If you immerse this 6-quark object in a ^{3}He or ^{4}He nucleus and then extract the finite size correction that you showed, can you reproduce the dip in the middle?
Miller: I haven't tried to do that.

Hynes: I have another question for you about electron scattering. A lot of data was taken at SLAC in the 70's on electron scattering from a hydrogen target going to excited states of the nucleon, which went up to very large momentum transfers. Have you applied your cloudy bags to fitting any of those form factors?
Miller: No, not yet. The calculations done at that time were, I think, by Walecka. They were very successful, and the mathematical elements are very similar.
Hynes: Walecka's was a potential model.
Miller: [The bag quark wave functions are similar to potential model wave functions.]

J. Vary (Iowa State Univ.): In making these arguments for why you should be concerned about quarks, unless I missed it, you didn't show any clear case among the data you presented where quarks would have been essential for the description of the nuclear phenomena. Is that correct?
Miller: Yes. As I said, the meson exchange picture is very successful. But look at the integrals I showed. You're in a region where I would say you better worry about it, because you don't have independent nucleons any more.
Vary: Let's backtrack a little bit. If you can't show in a particular model where you really need quarks, then you haven't killed the "old time" nuclear theory, which says that basically we can fit nucleon-nucleon data within a certain framework and carry it forward to calculate nuclear properites with it. That is essentially as successful as what you have done, so why should anybody be worried about quarks?
Miller: If you want to have a understanding of what's going on, I think you should worry about quarks, but we haven't demonstrated a specific effect that can be obtained by quarks and not by anything else.

J. Noble (Univ. of Virginia): It seems to me you didn't mention nucleon-nucleon scattering, in particular the one pion exchange component of it. As I understand it, your model has a rather soft pion-nucleon form factor in the time-like region. Now, I don't know what you say about the space-like region, although the space-like region is obviously important in computing the charge radius. In the old fits, people put in cut-offs on the pion-nucleon vertices, and they always used to get numbers on the order of a GeV, where the cut-offs were of the monopole variety.
Miller: Our cutoff is 800 MeV. The number that comes in is not one over the bag radius. The effective momentum is actually

$\sqrt{10/R}$ where R is the bag radius, and so 800 MeV or so is the actual momentum cutoff that comes in. I wrote down the form factor which is approximately $e^{-p^2 R^2/10}$, where p is the momentum transfer.
Noble: Are the data consistent with that small a cut-off?
Miller: Yes. [see the preprint by Thomas and Gersten].
M. Banerjee (Univ. of Maryland): In the work that Nien-chih Wei and I did, we found a way of extracting the $\pi N\Delta$ form factor in an almost model independent way, and we found that the initial slope, when expressed in the monopole form, is characterized by a mass not less than 10 pion masses.

C.-Y. Cheung (Carnegie Mellon Univ.): In all bag models the quarks are free inside the bag because of asymptotic freedom. Why do we feel comfortable using that feature and incorporating a bag model in calculating low momentum transfer processes?
Miller: It's a feature that QCD gives. In deep inelastic electron scattering, basically the quarks are free.
Cheung: That's true only at very high momentum transfer.
Miller: There're free inside the bag radius. The bag radius is 0.8 fm which is not all that large if the momentum that's associated with it is 800 MeV.
Cheung: That is so because you confine everything within a sharp boundary.
Miller: It's a phenomenological model and confinement is basically known from the data. There are other ways to get involved. There are also potential models, and I didn't have time to discuss the potential models, i.e., to treat the interactions between quarks and potentials. Potential models, at least as currently calculated, lead to very strong Van der Waals long range forces between two nucleons. This is a serious problem for potential models which does not occur in the MIT bag model because with the mechanisms of confinement in the MIT bag model no gluons escape the bag.

Cheung: One more question concerning $\pi d \to pp$. In the one-boson exchange model, it's the pion that carries momentum across. In the 6-quark model, is it true that it is the gluon that is responsible for momentum sharing?
Miller: Yes. Also in the quark wave-functions, which have their momentum given by the size of the bag.
Cheung: The momentum transfer may be high from a nuclear physics point of view (~500 MeV/c), but from the perturbative QCD point of view it is very low.
Miller: This is not a perturbative QCD calculation. We use an effective gluon-gluon interaction. All the complications built into the effective interaction are taken from the phenomenological work of Isgur and Karl. We have to see if the data are consistent with all experiments that we did. This model is new, but so far the same sizes are consistent with both the electromagnetic properties of the deuteron and this reaction.

P. Hwang (Indiana Univ.): I have a comment and a question. The comment is that it may be unfair to assert that there is no pion cloud in the MIT bag model, because you can calculate the πNN coupling constant from order α_s effects so that we do expect some pion cloud in this model.

My question is, in the introduction of your talk you mentioned several discrepancies in the old fashioned theory. To what extent do you think that, by introducing the quark degree of freedom, you are able to make improvements in the interpretation of nuclear saturation phenomena or of other discrepancies you mentioned?

Miller: I think it is fair to say that, in detail, there isn't much yet. But I think that people have gotten to the point where they've now made the standard calculations very reliably. No longer can we say we're messing up the many-body problem. These are the real results and you've got to live with them. There are real discrepancies here and, since you're asking for my opinion, I'm quite willing to stick my neck out and say that within 10 years we'll explain all these problems using quarks and that quarks will be a very important part of nuclei.

D. Koltun (Univ. of Rochester): You brought us back to the point I would like to bring up, which is the question of the successes and failures and the request for a little balance in such statements. For example, and I think this is not a quibble, if you conclude from a reasonable model, and I think your model is reasonable, that the radius of the nucleon is 0.8 fm, and you interpret this to mean that in nuclear matter, bags (or nucleons) are overlapping, then the question is, who is supposed to worry? What I mean by this is, is that a problem in conventional nuclear physics or a problem for quark bag models? I don't know the answer to it, but I think I know the direction. In the days when nuclear physics first encountered short distance problems of hard cores, a good 20 years ago or more, that was obviously a problem for nuclear structure, because nuclear structure as we understood it involved nucleons being able to pass over each other, and if you had a hard core they couldn't pass over each other. You had to adjust your thinking in radical ways to accomodate that, and that thinking was done right here at Indiana, as was mentioned earlier, when Bruckner and Watson were here. Now the question is, do we have the same sort of problem when bags overlap? I think the answer is no, partly because of asymptotic freedom. That is, there is no obvious dynamical problem if you happen to have six quarks, particularly if they have momentum (which you never mention) if they're passing over each other, and it may well be that a conventional nuclear view doesn't have to worry about the fact that they are overlapping. But what I do see is, the very simple models of QCD contained in the bag model don't yet know how to handle the question of what happens when six quarks are in the region which normally would be given to three quarks. So what I am saying is, I think you have to be a little more modest about the dynamical description you're dealing with, which is very, very simple and, I think, not yet able to deal with what you might call "surface energy problems" of distorted bags and things like that.

Miller: I did not say that we knew how to calculate the 6 quarks properly. That's why we went to the phenomenological shell model. This is an intermediate stage. Someday we may be able to calculate the spectroscopic factors for the shell model using QCD. That's certainly a reasonable hope. What we have to learn is how to do the many body problem with the quarks. That's our problem and the reason why nuclear matter and other things are not explained in the conventional model. And so this is something that we have to do; and if we do it, in my opinion, we'll answer all these problems.

Fritz Coester (Argonne National Lab.): My comments are related to the same subject. The results of Day cannot be interpreted as a failure of the conventional "nucleon model" to explain nuclear matter, they merely establish that three-body forces are needed. Dynamical models can be classified by specifying the active degrees of freedom. As I understood your talk the active degrees of freedom in your model are nucleons, pions, deltas and the dibaryon - not quarks. Quarks are used to determine the parameters of the model just as mesons are used to determine the nucleon-nucleon potentials in the conventional model.

Miller: We calculated the matrix elements in which one starts with two quarks, a potential interaction which fits the spectroscopy of the negative parity baryons, and we just take a quark matrix element. This is a number that comes into the calculation, so we definitely are dealing with the quark degrees of freedom.

Coester: You use the quark degrees of freedom to get the properties of other things that you afterward use. You don't have a dynamical model of the nucleus consisting of quarks.

Miller: That's certainly true in 1981, but that's the direction we have to go.

Peter Barnes (Carnegie-Mellon Univ.): In your talk you highlighted the physics at short distances and the possible role of quarks. To investigate this I am wondering why you pick a problem as complicated as a pion production reaction. In discussing the possible role of quarks why not focus on the nucleon-nucleon short range repulsion, which has been studied more thoroughly?

Miller: A lot of people, including Weber who will talk later on, have been worried about the nucleon-nucleon problems, and maybe that would be more fundamental. We are talking about pion production, and pion production is a very short distant process and also happens to be the subject of this conference. And the fact of the matter is, it is a process that wants to occur when nucleons are 0.4 fm apart.

Barnes: I have the impression that the quark model is not yet very successful in treating the short range nucleon-nucleon repulsion. Here in the pion production reaction the quarks appear in a small piece of the diagram. If the meson exchange parts are not treated correctly you may not understand what role the quarks play.
Miller: In the diagram I calculate there are no meson exchange parts. We have a short distance problem, and we have to use short distance physics. If you think of nucleons as being fairly large, it's hard to treat them as independent when they are 0.4 fm apart.

I want to say something about nucleon-nucleon forces. There is a serious question here. If there is some real physics reason which says that there is a hard core such that nothing is inside the hard core, not nucleons but really nothing, then the old picture of nuclear physics will come back. But if there is an interaction which gives you a hard core which prevents nucleons from getting inside, but leaves room for quarks on the inside, then you have to go to this kind of picture.

L.S. Kisslinger (Carnegie-Mellon Univ.) There is a secret that many nuclear physicists don't seem to know, and that is that the short range part of the strong interaction problem is easier than the long-range part. Particle physicists have now known this for 15 years, and we should also try to become familiar with these ideas.

Jerry has talked about a model which is applied in a certain energy region, for which the diagram which he showed seems to be perfectly set up. At that point there is a long range part and a short range part in $\Lambda N \to NN$. The quark interaction has been determined in that single momentum region, but what we're doing is making use of the baryonic hyperfine structure interaction which has been determined by baryonic spectroscopy, so it is again a unifying picture which seems to be quite adequate for this region. This is not brought out, in a sense, using this picture. A crucial question, which Jim Vary asked is: for the understanding of what experimental data is it necessary to introduce quark degrees of freedom? We have been discussing this a lot. Dan Koltun's question is very closely related to it. My own feeling, even though I love pions, is that electromagnetic interactions will be most useful for those crucial questions. For the two-body problem, we only have one bound state, the deuteron. As for the predictions of meson currents which have been used now for about 30 years, they now seem to fail as one is going to higher momentum. There are several publications now that indicate that one needs something that is somewhat more confined than those contributions. That's not a firm result, because if you go back to the mesons you can always throw in another boson. A ρ meson, for example, has a range of .25 fermi. Everyone knows that one is inside the bag at that point. Throwing in extra operators, although we call them ρ's or ω's, doesn't change the fact that it is an almost arbitrary procedure that has been used in nuclear physics. The meson picture is a non-converging theory that we have been working with for 30 years.

You are seeing now some places where the old method is failing. This new approach, I think, will be a converging procedure and is just starting. I think you will find as you get more familiar with it, that it is a very satisfactory way to do very short range interactions, because it is easier to do, and probably because it it correct.

Vary: Lenny knows that I sometimes ask questions for which I have an answer in mind. I was giving Jerry a chance and letting it sift in for a little bit. I think that there are certain statements that can be made which are far less bold than what Jerry stated about what is going to happen in the next 10 years. One statement that is clear is that, if elementary particle theorists claim they see quarks in deep inelastic scattering from nucleons, then, certainly, we should be able to see quarks in deep inelastic scattering from nuclei. The interesting question is, to what extent does deep inelastic scattering from nuclei by electrons show whether nucleons coalesce and form 6 quark objects, 9 quark objects and 12 quark objects? I have done some initial analysis in collaboration with Hans Pirner of Heidelberg [Phys. Rev. Lett. <u>46</u>, 1376 (1981)]. The model we build successfully describes, to a large extent, all the deep inelastic scattering data of electrons with ^{3}He, which is the most extensive set of data available to us. It involves certain model assumptions which we could debate at great length, but one simple conclusion that we have reached is that the point at which it becomes essential to think of nucleons as coalescing and forming multiquark objects within the nucleus is probably when their distance of closest approach is of the order of 0.9 fm. This means that a radius of the order of 0.4 to 0.5 fm would be the touching radius for this critical transition to occur. That doesn't necessarily contradict a lot of things you said, but I do believe there is at least direct evidence <u>at high energies</u> for the need to treat quark degrees of freedom <u>in nuclei</u>.

THEORIES OF PION PRODUCTION IN NUCLEON-NUCLEON COLLISIONS*

M. Betz, B. Blankleider, J.A. Niskanen[†] and A.W. Thomas
TRIUMF, 4004 Wesbrook Mall, Vancouver, B.C., Canada V6T 2A3

ABSTRACT

We first present an historical summary of the developments of the theory of pion production in nucleon-nucleon collisions, leading to the present time. Then the recent progress in each of three promising approaches to this problem are reviewed in detail. Finally we discuss problems that remain and suggest some possibilities for future research.

I. INTRODUCTION

As we shall see in Sec. II, the theoretical challenge of understanding pion production in nucleon-nucleon (N-N) collisions has engaged many physicists over the past thirty years. The issues raised are fundamental to our understanding of nuclear forces and the structure of the nucleus itself. Our aim in this brief review is first to put the modern calculations of pion production in perspective by examining what was done in the past. Then we report in detail on the most recent developments obtained in coupled channels calculations of NN scattering and $pp \to d\pi^+$ (Sec. III) and of the reaction $pp \to pn\pi^+$ (Sec. IV). In Sec. V we report on the few-body theory and its results in all of these systems.

It is obvious that in putting together three rather different models in a short paper to meet a conference deadline, the manuscript may still have rough edges. Nevertheless, we hope this lack of polish may be compensated to some extent by the fact that the information really is up-to-date! In the final section we try to point out the most important remaining problems of which we are aware, and thereby provide some direction for future work.

II. HISTORICAL DEVELOPMENT OF THE THEORY

Within a few years of the discovery of the pion (1947), considerable progress had been made towards a theoretical understanding of pion production. In 1951 Chew, Goldberger, Steinberger and Yang[1] suggested that the matrix element for pion production might be calculated in impulse approximation

$$M_{fi}^{IA} = <\psi_f|T.|\psi_i> , \qquad (2.1)$$

where ψ_f and ψ_i were the deuteron and initial two-proton wave functions, and T was a phenomenological operator whose general form was

---

*Invited talk presented by A.W. Thomas at the Workshop on Pion Production at IUCF, October 22-24, 1981.
[†]Address from 1st November 1981: Research Institute for Theoretical Physics, University of Helsinki.

taken from meson theory. The same approach was used by Geffen,[2] but he used Jastrow type wave functions and adjusted the two parameters in T to fit the experiment. In spite of the degree of phenomenology involved, some of his conclusions still hold. For example, Geffen found that the short range behaviour of both the initial and final N-N wave functions was important, as also was the deuteron d-state.

Until the work of Chew,[3] Low[4] and Wick[5] the relationship of even the simplest approximation for pion production to basic field theory was poorly understood. In view of the location of this conference, it is worthwhile to note that the first application of their theory to pion production was made at Indiana by Lichtenberg.[6] (In fact, much of the early work on the theory of pion production by Aitken, Brueckner, Mahmoud and Watson was also carried out at Indiana University.) Essentially what Lichtenberg did (in 1955) was to provide a rigorous derivation of the impulse approximation

$$M_{fi}^{IA} = \langle \psi_f | (V_{1k} + V_{2k}) | \psi_i \rangle , \qquad (2.2)$$

where V_{jk} had the now familiar (static) form

$$V_{jk} = i\left(\frac{f}{m_\pi}\right)\left(\frac{4\pi}{2\omega_k}\right)^{1/2} u(k)\vec{\sigma}\cdot\vec{k}\, e^{i\vec{k}\cdot\vec{r}_j}\, \tau_{jk} \qquad (2.3)$$

and f was the renormalised NNπ coupling constant.

Clearly Eq. (2.2) omits all effects of pion rescattering. It is interesting to note that at the time of publication Lichtenberg "[did] not know how to take [rescattering] into account consistently". We also observe that essentially all attempts to answer this question (most of which belong in the mid to late 70's as we shall soon discuss) have started from the same Chew-Wick model for πN dynamics. (In view of the recent progress in understanding πN dynamics in (e.g.) the cloudy bag model[7] — as reported at this meeting by G.A. Miller — it would be of interest to re-examine these derivations.)

The results obtained by Lichtenberg were not in good agreement with experiment, and he recognised that this was most probably caused by the omission of πN rescattering through the (3,3) resonance. The importance of such effects had already been demonstrated by Aitken, Mahmoud, Henley, Ruderman and Watson,[8] using an earlier meson theory formulation of N-N scattering by Brueckner and Watson.[9] Indeed, Aitken et al. correctly attributed the $(1 + 3\cos^2\theta)$ form of the pp $\to$ π^+d differential cross section to the (3,3) resonance — even though the latter had only been discovered a year or so before.

By 1957 Lichtenberg[10] had included the effect of pion rescattering through the (3,3) resonance,

$$M_{fi} = \langle \psi_f | H_1 | \psi_2^{(+)} \psi_i \rangle + \langle \psi_f | H_2 | \psi_1^{(+)} \psi_i \rangle . \qquad (2.4)$$

Here $\psi_j^{(+)}$ is the pion wave scattered from nucleon i before being absorbed (through H_j) on nucleon $j (j \neq i)$, that is

$$\langle \vec{k} | \psi_j^{(+)} \rangle = \delta(\vec{q}-\vec{k}) + \frac{\langle \vec{k} | T_j | \vec{q} \rangle}{(\omega_q - \omega_k + i\varepsilon)}\, e^{-i\vec{k}\cdot\vec{r}_j} \qquad (2.5)$$

The results of the new calculation were in fairly good agreement with the existing data on both the angular distribution and energy dependence of the pp → π^+d cross section. Similar results were obtained by Durney[11] at about the same time.

The work of Lichtenberg and Durney marks the end of the first phase of calculations of pion production. Some basic theoretical problems had been overcome, the effects of the (3,3) resonance included, and reasonable agreement with data obtained.

In 1960 Woodruff[12] published what we might consider the first modern paper on pion production in N-N collisions. Using a technique developed by Zachariasen[13] for deuteron photodisintegration, he derived a formally correct expression, to second order in the πN interaction. His arguments relied essentially on the topological properties of the diagrams in Chew-Wick field theory. The calculation itself was technically very complicated, involving πN rescattering in all s- and p-wave spin-isospin channels, and using the Gammel-Thaler potential for the N-N interaction. For p-wave pion production Woodruff's results agreed rather well with experiment, confirming Lichtenberg's earlier result that the (3,3) resonance and the deuteron d-state were the most important ingredients. The s-wave results, on the other hand, were not so satisfactory.

Some seven years later, Koltun and Reitan[14] (also at Rochester) re-examined the question of s-wave pion production in the light of the failure encountered by Woodruff. Their calculation formed the basis for the calculation of Thomas and Afnan,[15] in which the sensitivity of the s-wave pion production cross section to the deuteron d-state was investigated. Of course, the early 70's were innocent days, and one certainly does not expect to learn (say) P_D from pion production[16] now! Nevertheless, these investigations did explain why, for example, soft pion theory failed for pp → π^+d. They also prepared the way for more detailed analysis of data by the LAMPF group of Goplen, Gibbs and Lomon.[17] Similar calculations, based on LSZ reduction techniques were carried out by Lazard, Ballot and Becker.[18] Like Koltun and Reitan the latter authors were as much interested in developing a theoretical description for nuclear production and absorption, as in the N-N production itself.

The early 70's mark the end of phase two in the development of the theory. A great deal of numerical sophistication had been achieved, but there had been little conceptual change since 1960. That is, the calculations were essentially non-relativistic (except for pion kinematics) often involving some static approximation in intermediate propagators. Furthermore, they relied on first-order perturbation theory in the strong (3,3) scattering amplitude—at most one rescattering of the pion. With the rapid accumulation of pion production data in the 70's many more calculations of the same type were performed. In some cases there was even a valid theoretical reason for the calculation. That is to say, some were tests of operators intended for nuclear pion production, some tested the importance of relativistic corrections, and others included new processes such as ρ-meson exchange between the intermediate N and Δ. However, with a few notable exceptions like these, the progress made during the last decade was to abandon perturbative approaches based on one rescattering.

In the early 70's new approaches to pion production in N-N collisions began independently in three widely separated places. As a result of this work, calculations involving at most one pion rescattering can never again be considered acceptable.

In Helsinki, Green, Niskanen and co-workers[19-21] realized that multiple rescattering of the pion could be important. They also realized that since the delta was even more massive than the nucleon, it might be fruitful to treat it on the same footing. Thus in a very natural way the Helsinki group was led to a set of coupled differential equations involving NN, NΔ and $\Delta\Delta$ channels—hence its name, the Coupled Channels Method (CCM). From a purist's point of view one can raise objections to the CCM. The channel coupling potentials are static, the theory is essentially non-relativistic, and three-body unitarity is only approximately guaranteed. Nevertheless, the CCM has several strengths which more than compensate. Because of the familiarity of potential models in nuclear physics, it was easily adaptable to more general nuclear problems—indeed this approach was in many ways the forerunner of the microscopic Δ-hole calculations. Furthermore, as a potential model it readily permitted the inclusion of heavier mesons—such as ρ-exchange (although the parameters may have been a little unrealistic). It permitted an evaluation of the effect of pion production on N-N elastic scattering in a consistent way (and this is the major emphasis in Sec. III dealing with the CCM). Most importantly from our present viewpoint, it allowed a complete summation of the multiple scattering series for pion production, and found the higher order effects to be important (of order 20% in some amplitudes - see e.g. Sec. IV).

Meanwhile, at Rochester Myhrer and Koltun[22] applied the non-relativistic Faddeev theory to πd elastic scattering in the resonance region—see also the thesis work of Doolen.[23] In addition, Mizutani and Koltun[24,25] worked very hard using diagrammatic techniques and the Feshbach formalism to sort out the worrisome question of double counting in πd elastic scattering and absorption. That is, they obtained rules for separating what was already included in the N-N wave function, and what could legitimately be calculated as higher order contributions in pion production.

Also in the early 70's, but at the opposite end of the earth, Afnan and Thomas[26] proposed a simple ansatz, whereby all three important channels, NN, πd and NΔ could be incorporated in one set of exactly unitary, coupled, three-body integral equations. However doing this required that one nucleon be treated somewhat asymmetrically as an Nπ bound-state (N'). This was justified qualitatively by Thomas[27] by generalising the diagrammatic techniques of Woodruff, to which we referred earlier. Indeed Thomas obtained identical equations to those of Mizutani and Koltun for the effect of absorption on πd elastic scattering. However, it was soon realized that the asymmetric treatment of the N and N' in the Afnan-Thomas model led to significant undercounting.

Nevertheless the idea of coupled integral equations of the Afnan-Thomas type—incorporating *exact* two- and three-body unitarity—was so attractive that much effort has been devoted over the past four years to rigorously establishing such equations.[28,32] This work has been successfully accomplished by a number of groups (for a more detailed review we refer to the Physics Report of Thomas and Landau[33]) and numerical results have become available in the last year or so. The most recent results from the calculations of Blankleider and Afnan,[34] will be reviewed in Sec. V together with a brief summary of the theory.

Let us briefly summarise the features of these few-body calculations. The overwhelming advantage is that they incorporate in one consistent set of <u>exactly</u> unitary equations <u>all</u> the important channels — namely NN, $\overline{N\Delta}$ and πD — to all orders. Moreover, quite sophisticated relativistic kinematics corrections are easily included. On the other hand, such equations are most easily derived in a Tamm-Dankoff approximation, keeping intermediate states with at most one pion. Thus terms in which a pion is produced at one nucleon and travels backwards in time to rescatter do not usually appear. Such terms were certainly non-negligible in the calculations of Woodruff, Koltun and Reitan and others. In the CCM such terms double the strength of the NN-NΔ potential. Heavy meson exchange does not appear readily in these transition potentials either — although it can be included. We shall return to the discussion of the old controversy over the range of the NNπ vertex function versus ρ-exchange in the final section.

In the past couple of years the group of Betz, Coester and Lee at Argonne,[35,36] proposed a new coupled channels model. The idea was to start with a Hamiltonian somewhat like the Lee model, in which a bare nucleon and delta appeared, with a ΔNπ vertex but no NNπ vertex. This idea will be described in somewhat more detail in Sec. IV. In essence the intention was to generate a relatively simple and unambiguous description of the virtual NΔ component of the N-N force for application in nuclear matter (see also Sec. III).[37,38] Unknown channel coupling potentials were initially taken to have phenomenological, separable forms for simplicity. More recently, at TRIUMF, these channel couplings have been calculated in a OBE model, with the aim of making predictions for the reaction pp $\to$ pnπ^+. Such calculations are of considerable importance in the planning (and future analysis) of the extensive experimental programs now underway at TRIUMF and LAMPF. The initial results of this investigation have also been reported in Sec. IV.

III. THE COUPLED CHANNELS METHOD

Conventional potential models (CPM) of NN-scattering work in the two nucleon space with a potential. A generalisation is to expand the baryon space to include the internal excitations of nucleons, Δ's and N^*'s. The configurations including these "isobars" are coupled to the NN state by coupling potentials derivable from the meson exchanges between the two baryons. Thus the name of the model — the coupled channels model (CCM).[19-21] This model has more structure (with parameters, in general, obtainable from other sources) than conventional potential models, but is nevertheless much simpler calculationally than true three-body models. In particular, after nearly a decade of NΔ CCM calculations, it is still not widely realised that, even though the CCM interaction can be phase equivalent to a CPM, it is <u>not</u> the same when off-shell effects are considered in nuclear matter[37,38] and in reactions such as pp → $d\pi^+$.

In free scattering, the inclusion of (say) NΔ states with a width in the CCM brings in the possibility of reactions and quite naturally leads to a prediction of inelasticities (complex phase shifts). In CPM's these should be calculated <u>after</u> the solution of the Schrödinger equation or its equivalent, or by adding a purely phenomenological absorptive potential. Whilst the low energy ($T_{lab} \leq$ 400 MeV) behaviour of these phases can be the same in the two models, it is very hard to see how one could get a resonance-like behaviour at or above the NΔ threshold in a CPM—whereas such behaviour is unavoidable in the CCM.

This behaviour is very clearly pronounced in the $^3F_3$ partial wave, which was predicted to be very important in pp → $d\pi^+$ (because of the NΔ admixture).[21] The "bump" in the phase shift (δ) at 600 MeV can well be reproduced by the CCM and explained easily by second order perturbation arguments. In particular, the coupling to higher energy excitations is attractive. As the energy approaches the excitation the energy denominator decreases, and the attraction increases, until above the excitation energy the energy denominator changes sign and what used to be strong attraction changes into repulsion. Of course, the width of the excitation acts as a smoothing and smearing agent. This behaviour of slow rise (without NΔ, $\delta(^3F_3)$ would be a decreasing function of energy), and then a nosediving decrease, is shown by the experimental $\delta(^3F_3)$—see Fig. 1. In addition to the Δ-nucleon mass difference a repulsive contribution of about 80 MeV from the centrifugal force in $^5P_3$(NΔ) state <u>and</u> the NΔ diagonal potential determine the exact position of the bump. In fact this might be used as a piece of experimental data in determining the NΔ interaction. Another place where the effect of the NΔ is very pronounced is ε_2. There again the contribution of NΔ to the tensor force acts counter to the pure OPE and gives rise to behaviour qualitatively similar to that for $\delta(^3F_3)$.

In view of these arguments, it is interesting to note that the partial waves where NΔ configurations are most important (strong coupling to NΔ and a considerable decrease in centrifugal barrier) are $^1D_2$, $^3F_3$ and $^1G_4$. These are precisely the channels in which the

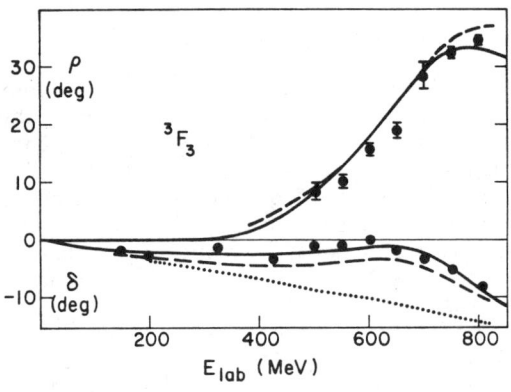

Fig. 1. A comparison of the OBE model calculations using the coupled channels method,[41] with (dashed) and without (dotted) NΔ coupling, with the phase shift analysis of Arndt and Verwest (solid line).[42]

so called dibaryon resonances have been reported. Moreover, the energies are very near those expected from a CCM calculation. Thus one is easily led to thinking of these resonances as mere reflections of an intermediate N∆ state. Surely the effect of N∆ configurations is not a "constant" or "smooth" background as assumed in some dibaryon analyses.

So far the CCM has been applied to NN scattering[20,39] and pp → dπ^+,[21,40] phenomenologically fitting the real parts of experimental phase shifts. The old amplitudes for pp → dπ^+ calculated in Ref. 21, and later used in Ref. 40, provided what were for a longtime the only predictions for asymmetries with a polarized beam and target or for the deuteron polarization. Qualitatively the prediction was correct while in detail there were discrepancies. However, with time the experimental results[43] have slowly approached theory. In particular, the experimental parameter A_{zz} at 580 MeV in the parametrization

$$\sigma_{ij} \text{ (polar)} = \sigma_{00}(1 + P_i A_{io} + P_j A_{oj} + P_i P_j A_{ij}) \quad (3.1)$$

has recently changed in shape and in normalization[44] towards the theoretical values.[45] Today there is a constant difference of about 0.2, which can perhaps be cured by changes in only one production amplitude—presumably the s-wave pion amplitude where the s-wave rescattering mechanism has some ambiguity. However, this s-wave rescattering strongly influences the old polarization parameter λ_0 in the region around 650 MeV where new, reliable data do not exist. To fix this ambiguity, an experiment to fill this gap would be most useful.

Perhaps the most disconcerting problem experienced in the CCM is that above about 500 MeV, after the inclusion of all relevant NN partial waves, the differential cross section has the wrong curvature as a function of $\cos^2\theta$. In particular, the coefficient of $\cos^4\theta$ (γ_4) is positive. It is possible that both this problem and the residual problems with other polarization parameters (other than A_{zz}) at the 10% level, could be related to the omission of the non-resonant p-wave rescattering in the CCM. (Note, however, that the few-body calculations discussed in Sec. V did include these other πN partial waves, but found no significant improvement!). Finally, on a more positive note, we observe that the fit to A_{zz} at lower energies (e.g. 440 MeV) is nearly perfect.[44,45]

IV. THE NN → NNπ REACTION

At laboratory energies above roughly 500 MeV, the three-particle channel provides the dominant contribution to the nucleon-nucleon inelastic cross-section. The mechanism of pion production can therefore hardly be considered fully unravelled until a satisfactory understanding of the NN → NNπ reaction is achieved. Measurements for appropriately chosen final state geometries allow the mapping of pion production amplitudes in regions of phase space not accessible to the pp → πd reaction. In contrast to the latter, the NN → NNπ process gives information about transitions to both spin singlet and triplet configurations of the nucleon pair.

In most of the energy range relevant to nuclear physics, the reaction is known to be dominated by the process NN → N∆ → NNπ and constitutes a unique tool for the phenomenological determination of the NN → NΔ partial wave amplitudes. This will of course require abundant data, including differential cross-sections, spin asymmetries, spin-spin correlations and spin transfer coefficients for fully determined kinematics. Since the angular distribution of decay products is known to provide information about the state of polarization of a decaying particle,[46] angular distribution measurements (in both polar and azimuthal variables) are roughly equivalent to detecting the polarization of the Δ. This makes experiments with noncoplanar geometries particularly desirable.

The study of the NN → NNπ reaction should shed light on the controversial issue of dibaryon resonances.[47] These resonances are believed to have large partial widths for decay into the three-body continuum. Therefore, if they indeed exist, they are most likely to show up as distinctive signals in this reaction. Only by a detailed study of the three-body channel can one hope to separate a true s-channel resonance from a pseudo-resonance phenomenon due to the resonant behaviour of the πN subsystem.

Because of the additional complexity due to the presence of three particles in the final state, as well as the relative lack of data, theoretical investigations of the process NN → NNπ have not, at this time, reached the same level of sophistication as those of the πd → πd and πd → pp reactions discussed in other sections. Several calculations[48-51] based on the single pion (or pion + ρ-meson) rescattering picture have been published. They provide a generally satisfactory description of the averaged features of the reaction, such as the energy dependence of integrated cross-sections and the shape of inclusive spectra. One such approach also succeeds in reproducing the experimental cross-section for fully determined geometries corresponding to small momentum transfer. These calculations are useful for exploring the general trends of the data; they confirm that the Δ production picture, suggested in this context by Mandelstam[52] is essentially valid.

However, these approaches can hardly be expected to account for the finer details of the reaction and cannot be relied upon to predict the behaviour of individual partial waves. For this purpose, it is necessary to construct dynamical models which describe all possible reactions of the baryon-number-two system in a unified way, and satisfy the constraints imposed by unitarity. This requires the specification of the coupling between at least all open channels and, in general, the solution of a system of coupled equations for the relevant amplitudes. A variety of such models are nowadays available, some of which are dealt with in other sections. As mentioned above, however, most calculations published to date are restricted to two-body reactions. Only the three-body, or bound-state model of Ref. 53 has been used to calculate NN → NNπ observables — see also Ref. 54. In the following, we discuss one such model for which calculations are in progress and preliminary results are available.

The spirit of this work is similar to that of Refs. 35 and 36, although the specifics are rather different. The basic degrees of freedom in the model are N, Δ and π. The Hamiltonian is

$$H = H_0 + V + h = H_0 + U \qquad (4.1)$$

where H_0 describes non-interacting particles, V is a baryon-baryon interaction performing the transitions $NN \to NN$, $NN \to N\Delta$ and $N\Delta \to N\Delta$, and h stands for the $\Delta \rightleftarrows N\pi$ vertex. Note that the explicit introduction of an $N \to N\pi$ vertex is avoided at this stage. The motivation for this approach is to maintain the unitarity structure of the model without having to face the problems of renormalization, vertex corrections and NN propagator dressings. Of course, a model more firmly grounded in field theory would include these effects (see Sec. II and V). The success of the CCM (similar in this respect to the present one) developed by Niskanen for $pp \to \pi d$ (Sec. III) suggests that they are not of great numerical importance. Non-resonant πN partial waves are omitted in (4.1). Appropriate two-body interactions could be added, but they have been neglected in the calculations performed so far.

Scattering equations corresponding to the Hamiltonian (4.1) can be formulated in standard fashion. One obtains:

$$T_{N,N\pi} = T_{N,\Delta} \ \tilde{G}_\Delta \ h_{\Delta,N\pi}[1 + G_0 \tau^{(\pi)}] \qquad (4.2)$$

$$T_{B,B} = \tilde{V}_{B,B}[1 + \tilde{G}_B \ T_{B,B}] \qquad (4.3)$$

$$\tilde{V}_{B,B} = V_{B,B} + \{h_{\Delta,N\pi} \ G^B \ h_{N\pi,\Delta}\}_C \qquad (4.4)$$

$$(G^B)^{-1} = G_0^{-1} - U^{B,B} \qquad (4.5)$$

$$\tilde{G}_N = G_0 \ ; \quad \tilde{G}_\Delta^{-1} = G_0^{-1} - \{h_{\Delta,N\pi} \ G_0 \ h_{N\pi,\Delta}\}_D . \qquad (4.6)$$

In these equations, the subscripts N, Δ, B, $N\pi$ stand for projections onto the NN, $N\Delta$, $NN \oplus N\Delta$ and $NN\pi$ subspaces, respectively. The corresponding superscripts denote the orthogonal projections. The label C(D) denotes the connected (disconnected) part of an operator. The NN T-matrix in the presence of a spectator pion is denoted by $\tau^{(\pi)}$. The second term of (4.4) can be broken up in two pieces, according to whether the intermediate pion undergoes interactions or not. Contributions to these two pieces are shown in Fig. 2a and b. Keeping only the piece in which the pion remains a spectator throughout, one may rewrite (4.4):

$$\tilde{V}_{B,B} = V_{B,B} + \{h_{\Delta,N\pi} \ G_0 \ h_{N\pi,\Delta}\}_C + h_{\Delta,N\pi} \ G_0 \ \tau^{(\pi)} \ G_0 \ h_{N\pi,\Delta} . \qquad (4.7)$$

One might wish to calculate $\tau^{(\pi)}$ consistently within the model, namely from an effective interaction of the form (4.7), but evaluated at the subenergy of the NN pair in the presence of the spectator pion. However, in order to avoid generating an infinite hierarchy of equations, we have chosen to calculate it independently from a two-body potential (e.g. fit to scattering data below the pion production threshold). The resulting equations (4.2 - 4.7) are depicted in Fig. 2c.

It remains to specify some model for the baryon-baryon interactions V. Following conventional wisdom, a one-boson-exchange (OBE) model is adopted. We remark that the $N\Delta \to \Delta N$ pion exchange with $NN\pi$ intermediate state should not be included in V, since it is generated dynamically by the $\Delta \rightleftarrows N\pi$ vertex and appears automatically in $\tilde{V}$ (see

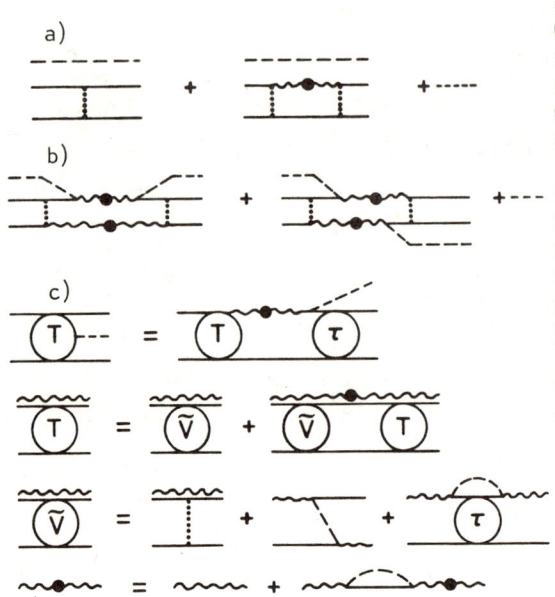

Fig. 2. a) contributions to G^B in which the pion remains spectator; b) contribution to G^B in which the pion undergoes interactions; c) the scattering equations (4.2-4.7). Solid, wiggly and dashed lines stand for nucleons, deltas and pions respectively. A solid and a wiggly line appearing together indicate that the particle can be either a nucleon or a delta. The interaction V is represented by a dotted line.

Fig. 2c). Heavy boson exchanges, as well as pion exchange with an intermediate $\Delta\Delta\pi$ state should in principle be included in V, but they are omitted in the calculations to be discussed below.

In these calculations, the NN interactions in intermediate NNπ states are neglected (i.e. $\tau^{(\pi)} = 0$) and so is the $N\Delta \to \Delta N$ pion exchange [i.e. the 2^d term of (4.7)]. In view of the first approximation, one cannot expect a satisfactory description of pion production below roughly 600 MeV where the coupling to the πd channel is important. For this reason, we shall present calculations at 800 MeV, where this approximation should not be too bad in most regions of kinematics. The second approximation amounts to keeping one πN rescattering only (albeit with a consistent description of NN scattering in the initial state). As emphasized in other sections, this approximation is undesirable and should be removed. Work along these lines is in progress. Calculations of pp $\to \pi d$[36,55] indicate that one can expect corrections of about 20% to amplitudes from higher order rescattering terms. The effect of the second term of (4.7) on NN elastic scattering has been investigated in Ref. 36 and found to be generally rather small.

After these simplifications, Eqs. (4.2-4.7) reduce to

$$T_{N,N\pi} = \Omega^+_{N,N} V_{N,\Delta} \tilde{G}_o h_{\Delta,N\pi} , \qquad (4.8)$$

$$\Omega^+_{N,N} = 1 + \overline{V}_{N,N} G_o \Omega^+_{N,N} , \qquad (4.9)$$

$$\overline{V}_{N,N} = V_{NN} + V_{N,\Delta} \tilde{G}_\Delta V_{\Delta,N} . \qquad (4.10)$$

That is, the initial state distortion is generated by an effective NN potential including, in addition to OBE contributions, the NΔ box diagram in which the intermediate Δ is dressed by its coupling to the πN channel.

The calculations presented here employ the $\Delta \rightleftarrows N\pi$ vertex of Ref. 36, which is determined by fitting πN scattering in the (3,3) channel. A form factor $\beta^4/(\beta^2 + k)^2$ is used, with $\beta = 358$ MeV/c. For the baryon-baryon interaction V, a OBE model is adopted, with π-, ρ-, ω-, σ-, δ-, η-, φ-meson exchanges for the NN → NN transition and π,ρ exchanges for the NN → NΔ transition. No NΔ → NΔ interaction is included. The particular forms of OBE interactions used here are the same as in Ref. 38. As in the latter work, the high momentum components of the matrix elements $\langle \vec{q}'|V|\vec{q}\rangle$ are damped by multiplying them by a form factor

$$F_\alpha^2(\Delta^2) = \left(\frac{\Lambda_\alpha^2 - m_\alpha^2}{\Lambda_\alpha^2 - \Delta^2}\right)^n \quad (4.11)$$

where $\Delta^2 = (E_{q'} - E_q)^2 - (\vec{q}' - \vec{q})^2$ and $n = 2(3)$ for scalar (vector) meson exchange [retardation effects are ignored in the calculation of the NN → NΔ interaction, i.e. $\Delta^2 = -(\vec{q}' - \vec{q})^2$]. In the calculations presented here, the parameters of the OBE interactions (coupling constants and cutoffs) have been taken from Ref. 38, with the exception of the cutoffs of the NN → NΔ interaction for which a few sets of values were tried. The results of sample calculations of differential cross sections for kinematically complete measurements of pp → pnπ$^+$ at 800 MeV are shown in Fig. 3. Isospin coupling coefficients give a ratio of 3 to 1 in amplitudes for the processes pp → nΔ^{++} and pp → pΔ$^+$. The experimental geometries were chosen in such a way as to further enhance the former, by requiring larger momentum transfer for the latter. Since calculations show that the amplitudes for the two processes do not interfere much, one can think of the reaction as proceeding mostly via pp → nΔ^{++}.

The calculation reproduces the trends of the data[50] such as the large bump corresponding to an invariant mass of the pπ system close to the Δ mass, and the falloff with increasing momentum transfer. The bump at small proton momentum is due to NN final state interactions which are not included in the present calculation. The NN → NΔ interaction of Ref. 38, with $\Lambda_\pi = \Lambda_\rho = 1200$ MeV/c is found to overestimate the cross-section, especially at large momentum transfer. Increasing the ρ cutoff to 1600 MeV/c brings both the shape and the magnitude of the cross section into better agreement with data. However, essentially the same result can be obtained by decreasing the π cutoff to 700 MeV/c and neglecting ρ-exchange altogether.

Figure 3 also illustrates the importance of NN initial state interactions, which lower the cross section by a significant amount. This effect is due mostly to absorption and provides further evidence for the inadequacy of a perturbative treatment of pion production.

Analyzing powers for the same reaction are shown in Fig. 4, together with the preliminary data of Ref. 56. The simple isobar model is clearly not adequate at the relatively large momentum transfers for which the data was taken. This was already apparent in the cross section calculation. It is likely that NN and non-resonant πN final state interactions play a significant role here.

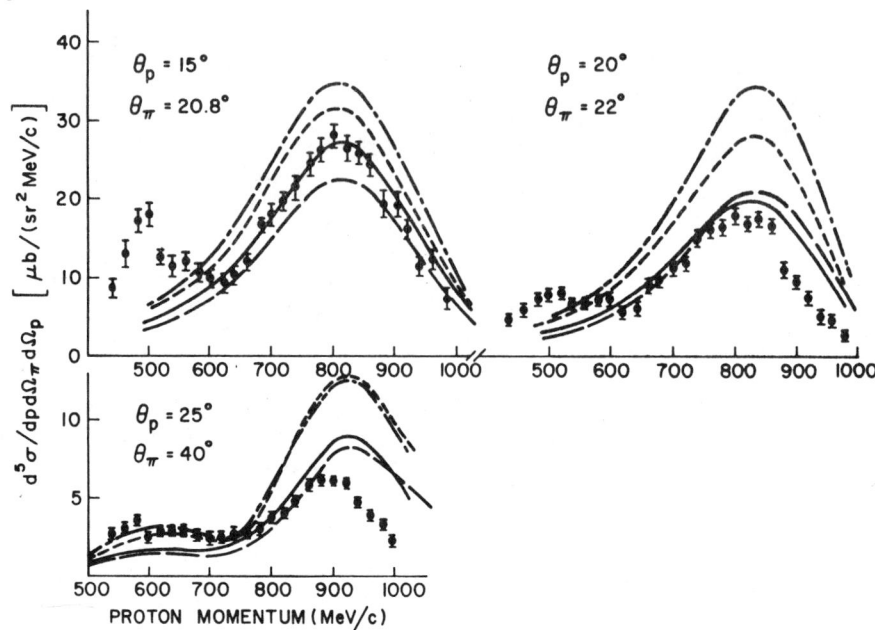

Fig. 3. Differential cross sections for kinematically complete measurements[50] of $pp \rightarrow pn\pi^+$ at $T_L = 800$ MeV. The geometries are coplanar; θ_p and θ_π are the laboratory angles of the proton and the pion, respectively. The data is plotted versus the lab proton momentum. The parameters of the $NN \rightarrow N\Delta$ transition operator corresponding to each curve are (in MeV/c): a) --- $\Lambda_\pi = \Lambda_\rho = 1200$; b) —— $\Lambda_\pi = 1200$, $\Lambda_\rho = 1600$; c) —·— $\Lambda_\pi = 700$, no ρ-exchange; d) ——— $\Lambda_\pi = 700$, no ρ-exchange, no initial state interactions. The momentum transfers for the $pp \rightarrow n\Delta^{++}$ process corresponding to the three geometries are roughly 300, 360 and 460 MeV/c.

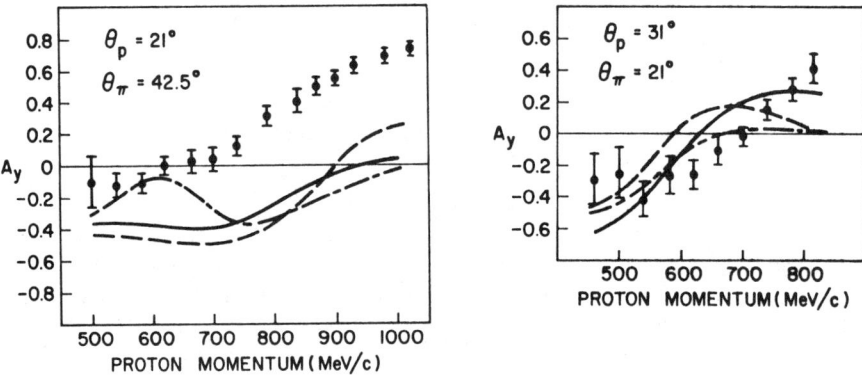

Fig. 4. Analyzing powers for kinematically complete measurements of $pp \rightarrow pn\pi^+$ at $T_L = 800$ MeV.[56] The meaning of each curve is the same as in Fig. 3. The momentum transfers for the $pp \rightarrow n\Delta^{++}$ process corresponding to the two geometries are roughly 400 and 550 MeV/c.

To close this section, we briefly comment on the NN phase shifts and inelasticities predicted by our model. If the parameters of the NN → NΔ interaction are chosen to give a satisfactory description of the pion production data, the inelasticities in the $^1D_2$ and $^3F_3$ partial waves are in good agreement with the phase-shift analysis of Arndt et al.[42] The $^1D_2$ phase shift is also well reproduced, but the kink in the $^3F_3$ phase shift is not so pronounced.

The calculated inelasticity in the $^1S_0$ wave is small in contrast with the phenomenological one. The results for the 3P partial waves are rather sensitive to the choice of parameters. In particular, in the $^3P_0$ channel, the inelasticities corresponding to the model with π- and ρ-exchange are large (in agreement with the analysis of Arndt et al.) while those calculated using the model with π-exchange only are small (in agreement with the analysis of Bugg[57]).

It is our hope that the next generation of measurements of NN → NNπ will allow us to discriminate between these alternatives.

V. THE FEW-BODY APPROACH

A. Theoretical Development

It is clear that for an adequate description of N-N scattering above the pion production threshold one needs to include the coupling to the πNN channels. For example, one needs to couple N-N to both N-Δ, resulting in the well-known resonant behaviour, and π-d, resulting in a simultaneous description of NN → πd. Similarly π-d scattering is coupled to both N-N and N-Δ channels. If in addition one believes that the one-pion approximation is adequate for energies below the two-pion production threshold then one is left with the task of describing a system consisting explicitly of at most 3 bodies. As 3-body scattering problems are able to be solved exactly it is not surprising that recent years have seen attempts to also describe the πNN-NN system in an exact fashion.

Indeed the first work in this area was a pure 3-body description of the πNN system by Afnan and Thomas (AT).[26] In their model, the so called bound state model (BSM), they allowed for coupling to the N-N channel by constructing a bound state in the π-N P_{11} channel and then identifying this bound state with a nucleon. Effectively this means that the pion is never really destroyed, and the three-body description of πNN-NN follows. In particular, AT solved the Alt-Grassberger-Sandhas (AGS)[58] equations at low energies and obtained, with the one set of integral equations, the coupled amplitudes for πd → πd, πd ↔ set of integral equations, the coupled amplitudes for πd → πd, πd ↔ NN, and NN → NN. Although their calculation provided a good description of each of the above reactions at low energies, there were serious theoretical problems arising from the fact that only one of the nucleons (the bound state) could emit the pion. Thus, for instance, the process of Fig. 5a is included in their model, while Fig. 5b is not.

Fig. 5. Examples of contributions to π-d elastic scattering that a) are included, b) are not included in the BSM.

Although the problems of the BSM discouraged its use for calculations retaining all physical channels, there have been a number of calculations that may be considered as restricted cases of the BSM. Firstly there are the three-body calculations that neglect pion absorption altogether in π-d elastic scattering,[59-65] so that the above problems do not arise. Secondly, there are the analyses of N-N elastic scattering[53,66,67] that use the BSM but do not couple to the π-d channel, and thereby lose the possibility of describing NN → πd. The main effort in this area has come from Silbar and Kloet who also attempt to compensate for the undercounting problems of the BSM by adding one-half a static N-N potential to the Born term. More recently Araki et al.[68,69] used the BSM with couplings to all physical channels, but only $J^\pi = 3^-$ results have been given.

Deficiencies of the BSM were recognized at an early stage, and there has since been steady progress in the description of the πNN-NN system from various more fundamental approaches based in field theory.[24,25,27-34,70] Practical three-body-like equations have now been derived by Blankleider and Afnan,[34] and independently by Avishai and Mizutani.[30] The main features of these equations, from now on called "the πNN equations" or "few-body equations", are that they satisfy two- and three-body unitarity—dressings of both nucleon propagators and πNN vertices are explicitly included, and they contain terms that may be identified with heavy meson exchanges.

Although the derivations of these equations are quite involved, an intuitive derivation, which involves "correcting" the undercounting problem in the BSM, has recently been given by Blankleider and Afnan.[34] As this is a good way to introduce the uninitiated reader to the πNN equations, we briefly reproduce the argument here.

The AGS equations for the BSM are

$$U^{(o)}_{\alpha\beta} = \bar{\delta}_{\alpha\beta} G_o^{-1} + \sum_{\gamma=1}^{3} \bar{\delta}_{\alpha\gamma} t_\gamma G_o U^{(o)}_{\gamma\beta}, \qquad (5.1)$$

where $\bar{\delta}_{\alpha\beta} = 1 - \delta_{\alpha\beta}$, $G_o = (E-H_o)^{-1}$ is the free Green's function for the πNN system, and the superscript in $U_{\alpha\beta}$ indicates that this is a pure three-body amplitude. The labels α, β, γ denote the "spectator + interacting pair" channels, with t_γ describing the two-body interaction in channel γ. Labelling the particles, 1 and 2 for the nucleons and 3 for the pion, Eq. (5.1) may be written

$$U^{(o)}_{\alpha\beta} = \bar{\delta}_{\alpha\beta} G_o^{-1} + \bar{\delta}_{\alpha 3} t_3 G_o U^{(o)}_{3\beta} + \sum_{i=1}^{2} \bar{\delta}_{\alpha i} t_i G_o U^{(o)}_{i\beta}, \qquad (5.2)$$

where t_3 is the N-N amplitude and t_i is the π-N_i amplitude. It can be shown that in the P_{11} channel, t_i may be written in terms of a pole part (t_i^P) and a non-pole part (t_i^{NP})

$$t_i = t_i^P + t_i^{NP}, \qquad (5.3a)$$

$$t_i^P = f(i)g(i)f^+(i) \qquad (5.3b)$$

where $g(i)$ and $f(i)$ are the dressed nucleon propagator and πNN vertex for the ith nucleon. For other channels we may take $t^P = 0$. Equation (5.2) may then be written diagrammatically as shown in Fig. 6.

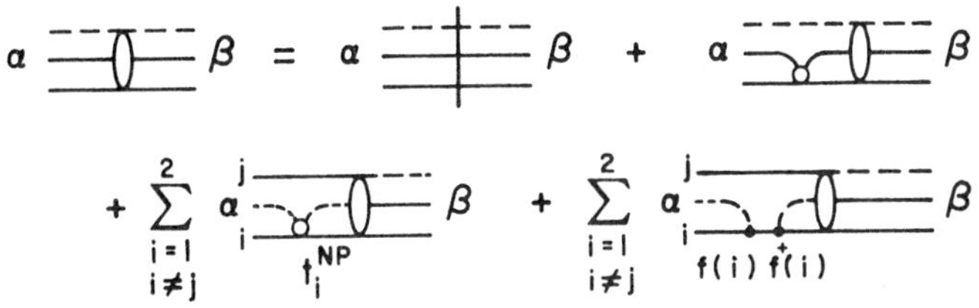

Fig. 6. The AGS equations in the πNN BSM. Refer to Eq. (5.2).

The unsymmetrical nature of the last diagram in Fig. 6 is apparent, and this may be remedied by adding the term shown in Fig. 7. Not only does the inclusion of this term allow either nucleon to emit a pion, but it also contributes to the dressing of the jth nucleon—thereby putting both nucleons on the same footing.[34] With the added term the equations become

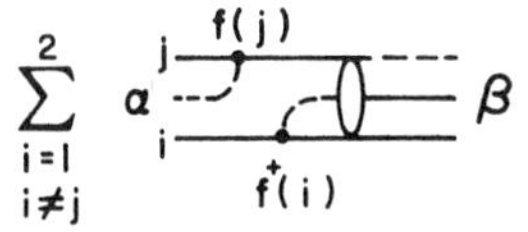

Fig. 7. The "correction" term to the BSM equations represented in Fig. 6.

$$U_{\alpha\beta} = \bar{\delta}_{\alpha\beta} G_0^{-1} + \sum_{\gamma,\rho} \bar{\delta}_{\alpha\gamma} B_{\gamma\rho} G_0 U_{\rho\beta} , \quad (5.4a)$$

$$B = \begin{bmatrix} t_1 & V_{12} & 0 \\ V_{21} & t_2 & 0 \\ 0 & 0 & t_3 \end{bmatrix} , \quad (5.4b)$$

$$V_{ij} = \bar{\delta}_{ij} f(i) g f^+(j) , \quad (5.4c)$$

where there is no particle label on g as the two nucleons are now identical. Equations (5.4) are in fact the correct πNN equations and differ from the BSM only in the non-diagonal terms V_{ij}. However, in order to guarantee three-body unitarity one needs to explicitly dress the nucleon propagators and πNN vertices according to the following equations[34]

$$g = g_0 + g_0 \sum_{i=1}^{2} \left[f_0^+(i) G f_0(i) + f_0^+(i) G t_i^{NP} G f_0(i) \right] g , \quad (5.5a)$$

$$f(i) = f_0(i) + t_i^{NP} G f_0(i) , \quad (5.5b)$$

where g_0 and $f_0(i)$ are the undressed quantities and G is the three-particle propagator. Equations (5.5) are represented diagrammatically in Fig. 8.

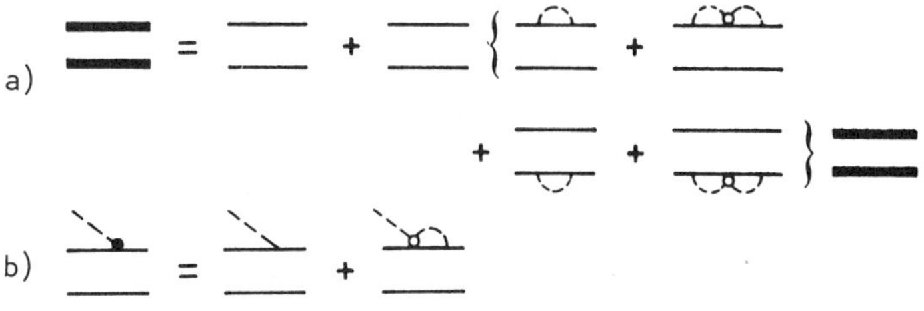

Fig. 8. a) dressing of the two-nucleon propagator—see Eq. (5.5a); b) dressing of the πNN vertex—see Eq. (5.5b).

In order to facilitate the solution of Eq. (5.4) it is usual to make the approximation of separable interactions $t_i^{NP} = |\phi_{\Delta i}\rangle \tau_{\Delta i} \langle \phi_{\Delta i}|$, and $t_3 = |\phi_d\rangle \tau_d \langle \phi_d|$, where "d" represents all N-N interactions, and "Δ" all π-N interactions. One may then relate the AGS amplitudes $U_{\alpha\beta}$ to the physical amplitudes $X_{\alpha\beta}$ in the usual way,[34] so that Eqs. (5.4) (after antisymmetrization) become

$$X_{\alpha\beta} = Z_{\alpha\beta} + \sum_{\gamma} Z_{\alpha\beta} \tau_\gamma X_{\gamma\beta}, \qquad (5.6)$$

with α = N, Δ or d, and $\tau_N = g/2$. Of course $Z_{dd} = 0$, and in general the fully antisymmetrised Born terms $Z_{\alpha\beta}$ are related to the BSM Born terms of AT by

$$Z_{\alpha\beta} = C_{\alpha\beta} Z_{\alpha\beta}^{AT}, \qquad (5.7)$$

where

$$C = \begin{bmatrix} 4 & 2 & 2\sqrt{2} \\ 2 & 1 & \sqrt{2} \\ 2\sqrt{2} & \sqrt{2} & 0 \end{bmatrix}, \qquad (5.8)$$

and α, β run over N, Δ, d in that order.

Although Eq. (5.6) was presented in Refs. 29, 30 and 34, earlier attempts at deriving the πNN equations had some success[24,70] in that they managed to produce partially coupled equations for the reactions $\pi d \to \pi d$, $\pi d \to N\Delta$, $\pi d \to NN$. In particular, this set of equations was equivalent to Eq. (5.6), with the NN $\to$ NN amplitude X_{NN} not calculated self-consistently. These equations were used in several calculations,[24,71,72] the most extensive of which was due to Rinat et al.[71,72] These authors used a phenomenological potential for X_{NN} and investigated the effect of absorption in $\pi d \to \pi d$.[71] Although their model was able to reproduce most π-d elastic data,[71] the corresponding predictions for pp $\to \pi^+$d were disappointing.[72] Because of the complicated nature of the calculation, it is not clear whether this was caused by the lack of a self-consistent construction of X_{NN}, or some other reason.

The first attempt at solving the full set of πNN equations (5.6) was carried out by Afnan and Blankleider[73] at a pion laboratory energy T_π = 48 MeV. In their calculation they used relativistic pion kinematics (RPK) (relativistic pion, non-relativistic nucleon),[59] and did not implement explicit dressings of the propagators or πNN vertex. However, very encouraging results were obtained for all the reactions in Eq. (5.6) and their analysis (which forms the basis for the next sub-section) has now been extended to the whole energy region $T_\pi \leq$ 256 MeV. A theoretically more ambitious calculation has been performed by Fayard et al.[27] In this work the authors used fully relativistic kinematics with the usual Blankenbecler-Sugar reduction, and also they included the dressings of Eqs. (5.5). However, they only solved the set of equations for πd elastic scattering and restricted their discussion to the effect of absorption in π-d elastic scattering. In particular their results for NN $\leftrightarrow \pi$d were not presented. More recently Fayard et al. have presented some limited results for both NN $\leftrightarrow \pi$d and NN $\to$ NN.[75]

At the present time the only available, detailed calculation of πd $\to \pi$d, NN $\leftrightarrow \pi$d and NN $\to$ NN using the fully coupled πNN equations is the one of Blankleider and Afnan (BA).[34] Their results for the production (or absorption) reaction NN ↔ πd will be reviewed below—although in view of the couplings, some mention of the results for other channels is also necessary.

B. Results

In order to solve Eq. (5.6) one needs to specify as input separable N-N and π-N interactions in all included partial waves, as well as the form factor for the πNN vertex. A most important aspect of the assumed input is the construction of the P_{11} interaction. For this Blankleider and Afnan (BA)[34] took two-term separable potentials

$$V_{P_{11}} = |U_1\rangle \lambda_1 \langle U_1| + |U_2\rangle \lambda_2 \langle U_2|, \qquad (5.9)$$

where the form factors $U_i(k)$ are of the usual Yamaguchi type

$$U_i(k) = \frac{c_i k^{n_i}}{(k^2 + \beta_i^2)^{m_i}}, \quad i = 1,2 \qquad (5.10)$$

and λ_i, c_i, β_i, n_i, m_i are chosen by fitting to the P_{11} phase shifts and scattering length, and by requiring a pole in the corresponding P_{11} t-matrix at the nucleon mass. The correct asymptotic behaviour of the wave functions requires that $n_1 = 1$, $n_2 \geq 1$ and $2m_i > n_i$. In order to test the model's sensitivity both to the off-shell description of the P_{11} channel and to the (experimentally uncertain) value of the P_{11} scattering length, various choices of the form factors in Eq. (5.10) were constructed.[76]

As the constructed P_{11} t-matrix has a pole part, $t_{P_{11}}^P(k) = f^2(k)/(E-m_N)$, the P_{11} non-pole part is given by $t_{P_{11}}^{NP} = t_{P_{11}} - t_{P_{11}}^P$. We note that the constructed P_{11} interaction enters the πNN equations in two parts. Firstly $t_{P_{11}}^{NP}$ is included on an equal footing with non-P_{11} π-N interactions while $t_{P_{11}}^P$ enters separately through the πNN vertex function $f(k)$. As BA did not implement any dressings, their vertex function is energy independent and it is expected that three-body unitarity is violated to some extent. The only other interaction to be varied in this calculation is in the deuteron channel. For this BA used the unitary pole approximation (UPA) to the Reid soft core (RSC) (P_d = 6.56%) and Bryan-Scott (BS) (P_d = 5.36%) potentials. We note that the calculation retains two-body partial waves up to D-wave N-N and P-wave π-N interactions.

The calculations are very sensitive to the choice of P_{11} interaction, but the best results are obtained with the P_{11} form factors "Bφ8" of Ref. 34, and with the Reid soft core—UPA deuteron. Unless otherwise stated, all results are with this choice of input.

1. Low energy predictions for NN → πd

At energies close to pion production threshold only s- and p-waves should contribute, so that the differential cross section is given by

$$\frac{d\sigma}{d\Omega} = \frac{1}{32\pi}(\gamma_0 + \gamma_2 \cos^2\theta), \qquad (5.11)$$

where

$$\gamma_0 = 2|a_1|^2 + 2|a_0|^2 + |a_2|^2 + 2\sqrt{2}\,\text{Re}(a_0 a_2^*)$$
$$\gamma_2 = 3(|a_2|^2 - 2\sqrt{2}\,\text{Re}(a_0 a_2^*))\;, \tag{5.12}$$

and a_0, a_1, a_2 are the $J^\pi = 0^+$, 1^-, 2^+ NN $\to \pi d$ rotationally invariant amplitudes in the convention of Mandl and Regge.[77] Clearly a measurement of $d\sigma/d\Omega$ alone is not enough to determine the three amplitudes. The corresponding total cross section is given by

$$\sigma_T = \frac{1}{4}(|a_1|^2 + |a_1|^2 + |a_2|^2)$$
$$\sigma_T = \alpha\eta + \beta\eta^3\;, \tag{5.13}$$

where the last expression is due to Gell-Mann and Watson and Rosenfeld,[78] and expresses the total cross section in terms of an s-wave part (α) and a p-wave part (β), with the basic momentum dependence ($\eta = k_\pi/m_\pi$) separated out. It has been a hope that the parameters α and β are essentially constant at small energies so that an analysis of low energy cross sections would yield values for α and β and thus some information on the amplitudes themselves. Indeed within the accuracy of all low energy experiments, the data has been consistent with a two parameter fit of the form of Eq. (5.13) and commonly quoted values are tabulated in Table I. The first indication that α might not be constant came from the calculation of Afnan and Thomas,[26] who found that α markedly decreases with energy. (Although Lichtenberg[10] had remarked 17 years earlier that it need not be constant.) In fact they suggested that this behaviour could explain the discrepancy in the experimental value of Rose,[79] and the higher energy one of Crawford and Stevenson.[80]

Table I. Some experimental values of α and β — refer to Eq. (5.13).

Group	α µb	α µb
Crawford and Stevenson[80]	138 ± 15	1010 ± 80
Rose[79]	240 ± 20	520 ± 200
Richard-Serre et al.[81]	180 ± 20	950 ± 150

The present calculation[34,76] agrees with Afnan and Thomas on the decreasing behaviour of α. However, the low values at threshold ($\sim 140\mu b$) are inconsistent with Rose's experimental result. We note that Fayard et al.[74] also obtained low values for α. In addition, BA's calculation gives an approximately constant value of β ($\sim 1000\mu b$), which precludes us from using similar arguments to Afnan and Thomas to explain the discrepancy in the experimental values of β. Thus at threshold, accurate experiments are badly needed, and it is encouraging that some effort is now being made[82] to cover this long neglected region.

2. Medium energy predictions for NN → πd

The dependence of the few-body model on the assumed input is greatly restricted, as one needs to describe as accurately as possible all the reactions πd → πd, NN ↔ πd and NN → NN at the same time. However, for the present calculation all heavy-meson exchanges are neglected, so one does not expect to fit the N-N scattering data. On the other hand, indications are that ρ-exchange might play a small role in πd → πd,[71,83] and also in NN → πd if one judiciously chooses the range of the πNN vertex.[84]

Using the optimum choice of input mentioned earlier, BA obtain the differential cross sections for πd → πd and pp → π⁺d as illustrated in Fig. 9. For π-d elastic scattering we show

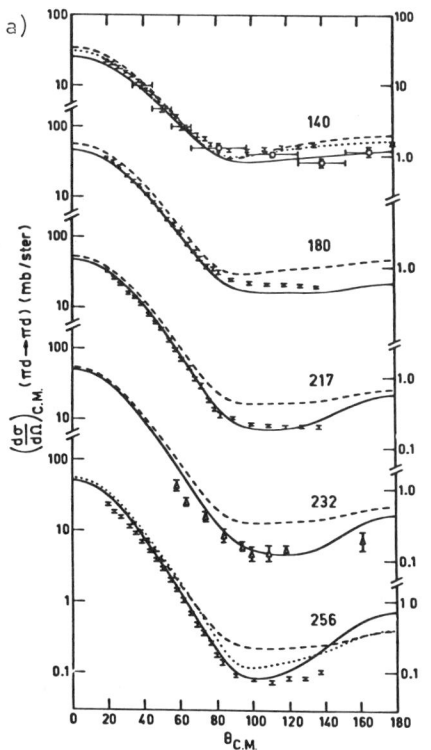
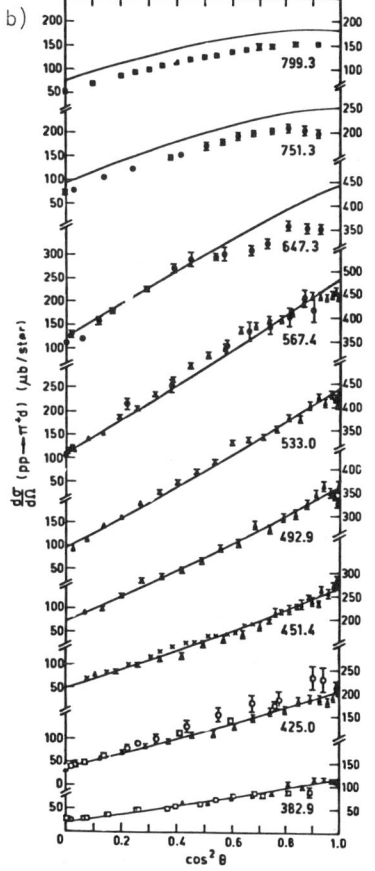

Fig. 9. Simultaneous calculations of the differential cross sections for the coupled processes: a) πd → πd: for various pion lab energies we show the full calculation (—), no absorption (--) and no absorption or P_{11} rescattering (··). The experimental data are those of Cole et al.,[85] Gabathuler et al.,[86] Holt et al.[87] and Pewitt et al.[88] b) pp → π⁺d; curves are labelled by the proton lab energy. The experimental results are those of Axen et al.,[105] Dolnick,[89] Aebischer et al.,[90] Richard-Serre,[81] Nann et al.[91] and Hürster et al.[92] The data of Hürster was scaled since it was normalized to 1 at zero degrees.

results that contain no P_{11} input (dotted curves), no absorption (i.e. no t^P) (dashed curves), and full P_{11} input (solid cuvres). A general feature of the effect of absorption in the BA calculation is the lowering of the cross section, especially at the backward angle dip. In particular, absorption in the model is instrumental in reproducing the notorious dip at 256 MeV. The differential cross sections for pion production fit experiment rather well except at energies close to T_p = 800 MeV, where one expects problems with the RPK approximation. For N-N elastic scattering BA also obtain reasonable results, which are consistent with a lack of short range repulsion in the N-N generated interaction.

Having established the basic success of the model, we examine more closely some aspects of the pion production part of the calculation. The multiple scattering nature of the model enables one to easily test the common practice of calculating at most one-pion rescattering in pp → πd. In particular by iterating Eq. (5.6) one obtains successive terms of the multiple scattering series (MSS) and its convergence may be examined. Using BA's amplitudes we investigate the MSS in the dominant $J^\pi = 2^+$ channel for T_p = 567 MeV. All forementioned N-N and π-N partial waves are retained, although it is of course the N-Δ(P_{33}) in relative s-state that dominates. Starting with the Born term (direct production), successive terms in the MSS for the t-matrix $T^J_{\ell's';\ell s}$ (pp → π^+d, ℓ'=s'=1, ℓ=J=2) are tabulated in Table II. Firstly we note that the MSS does not converge until about

Table II. Iteration of Eq. (5.6) in the $J^\pi = 2^+$ channel, generating the MSS for pp → π^+d at T_p = 567 MeV.

Iteration	MSS for $T^J_{\ell's';\ell s}$ (fm^2)
0	-0.0087
1	-0.0299 - 0.0134i
2	-0.0355 - 0.0292i
3	-0.0297 - 0.0349i
4	-0.0234 - 0.0378i
...	...
10	-0.0194 - 0.0327i

the 10th iteration. In addition, comparing the direct plus one-rescattering pion production (iteration 1), with the full solution, we see that these amplitudes would lead to differences of about 15% in the cross section. However, as the magnitude is distributed differently between real and imaginary parts in the two cases, we would expect much larger differences to appear in the polarization observables.

Another commonly made assumption regards the number of partial waves that contribute to pp → π^+d. At first it was hoped that deviations from a straight line in plots of dσ/dΩ vs. $\cos^2\theta$ might show evidence for d-wave pions in pp → π^+d. Although this has since been confirmed, deviations from straight lines are small below the 3-3 resonance. More apparent evidence has come from polarization experiments.[93] However, it is not easy to separate the different higher partial wave contributions from experimental results. Indeed Niskanen[21] found that d-wave pion production in the $J^\pi = 3^-$ channel was very important in his model. As BA retained up to h-wave pions in their calculation, they were in a good position to test higher partial wave contributions. In fact they found that both the $J^\pi = 3^-$ and the $J^\pi = 4^+$ (f-wave pion) partial amplitudes had a significant effect on $d\sigma/d\Omega$. However, the inclusion of both resulted in a straight line which in turn, in some simple analyses, would be taken as evidence for just s- and p-waves! This illustration should caution those who analyse data without retaining amplitudes up to the 4^+. On the other hand, still higher partial waves were found to have a significantly smaller effect (perhaps not negligible though) on $pp \to \pi d$ observables.

A more exacting test of a model is provided by polarization observables. In the case of $pp \to \pi^+ d$ there is a good deal of data with the incident proton polarized (Ayo), there are the SIN-Geneva[43] measurements having both initial protons polarized (Axx, Ayy, Azz, Azx), and some preliminary data on iT_{11} in $\pi^+ d \to pp$ has also been recently reported.[94] Again we use BA's amplitudes, and on comparison with experiment we find that the predicted Ayo is too large across the whole energy range—see Fig. 10. On the other hand, of all the Aij we find Ayo to be the most model-dependent. In particular we find Ayo(90°) is dominated by interference terms between the large amplitudes of $J^\pi = 2^+$ and 3^- and the small ones of 0^+ and 1^-, so that any mechanism which affects only these smaller amplitudes would change Ayo(90°) without affecting the cross section greatly. Such a mechanism is provided, for example, by the deuteron d-state components and their effect on Ayo is displayed in Fig. 10. The results for other Aij parameters are similar to those of both Niskanen[21] (see also Sec. III) and Rinat et al.[72] so that there are now some well established and substantial disagreements between theory and experiment in $\vec{p}\vec{p} \to \pi^+ d$. As yet the source of these discrepancies has not been established. Nevertheless, the overall success of the few-body approach in describing the πNN-NN system with a minimum number of parameters has been remarkable.

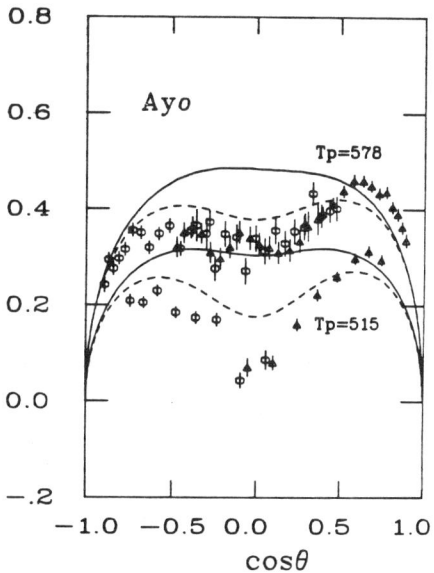

Fig. 10. The analyzing power Ayo for $pp \to \pi^+ d$. Solid curve corresponds to our usual choice of UPA to RSC deuteron (Pd = 6.56%), and dashed curve corresponds to the UPA to BS (Pd = 5.36%). Experimental points are from the SIN-Geneva group.[43]

VI. SUMMARY AND FUTURE OUTLOOK

In this brief review we have covered a great deal of territory, from the beginnings of the theoretical attempts to understand meson production to the present time. We have seen that the 70's led to a major advance in which a complete, consistent set of integral equations was derived, including all orders of multiple scattering, and self-consistently coupling all important channels. The fixed scatterer approximation for the nucleon and delta, which was essential in the 50's and 60's has also been left behind, and one can even include some relativistic corrections.

For the problem of N-N elastic scattering (Secs. III, IV and V) we have seen that the coupling to $N\Delta$ and πd produces precisely the type of rapid variation observed in the $^1D_2$ and $^3F_3$ scattering amplitudes. It seems unlikely that one needs exotic, dibaryon states in addition. Nevertheless, the existence of exotic quark states, for example, would be of fundamental importance, so perhaps one should keep an open mind for a while yet.

In the time honoured process of $pp \to \pi^+ d$ (Secs. III and V) there is now considerable experimental activity, so that we should soon know much more about both its threshold behaviour, and its spin dependence. The agreement with existing differential cross section data, both in the CCM and the few-body approach is rather good below 600 MeV, but deteriorates by 800 MeV. More worrying are the discrepancies with some polarization observables, although at the present time we have essentially no systematic understanding of why one observable is reproduced better than any other.

When it comes to three-body final states $NN \to NN\pi$ (Sec. IV), the game is just beginning. We have seen that the agreement with data is rather good for low momentum transfer to produced Δ, however, this gets worse as the Δ-momentum rises. Of course, the calculation reported here was somewhat preliminary in that it omitted the $N\Delta$-ΔN transition potential. We look forward to the time when complete multiple scattering results, based on the full integral equations of Sec. IV can be completed. Once again, the study of polarization observables will be very important.

So much for past theoretical achievement and the comparison with data. In what follows we discuss a few of the key theoretical issues which remain undecided, and which therefore provide the material for future research. In our view the most critical issue is the choice of theoretical model, at the nucleon level. That is, should one begin with a Lagrangian involving heavy mesons such as ρ, ω ... as well as the pion, or should one just deal with pions? The first approach is essentially that of classical nuclear physics with point-like nucleons and very hard meson-baryon vertex functions.[95] It has been used by many groups, notably Brack, Riska and Weise (BRW)[96] and the CCM of Sec. III. (Although it has recently been suggested by Niskanen[97] that the hard form factors lead to inconsistencies in the description of the Δ!) There is no doubt that with a hard pion cutoff ($\Lambda_\pi \sim 1.5$ GeV) the ρ-meson plays a critical role in reproducing the experimental $pp \to \pi^+ d$ cross section. For example, Niskanen finds that a 10% decrease in $f_{NN\rho}$ increases the peak $\sigma_{tot}(pp \to \pi^+ d)$ by more than 20%.[21]

The alternative approach is based upon more recent developments in our understanding of the structure of the nucleon itself. For example, in the cloudy bag model (CBM) the nucleon consists of a fairly large MIT bag (R ~ (0.8-0.9) fm), containing quarks, to which the pion is coupled in the minimal fashion necessary to guarantee chiral symmetry.[7,98-100] Because the nucleon bag (that is the pion source) is so large there is a <u>natural soft cutoff in the theory</u>. Indeed one is led to an almost indecently convergent theory of strong interactions,[100] in which, for example, the bare and renormalised coupling constants differ by less than 10%.[99] This model has been applied with great success to a number of problems ranging from the neutron charge distribution,[7,99] the magnetic moments of the nucleon octet,[101a] the violation of charge symmetry in N-N scattering[101b] and the nature of the P_{33} resonance itself.[7] In such a model there is essentially no philosophical reason for introducing the ρ-meson. Range considerations suggest it would make more sense to calculate such effects as $q\bar{q}$ fluctuations within the bag. Our studies of N-N scattering suggest that one can not distinguish between these two models using that system.[102] If anything, there is some support for the second picture. In pp → π^+d we saw that both the CCM using the first approach (Sec. III) and the few body model (Sec. V) using a very soft form factor, reproduced the existing data about equally well! A similar result was reported for NN → NNπ in Sec. IV.

Figure 11, which shows the driving term (V_2) in the CCM of Niskanen, illustrates our point. There are four separate contributions shown in Fig. 11, of which only one — the forward going pion, is included in the few-body calculations of Sec. V. Now in the absence of ρ-exchange, a very simple argument based on an effective Lagrangian suggests that the inclusion of the backward pion would increase the driving term by (50-100)%. Naively this means a factor of (2-4)

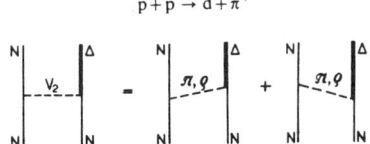

Fig. 11. The meson exchanges included in the transition potential V_2.

in the cross section, quite apart from the question of the pion form factor. How can both models agree with the data? The answer lies partially in the strong dependence on $f_{NN\rho}$ mentioned earlier, which essentially permits one to tune the calculation to σ_{tot}(pp → π^+d).

Of course, there is no theoretical reason for ignoring backward going pions in the few-body approach except that they are not necessary for unitarity. The incorporation of such processes into the few-body equations is an interesting problem for the future. However, the most interesting challenge would be to generalise the derivation of the few-body equations to the CBM Hamiltonian. In this regard we make one final remark on backward going pions. If one uses a renormalisable theory, like the CBM, as opposed to an effective Lagrangian, there are some interesting consequences. In particular, it is easily established using Table I of Ref. 99a and Fig. 6 of Ref. 99b that the $\Delta N\pi$ coupling constant would be some 30% smaller for Fig. 11b than 11a. This combined with the increase in mass of an off-shell delta would suppress the backward pion graph 11b by perhaps a factor of two. Clearly future calculations must take this into account.

Let us return to the question of N and Δ recoil, which was ignored in all the calculations of the 50's and 60's and is still omitted in many calculations of pion production. As stressed by Niskanen,[45] because the delta can be on-shell near T_{lab} = 600 MeV, one finds that the range of the NΔ wave function can be rather long. This is quite the opposite of what is found in static calculations like that of BRW, where the NΔ wave function peaks near 1 fm. Clearly then, the process of pion production will be shifted to larger inter-baryon separations in a careful treatment. This explains why, for example, Niskanen finds much less sensitivity to the deuteron d-state and the N-N short range correlations.

While this observation on the range of the absorption process is interesting per se, it is directly relevant to another recent mechanism proposed for pion production. Kisslinger has suggested, on the basis of a model strikingly similar in spirit to the CBM, that because pion production is a short ranged process[95,103] it should proceed dominantly through formation and decay of a six-quark state.[104] Certainly it will be of great interest to study the contribution to pion production from quarks in overlapping bags. Nevertheless, we hope to have made it clear that, first, pion production is not such a short range process. Second, the normal multiple scattering processes can explain a great deal of data with essentially no adjustment of parameters (although there is some off-shell sensitivity). In direct analogy to our earlier comments on dibaryons, any calculation which attributes a feature of pion production to multi-quark effects must first include the best possible description of the conventional mechanism. It is our hope that this brief review will help with the latter.

ACKNOWLEDGEMENTS

We all benefitted greatly from a stimulating visit by Prof. Avraham Rinat earlier this year. We are also indebted to Lorraine Gray for her rapid and efficient production of this manuscript.

REFERENCES

1. G. Chew et al., Phys. Rev. 84, 581 (1951); see also S. Matsuyama and H. Miyazawa, Prog. Th. Phys. 9, 492 (1953).
2. D.A. Geffen, Phys. Rev. 99, 1534 (1955).
3. G. Chew, Phys. Rev. 94, 1748, 1755 (1954); ibid. 95, 1669 (1954).
4. F.E. Low, Phys. Rev. 97, 1392 (1955).
5. G.C. Wick, Rev. Mod. Phys. 27 339 (1955).
6. D.B. Lichtenberg, Phys. Rev. 100, 303 (1955).
7. S. Théberge, A.W. Thomas and G.A. Miller, Phys. Rev. D 22, 2838 (1980); 23, 2106(E) (1981); G.A. Miller, A.W. Thomas and S. Théberge, Comm. Nucl. Part. Phys. 10, 101 (1981).
8. A. Aitken et al., Phys. Rev. 93, 1349 (1954).
9. K.A. Brueckner and K. Watson, Phys. Rev. 86, 923 (1952); ibid. 90, 699 (1953).
10. D.B. Lichtenberg, Phys. Rev. 105, 1084 (1957).

11. B. Durney, Proc. Phys. Soc. (London) 71, 654 (1958).
12. A.E. Woodruff, Phys. Rev. 117, 1113 (1960).
13. F. Zachariasen, Phys. Rev. 101, 371 (1956).
14. D.S. Koltun and A. Reitan, Phys. Rev. 155, 1139 (1967); ibid. 162, 963 (1968); Nucl. Phys. B4, 629 (1968).
15. A.W. Thomas and I.R. Afnan, Phys. Rev. Lett. 26, 906 (1971).
16. H.C. Pradhan and Y. Singh, Can. J. Phys. 51, 343 (1973).
17. B. Goplen, W.R. Gibbs and E.L. Lomon, Phys. Rev. Lett. 32, 1012 (1974).
18. C. Lazard, J.L. Ballot and F. Becker, Nuovo Cim. 65B, 117 (1970).
19. A.M. Green, Rep. Prog. Phys. 39, 1109 (1976).
20. A.M. Green, J.A. Niskanen and M.E. Sainio, J. Phys. G4, 1055 (1978).
21. J.A. Niskanen, Nucl. Phys. A298, 417 (1978).
22. F. Myhrer and D.S. Koltun, Nucl. Phys. B86, 441 (1975); Phys. Lett. 46B, 322 (1973).
23. G. Doolen, Purdue University thesis (1968) unpublished.
24. T. Mizutani and D. Koltun, Ann. Phys. 109, 1 (1977).
25. T. Mizutani, University of Rochester thesis (1976) unpublished.
26. I.R. Afnan and A.W. Thomas, in Proc. Int. Conf. on Few Particle Problems in the Nuclear Interaction, Los Angeles, 1972, ed. I. Slaus et al. (North-Holland, Amsterdam, 1973) p. 861; Phys. Rev. C 10, 109 (1974).
27. A.W. Thomas, Flinders University thesis (1973) unpublished; Proc. Int. Conf. on Few-Body Problems in Nuclear and Particle Physics, Université Laval (Les Presses de l'Université Laval, 1975) p. 287.
28. K. Kowalski et al., Phys. Rev. C21, 2122 (1980).
29. A.W. Thomas and A.S. Rinat, Phys. Rev. C 20, 216 (1979).
30. Y. Avishai and T. Mizutani, Nucl. Phys. A326, 352 (1979); ibid. A338, 377 (1980); ibid. A325, 399 (1981); Univ. Lyon preprint (1980).
31. A.T. Stelbovics and M. Stingl, Nucl. Phys. A299, 391 (1978); J. Phys. G4, 1371 (1978); ibid. 1389 (1978).
32. I.R. Afnan and A.T. Stelbovics, Phys. Rev. C 23, 1384 (1981).
33. A.W. Thomas and R.H. Landau, Phys. Rep. 58C, 121 (1980).
34. B. Blankleider and I.R. Afnan, Flinders Univ. preprint FIAS-R-76 (1981), to be published; Phys. Rev. C 22, 1638 (1980).
35. M. Betz and F. Coester, Phys. Rev. C 21, 2505 (1980).
36. M. Betz and T.-S.H. Lee, Phys. Rev. C 23, 375 (1981).
37. A.M. Green and P. Haapakoski, Nucl. Phys. A221, 429 (1974); A.M. Green and J. Niskanen, Nucl. Phys. A249, 493 (1975).
38. K. Erkelenz, Phys. Rep. 13, 191 (1974); K. Holinde and R. Machleidt, Nucl. Phys. A280, 429 (1977); K. Holinde et al., Phys. Rev. C 18, 870 (1978).
39. A.M. Green and M.E. Sainio, J. Phys. G5, 503 (1979).
40. J.A. Niskanen, Phys. Lett. 79B, 190 (1978); ibid. 82B, 187 (1979).
41. J.A. Niskanen, to be published.
42. R. Arndt and B. Verwest, in Proc. 5th Int. Symposium on Polarisation Phenomena in Nuclear Physics, eds. G.H. Ohlsen et al. (Santa Fe, 1980) AIPCP 69 (to appear); Texas A&M Univ. preprint DOE/ER/05223-29 (1980).
43. E. Aprile et al., in Proc. 8th ICOHEPANS, eds. D.F. Measday and

A.W. Thomas (Vancouver, 1979); Nucl. Phys. A335, 245 (1980); Nucl. Phys. A335, 245 (1980).
44. C. Lechanoine-Leluc and W. Leo, private communication to J.A.N.
45. J.A. Niskanen, in Proc. 5th Int. Symposium on Polarisation Phenomena in Nuclear Physics, eds. G.H. Olsen et al. (Santa Fe, 1980) AIPCP 69 (to appear).
46. M. Simonius, in Polarisation Nuclear Physics, Lecture Notes in Physics, vol. 30, ed. D. Fick (Springer-Verlag, Heidelberg, 1974).
47. For a recent review see the rapporteurs talks by D. Bugg and A.W. Thomas at the 9th ICOHEPANS (Versailles, 1981) to be published by Nucl. Phys.
48. B.J. Verwest, Phys. Lett. 83B, 161 (1979).
49. A. König and P. Kroll, Nucl. Phys. A356, 345 (1981).
50. J. Hudomalj-Gabitzsch et al., Phys. Rev. C 18, 2666 (1978).
51. F.H. Cverna et al., Phys. Rev. C 23, 1698 (1981).
52. S. Mandelstam, Proc. Roy. Soc. A244, 491 (1958).
53. W.M. Kloet and R.R. Silbar, Nucl. Phys. A338, 281 (1980); ibid. A338, 317 (1980); Phys. Rev. Lett. 45, 970 (1980).
54. J. Dubach, W.M. Kloet and R.R. Silbar, Phys. Rev. D 22, 2761 (1980); LASL preprint LA-UR-81-2491 (1981); and J. Dubach, W.M. Kloet, A. Cass and R.R. Silbar, Univ. of Massachusetts preprint (1981).
55. A.S. Rinat, Y. Starkand and E. Hammel, Nucl. Phys. A364, 486 (1981).
56. P.R. Bevington et al., Expts. 336/33, LAMPF Progress Report LA-8456-PR(1980)6.
57. D. Bugg, private communication.
58. E.O. Alt, P. Grassberger and W. Sandhas, Nucl. Phys. B2, 167 (1967);
I.R. Afnan and A.W. Thomas in Modern Three-Hadron Physics, ed. A.W. Thomas (Springer-Verlag, Heidelberg, 1977) chap. 1.
59. A.W. Thomas, Nucl. Phys. A258, 417 (1976).
60. A.S. Rinat and A.W. Thomas, Nucl. Phys. A282, 365 (1977).
61. V.B. Mandelzweig, H. Garcilazo and J.M. Eisenberg, Nucl. Phys. A256, 461 (1976).
62. J.M. Rivera and H. Garcilazo, Nucl. Phys. A285, 505 (1977).
63. N. Giraud et al., Phys. Rev. C 19, 465 (1979).
64. N. Giraud, C. Fayard and G.H. Lamot, Phys. Rev. C 21, 1959 (1980).
65. H. Garcilazo, Phys. Rev. Lett. 45, 780 (1980); Nucl. Phys. A360, 411 (1981).
66. V.S. Varma, Phys. Rev. 163, 1682 (1967).
67. W.M. Kloet et al., Phys. Rev. Lett. 39, 1643 (1977).
68. M. Araki, Y. Koike and T. Ueda, Prog. Th. Phys. 63, 335 (1980); ibid. 63, 2133 (1980).
69. M. Araki and T. Ueda, Osaka Univ. preprint OUAM 81-8-7.
70. A.S. Rinat, Nucl. Phys. A287, 399 (1977).
71. A.S. Rinat et al., Nucl. Phys. A329, 285 (1979); Phys. Lett. 80B, 166 (1979).
72. A.S. Rinat, Y. Starkand and E. Hammel, Nucl. Phys. A364, 486 (1981).
73. I.R. Afnan and B. Blankleider, Phys. Lett. 93B, 367 (1980).

74. C. Fayard, G.H. Lamot and T. Mizutani, Phys. Rev. Lett. $\underline{45}$, 524 (1980).
75. C. Fayard et al., abstract submitted to 9th ICOHEPANS, Versailles (1981).
76. B. Blankleider, Flinders Univ. thesis (1980), preprint #FIAS-R-72.
77. F. Mandl and T. Regge, Phys. Rev. $\underline{99}$, 1478 (1955).
78. M. Gell-Mann and K.M. Watson, Ann. Rev. Nucl. Sci. $\underline{4}$, 219 (1954);
A.H. Rosenfeld, Phys. Rev. $\underline{96}$, 139 (1954).
79. C.M. Rose, Phys. Rev. $\underline{154}$, 1305 (1967).
80. F.S. Crawford and M.L. Stevenson, Phys. Rev. $\underline{97}$, 1305 (1955).
81. C. Richard-Serre et al., Nucl. Phys. $\underline{B20}$, 413 (1970).
82. B.G. Ritchie et al., to be published;
P. Walden et al., TRIUMF expt. # 132.
83. E. Levin and J.M. Eisenberg, Nucl. Phys. $\underline{A292}$, 459 (1977).
84. B.D. Keister and L.S. Kisslinger, Nucl. Phys. $\underline{A326}$, 445 (1979);
L.S. Kisslinger, Zeit, f. Phys. $\underline{A291}$, 163 (1979).
85. R.H. Cole et al., Phys. Rev. C $\underline{17}$, 681 (1978).
86. K. Gabathuler et al., Nucl. Phys. $\underline{A350}$, 253 (1980).
87. R.J. Holt et al., Phys. Rev. Lett. $\underline{43}$, 1229 (1979).
88. E.G. Pewitt et al., Phys. Rev. $\underline{131}$, 1826 (1963).
89. C.L. Dolnick, Nucl. Phys. $\underline{B22}$, 461 (1970).
90. D. Aebischer et al., Nucl. Phys. B106, 214 (1976).
91. H. Nann et al., Phys. Lett. $\underline{88B}$, 257 (1979).
92. W. Hürster et al., Phys. Lett. $\underline{91B}$, 214 (1980);
H. Schmitt, private communication.
93. P. Walden et al., Phys. Lett. $\underline{81B}$, 156 (1979).
94. G.R. Smith et al., contribution to 9th ICOHEPANS, Versailles (1981).
95. G.E. Brown and W. Weise, Phys. Rep. $\underline{58C}$, 121 (1980);
E. Oset, H. Toki and W. Weise, Regensburg Univ. preprint (1981) to be published in Phys. Rep.;
V. Vento et al., Nucl. Phys. $\underline{A345}$, 413 (1980).
96. M. Brack, D.O. Riska and W. Weise, Nucl. Phys. $\underline{A287}$, 425 (1977);
O.V. Maxwell et al., Nucl. Phys. A348, 388, 499 (1980).
97. J.A. Niskanen, TRIUMF preprint, TRI-PP-81-31.
98. A.W. Thomas, TRI-PP-81-19 to be published in the Czech. J. Phys.
99. A.W. Thomas, S. Théberge and G.A. Miller, Phys. Rev. D $\underline{24}$, 216 (1981); Can. J. Phys. $\underline{59}$,   (1981).
100. L.R. Dodd, A.W. Thomas and R.F. Alvarez-Estrada, Phys. Rev. D $\underline{24}$,   (1981).
101. S. Théberge and A.W. Thomas, TRI-PP-81-34 to be published in Phys. Rev. D;
A.W. Thomas, P. Bickerstaff and A. Gersten, Phys. Rev. D $\underline{24}$,   (1981).
102. A. Gersten, invited paper at the 2nd Int. Conf. on Recent Progress in Many-Body Theories (Oaxtepec, Mexico, 1981), TRI-PP-81-2;
A. Gersten and A.W. Thomas, TRI-PP-81-28 (to be published).
103. M.A. Alberg et al., Nucl. Phys. $\underline{A306}$, 447 (1978).
104. L.S. Kisslinger, these proceedings; private communication.
105. D.A. Axen et al., Nucl. Phys. $\underline{A256}$, 387 (1976).

DISCUSSION

D. Koltun (Univ. of Rochester): Tony, I was struck by your remark that the relatively sophisticated unitary calculations tend to agree with each other even where they disagree with the data. The question is one of interpretation: Is it that we have the delta dominance correct, and we don't understand all of the background?
A. Thomas (TRIUMF): In all of the few body calculations which I described, all of the small partial waves are included. In the calculations of Niskanen only the s-waves are included.
Koltun: The point is, maybe they are not properly treated even though they're there, whereas the delta with its separability may be just easier to handle. It just gets harder with small partial waves, in the way that you indicated in your last few sentences.

R. Silbar (Los Alamos Scientific Lab.): Let me ask your opinion. Has anything ever been done well with πd backward scattering at 256 MeV? That has never gotten cleared up.
Thomas: It never comes out right.
Silbar: You didn't say why?
Thomas: I don't know why. If I knew why I'd say so.

F. Coester (Argonne National Lab.): Are you proposing to scatter two cloudy bags from each other and apply that to this problem?
Thomas: Sounds like a nice idea.

D.B. Lichtenburg (Indiana Univ.): Is your remark that there is no need for a dibaryon resonance a personal opinion? I know of a lot of people who still think very strongly that there are dibaryons. You say the controversy is dead now, or is there still a controversy?
Thomas: I didn't say there isn't such a thing as a dibaryon. I just said it doesn't appear to be necessary to explain the experiments we have at the moment.
Lichtenberg: But it's on the basis of experiments that people postulate it and still believe it.
Thomas: Sometimes ideas have been right for the wrong reasons.

L.S. Kisslinger (Carnegie-Mellon Univ.): Is that an isospin one dibaryon we're talking about?
Thomas: Yes.

SUMMARY: FUNDAMENTAL INTERACTIONS AND PROCESSES

Daniel S. Koltun
University of Rochester, Rochester, NY 14627

ABSTRACT

The subjects of the talks of the first day of the workshop are discussed in terms of fundamental interactions, dynamical theory, and relevant degrees of freedom. Some general considerations are introduced and are used to confront the various approaches taken in the earlier talks.

1. INTRODUCTION

The title of the general session had the word *fundamental* in it; I want to describe what we mean by fundamental. What I'm trying to do in to the summary is to relate the kind of things you have heard this morning and this evening, to the notion of fundamental in nuclear theory, and in particular, within the context of the subject of this workshop. One definition of fundamental which is often implied but rarely written down, is: "What interests *me* is fundamental-what interests *you* is a technical detail." But beyond that I'm going to talk about three aspects of fundamental, and in some sense in increasing order of fundamental, or decreasing, if the foundation is supposed to be underneath: interactions, dynamical theory, and relevant degrees of freedom.

2. WHAT DO WE MEAN BY "FUNDAMENTAL"?

a) Interactions. The first way that Nuclear Physicists often use the word "fundamental" in talking about problems like those of π-production or π-absorption is to try to separate the fundamental interaction—the central interaction, the interaction of interest—from the other parts of the problem — the nuclear physics, the reaction processes, and so on. This is expressed naturally in terms of the Distorted Wave Born Approximation (DWBA) in which we write a transition amplitude, e.g. for π-production, $pA \to \pi(A+1)^*$, as a matrix element

$$< \Psi^{(-)}[\pi, (A+1)^*] \mid H_i \mid \Psi^{(+)}[p, A]> \qquad (1)$$

of an interaction H_i. It is this operator, for the moment, that you are considering fundamental. There are other parts of the problem — namely the scattering processes and nuclear structure physics implied by $\Psi^{(\pm)}$ — not that they are not of interest, but you have other ways of thinking about them which are not necessarily considered so fundamental. In terms of the process that we all love and that has brought us here together (π-production and absorption), Hint is supposed to be the interaction involving single pions. In old-fashioned language it would often be written as some sort of a nucleonic current or hadronic current times the pion field,

$$H_i \sim \int j \cdot \phi_\pi \, . \qquad (2)$$

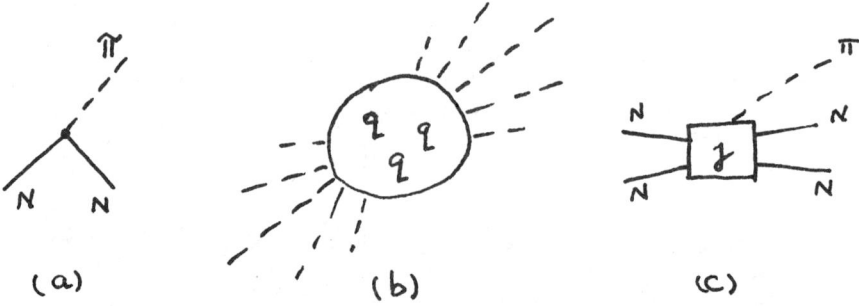

Fig. 1 Three models for the nucleonic current j of Eq. (2), as discussed in the text.

In Fig. 1 I've drawn some pictures for the kinds of things we've heard discussed (a,b) and one (c) we haven't heard very much discussed (which may be the sign of the times) for the way that this gets treated. The fundamental interaction could be the NNπ vertex (a) which we heard about this morning. I'll come back to that. This object in (b) is a peculiar representation of a p-wave pion on the surface of the bag. The third thing (c) represents the possibility that there might be dibaryons as resonant states, which couple both the two nucleon system to the two nucleon plus pion system carrying definite spin-parity. Any of those might be the thing that we consider, in this context, to be the fundamental interaction, all the rest being in some sense nuclear physics. Now, of course, we very quickly learn that even if we start with that point of view, we don't stick to it because as soon as we look further into it, we find, for the reasons that have been very well outlined by Tony Thomas in the preceeding talk, that the nuclear physics has fundamental aspects in just the same sense. The same interactions come in, and one can't avoid that. We heard discussions several times today about how whenever we are dealing with the so called short-range interactions in terms of N* excitations, or of the quark bag pictures where the bags are forced to overlap, those aspects are fundamental in the sense that we're describing. (Remember the ranges are not so terribly short, often not much shorter than a fermi.)

Another sense in which we can use the word fundamental includes *relativity*. How nucleons behave, for example, in this (p,π) reaction when the velocities involved are not so small becomes an important question, as also was alluded to earlier by Banerjee in his talk. I will come back to some of these things in a little bit more detail.

b) Dynamical Theory. The next level, above this level of simply trying to separate what we consider the physics we understand from the fundamental physics, or physics we hope to understand, is the dynamical theory in which all the parts of the amplitude, say that of Eq. (1), are consistently defined. That is, what do we mean by the wave-function for a nucleon scattering on a nucleus A? What do we mean by the wave-function for the pion scattering on an excited state of a nucleus in the final state? What are those things, and how are they defined in the same dynamical picture in which we're describing our H_i? That, of course, is the problem; that's where we soon discover that we can't totally separate the discussion of those two things, which brings us to the next level of fundamental.

c) Relevant degrees of freedom. The third level, which goes further than the dynamical theory but also has a lot to do with it, is: what are the *relevant* degrees of freedom which we use in levels (a) and (b); i.e., use in our description of the transition and use in a dynamical theory from which we get the description. We may follow two lines of thought; in the first we talk about nucleons, pions, deltas, rhos and maybe anti-nucleons, maybe even dibaryons, as separate degrees of freedom if in fact there is any convincing evidence that they are separate degrees of freedom. The other line of thought has been offered and argued this morning, that we must or may talk about quarks and anti-quarks, that that is a natural way to start because it's simpler and will lead to cleaner results. We must also include the bag itself, which is really a dynamical degree of freedom in such a theory, and that has to be remembered whenever one collects more than one nucleon in a nuclear system. And one must remember the degree-of-freedom that we like to call pi, particularly when we're producing π's in the lab, which are still something of a problem in quark-based models. Therefore one has to decide at the beginning, what are the relevant degrees of freedom--is the pion a degree of freedom or is it part of the quark picture? --for which you need a consistent theory?

What I'm going to do next is work backwards through these three cagetories I've defined: the degrees of freedom, the dynamical theory, and then our picture of the interaction itself, to summarize some of the things that we have learned today.

3. DEGREES OF FREEDOM

The key question is not what degrees of freedom are there in the universe, but what relevant degrees of freedom are there to the kinds of problems we're dealing with? There is no unambiguous answer to that, and I want to stress the notion that there is always a choice between generality and efficiency. Generality favors the notion of talking about quarks and QCD, because there are many other parts of physics for which we believe that may well be a successful description, and thats a good thing. So the scope of the phenomena discussed may be larger for some choices of degrees of freedom then for others. However, efficiency means, among other things, the ability to calculate, and, I leave it as an open question whether a general choice or a more limited choice is in fact efficient for a given purpose. For example, the quark and bag picture may be the more efficient choice to work with if we're talking about nucleons and deltas! However, if we want to talk about the π, as I remarked earlier, the π itself is a problem in the quark and bag theories unless its imposed as a separate degree of freedom. For nucleon-nucleon forces (that is something we may hear from some of the panelist about): which will we choose to work with? That's an open question: I will just make a few remarks about that.

One very important remark made this morning by Tony Thomas is that convergence certainly has a lot to do with the question of what I'm calling efficiency--the question of a choice of degrees of freedom. First of all, if you can demonstrate that one approach converges and another doesn't, there isn't much choice at all. It is not usually the case that you can prove that, but if you can prove that one method of calculation converges very much faster than another method, you do have a definite advantage and thats the kind of case I'd like to see made for the rival theories. Now, let me expand on the notions of efficiency, convergence, etc., by taking two examples that were discussed this morning. One was how to calculate nuclear matter and the other one was how to calculate π-production in proton+proton collisions.

Let us start with nuclear matter. I shall discuss four schemes for treating the degrees of freedom. Some of these schemes were summarized this morning by Gerry Miller, some of them were summarized by Tony Thomas this evening. I want to do it in outline:

a) $N + N + N + \ldots$

$$H = H_o + V^{(2)} + V^{(3)} + \ldots + V^{(n)} + \ldots$$

The first one is what we would call conventional nuclear physics, that is, there are only nucleons; those are the degrees of freedom. But we expand the Hamiltonian to include not only the one we were given this morning — the two body force — but allowing also three-body forces, four-body forces etc., for whatever reason. Is that a convenient way to do nuclear matter? Well, we don't know yet. Three-body forces can be included but only relatively limited versions of these have been used in calculations.

Certainly it would be true that if you had to go to very large numbers of nucleons to get sensible answers, or if you did not find that as you went up in $\underline{n}$ the many body forces didn't begin to decrease, i.e. that they didn't have any convergence, you would abandon that line of approach as a way of dealing with nuclear matter.

b) $N + N + \Delta + N + \ldots$

$$H = H_o + V(NN \leftrightarrow NN) + V(NN \leftrightarrow N\Delta) + \dots$$

The second scheme was mentioned this evening. Suppose you have nucleons and deltas and maybe some other baryonic excitations, and you have not only potentials of the usual sort here but you have transitions which involve two nucleons becoming, say, a nucleon and a delta, and you deal with a coupled channel theory. Is that a better approach? That may be; I don't know because it hasn't been carried out in all detail to the extent that nuclear matter with the two-body forces has been carried out. Here the convergence question would be: how many channels in fact will you need in order to get a successful calculation?

c) $N + N + \pi + \rho + \omega + \sigma + \dots$?

$V_{NN\pi}$, $V_{NN\rho}$, etc.

Going quickly to the third, suppose that instead of talking about deltas, you talk about various sorts of mesons, as has been done in literature in the last 8 years in terms of semi-classical calculations (or others) and you have vertices for these: coupling constants, perhaps with form factors, and so on. How many mesons do you need, and how often do they interchange? If they are interchanged a very large number of times, then you need a high degree of unitarity in the theory which means that you need a rather complicated way of calculating.

d) $q + q + \bar{q} + \dots + \text{bag} + \dots + $ "fluctuations"

The last one is the quark bag model. For example, you might imagine that a model that tries to do nuclear matter with today's understanding of quarks and bags would get to the point where it had to decide how to deal with what I call fluctuations of the quark-anti-quarks in such a theory, which in different language would be thought of as pi's and rho's, except of course they'd be virtual pi's and rhos. Now the question of efficiency would come up here. If it turned out that fluctuations don't look very much like the physical pi's and rho's, then it might be a very efficient way to deal with the problem. If, however, the pi's and rho's that show up in nuclear matter are anything like real mesons (they're virtual of course) but anything like real mesons off mass shell or close to mass shell, then of course the fluctuation picture might be a very poor way of doing the nuclear matter problem. I don't know the answer to which one of these is the best way to go, but I think this is the way at least one might discuss what's a better theory, and these are the kind of questions that one would like to see answered.

The same questions can be raised for the second example, the calculation of the $pp \rightarrow \pi^+ d$ process. We may treat this in terms of N and Δ, as discussed by Thomas:

a) $N + N \to N + \Delta \to d + \pi$,

or in terms of a six-quark intermediate state, as suggested by Miller:

b) $N + N \to 6q \to d + \pi$.

We are more familiar with the successes (and perhaps, the failures) of method (a), which is now calculated to high accuracy, on its own grounds, than we are for (b), which is at present somewhat schematic.

4. DYNAMICAL THEORY

On to the dynamical theory, which is the second level in our discussion of fundamentals. Dynamical theories usually start with quantum field theory, at least one says they start with quantum field theory. One doesn't know necessarily which complete field theory and exactly how to solve it. One doesn't intend to solve the complete field theory, but ususally it is reduced to a more limited theory with a restricted number of degrees of freedom. (That's the matter we just discussed in Sec. 3.) How it is reduced is a highly technical problem of no small importance. For example, if one uses the so-called non-relativistic field theories, usually you approach it as a reduction of some other relativistic theory, and how you do that is a problem and can be ambiguous. We heard from Thomas about, for example, proceeding with dynamical theories involving coupled channels of the sort I just mentioned where, say, a $\pi + N \to N$, or $\pi + N \to \Delta$ and how you can build a dynamical theory that way. In that case the elementary objects are the π, the N, the Δ and so on; they describe the degrees of freedom. The coupling and the rules for handling coupling give the dynamical theory, and the constraints you get from unitarity. Gerry Miller told us about how to think in terms of the quark and bag models of quantum chromodynamics. In fact one can think of the results so far (in the cloudy bag model at least) which have to do with π scattering on nucleons, as being in a sense the old Chew-Low (or, more properly, Chew) theory in which there is real new physical content. The new physical content is the introduction in the model of finite size to the bare source of the pions. It's a very nice theory, I think, from that point of view. But of course the degrees of freedom are different here and the rules of calculation are different, from those of a coupled channel theory based on π, N, and Δ.

May I just remind you, particularly those of you, like myself, who have worked in other pastures of nuclear physics, that the choice of degrees of freedom and the concurrent choice of a dynamical theory is, in the many-body theory or many-body spectroscopy as we used it in nuclear physics, called the *model space question*, and a lot of things we learned in those days about the fact that these are coupled problems are still relevant here. When you choose a model space, you choose a set of degrees of freedom and at the same time you put some constraints on the kind of dynamical theory you can write down.

5. FUNDAMENTAL INTERACTIONS

The particular interaction which got the most discussion today was in fact the πNN vertex, the simplest of the possible interactions. I just remind you that you heard three times, three different aspects of this, which come from, if you like, different ways of looking at the dynamics. From Tony Thomas we heard about the problems of counting and separating and dealing with that vertex and how one uses that in the coupled channels of few-body approach to scattering in the two-nucleon-plus-π case. From Banerjee we heard about particular aspects that had to do with relativity. In a Feynman diagram theory one makes particular use of the relativistic structure to simplify or regroup the diagrams. And then from Gerry Miller we heard, of course in terms of the cloudy bag theory, how one could get a vertex function for N ↔ Nπ (also for Δ ↔ Nπ, etc.). Now, it is an old idea in meson theory that the πNN vertex function which appears in π-absorption and in πN scattering (e.g. Chew-Low) theory, also governs the nucleon-nucleon interaction at long range (OPE). The cloudy bag model of course maintains that that should be the case. The case isn't completely made because the nucleon-nucleon part of that is not yet done.

I want to make a technical remark about the work of Banerjee and Walker which you heard this morning. Their method reminds me of something that a number of us in a few-body business did study at one time, which was the Taylor[1] method of how to group Feynman diagrams. Taylor's methods are quite old and somewhat neglected — maybe because they were written for the wrong audience. But, they are a way of dealing with few-body theory and, in fact, they turn out to be a way in which you can introduce absorption and production into such a theory. The rules Taylor derived have been discovered separately by Thomas and by Mizutani, each in thesis work devoted to the problem of πd scattering and absorption. Taylor found what are called "cutting rules". Cutting rules have to do with how you cut up a Feynman diagram, by cutting on the particle lines (they would be nucleon lines in this case). The method given by Banerjee and Walker of treating relativity has some similarity to the Taylor cutting rules, although I do not believe he dealt explicitly with momentum dependent coupling as they have. In non-relativistic physics we would call it non-local coupling. It's a way of making the theory more local and therefore in many ways more attractive, and it seems like a nice thing to do. The point is that what Banerjee and Walker have got into certainly is intelligent; the question is whether the grouping is efficient. The efficiency we don't yet know, that's the point, I think, of Thomas' remarks after Banerjee's talk.

6. FUNDAMENTAL PROCESSES

I want to remind you that the title of the whole discussion we've had this morning and this evening includes not just *fundamental interactions* but also *fundamental processes*, and there is a reason for that. We had a talk this morning by Jones on the processes NN → πd and NN → NNπ. The fundamental interaction that we've been dealing with, as we said several times, is central to understanding π-nucleon scattering and the nucleon-nucleon force. But why is it, if that's the case, that we spend all of our time talking about these π-production reactions? I think the reason is that it has long been the general point of view that the subject of absorption and production has everything in it. That is, it has many of the complications of all of nuclear physics: reaction mechanisms, nuclear structure etc., that we have to understand, and of course if we don't understand these in this context we have very little confidence in going on to further things such as π-production on complex nuclei. So that's the reason that we must have considerable information, both experimental and theoretical, about these particular NN reactions. We had a very good review from Tony Thomas of what the theoretical situation is and what the history has been for that subject. Tony reminded us that the elastic π-scattering and the elastic nucleon scattering are strongly coupled channels in this particular problem and they both have to be included systematically here. Therefore, when we go back to the earlier catagories of what degrees of freedom do you choose and what dynamics do you choose, your choice is affected by whether it's likely to be adequate in talking about one particular domain. In other words you can't narrow your domain so much that you simply can't do the physics in that very narrow domain. In the present context, this means a coupled channels treatment is required.

A last remark: it is in the context of NN → NN and NN → πd that the question of possible <u>dibaryon</u> states has been raised. This has not been much discusssed today. If, in fact, there are some exotic states playing a dynamical role here, it would be interesting to know if that fact leads to any interesting predictions for exclusive pion production on nuclei:

$$pA \rightarrow \pi(A+1)^* \quad .$$

REFERENCES

1. J. G. Taylor, *Nuovo Cimento Supp.* **1**, 857 (1965), Phys. Rev. **150**, 1321 (1966).

PANEL DISCUSSION

Participants: F. Coester, Argonne National Laboratory
J. A. Niskanen, University of Helsinki
H. J. Weber, University of Virginia

Summary by

Daniel S. Koltun
University of Rochester, Rochester, NY 14627

ABSTRACT

What follows is a paraphrase and summary of the remarks by the participants in the panel discussion, and some of the open discussion.

J. Niskanen made a number of remarks about the coupled-channels approach to pp → π^+d, showing some of the most recent results of calculation. Variation of some interaction parameters can improve the shape of the theoretical integrated cross section, but differential cross sections are harder to adjust through the Δ region. The newer spin-dependent data give much greater constraints than differential cross sections. Around 600 MeV the discrepancies between experiment and theory remain at 10-20% and are not easily reduced. "It's sort of childhood's end in the field of pion production in two-nucleon systems."

Niskanen then discussed several areas where new theoretical treatment of the coupled channels problem could be improved: Δ-N interaction, Δ-width, small p-waves (πN), inclusion of centrifugal barrier effects in the rescattering method. With regard to exotic effects in the πd or pp system: six quarks, dibaryons, and the like, Niskanen argued that the "old" coupled channel and related methods should be taken to the limits of their validity to look first for non-exotic explanations for any "interesting" partial wave effects.

F. Coester addressed three questions: has the traditional model (of nuclear structure) failed? what are the necessary degrees of freedom for nuclear physics? and, what is the role of relativity in all this?

The *traditional model* refers to a Schrödinger equation for nucleons, where the Hamiltonian may include a modest three-body interaction, and possibly a small four-body force. Then the test would be to account simultaneously at least for ^{3}H, ^{3}He, ^{4}He, and nuclear matter. So far, reasonable three-body forces have failed to pass this test, "The present situation on that is that it indeed is still a possibility (even discounting the electromagnetic thing to which I come in a minute) that this model will fail."

Additional degrees of freedom can show up in two ways: either they constrain the form of the nuclear Hamiltonian or other operators (as in meson-theoretic potentials), or they may enter explicitly as degrees of freedom in the nuclear model. At the level of the kind of reactions induced by pions, it is almost compelling to introduce both pions and deltas explicitly. It is still a question whether quarks should enter the nuclear model as active degrees of freedom in the same sense, or as implicitly determining the underlying dynamics, in the first sense.

Coester went on to discuss the problems encountered in producing a workable many-body theory starting from relativistic quantum fields. Although it is difficult to produce a complete calculational theory, one can learn a lot about the structure of the resulting theory by using the constraints induced by Lorentz invariance. When this is applied to the electromagnetic properties, one finds it necessary to include many-body charge and current operators which are consistent with the dynamics. This kind of consistency has not yet been put into treatments of, say, the electromagnetic properties of ^{3}He, and it is not yet known whether this can be done.

The cloudy bag model is non-relativistic. One can, however, treat it as a quark model for the source of pions, and then go over to a model with explicit pion, nucleons, (without explicit quarks) which then can be made relativistic.

H. J. Weber began his remarks on quark models with the observation that the cloudy bag or chiral bag models have succeeded in reconciling, in a convergent calculation, the constituent quark model with the Chew-Low model of the Δ. However, to go further in the direction of reconciling the quark model with other features of particle and nuclear physics, it seems to be necessary to consider what Weber calls a quark-molecular approach, starting from a quark-quark potential. This method has been pursued to obtain the nucleon-nucleon force from a six-quark model with gluon exchange. However, this seems to give wrong results, as has been summarized recently by Greenberg and Lipkin. One problem in perturbative QCD can be seen as the inability of the color-singlet vacuum to allow a single colored object: quark or gluon, to propagate from one bag to the other.

What is required is a nonperturbative method of treating exchange, using a strong coupling picture with a $q\bar{q}$-condensate as the exchanged object. This leads, at long range, to the one-boson-exchange results. At shorter range, one has overlapping bags, but with a surface separating the color-singlet clusters of three quarks. One does not expect this surface, which is the QCD vacuum, to disolve at low energy or low momentum transfer. The $q\bar{q}$-condensate in the NN problem acts to restore the missing chiral symmetry to the quark model, much as the chiral bag model does for one nucleon.

In the discussion, questions were raised about how much of the results could be calculated, and how much were postulated. The details of the vacuum state and the condensate are not calculated, but the couplings to the exchanged $q\bar{q}$-condensate (mesons) are calculated. Deformation of bags is not included.

A general free-for-all discussion of quark degrees of freedom followed, and is probably still going on.

II Pion Production and Absorption in Nuclei: Experimental Results

Top: B. Höistad, M.A. Pickar, M.C. Green, E.G. Auld
Middle: N. Willis, Y. Le Bornec, P. Couvert
Bottom: E. Rössle, G.J. Lolos, H. Nann
Photos by Kent Berglund

AN OVERVIEW OF EXPERIMENTAL TRENDS AND SYSTEMATICS IN PION PRODUCTION AND ABSORPTION IN NUCLEI[+]

B. Höistad

Physics Department, University of Texas at Austin
Austin, Texas 78712, USA

ABSTRACT

An experimental overview of the (p,π) reaction on nuclei is given, emphasizing systematic behaviour in the data in general. Specific features in the data which point to nuclear structure dependence as well as the influence from the reaction dynamics are also discussed.

INTRODUCTION

The history of the (p,π) reaction is rich in experimental achievements, but short of significant theoretical progresses. There are few fields in nuclear physics were so much data are available but so little is known about the basic reaction mechanism. This is of course merely a reflexion of the complexity of the reaction process rather than a lack of theoretical work. In the early days of the (p,π) experiments we learned that a limited sample of data does not give enough constraints on proposed reaction models, since these allowed too much freedom in parameter choice etc. to make the interpretation of a qualitative fit to the data conclusive. It was therefore realized that progress in our understanding of the (p,π) reaction requires a comprehensive data base to guide further theoretical work. Experimental data are now available from a large variety of nuclear transitions, including angular distributions of differential cross section as well as analyzing power. Furthermore the reaction is explored at energies very close to the threshold up to 800 MeV, although the major part of the data exist below 250 MeV.

In the following some characteristic features of the data will be discussed rather than reviewing the experimental situation as a whole. For a comprehensive review of this field the reader is referred to Ref. 1. As the title of this talk indicates, we shall address the question of general trends in the (p,π) data. However, it should be pointed out that there are vast differences in the near threshold and the high energy behaviour of the

---

[+] Work supported by the U.S. Department of Energy

0094-243X/82/79105-25 $3.00 Copyright 1982 American Institute of Physics

(p,π) reaction. It is therefore neccessary to treat these energy regions separately in order to find any systematics in the data. No comparison between existing theoretical predictions and experimental data will be made, simply because this would not lead to any conclusive answer regarding the reaction mechanism. The situation will hopefully change dramatically in this respect when the full results from the new theoretical approaches, to be discussed in this workshop, are available.

THE (p,π) REACTION AT LOW ENERGIES

General features in the low energy data

Before we start the discussion of the (p,π) data a brief reminder about the momentum transfer q involved in the reaction, could be helpful. From the presentation of q in Fig. 1, it is noted that the reaction can only take place at a large momentum transfer to the residual nucleus. Fig. 1 also shows how the range in q varies with incident energy in a fixed angular interval. From this it is clear why the low energy data usually cover large scattering angles, while the high energy data are limited to forward angles.

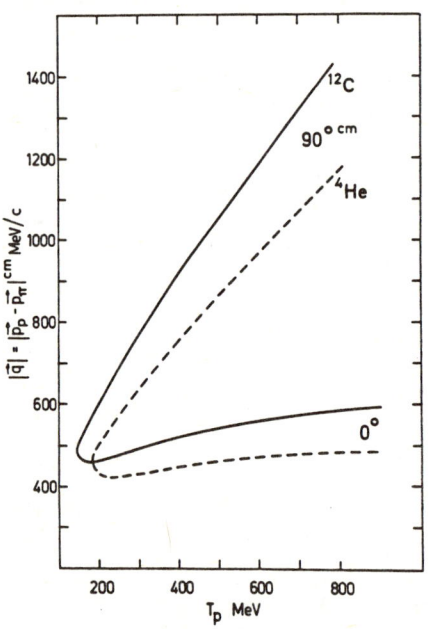

In order to demonstrate some characteristics in the near threshold energy data, we show in Fig. 2 angular distributions from the ground state transition with ^{3}He, ^{16}O, and ^{90}Zr as targets. The featureless slope in the ^{3}He(p,π)^{4}He g.s. data [2] is the most common shape of the angular distributions from transitions in the 1s-1p shells. The slope of the distribution from this reaction is however less steep than generally found for transitions with e.g. ^{9}Be, ^{10}B and ^{12}C as

Fig. 1. The momentum transfer in the (p,π) reaction on ^{12}C and ^{4}He leading to the ground state of ^{13}C and ^{5}He.

target nuclei. The ^{16}O(p,π)^{17}O g.s. data [3] shows a very pronounced minimum around 90 deg. Such a minimum occurs frequently for transitions in the 2s-1d shell. The depth and

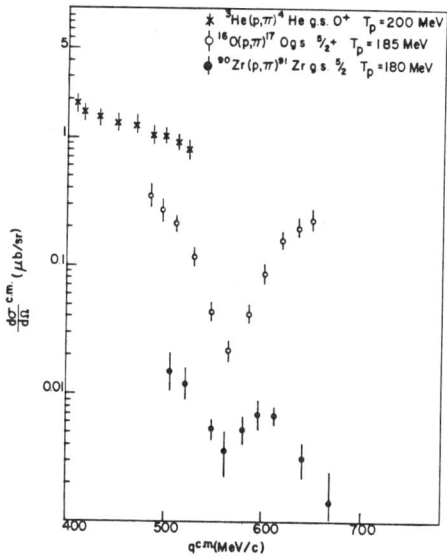

Fig. 2. Angular distributions ($\Theta_\pi \sim$ 20-130 deg.) from the (p,π^+) reaction on ^{3}He, ^{16}O and ^{90}Zr.

position of this minimum can vary with incident energy and transition involved. For heavy nuclei there are still very little data, and it would be premature to try to deduce any systematics. The data from the ^{90}Zr$(p,\pi)^{91}$Zr g.s. transition [4] illustrates merely that structure in the angular distributions can also be found with heavy targets.

Besides the variation in the shape of the angular distributions from different transitions, we also note from Fig. 2 a strong dependence in the cross section on the nuclear size. From ^{3}He to ^{90}Zr the cross section decreases about two order of magnitudes. (Preliminary data indicate that this drop in cross section can be even larger at higher energies.) Such a behaviour is in fact expected, at least qualitatively. Assuming the reaction takes place at the nuclear surface, the angular momentum missmatch increases for the transitions in He, O and Zr, implying that the probability for the reaction is reduced correspondingly. If instead the reaction preferably occurs for small values of the impact parameter, in order to reduce the angular momentum missmatch, then the stronger absorption for heavier nuclei would cause the cross section to drop. Either argument leads to a decrease in the cross section with the nuclear size, which basically is a consequence of the large momentum transfer in the reaction.

Negative pion production is generally suppressed compared to π^+ production, at least at forward angles. Also the π^- angular distributions are rather isotropic at threshold energies with only a few exceptions detected as yet. As a representative example we show in Fig. 3 the angular distribution from the ^{12}C$(p,\pi^-)^{13}$O g.s. reaction, and compare that with the one from the ^{12}C$(p,\pi^+)^{13}$C g.s. reaction [5]. Owing to the isotropic π^- angular distributions it is generally believed that the information about nuclear structure from these data are limited. However, the π^- measurements plays a fundamental role in the analysis of the reaction mechanism.

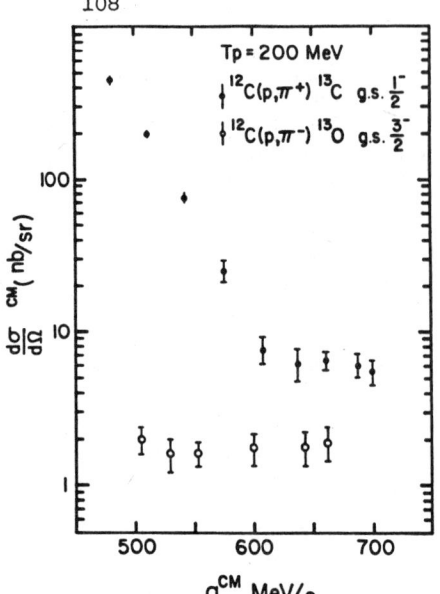

Fig. 3. Angular distributions (θ_π = 25-155 deg.) from the $(p,\pi^\pm)$ reactions on ^{12}C.

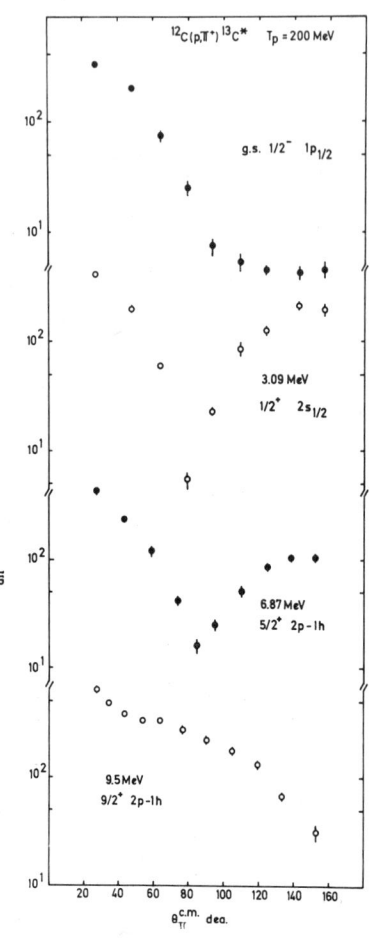

Fig. 4. Angular distributions from transitions in the $^{12}C(p,\pi^+)^{13}C$ reaction.

Nuclear structure dependence

We now focus our attention on the influence from the nuclear structure on the shape of the angular distribution. Let us therefore study some transitions which differ with respect to the nuclear structure involved. In Fig. 4, the angular distributions from some transitions in the $^{12}C(p,\pi)^{13}C^*$ reation [6] are presented. Different distributions are selected where the final state in ^{13}C is described by either 1 particle - 0 hole or by a 2 particle - 1 hole configuration. The ground state and 3.09 MeV state are largely given by a neutron in the $1p_{1/2}$ and $2s_{1/2}$ shells, respectively, while the 6.86 MeV and 9.50 MeV states have most of their strengths in the $|1d_{5/2} \times 2^+ + 2s_{1/2} \times 2^+|$ and $|1d_{5/2} \times 2^+|$ configurations, respectively. Among the distributions in Fig. 4, we find shapes, with a deep minimum as well as with a plain slope. Since both the single particle states and the 2p-1h states are represented in the two types of distributions, it can merely be concluded that the nuclear structure must be involved in determining the shape of the distribution, but it is not at all apparent how this is accomplished.

Another example is given by the transitions to the ground state (5/2$^+$) and 3.84 MeV (5/2$^-$) state in ^{17}O, whose angular distributions [3] are compared in Fig. 5. Both the shape and magnitude of the cross section are significantly different in this case. Note that, since these states have the same spin, the difference in the cross section cannot be explained by a difference in angular momentum missmatch or spin statistics. It is therefore likely that the specific properties of the final nuclear state cause the differences. The ground state is a single particle state ($1d_{5/2}$) while the 3.84 MeV state is a 2p-1h state with the hole in the $1p_{1/2}$ shell and the two particles in the $1d_{5/2}$ shell or mixed in the s-d shells. It should be emphasized that in a one-step reaction process only the ground state can be populated, but in a two step process (or equivalently a two nucleon process) both states can be reached. From the example given above one might therefore get the impression that the (p,π^+) reaction is dominantly a one-step process, but as we shall see in the following this is not supported by the data in general.

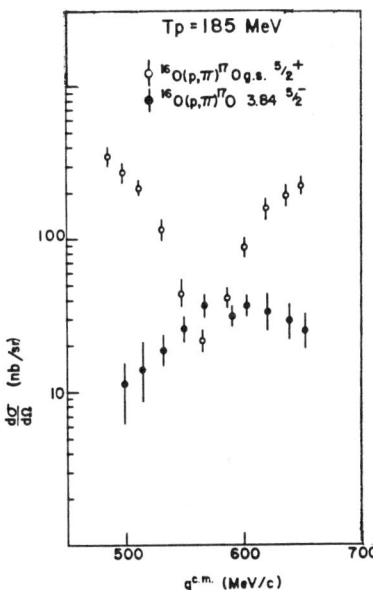

Fig. 5. Angular distributions from transitions in the ^{16}O(p,π^+)^{17}O reaction.

The dominant cross section from the transition to the ground state over that to the excited state at 3.85 MeV in ^{17}O brings up the question of selectivity in the (p,π) reaction. Let us examine some representative spectra in order to see if any preference for specific final states is prevailing. As a first example the ^{16}O(p,π)^{17}O* spectrum [3] is chosen, shown in Fig. 6, which is rich in structure. Due to the high level density beyond 5 MeV excitation energy no attempt will be made to identify levels in that region. The spectrum below 5 MeV shows a clear preference for transitions to single particle states. The $1d_{5/2}$ ground state and the $2s_{1/2}$ state at 0.87 MeV dominate the spectrum. The 1/2$^-$, 5/2$^-$ and 3/2$^-$ states at 3.06, 3.84 and 4.55 MeV are all very weakly populated. The structure of those states are given by

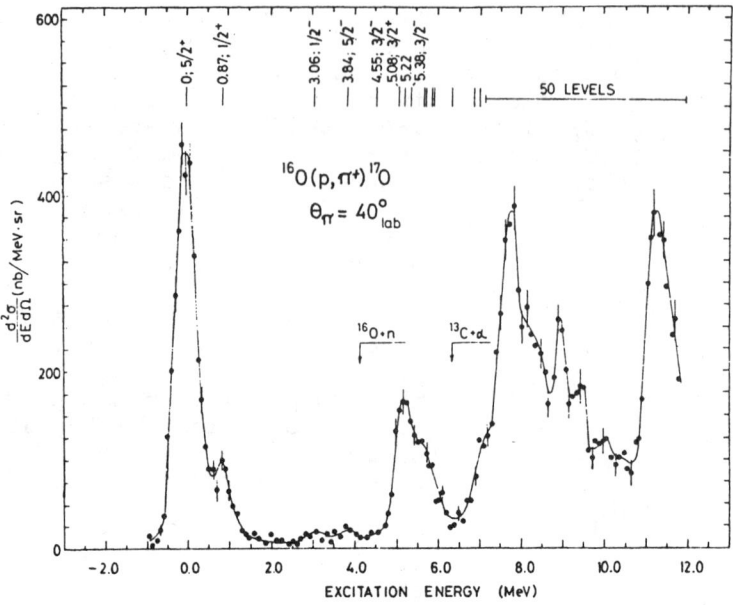

Fig. 6. A pion spectrum from the $^{16}O(p,\pi^+)^{17}O$ reaction obtained at 185 MeV.

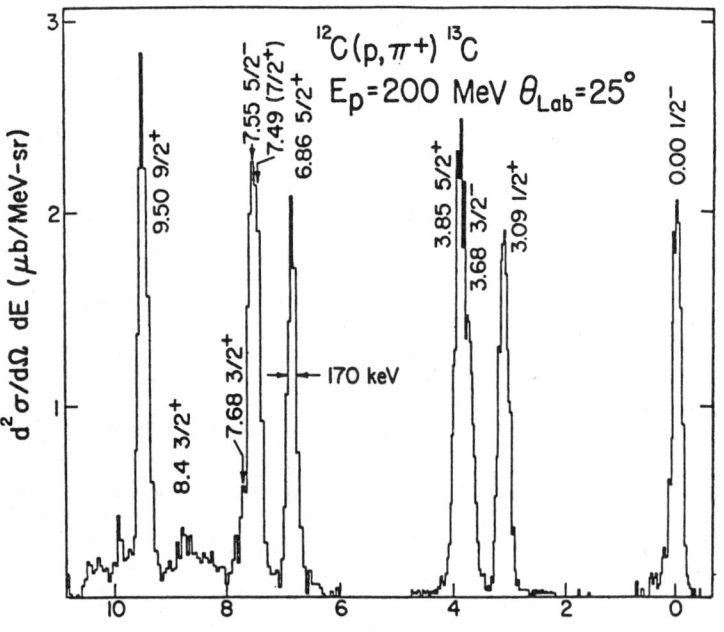

Fig. 7. A pion spectrum from the $^{12}C(p,\pi^+)^{13}C$ reaction.

2p-1h configurations, with the hole in the $1p_{1/2}$ shell and the two particles in the s or d shells. Another example is given in Fig. 7 where the $^{12}C(p,\pi)^{13}C$ spectrum [7] is shown. The single particle states in ^{13}C are found at the g.s. ($1p_{1/2}$), 3.09 MeV ($2s_{1/2}$) and 3.85 MeV ($1d_{5/2}$). The other states in the energy region up to 9.5 MeV can be understood as a single particle in the $1p_{1/2}$, $2s_{1/2}$ or $1d_{5/2}$ shell coupled to the 2^+ core of ^{12}C. In this spectrum there is no preference for the single particle states being populated via the (p,π) reaction. On the contrary, the 2p-1h states seem to be equally strongly excited. More examples showing this dual picture can easily be found. For example the $^{10}B(p,\pi)^{11}B$ spectrum [7] shows some similarities with the low energy stripping reaction, thus indicating selectivity to the single particle states. However, in the $^{12}C(\pi,p)^{11}C$ reaction [8], where the mirror states of ^{11}B are reached, the $|2^+ \times 1p_{3/2}^{-1}|$ states are populated as strongly as the states formed by a simple removal of a nucleon in the p-shell. Thus, again we find the picture ambiguous, and no general trend is apparent. Evidently there is no simple selectivity in the (p,π) reaction which can be decerned by observing the relative strength with which different final nuclear states are populated. The data seem to indicate a complicated interplay between the reaction process and the nuclear structure. The different diagrams in the complex reaction process might be sensitive to the nuclear structure in different ways.

Several examples from the low energy (p,π) data have already been given which show that the nuclear structure affects both the magnitude and shape of the differential cross section. From those examples it is however by no means clear if the shape of the angular distributions is mostly determined by a nuclear wave function or the dynamics of the reactions process or both. If for example the relative contributions from different diagrams in the reaction process is determined by the nuclear structure involved, different shapes of the angular distributions can still be directly related to the dynamics of the reaction rather than the nuclear wavefunction involved. In an attempt to find out the role of the nuclear wavefunction in defining the shape of the angular distributions, we examine the distributions from different transitions which all involve a $1p_{3/2}$ neutron wavefunction. In the $^{12}C(\pi,p)^{11}C$ g.s. transition [8], a neutron is removed from the $1p_{3/2}$ shell. This is also the case for the $^{13}C(\pi,p)^{12}C^+$ reaction [8] in the transition to the 2^+ level at 4.4 MeV. The third transition we consider is from the $^{10}B(p,\pi)^{11}B$ g.s. reaction [9], which involves a capture of a neutron to the $1p_{3/2}$ shell. If the wavefunction of the $1p_{3/2}$ neutron to a large extent determines the shape of the angular distributions, one would expect these to be rather alike. As appears from Fig. 8, no obvious similarities can be found. The minimum in the $^{12}C(\pi,p)^{11}C$ distribution does not occur in the other two. Consequently no evidence is found that this minimum is due to the $1p_{3/2}$ neutron wavefunction.

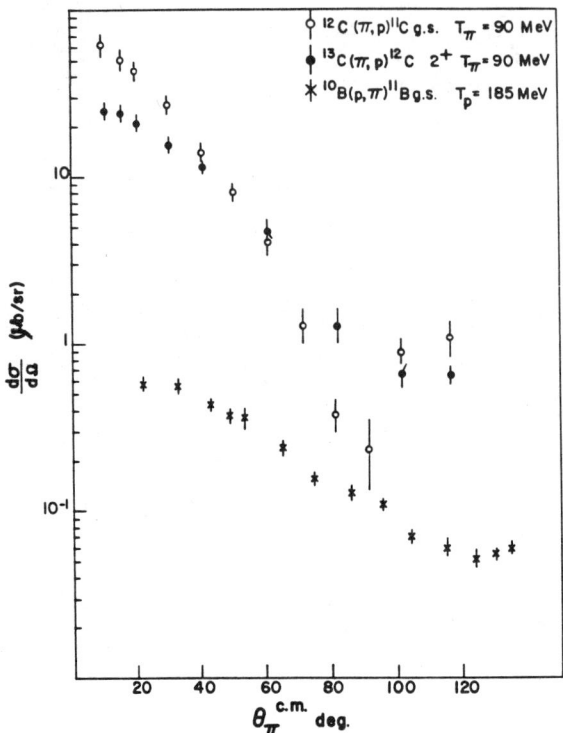

Fig. 8. Angular distributions from transitions which involves a capture or removal of a neutron in the $1p_{3/2}$ shell.

The energy dependence

A clue to the origin of the shape in the angular distributions is perhaps easiest found in the energy dependence of the reaction. For example, a minimum associated with the nuclear momentum distribution should largely remain at a fixed q when the incident energy varies, while minima due to the reaction mechanism should affect the shape with the characteristic energy dependence of the basic reaction amplitude.

The angular distribution from the $^{12}C(\pi,p)^{11}C$ g.s. reaction discussed above has been measured at the pion energies $T_\pi=50$ MeV[10], 90 MeV[8], and 180 MeV[8], and these are shown in Fig. 9. We observe that the minimum in the distribution recorded at 90 MeV and the shoulder seen at 50 and 180 MeV occurs approximately at the same pion angle, and consequently at very different momentum transfer. It is therefore unlikely that this structure in the angular distribution is related to the nuclear wavefunction. The distribution at 50 MeV might also contain a minimum around 90 deg., but better data in this angular region is needed to verify this. It should be noted that the minimum seen at $T_\pi = 90$ MeV seems to vanish at higher energies, where the effects from pion distortion is large. In fact all

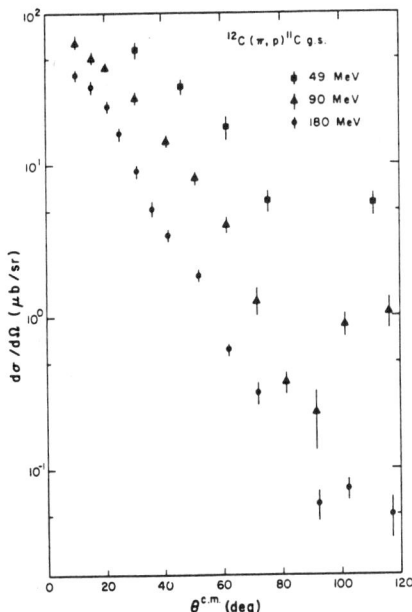

Fig. 9. Angular distributions from the $^{12}C(\pi,p)^{11}C$ g.s. reaction obtained at different energies.

data available as yet in the (3,3) resonance region show rather featureless slopes. The deep minima in the distributions are only seen in the near threshold data. This means that distinct minima are absent in a region dominated by the strong p-wave pion rescattering. However, more data is needed around the resonance energy to confirm if this is generally true.

The most dramatic energy dependence in the (p,π) data is observed in the $^{40}Ca(p,\pi)^{41}Ca$ g.s. reaction, which is measured at several energies from very close to the threshold up to 200 MeV [11]. These data are shown in Fig. 10 a. Both the position and the depth of the minima in the angular distributions change substantially in this energy interval. Moreover, a second minimum is developed above $T_p \sim 180$ MeV. Since the pion energy varies by a large factor in these data it is plausible that the changes in the distributions are due to the variations in the pion distortion. The strength of the p-wave as well as the relative contribution from the s- and p-wave pion rescattering changes significantly in the interval $T_\pi \sim 5-50$ MeV.

In order to find some systematics in the energy variation of the angular distributions, one approach is to examine how the position of the minima varies with energy for different nuclear transitions. From the distributions presented in Fig. 10a-c, we make the following observations. The first minimum in the $^{40}Ca(p,\pi)^{41}Ca$ g.s. distribution moves to larger angles with increasing energy. For the $^{12}C(p,\pi)^{13}C^*$(3.85 MeV) transition [6] the minimum moves in the opposite direction. A third case is found in the $^{28}Si(p,\pi)^{29}Si^*$(2.03 MeV) transition [3,12] where the position in the minimum remains at fixed angle in the energy interval considered. Any systematic behaviour in these distributions might appear remote. However, some order could be brought into the data if the momentum transfer at the minimum q_{min} is displayed versus the pion momentum p_π. In such a plot [7], shown in Fig. 11,

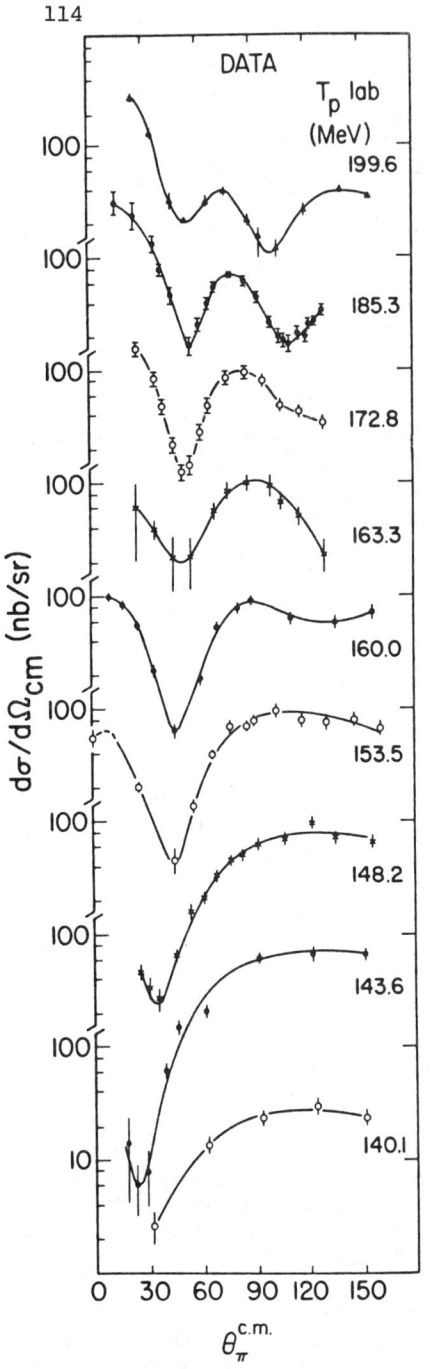

Fig. 10a. Angular distributions from the $^{40}Ca(p,\pi^+)^{41}Ca$ g.s. reaction.

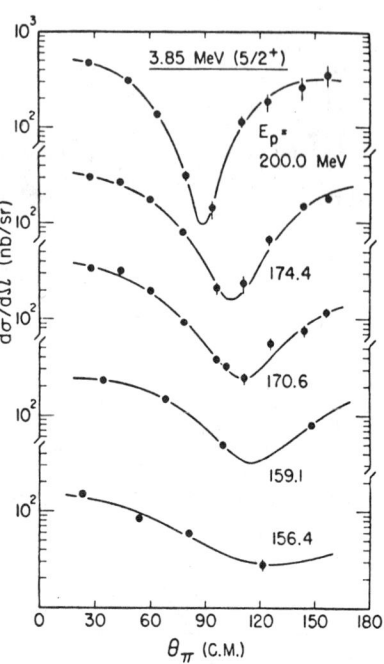

Fig. 10b. Angular distributions from the $^{12}C(p,\pi^+)^{13}C^*$(3.85 MeV) reaction.

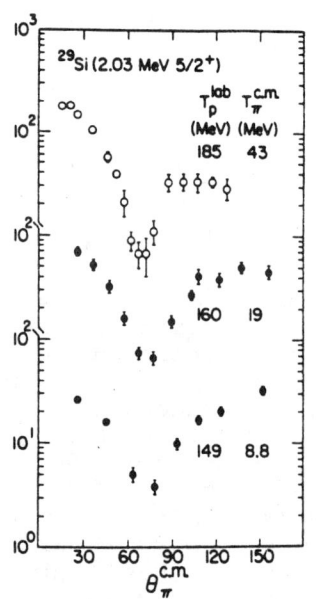

Fig. 10c. Angular distributions from the $^{28}Si(p,\pi^+)^{29}Si^*$ (2.03 MeV) reaction.

q_{min} is found to vary linearly (within the accuracy of the data) with p_π. Moreover, these three transitions, give rise to the same slope, which is perhaps more remarkable. It should, however, be pointed out in this context that if a larger energy interval is considered by using the $^{12}C(\pi,p)^{11}C$ g.s. data at T_π = 50, 90 and 180 MeV, the position of the minimum or shoulder also follows a linear but somewhat steeper slope, which is simply a kinematical consequence of the fact that the minimum occurs at fixed angle.

The magnitude of the slope should somehow be related to the reaction mechanism. For example, neglecting distortion effects, a minimum caused by a wavefunction should show up as a horizontal line in the q_{min} plot. It is therefore interesting to examine if the nuclear structure has any influence on the energy variation of q_{min}. The nuclear transitions discussed above only involve final states with dominant single particle configuration. An example of a transition to a 2p-1h final state, which also contain a clear minimum in the angular distribution, is the population of the $5/2^+$ level in ^{13}C at 6.86 MeV. The energy variation in the position of this minimum is also included in Fig. 11. We observe immediately that the slope of this linear curve is different from that given by the transitions to the single particle states. Assuming that the slope of the curves is largely determined by reaction dynamics, we thus have found still another feature in the data which indicate a coupling between the nuclear structure and the reaction dynamics.

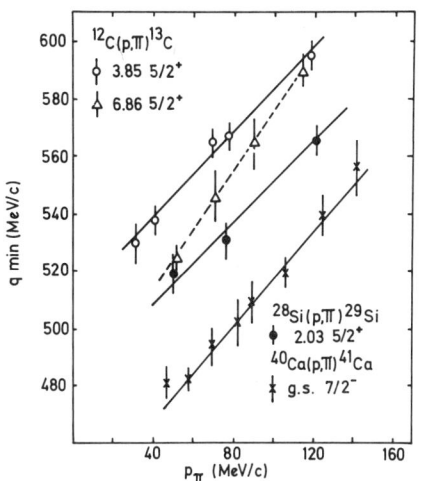

Fig. 11. The momentum transfer versus pion momentum at the minimum in angular distributions from transitions in C, Si and Ca.

The energy variation of the total cross section for specific transitions should further elucidate the nuclear structure dependence in the (p,π) reaction. In Fig. 12 the integrated cross section for some transitions in the $^{12}C(p,\pi)^{13}C$ reaction [6] is presented as a function of pion energy. The single particle final states give rise to very similar energy variations.

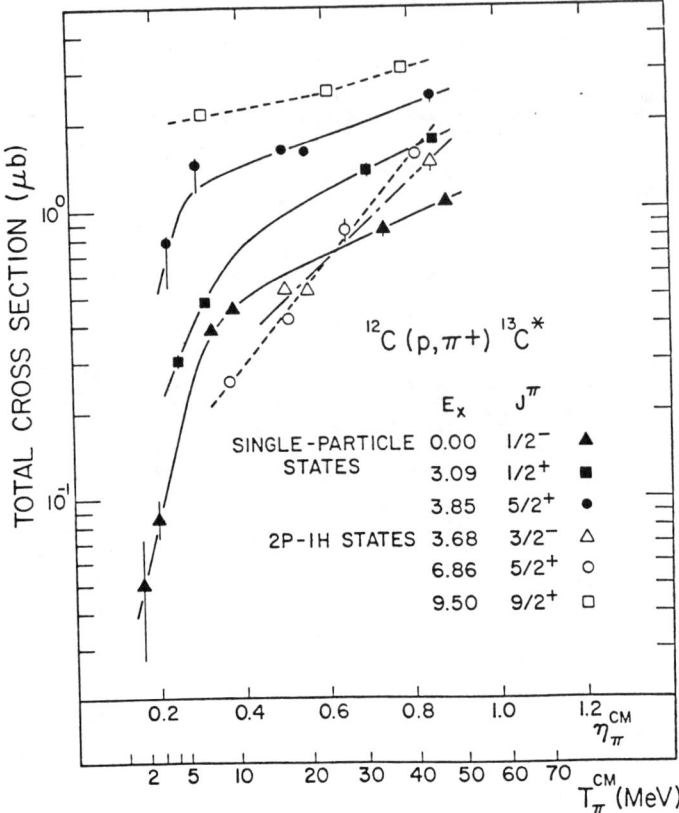

Fig. 12. The total cross section versus pion energy for some transitions in the $^{12}C(p,\pi^+)^{13}C^*$ reaction.

On the other hand, the 2p-1h state at 6.86 MeV does not conform to this pattern but deviates also in this variable from the 1p-0h trend. However, the transition to the 2p-1h state at 9.5 MeV shows a remarkable similar energy variation as those leading to single particle final states. Obviously, the total cross section for individual nuclear transitions also show clear sensitivity to the nuclear structure, but there is no simple general distinction in the data which gives a hint of the character of this nuclear structure dependence.

THE (p,π) REACTION AT HIGH ENERGIES

Many aspects of the (p,π) reaction have now been investigated experimentally far above the energy threshold. At 800 MeV (p,π$^+$) data from several light nuclei up to ^{16}O are available, but for the (p,π$^-$) reaction, data from only one target nucleus (^{9}Be) exist [13]. In the energy interval, between 250 and 800 MeV some data exist from a few light nuclei [14]. Asymmetry measurements at various energies have just recently been initiated. As already mentioned in the introduction there are very few direct similarities between the high and low energy data. This is of course not surprizing since the pion-nucleus interaction differs at ∼ 50 MeV and ∼ 650 MeV, which certainly has a large impact on the (p,π) reaction. It is however by no means clear how the change in the pion-nucleus interaction will be reflected in the data. In particular since we have only a vague idea about the reaction mechanism even at low energies.

Let us now examine the high energy (p,π) data and focus on the response from the nuclear structure. As a first example the spectrum from the ^{16}O(p,π)^{17}O reaction obtained at 800 MeV [15] (see Fig. 13) is chosen, since this spectrum was also included in our discussion of the

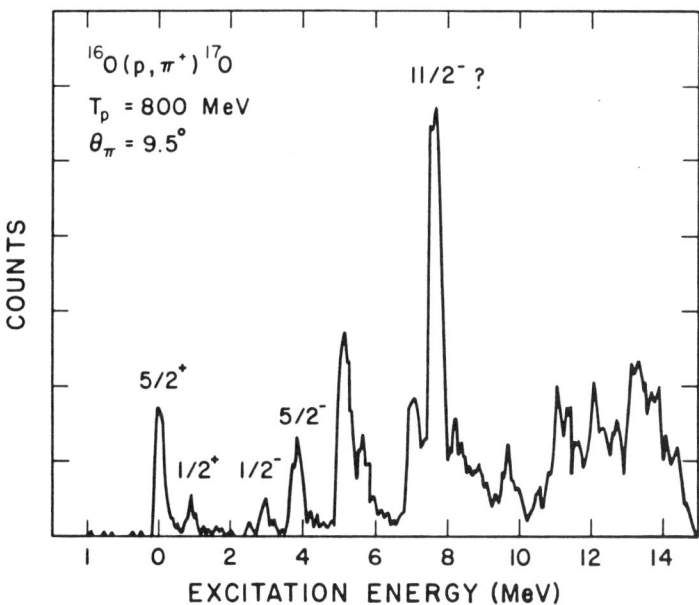

Fig. 13. Pion spectrum from the ^{16}O(p,π$^+$)^{17}O reaction.

low energy data. One characteristic of this spectrum, which is immediately noticed in the comparison with the low energy data, is that the single particle states are no longer the dominating peaks. The 2p-1h states at 3.06 and 3.84 MeV are about as strongly populated as the ground state ($1d_{5/2}$) and the 0.87 MeV state ($2s_{1/2}$). The only coarse selection rule appears to be that high spin states are preferably populated. This fact has probably the trivial explanation in spin statistics and angular momentum missmatch. Consequently any clear nuclear structure dependence is difficult to deduce from this spectrum alone. A comparison between the angular distributions from transitions to the single particle state $1d_{5/2}$ and the 2p-1h state at 3.84 MeV, which both have $J = 5/2$, should be more informative. Those distribution are shown in Fig. 14. Both distributions exhibit a featureless slope and differ only slightly in the magnitude of the cross section. This behaviour is drastically different from that, given by the low energy data (see Fig. 5). Moreover, the high energy data show much less sensitivity to the nuclear structure. Only the magnitude of the cross section is affected somewhat. The same trend, with few exceptions, can in fact be seen in all high energy data recorded so far. The large variety in the shapes of the distributions seen in the low energy data is not prevailing at high energies. This is further illustrated in Fig. 15 where the $J = 1/2$ levels at 0.87 MeV ($2s_{1/1}$) and 3.05 MeV (2p-1h) can be compared. Although in this case there are in fact some minor differences also in the shape of the distributions. This example constitutes one of the exceptions where the influence from the final nuclear state is seen rather clearly. Another example in this category is found in the transitions to the $J = 5/2$ levels at 3.85 and 6.87 MeV in ^{13}C [16]. Even including the last two examples it is generally true that in order to find any nuclear state dependence in the high

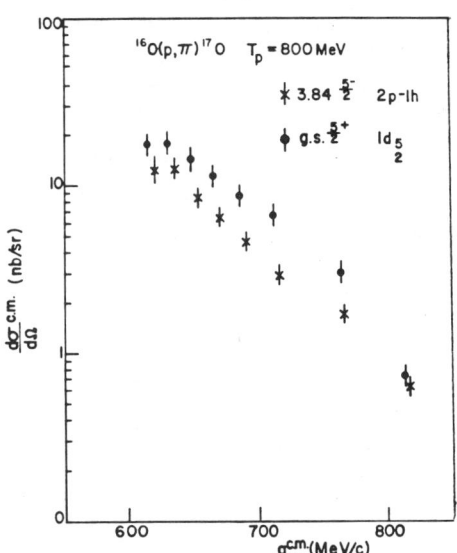

Fig. 14. Angular distributions from transitions in the $^{16}O(p,\pi^+)^{17}O^+$ reaction.

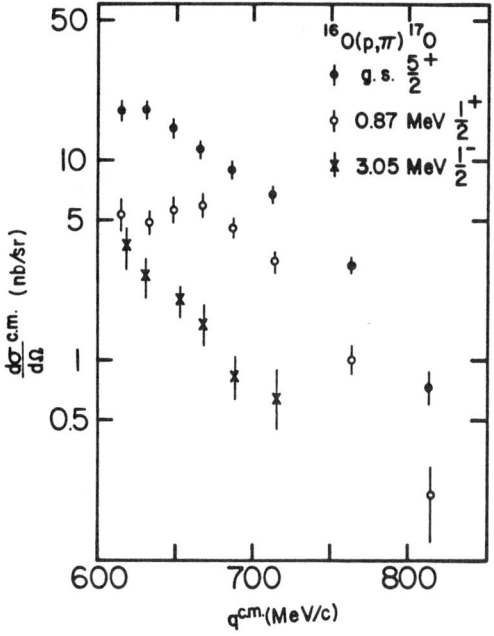

Fig. 15. Angular distributions from transitions in the $^{16}O(p,\pi^+)^{17}O^*$ reaction.

energy data it is necessary to examine the angular distributions with respect to small details.

The difference between the (p,π^+) and (p,π^-) reactions has been investigated with ^{9}Be as target nucleus [17]. A spectrum from that measurement is shown in Fig. 16. The choice of an isospin 1/2 target nucleus, implies that the residual nuclei are isobaric analogs and consequently the difference in the (p,π^+) and (p,π^-) cross section is mainly due to the reaction mechanism. It is therefore interesting to observe in the ^{9}Be$(p,\pi^{\pm})$ spectra that the relative strength with which the final states are populated differs significantly in ^{10}Be and ^{10}C. This indicates that the nuclear structure affects the positive and negative pion production in different ways.

DIFFERENCES IN THE ENERGY DEPENDENCE OF THE $(p,\pi^{\pm})$ REACTIONS

Data from the ^{9}Be$(p,\pi^+)^{10}$Be reaction exist at some energies from 185 to 800 MeV, and can thus be used to study the energy variation in the cross section. In Fig. 17 the angular distributions from the ground state transition obtained at three energies are presented as a function of momentum transfer. In these distributions the energy-dependence is only appearing in the magnitude of the cross section. The slope remains the same from 185 MeV all the way up to 800 MeV. This is also observed for other transitions where the distributions consist of plain slopes.

In view of the energy independent slope in the angular distribution from the ^{9}Be$(p,\pi^+)^{10}$Be reaction, it is rather surprizing to find large variations with energy in the shape of the distributions from the ^{9}Be$(p,\pi^-)^{10}$C transition, which appears from Fig. 18. The distribution at 200 MeV [18] is isotropic out to about 120° where a backward rise starts, while the 225 MeV [19] data show a minimum at ~ 120°, and the 800 MeV [17] data exhibit a plain relatively steep slope.

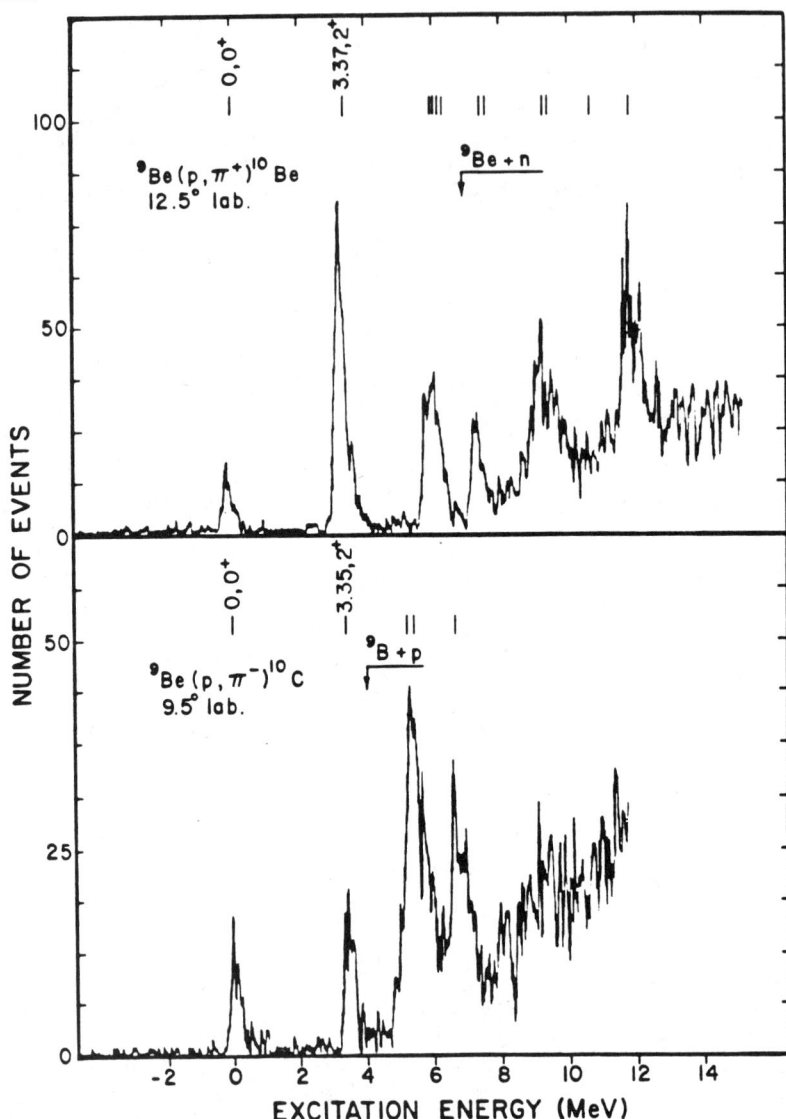

Fig. 16. Pion spectra from the ^{9}Be(p,$\pi^{\pm}$) reactions.

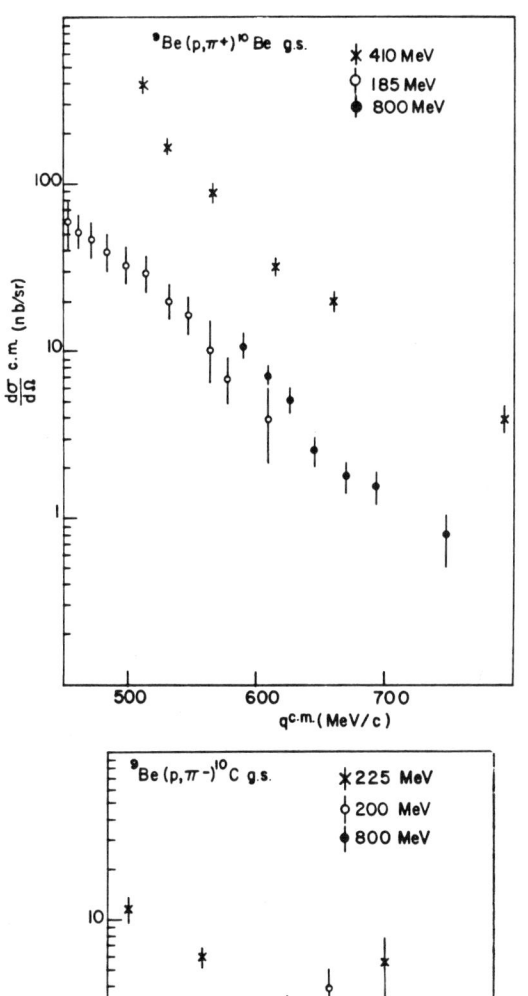

Fig. 17.
Angular distributions obtained from the ^{9}Be$(p,\pi^+)^{10}$Be g.s. reaction at 185 MeV (Θ_π^{lab} = 15-135 deg.), 410 MeV (Θ_π^{cm} = 6 - 70 deg.) and 800 MeV (Θ_π^{lab} = 9.5 - 27.5 deg.).

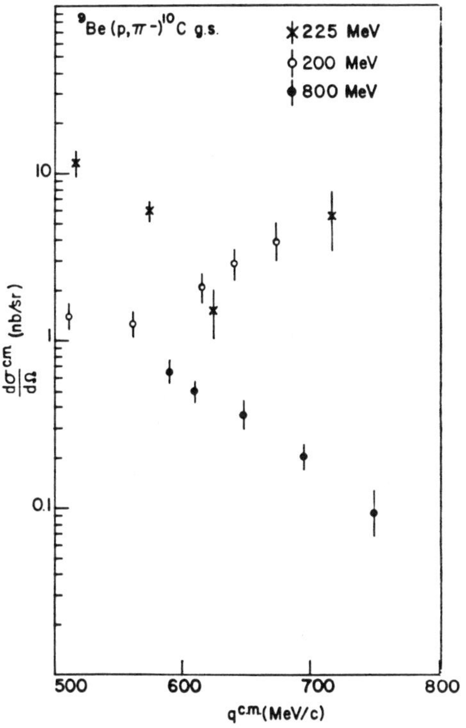

Fig. 18
Angular distributions from the ^{9}Be$(p,\pi^-)^{10}$C g.s. reaction at different energies. The data at 225 MeV are arbitrarily normalized.

Some idea of the energyvariation in the total cross section for a specific transition can be obtained by examine how the differential cross section varies at a fixed momentum transfer q. Since the dependence on the nuclear form factor thereby is minimized, the energy variation of the cross section should largely be due to the reaction mechanism. In Fig. 19 such data from the target nuclei ^{12}C and ^{9}Be are presented for both π^+ and π^- production 5,8,14,18,20,21, 22 at angles corresponding to q^{cm} = 600 MeV/c. The energyvariation of the (p,π^+) reaction presents no surprizes. The enhancement around 300 MeV could easily be associated with the (3,3) resonance. The energyvariation in the (p,π^-) cross section is much harder to understand. A minimum around 600 MeV, which is indicated by the data, has no straight forward explanation. More accurate data is however needed to deduce the true energy behaviour. In particular since the π^- angular distributions might vary quite significantly in this energy region and the cross section at a fixed q can be a poor indicator of the total cross section. Nevertheless it can be safely concluded that there is a distinct difference in the energy-variation of the (p,π^+) and (p,π^-) reactions.

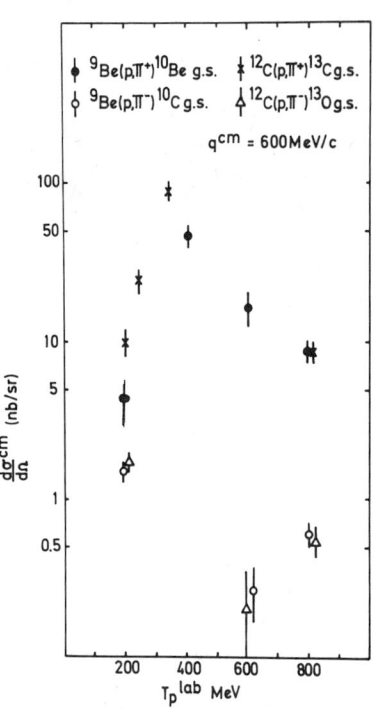

Fig. 19. The differential cross section from π^+ and π^- production at q = 600 MeV/c versus incident proton energy. The data points from the ^{12}C(p,π^+)^{13}C g.s. reaction at T_p = 240 and 330 MeV are calculated from the data on the reversed reaction at T_π = 90 and 180 MeV.

In the presented data we have seen large differences between the (p,π^+) and (p,π^-) reactions. The magnitude of the differential cross section as well as its angular dependence and energy dependence show different signatures for the two reactions. Moreover, the nuclear structure affects the reactions differently. Consequently one is inclined to believe that the leading reaction diagrams for the (p,π^+) and (p,π^-) reactions are not the same.

THE ANALYZING POWER IN THE (p,π) REACTION

The first experimental results on the analyzing power A_y in the (p,π) reaction [23] appeared at 200 MeV from TRIUMF and comprised of angular distributions for some transitions in ^{9}Be and ^{12}C. These data showed a uniform and simple pattern with striking similarities with the analyzing power from the elementary $pp \to d\pi$ reaction. All transitions studied gave negative values on A_y, distributed along a curve with a clear minimum around 60 deg (see Fig. 20). This curve is recognized from the analyzing power of the $pp \to d\pi$ reaction at low energies ($T_p \sim 350$ MeV). It is assumed that kinematical arguments are valid to explain the 30 deg. shift to smaller angles in the position of the minimum in the (p,π) data. This observation led early to the interpretation that the $pp \to d\pi$ reaction might be considered as a subprocess in the exclusive positive pion production on nuclei. However, recent data from a larger sample of nuclear transitions show much more variations in the shape of the A_y distributions, thereby revealing that the nuclear structure dependence of A_y is by no means negligible, which makes it doubtful whether the simple $pp \to d\pi$ process survives as a separate amplitude in the (p,π) reaction on nuclei. We demonstrate the nuclear state dependence in Fig. 21, which shows the angular distributions of A_y from two transitions leading to the $2s_{1/2}$ state in ^{13}C($E_x = 3.09$ MeV) and ^{17}O($E_x = 0.87$ MeV) respectively [24]. The distribution from the transition in O deviates significantly from that in C, in spite of fact that the nuclear structure of the final states is largely the same.

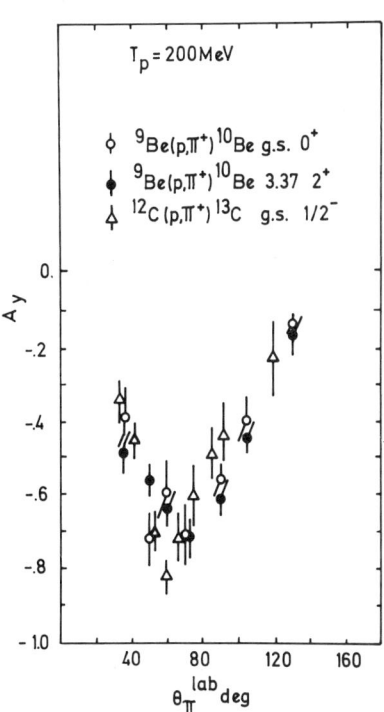

Fig. 20, Analyzing power for some transitions from the (p,π) reaction on ^{9}Be and ^{12}C.

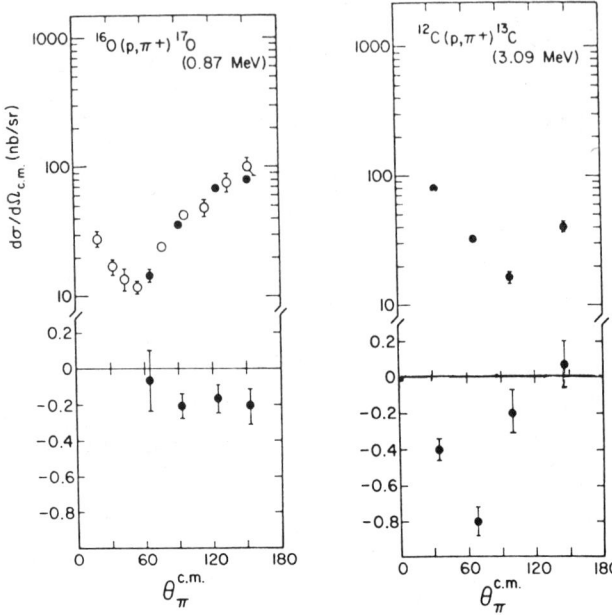

Fig. 21. The differential cross section and analyzing power at 159 MeV for the transitions to the $2s_{1/2}$ states in ^{13}C and ^{17}O.

Evidently, A_y can be sensitive to subtle details in the nuclear wavefunction. Other examples could be given indicating the nuclear structure dependence, but it is too early to discuss the data in terms of systematics. More data will appear in the near future and surprizes of different kinds are probably yet to come.

Regarding the analyzing power for the negative pion production very little is known. The experimental situation is currently being improved with new experiments in progress, which results are discussed elsewhere in this workshop. Let us only mention the first result which appeared for the $^9Be(p,\pi^-)^{10}C$ reaction at 200 MeV from IUCF [18]. This result, presented in Fig. 22 illustrates both the existence of specific features of the reaction mechanism for π^- production as opposed to the one for π^+, and a sensitivity to the nuclear structure, which probably to some extent is responsible for the somewhat peculiar shape of angular distribution of A_y in this case. We note that the analyzing power goes from negative to positive values at about the same angle as the differential cross section starts the rise in the backward direction.

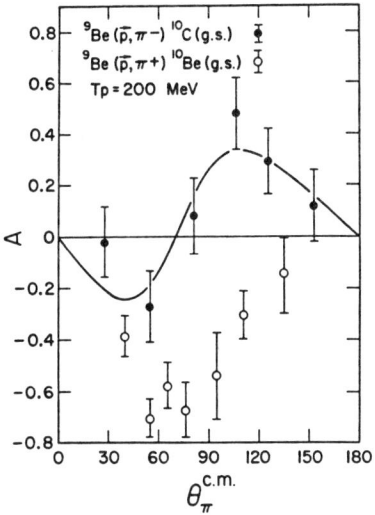

Fig. 22. The analyzing power for the ground state transitions in the $(p,\pi^{\pm})$ reactions on Be at 200 MeV.

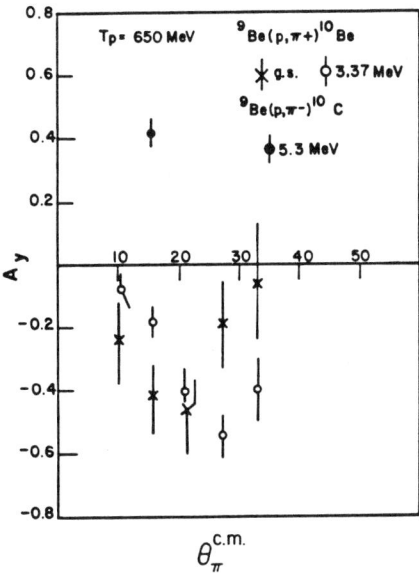

Fig. 23. The analyzing power for some transitions in the $(p,\pi^{\pm})$ reactions on ^{9}Be at 650 MeV.

Finally it should be mentioned that measurements of the analyzing power at high energy from some light nuclei have just started at LAMPF. Very preliminar results from A_y in the $^9\text{Be}(p,\pi^{\pm})$ reactions [25] can now be shown. In Fig. 23, A_y from the transition to the ground state and first excited state in ^{10}Be is presented as well as one datum point from the negative pion production. The angular distributions of A_y show a clear minimum for the g.s. transition and probably also for the 3.37 MeV state. The positions of the two minima are somewhat displaced indicating a weak final state dependence. We note that these data do not show any clear traces from the pp -> dπ reaction, which analyzing power is positive at all angles in this region. However, it should be emphazized that both the pd->tπ and the p^{3}He -> ^{4}Heπ reactions also show deep minima at forward angles.

THE ENERGYVARIATION IN THE ANALYZING POWER

Our experimental knowledge about the energy dependence of A_y is still in its infancy. The very first results from the energy variation in A_y show a dramatic dependence on the incident energy between 200 and 250 MeV, in particular for the transition $^{12}\text{C}(p,\pi)^{13}$ g.s. [26] for which the data are shown in Fig. 24. A distinct maximum in A_y shows up at 225 MeV which is not even hinted at in the 200 MeV data. One might guess that this rapid variation in A_y with energy is connected to the change in the pion-nucleon interaction. However, the nuclear structure also has to be involved in some

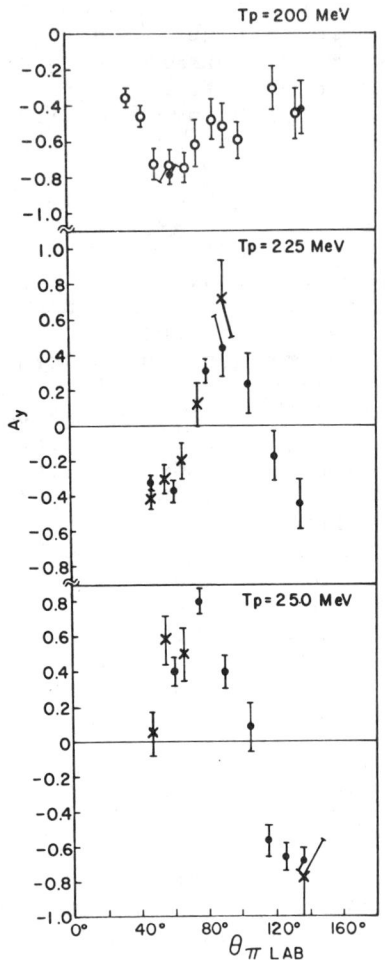

Fig. 24. The analyzing power for the $^{12}C(p,\pi^+)^{13}C$ g.s. reaction at 200, 225 and 250 MeV.

fundamental way since other transitions in ^{12}C and in ^{9}Be show a slightly different and less pronounced energyvariation. One should also make the remark that the variations in A_y with energy is not easily reconciled with that from the $pp \rightarrow d\pi$ reaction. Simple final state interaction, e.g. proton distortion should contribute to the asymmetry in the (p,π) reaction, but considering the nuclear structure dependence as well as the rapid energyvariation we find it more likely that A_y is closely linked to the production process itself.

It is interesting to study if the energyvariation in A_y follows some characteristic changes in the differential cross section. In Fig. 25 the angular distributions from the transitions in ^{12}C at $T_p=$ 225 and 250 MeV [26] are therefore presented. The distribution at 200 MeV is already given in Fig. 3. As observed these distributions show a very weak and smooth energydependence, in sharp contrast to the situation for A_y. The strong energydependence evidently exists in the individual spin terms of the cross section but is canceled in the spin average.

Although the existing data on the energyvariation of A_y is far too scarce to deduce the true behaviour, a very coarse picture might nevertheless be discerned. Using various data on A_y from light nuclei at 200, 225, 250, 650 and 800 MeV as well as available experimental information on the $pd \rightarrow t\pi$ reaction, and assuming the observed similarity between A_y from $pd \rightarrow t\pi$ and $pA \rightarrow (A+1)\pi$ remains in general at high energies, we dare to make a guess how the shape in the angular distribution changes between 200 and

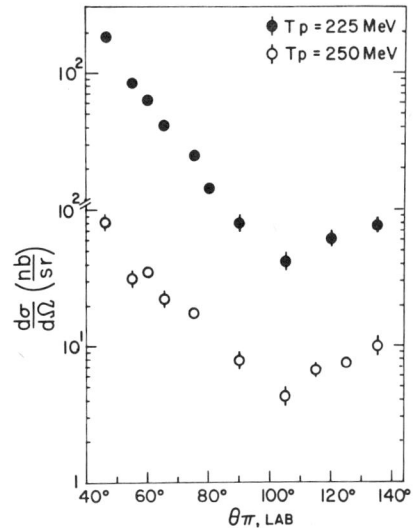

Fig. 25. The differential cross section for the $^{12}C(p,\pi^+)^{13}C$ g.s. reaction at 225 and 250 MeV.

800 MeV. The 60 deg. minimum in A_y at 200 MeV moves gradually towards smaller angles with increasing energy. A maximum around 90 deg. is building up rapidly just above 200 MeV, thereby also forming a negative second minimum at some large angle. With increasing energy the maximum gets broader and the second minimum more shallow, until at some energy, perhaps around 550 MeV, A_y is positive at all angles beyond the maximum. This scenario might be the qualitative trend in the A_y data, although strong modulations due to the specific nuclear structure involved are certainly present.

Regarding the energy-variation of A_y for negative pion production, nothing can essentially be said as yet. Only one datumpoint exist at high energy (650 MeV) and is from the $^9Be(p,\pi^-)^{10}C$ reaction (see Fig. 22). All final states in ^{10}C yields positive values, which is in sharp contrast to A_y for π^+ production. Indications of a difference in the two production modes are thus present in A_y also at high energies.

From the data on the analyzing power presented above, it is indicated that a lot of valuable information about the reaction mechanism is contained in A_y. The analyzing power has been a crucial parameter to measure in the elementary pp -> dπ reaction in order to improve our understanding of this process. In the same way one can anticipate data of A_y from the (p,π) reaction on nuclei to play a fundamental role in the search for the basic production amplitude. The present access to intense polarized beam at many laboratories will grant that future data are likely to include angular distributions from A_y as well as the differential cross section.

CONCLUSIONS

We have discussed some features in the (p,π) data which are of specific importance to know in order to advance our understanding of the reaction process. Special attention has been paid to find any significant systematics in the data. This search, however, has been met with limited progess, simply because there are very few general trends in the data which are nontrival. The data exhibit a variety of different features for which it is hard to find the connecting link. What appears to be a clear indication of a certain reaction process in one set of data is contradicted by another set. In the low energy data we have seen a large sensitivity to the nuclear structure, but not in the direct way, i.e. so that the shapes of the angular distributions are mainly determined by a nuclear wavefunction. The reaction dynamics, which partly is contained in the off-shell pion rescattering, certainly affects the data, but not in a transparant way. Much of the (p,π) data, including angular distributions of differential cross section and analyzing power as well as their energydependence, suggest an interplay between the details of the reaction process and the nuclear structure. Different nuclear transitions may be dominated by different reaction diagrams. Such a complexity in the reaction process, hinted at in the data, can of course only be resolved by detailed theoretical calculations.

At high energy the (p,π) reactions is much less sensitive to nuclear structure than at low energy. The reaction process and perhaps also general distortion effects evidently dominate the picture, and prevents the nuclear structure from being seen clearly in the data. Whether such a circumstance simplifies or complicates the description of the (p,π) reaction remains to be seen.

The new data on the analyzing power look very promissing, in the sense that these may turn out to contain a lot of information about the reaction mechanism. More data in this field is, however, needed to be able to judge how selective those results are to different reaction models.

Finally it should be emphasized that the wealth of experimental data now qualifies for a vigorous attack of theoretical work.

REFERENCES

1. H.W. Fearing, Prog. Part. Nucl. Phys. $\underline{7}$, 113 (1981).
   B. Höistad, Adv. Nucl. Phys. $\underline{11}$, 135 (1979).
   D.F. Measday and G.A. Miller. Ann. Rev. Nucl. Part. Sci. $\underline{29}$, 121 (1979).
2. N. Willis et al., Journal of Phys. $\underline{G7}$, 195 (1981).
3. S. Dahlgren et al., Nucl. Phys. $\underline{A227}$, 245 (1974).
4. B. Höistad et al., Phys. Lett. $\underline{79B}$, 385 (1978).
5. B. Höistad et al., Phys. Lett. $\underline{94B}$, 315 (1980).
6. F. Soga et al., Phys. Rev. $\underline{C24}$, 570 (1981).
7. F. Soga et al., Phys. Rev. $\underline{C22}$, 1348 (1980).
8. R.E. Anderson et al., Phys. Rev. $\underline{C23}$, 2616 (1981).
9. S. Dahlgren et al., Physica Scripta $\underline{10}$, 104 (1974).
10. J.F. Amann et al., Phys. Rev. Lett. $\underline{40}$, 378 (1978).
11. R.D. Bent, AIP Conf. Proc. $\underline{54}$, 142 (1979).
12. T.P. Sjoreen et al., IUCF preprint No. 151 (1981).
13. B. Höistad, LAMPF Users Group Newsletter $\underline{12}$, 87 (1980).
14. M. Dillig et al., Nucl. Phys. $\underline{A333}$, 477 (1980).
15. B. Höistad et al., to be published.
16. H. Nann, this conf. proc.
17. B. Höistad et al., Phys. Rev. Lett. $\underline{43}$, 487 (1979).
18. T.P. Sjoreen et al., Phys. Rev. Lett. $\underline{45}$, 1769 (1980).
19. G.J. Lolos et al., Abstract E10 in contributed paper to the 9 ICOHEPANS, Versailles, 1981.
20. B. Höistad et al., to be published.
21. S. Dahlgren et al., Nucl. Phys. $\underline{A204}$, 53 (1973).
22. P. Couvert et al., Phys. Rev. Lett. $\underline{41}$, 530 (1978).
23. E.G. Auld et al., Phys. Rev. Lett. $\underline{41}$, 462 (1978).
24. T.P. Sjoreen et al., Phys. Rev. $\underline{C24}$, 1135 (1981).
25. B. Höistad et al., to be published.
26. G.J. Lolos et al., to be published in Phys. Rev. C.

RECENT DEVELOPMENTS AND RESULTS IN $(p,\pi^{\pm})$ AT IUCF*

M.C. Green
Indiana University Cyclotron Facility, Bloomington, Indiana 47405

INTRODUCTION

Proton-induced charged pion production has been studied at IUCF[1-11] for proton kinetic energies below 210 MeV (well below the nucleon-nucleon production threshold). Differential cross section and analyzing power angular distributions have been measured for a variety of nuclei where the residual nucleus is left in some discrete final state. Prof. Höistad has presented[12] an excellent overview of the currently published data from IUCF and other facilities. In the past few months we have taken a large amount of data on the carbon isotopes for both positive and negative pions with a new spectrometer specifically designed for pion production measurements. The first part of this presentation will be a brief discussion of the new spectrometer and its performance. In the second part, some recent experimental results will be discussed.

ADVANCES IN PION DETECTION AT IUCF

A magnetic spectrometer is ideally suited as the base for a pion detection system since the ratio of the pion mass to that of the proton (a prime candidate for background) is about seven. Target-to-detector flight time provides an additional constraint to separate charged pions from other background. In this energy region, π^+ and π^- production total cross sections have been found to be 100's and 10's of nanobarns, respectively. Thus data rates are generally limited by the solid angle and momentum bite of the spectrometer. In addition, for near threshold measurements, significant losses result from pion decay in flight from the target to the detectors. A spectrometer "optimized" for efficient pion detection should have a large solid angle, a large momentum bite and a short flight path.

Initial (p,π^+) measurements at IUCF were carried out with the QDDM spectrometer,[13] which serves as a general purpose spectrometer for charged particle reaction studies. Shown schematically in

---

*Work supported in part by National Science Foundation

Figure 1 is the QDDM as well as more recent IUCF pion spectrometers. The QDDM spectrometer is far from optimal with respect to the above criteria. The $D\bar{D}$ spectrometer[14] (2 dipoles, with second dipole field opposite from the first) was built and used by P. Pile for detection of very low energy pions.[15]

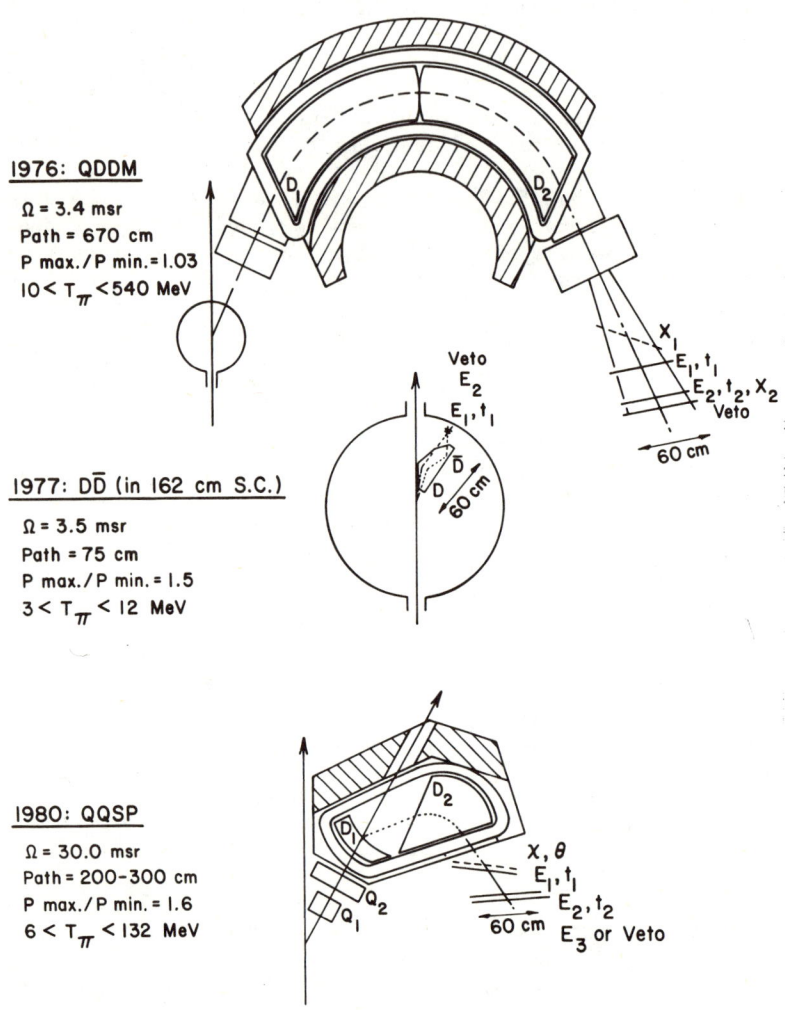

1976: QDDM

Ω = 3.4 msr
Path = 670 cm
P max./P min. = 1.03
10 < T_π < 540 MeV

1977: $D\bar{D}$ (in 162 cm S.C.)

Ω = 3.5 msr
Path = 75 cm
P max./P min. = 1.5
3 < T_π < 12 MeV

1980: QQSP

Ω = 30.0 msr
Path = 200-300 cm
P max./P min. = 1.6
6 < T_π < 132 MeV

Figure 1:   IUCF Pion Spectrometers

The most recent IUCF pion spectrometer, the QQSP, was specifically "optimized" for low to medium energy pion detection. The QQSP spectrometer is composed of two quadrupoles and two dipoles. The dipoles are excited by a single coil. Hence the "SP" stands for the Split Pole dipole. Figure 2 is more detailed diagram of the QQSP and its associated detectors.

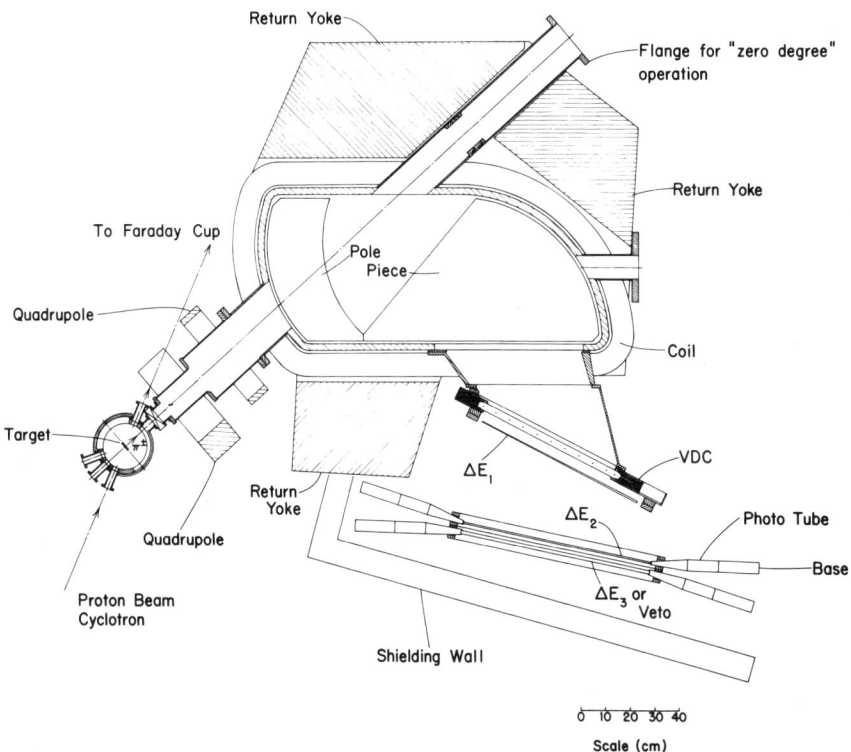

Figure 2: QQSP Spectrometer

The process of optimizing the design of the QQSP for maximum pion count rate efficiency inevitably introduced significant horizontal aberrations in the spectrometer optics. Shown at the top of Figure 3 is output from a computer raytrace simulation where

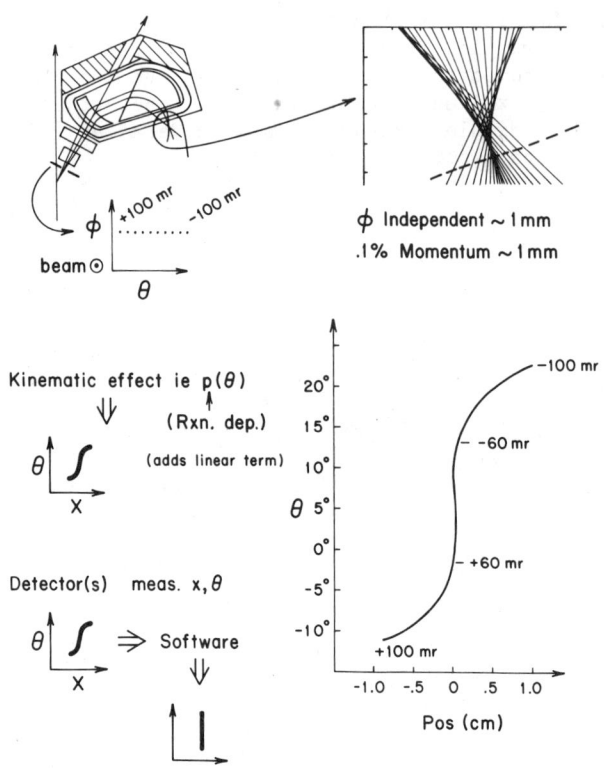

Figure 3: QQSP Horizontal Aberrations

pencil beams of equal momentum (magnitude) enter the spectrometer at equal horizontal angular intervals up to ±100 mr from the central ray. The fact that all these rays do not come to a focus at the same point is what is meant by the term aberration. A more quantitative picture of this effect is shown in the lower righthand side of Figure 3. Here the angle of the outgoing ray with respect to the normal of a reference line (dashed in upper righthand corner) is plotted versus the position at which the ray intercepts the line. If all the rays had focused at a single point, the resulting curve would have been a straight vertical line. For a given excited state in a residual nucleus, the outgoing particle's momentum is a function of the outgoing angle. This kinematic effect introduces to first order a linear term which in effect tilts the raytrace plot. In order to remove or correct these aberrations the detection system must measure both the position and angle of the outgoing ray.

The detector used to measure position and angle is a vertical wire drift chamber placed in the focal region of the spectrometer (Figure 4). The chamber's construction closely follows a design by Bertozzi et al.[16] The chamber is schematically shown in the lower portion of Figure 4. Charged particles passing through the chamber ionize gas molecules from which electrons then drift along electric field lines. As the electrons pass into the non-linear field (near the sense wires), they avalanche and deposit a pulse of negative charge. The arrival time of electrons at the wire is directly related to the distance from the ground plane to the region where the gas is ionized, since the electron drift velocity is constant over all but a small portion of the electron's path. The chamber is oriented so that at least three drift regions are crossed by an incoming particle. A following scintillator provides a a start signal for timing. The resulting three drift times are used to calculate the position and angle of the particle's trajectory. The scintillators following the VDC provide both energy and timing information used in background reduction.

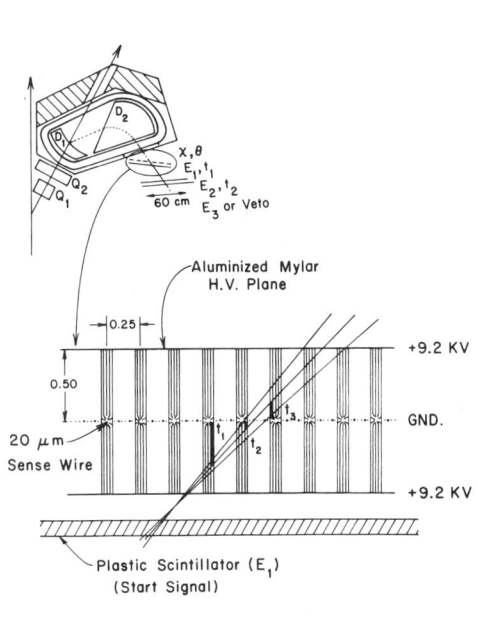

Figure 4: Vertical wire Drift Chamber measures both position and angle.

A typical pion spectrum (corrected for horizontal aberrations) from the reaction $^{12}C(p,\pi^+)^{13}C$, for a proton kinetic energy of 183 MeV and an outgoing pion lab angle of 28°, is shown in Figure 5. All of the known states up to 9.5 MeV in ^{13}C have been labeled. The 100 keV FWHM resolution corresponds to 2 mm displacement along the focal plane. The design goal for the QQSP magnet optics after software correction was 1 mm (FWHM), and intrinsic resolution of the VDC has been measured to be ~ 0.5 mm. We believe that the resolution measured so far has been limited by the energy

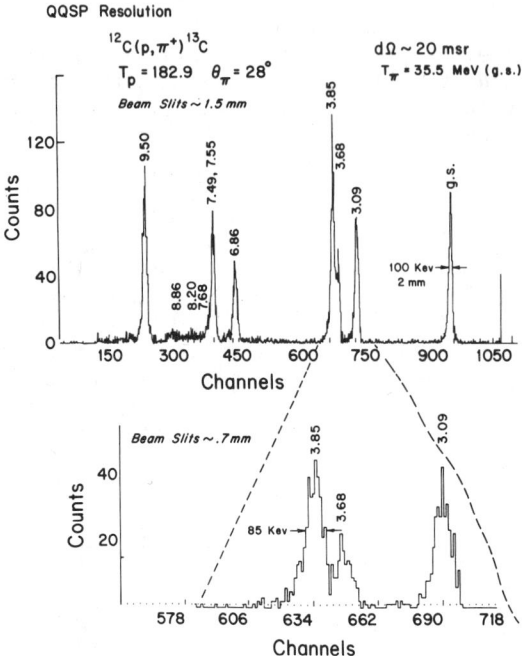

distribution of the incident proton beam. Shown in the lower half of Figure 5 is the 3-4 MeV excitation region for the same reaction. Here, the image and object slits of the beam analyzing magnet have been reduced by 50%. Some increase in resolution (FWHM ~ 85 keV) is seen, but this gain is at the expense of beam intensity which was reduced by more than 75% in this case.

Figure 5: Typical pion spectra from the QQSP.

RECENT RESULTS IN $(p,\pi^{\pm})$ AT IUCF

A. SIMPLE SPIN SYSTEMS

This past August a study (Exp. #130)[17] was begun of $A(p,\pi^{\pm})A+1$ where the reaction participants have sufficiently simple spins to allow one to "completely" determine the partial wave amplitudes from differential cross-section and analyzing power data. Using R-Matrix theory,[18] the energy dependence of the partial wave amplitudes expected from the Coulomb interaction of the incoming and outgoing reaction participants can be calculated. Comparisons of the energy dependence of the experimentally determined partial wave amplitudes to the calculated Coulombic energy dependence should leave a "residual" to be described by more complicated aspects of pion production.

Initial interest in this approach was generated by the suspicion that the energy dependence of this residual is rather weak. Marrs, Pollock, and Jacobs[3] found the total cross sections very near threshold for a number of light nuclei to scale in energy like phase space times a Coulomb penetrability factor. Also a preliminary R-Matrix analysis by the author of previously existing data for the $^{12}C(p,\pi^+)^{13}C_{gs}$ reaction could not rule out that below incident proton energies of 200 MeV the "residual" has no energy dependence.

This past summer additional data were taken for the $^{12}C(p,\pi^+)$ reaction at T_p = 190, 183, and 170 MeV and a complete analysis is now underway. Shown in Figure 6 are the data for $^{12}C(p,\pi^+)^{13}C_{gs}$ at T_p = 170 and 183 MeV from an on-line analysis. The pronounced positive analyzing power for the 170 MeV data in the backward hemisphere came as a surprise since previously only negative analyzing power values had been measured for $A(p,\pi^+)A+1$ below 200 MeV. With a 13 MeV increase in incident proton energy this feature disappears. The solid lines are fits to the data using an R-Matrix formalism as discussed earlier. The conclusion drawn is that the strong positive analyzing power naturally evolves from the 183 MeV data assuming <u>only</u> a Coulombic energy dependence. Data for $^{13}C(p,\pi^\pm)$ (at 170, 183, and 190 MeV) to simple spin states have also been taken and will be analyzed in a similar manner.

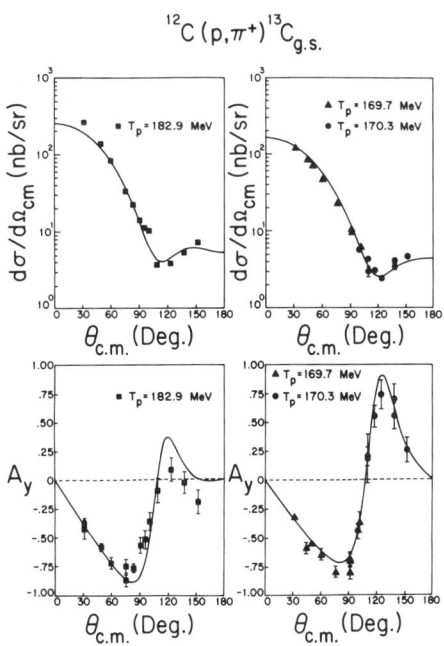

Figure 6: $^{12}C(p,\pi^+)^{13}C_{gs}$ cross section and analyzing power angular distributions for T_p = 183 and 170 Mev.

B. j-DEPENDENCE

Another experimental study (Exp. #142)[19] was begun at IUCF this past summer aimed at investigating the role played by two-nucleon processes in $(p,\pi^{\pm})$. If one examines the possible <u>free</u> N-N charged pion production processes; namely,

a) $p + p \rightarrow d + \pi^{+}$

b) $p + p \rightarrow p + n + \pi^{+}$

c) $p + n \rightarrow n + n + \pi^{+}$

d) $p + n \rightarrow p + p + \pi^{-}$

one notices that, in contrast to (p,π^{+}), where the pion can be produced by interaction with either a neutron or a proton, there is only one two-nucleon process which can contribute to (p,π^{-}). Assuming these fundamental NN processes play a dominant role in nuclear pion production, we[20] were led to predict a simple j-dependence for (p,π^{-}) analyzing powers, and a scaling of the (p,π^{-}) cross-section for an isotopic series of targets. Analysis of the data for the scaling test is now underway. Results from the on-line analysis of $^{12,13,14}C(p,\pi^{-})$ analyzing powers (see Figure 7) show a clear j-dependent difference in the sign of the analyzing power, at least at forward angles. The $^{13}C(p,\pi^{-})^{14}O_{g.s.}$ and $^{14}C(p,\pi^{-})^{15}O_{g.s.}$ transitions corresponding to a j transfer of 1/2 [and thus requiring the incident proton to interact with $p_{1/2}$ target neutron in the two-nucleon process (d) above] have a generally positive analyzing power; whereas the $^{12}C,(p,\pi^{-})^{13}O_{g.s.}$ transition, corresponding to a j-transfer of 3/2 (i.e., interaction with a $p_{3/2}$ neutron), has a negative analyzing power.

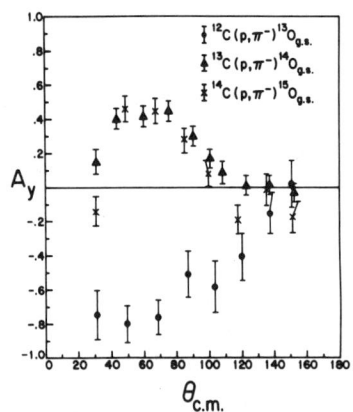

Figure 7: $C(p,\pi^{-})$ Analyzing power distributions to the final states: $^{13}O(gs, 3/2^{-})$, T_p=205 MeV, j=3/2^{-}; $^{14}O(gs, 0^{+})$, T_p=184 MeV, j=1/2^{-}; $^{15}O(gs, 1/2^{-})$, T_p=184 Mev, j=1/2^{-}; where for all cases $T_\pi \sim$ 30-40 MeV (preliminary on-line analysis).

This j-dependence of $A_y(\theta)$ may be understood from the point of view of two-nucleon processes, process (d) listed above, as follows. For nuclear final states where the two residual protons are coupled to spin zero, angular momentum and parity considerations require 1) the interacting proton and neutron to be in a spin triplet state, and 2) the struck neutron to be in a state within the target nucleus of uniquely defined spin and parity. Suppose now that distortions or other processes cause a preference for the reaction product to emerge on the same (or, alternatively, the opposite) side of the nucleus as the projectile. (It is a general property of distortions to introduce such a "sidedness" to reactions.) We note also that near-threshold pion production requires the Fermi motion of the struck nucleon to be toward the beam. With these general conditions, simple angular momentum coupling arguments lead to a prediction of a j-dependence in the (p,π^-) analyzing power (i.e., a difference in sign depending on whether the struck nucleon has $j=\ell+1/2$ or $j=\ell-1/2$), while no such j-dependence would be expected in general for (p,π^+), where conditions 1) and 2) above are not satisfied and the spin system is far less constrained). As these predictions for (p,π^-) analyzing powers follow directly from the assumed dominance of fundamental N-N processes in nuclear pion production, the new data in the figure support the validity of the two-nucleon mechanism.

FUTURE PLANS

There is yet a significant number of data tapes to be analyzed from the most recent experiments with the carbon isotopes. Once cross section and analyzing power angular distributions have been extracted, distributions from other simple spin states will be reduced to partial wave amplitudes providing additional information on the question of the dominance of phase space and Coulomb effects in (p,π^-) as well as (p,π^+). Completion of the analysis will also determine the extent to which the cross section angular distributions for (p,π^-) on ^{13}C and ^{14}C scale and the magnitude of the overall scale factor.

There are currently plans to take more data aimed at exhibiting the two-nucleon character of pion production. In addition to acquiring more data on the j dependence and scaling effects, it is planned to study specific final states where a direct comparison of (p,π^+) to (p,π^-) can be made by assuming charge symmetry and a two-nucleon production mechanism. The final state is chosen so as to "force" the π^+ reaction to be a two-nucleon process of the type (c). A simple prediction for such states is that a forward peaking (for example) of the cross section for (p,π^+) will correspond to a backward peaking in the charge symmetric configuration for (p,π^-). The analyzing power distributions may also be simply related (i.e., by a reflection about $90°_{cm}$ and a change in sign).

With its large momentum acceptance, the QQSP is an excellent instrument for studies of pion production in the continuum, and such plans for inclusive measurements are underway. In addition, there are plans to install proton detector telescopes outside the QQSP scattering chamber in order to perform (p,pπ) coincidence measurements. Here, it is hoped to establish the magnitude of the cross section, and if large, to use the kinematic specificity of the three body final state to investigate features of the pion production mechanism.

ACKNOWLEDGMENTS

K. Solberg worked diligently on the development the vertical wire drift chamber. L.C. Welch developed new Camac acquisition software partly in response to the particular needs of the QQSP project. The efforts of these and other IUCF staff who helped to make the QQSP project a success are greatly appreciated.

REFERENCES

1. R.D. Bent, P.T. Debevec, P.H. Pile, R.E. Pollock, R.E. Marrs, and M.C. Green, Phys. Rev. Lett. 40, 495 (1978).
2. P.H. Pile, R.D. Bent, R.E. Pollock, P.T. Debevec, R.E. Marrs, and M.C. Green, Phys. Rev. Lett. 42, 1461 (1979).
3. R.E. Marrs, R.E. Pollock, and W.W. Jacobs, Phys. Rev. C20, 2308 (1979).
4. R.E. Marrs and R.E. Pollock, Phys. Rev. C20, 2446 (1979).
5. F. Soga, R.D. Bent, P.H. Pile, T.P. Sjoreen, and M.C. Green, Phys. Rev. C22, 1348 (1980).
6. B. Hoistad, P.H. Pile, T.P. Sjoreen, R.D. Bent, M.C. Green, and F. Soga, Phys. Lett. 94B, 315 (1980).
7. T.P. Sjoreen, M.C. Green, W.W. Jacobs, R.E. Pollock, F. Soga, R.D. Bent, and T.E. Ward, Phys, Rev. Lett. 45, 1769 (1980).
8. W.W. Jacobs, A.G. Drentje, P.H. Pile, P.P. Singh, T.P. Sjoreen, and S.E. Vigdor, Phys. Lett. 94B, 319 (1980).
9. F. Soga, P.H. Pile, R.D. Bent, M.C. Green, W.W. Jacobs, T.P. Sjoreen, T.E. Ward, and A.G. Drentje, Phys. Rev. C24, 570 (1981).
10. T.P. Sjoreen, P.H. Pile, R.E. Pollock, W.W. Jacobs, H.O. Meyer, R.D. Bent, M.C. Green, and F. Soga, Phys. Rev. C24, 1135 (1981).
11. T.P. Sjoreen, P.H. Pile, R.D. Bent, M.C. Green, J.J. Kehayias, R.E. Pollock, F. Soga, M.C. Tsangarides, and J.G. Wills, Phys. Rev. C. 24, 2569, (1981).
12. B. Hoistad, contribution to this conference.
13. R.D. Bent, P.H. Pile, R.E. Pollock, and P.T. Debevec, Nucl. Instrum. Methods 180, 397 (1981).
14. P.H. Pile and R.E. Pollock, Nucl. Instr. and Meth. 165, 209 (1979).

15. P.H. Pile, "Near Threshold Positive Pion Production by Protons on Nuclei", (Ph.D. Thesis), IUCF Internal Report 78-9
16. W. Bertozzi, M.V. Hynes et al., Nucl. Instr. and Meth. $\underline{141}$, 457 (1977).
17. IUCF Experiment 130 Spokesman: M.C. Green
18. A.M. Lane and R.G. Thomas, Rev. Mod. Phys. $\underline{30}$, 257, (1958).
19. IUCF Experiment 142 Co-spokesmen: W.W. Jacobs and T.G. Throwe
20. S.E. Vigdor private communication.

NEAR THRESHOLD PROTON INDUCED NEUTRAL PION PRODUCTION FROM DEUTERIUM

Mark A. Pickar
Indiana University Cyclotron Facility,
Bloomington, IN 47405

ABSTRACT

An experiment wherin the angular distribution of the cross section and analyzing power for the reaction $^2H(p,\pi^\circ)^3He$ was measured in the extreme threshold region will be discussed.

INTRODUCTION

One aspect of the (p,π) reaction that has seen no really significant advances in the past twenty years, has been our understanding of the role played by s wave pion production. The most recent calculations for $pp \leftrightarrow d\pi^+$[1] differ little from the earliest calculations[2] in the way s wave pion production is handled. The calculations indicate that this type of production does not begin to dominate until $k_\pi/m_\pi \equiv \eta < 0.5$. Hence it is data in this region that is most useful in testing models for s wave pion production. Unfortunately there is very little detailed information available for $NN \leftrightarrow NN\pi$ in this energy region[3,4]. In fact, there is presently more data available for (p,π) on heavy targets in this region than there is for $pp \leftrightarrow d\pi^+$ [5,6,7]. Extraction of information on production mechanisms from such systems, however, is complicated by uncertainties in how to handle the wavefunctions in the entrance and exit channels. Studies of pion production from few nucleon systems should thus prove more useful in investigating reaction mechanisms, as the initial and final state wavefunctions are far better understood. Unfortunately, there has been essentially no data available until most recently on such systems. In the following is discussed an experiment investigating pion production from deuterium in an energy region extremely close to threshold.

THE EXPERIMENTAL TECHNIQUE

The reason one chooses the reaction channel in which a neutral pion is emitted is easily understood upon an examination of the kinematics (Fig. 1). Near threshold the energy of the emitted pion is small, as is its emission angle, making the pion, charged or uncharged, a poor choice as the principal probe. However, the range of energies of the recoil are sufficient to allow a detailed study of the reaction, provided one has a spectrometer of moderate resolution capable of making zero degree measurements. The fact that the angles of emission are small is here a benefit, as then the angular acceptance of the detector need not be too large in order to obtain an efficiency close to unity. Further, the principal decay mode of the associated $\pi^°$ makes it a very effective coincidence condition with which to cut down background.

The apparatus with which the experiment was performed is illustrated in Fig. 2. The major component of the system was the QDDM spectrograph, which has been used extensively in other reaction studies at IUCF. A number of modifications, however, had to be made to this device in order to perform the experiment. A zero degree measurement requires stopping the beam within the spectrometer itself. A special Faraday cup was designed for this purpose. This, of course, generates a large background in the focal plane. As the standard focal plane arrangement was unable to handle these high rates, it was replaced with a scintillator hodoscope with ΔE, E, and Veto planes. This device was divided into 11 bins, each having a width of less than 400 kev for the recoils of interest. In order to cut down background from the carbon in the CD_2 targets used, and from the cup, a coincidence was required with either or both of two large photon detectors placed on each side of the target. These photon detectors were constructed of lead glass blocks and hence were essentially insensitive to any radiation other than high energy photons and electrons. To get increased efficiency the standard scattering chamber was replaced with one of much smaller dimensions, permitting the detectors to be positioned closer to the target. The solid angle of these detectors was well defined by thick lead collimators. Throughout the experiment the target thickness and beam polarization were monitored by range telescopes placed on each side of the beam.

The events associated with the reaction $^2H(p,\pi^°)^3He$ were identified by:
1) A coincidence between the ΔE and E elements of the hodoscope with either of the photon detectors, and no signal from the Veto element.
2) Recoil time of flight against the cyclotron RF.
3) Recoil time of flight against the photon detectors.
4) Photon time of flight against the cyclotron RF.
5) Recoil pulse height in the hodoscope elements.

The spectra resulting from these conditions contained essentially no background.

Extraction of cross sections and analyzing powers from these data is a straightforward although somewhat tedious task. It is very important near threshold to know the beam energy as precisely as possible. This was accomplished by measuring the width of the recoil energy spectrum. The angular acceptance of the QDDM was not defined by its slits for much of the experiment, and hence had to be accurately determined during the experiment itself. The hodoscope had to be accurately calibrated. The efficiency of the photon detectors for detecting neutral pions in a given geometry had to be carefully calculated. Finally corrections had to be computed for distortions induced by such effects as finite target thickness, a finite spread in beam energy, and slit scattering in the photon collimators.

RESULTS

In Figures 3 and 4 are illustrated the results of a first order analysis on only a small portion of the data. A full analysis of the data is near completion and indicates these first order results are not changed substantially by the inclusion of higher order corrections, the principal effect being a slight change in overall normalization.

The most striking aspect of these data one first observes is how quickly the cross sections (Fig. 3) depart from isotropy as one rises above threshold. For a pion energy of 0.46 Mev one observes an ~ 2:1 ratio between the 0° and 180° cross sections, rising to ~ 4:1 for 1.40 Mev pions, and to ~ 6:1 for 2.71 Mev pions. Next one sees that the cross section at back angles varies much less rapidly with energy than does the cross section at forward angles. In contrast with the cross section, the analyzing power (Fig. 4) appears to vary little with either energy or angle. It is small, but significantly different from zero, staying fixed at about -0.15 over the energy rang 0.5-2.7 Mev.

In the near threshold region, where only s and p wave pion production might be expected to be significant, one expects the spin dependent cross section to be well described by an expansion in a few Legendre polynomials, i.e.,

$$d\sigma/d\Omega|_\alpha = a_0 P_0(x) + a_1 P_1(x) + a_2 P_2(x)$$
$$+ \vec{P}_\alpha \cdot \hat{n} \{ b_1 P_1^1(x) + b_2 P_2^1(x) \}$$

with
$$x \equiv \cos\theta$$
$$d\sigma/d\Omega|_0 = a_0 P_0(x) + a_1 P_1(x) + a_2 P_2(x)$$
$$A_y \, d\sigma/d\Omega|_0 = b_1 P_1^1(x) + b_2 P_2^1(x) .$$

The strengths of the coefficients a and b and their variation with pion momentum give one a better measure of the behavior of the various transition amplitudes than does an examination of the cross sections and analyzing powers alone.

In Figures 5 and 6 are illustrated the preliminary results of a full analysis on the data. The uncertainties indicated are statistical only and, unless otherwise indicated, are no larger than the size of the symbols used. The curves drawn through the coefficients illustrate the expected momentum dependence in that asymptotic region where

s wave production >> p wave production >> d wave production ,

and assume that the momentum dependence of a given amplitude is determined purely by phase space. In such a region one expects,

$a_0 >> a_1 >> a_2$, $b_1 >> b_2$

$a_0 \propto p^2$
$a_1 \propto p^3$ $\quad b_1 \propto p^2$
$a_2 \propto p^3$ $\quad b_2 \propto p^3$, where $p \equiv$ pion momentum in the center of mass.

The data appear to be reasonably consistent with this behavior, however there are significant deviations. One first notes that for $p_\pi > 15$ Mev/c, the coefficient a_1 is not much less than a_0, that in fact it very rapidly becomes comparable to a_0. Even though the coefficient a_2 is small, it is still larger than one would naively expect using $l_{max} \cong kR < 0.3$ as an estimate of the role played by emission of pions with angular momenta greater than 0. The fact that emission of such pions is important this close to threshold is further substantiated by the size of the coefficients b_1 and b_2. Based on these results one can not help but conclude that the production of pions with angular momenta greater than 0 plays a significant role extraordinarily close to threshold.

This fact somewhat complicates the problem of extracting information about s wave pion production directly from the data. However, one can immediately make two observations. First, s wave production appears to be small and dominant only close to threshold because of the speed with which p wave, or higher, pion production is observed to contribute to the anisotropy of the cross section. Second, the momentum dependence of s wave pion production does not appear to be described completely by phase space alone. One sees this most clearly upon examining the momentum dependence of a_0. The coefficient a_0 is a sum of the squares of the amplitudes describing the reaction, i.e.,
$$a_0 \sim c_0|s|^2 + c_1|p|^2 + c_2|d|^2 + \ldots,$$
where the coefficients c are all nonnegative. If the possible transitions were described by phase space alone one would find
$$a_0 \sim d_0 p + d_1 p^3 + d_2 p^5 + \ldots, \quad p \equiv \text{pion CM momentum},$$
and the coefficients d all nonnegative. Thus, such a model predicts the coefficient a_0 to curve <u>upward</u> away from a line proportional to p_π as the higher order partial waves start to contribute. What one sees in the data, however, is that the data appear to curve slightly <u>downward</u> away from a line proportional to p_π as p_π increases. Further, one also observes that p wave, and higher wave, pion production plays an important role because of the strength of the coefficients other than a_0, particularly a_1. Hence one knows that there must be contributions to a_0 from other than s wave transitions. Thus, if those effects are subtracted out, one finds evidence that the momentum dependence of the s wave transition cannot be described by phase space alone, that is, that it appears to curve <u>downward</u> away from that behavior determined by phase space. This kind of behavior was predicted in a Fadeev type calculation done by Afnan and Thomas for $pp \to \pi^+ d$,[8] and some support for this behavior was found in a fit to the data performed by Spuller and Measday[9]. The data discussed here, however, appear to be the first direct evidence of such an effect.

In conclusion, we find that the emission of pions with angular momenta other than 0 plays an important role in $pd \to \pi^0 \, ^3\text{He}$ extraordinarily close to threshold, and that s wave production appears to vary somewhat differently than phase space. It is hoped that these unique data will be of use in future theoretical studies of this intriguing problem.

ACKNOWLEDGEMENTS

I would like to thank my collaborators R. E. Pollock, H. O. Meyer, A. D. Bacher, and G. T. Emery for their help and support in this rather difficult experiment. I would also like to thank the IUCF technical staff for their rather extensive efforts in making this experiment possible.

REFERENCES

1. O. V. Maxwell, W. Weise and M. Brack, Nuc. Phys. A348(1980)388.
2. A. E. Woodruff, Phys. Rev. 117(1960)1113.
3. C. M. Rose, Phys. Rev. 154(1967)1305.
4. P. Walden, D. Ottewell, E. L. Mathie, T. Masterson, G. Jones, R. R. Johnson, H. Haynes and E. G. Auld, Phys. Lett. 81B(1979)156.
5. R. E. Marrs, R. E. Pollock, and W. W. Jacobs, Phys. Rev. C20(1979)2308.
6. P. H. Pile, thesis (unpublished).
7. P. H. Pile, R. D. Bent, R. E. Pollock, P. T. Debevec, R. E. Marrs, T. P. Sjoreen and F. Soga, Phys. Rev. Lett. 42(1979)1461.
8. I. R. Afnan and A. W. Thomas, Phys. Rev. C10(1974)109.
9. J. Spuller and D. F. Measday, Phys. Rev. D12(1975)3550.

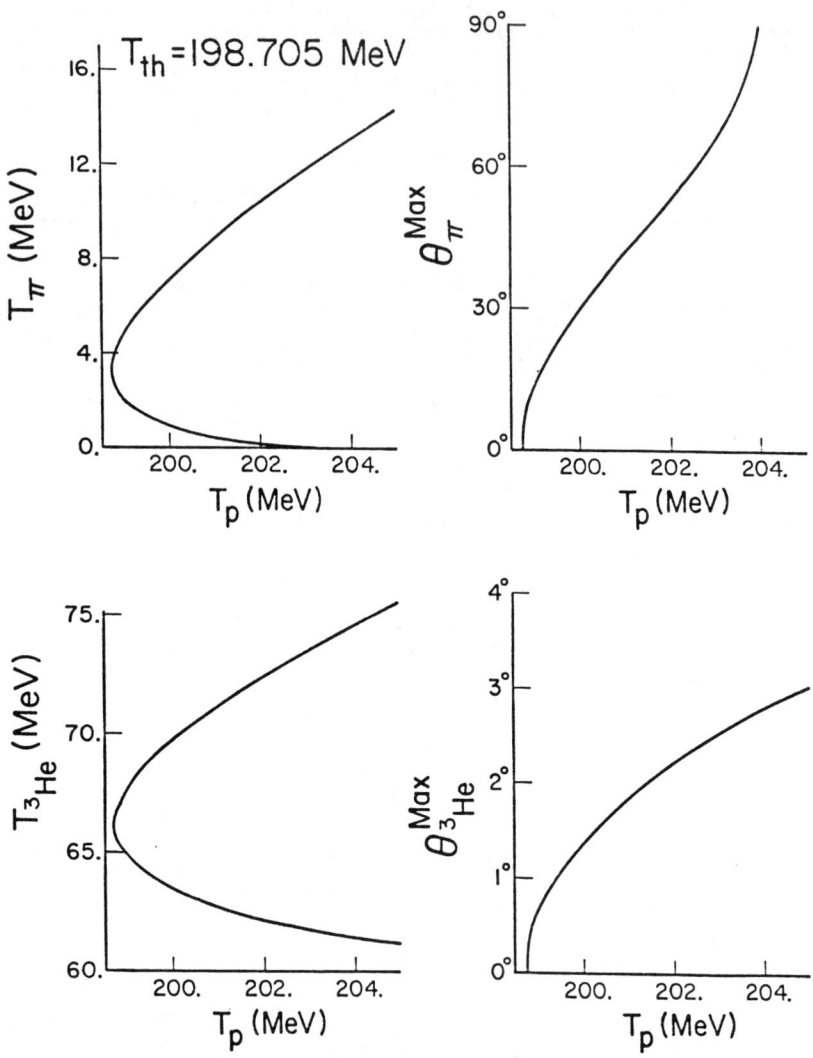

Fig. 1 Diagram illustrating the near threshold kinematics for the reaction $pd \to \pi^0\, ^3He$.

149

Principal Features of the Geometry

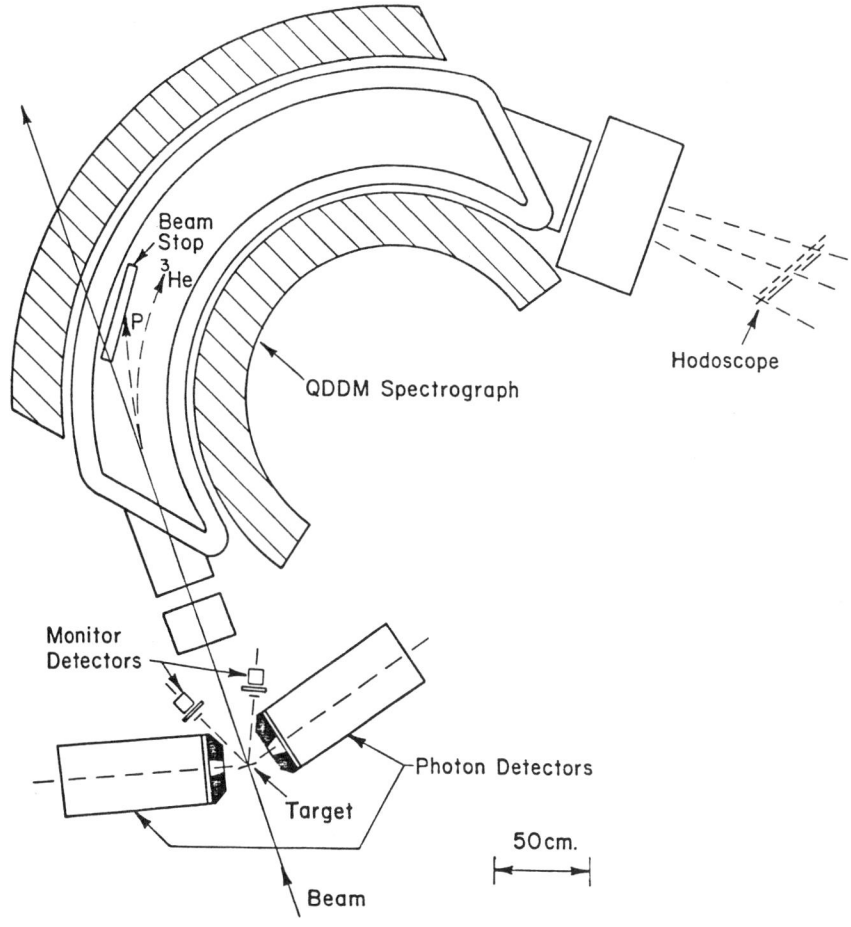

Fig. 2 Diagram illustrating the principal features of the experimental setup.

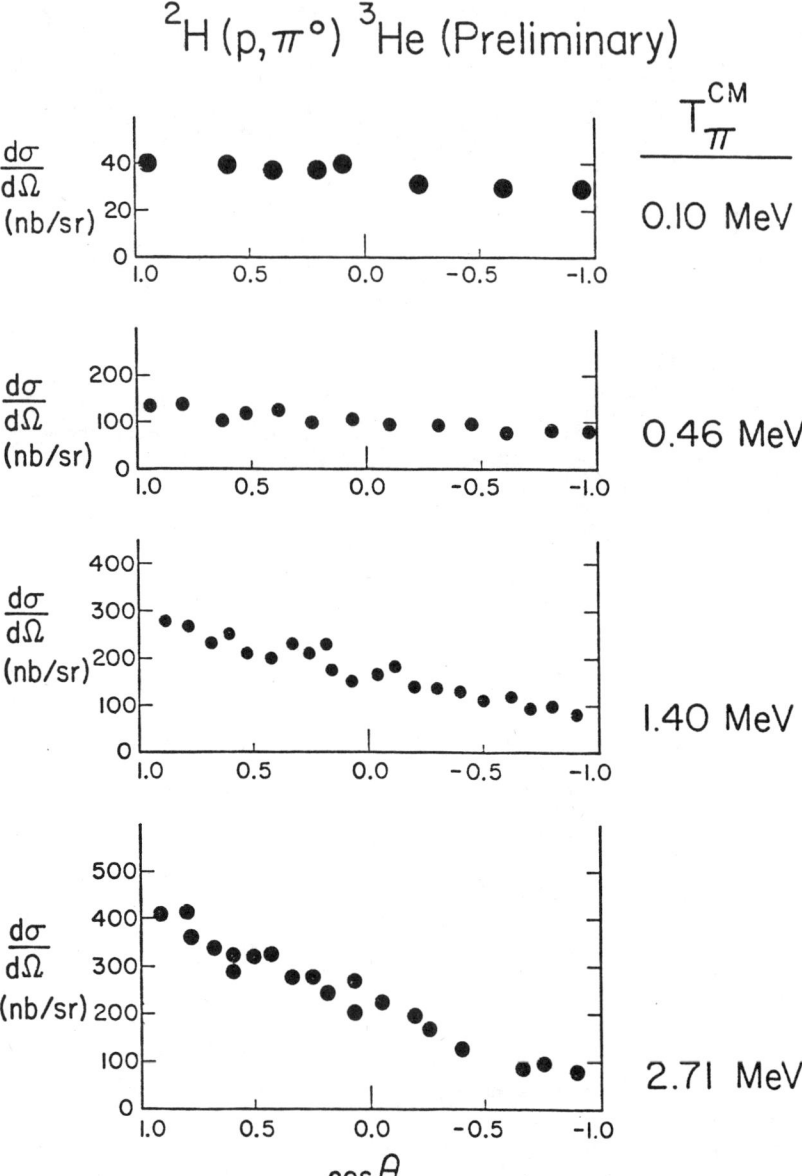

Fig. 3 Pion differential cross section as determined by a first order analysis of a small fraction of the data.

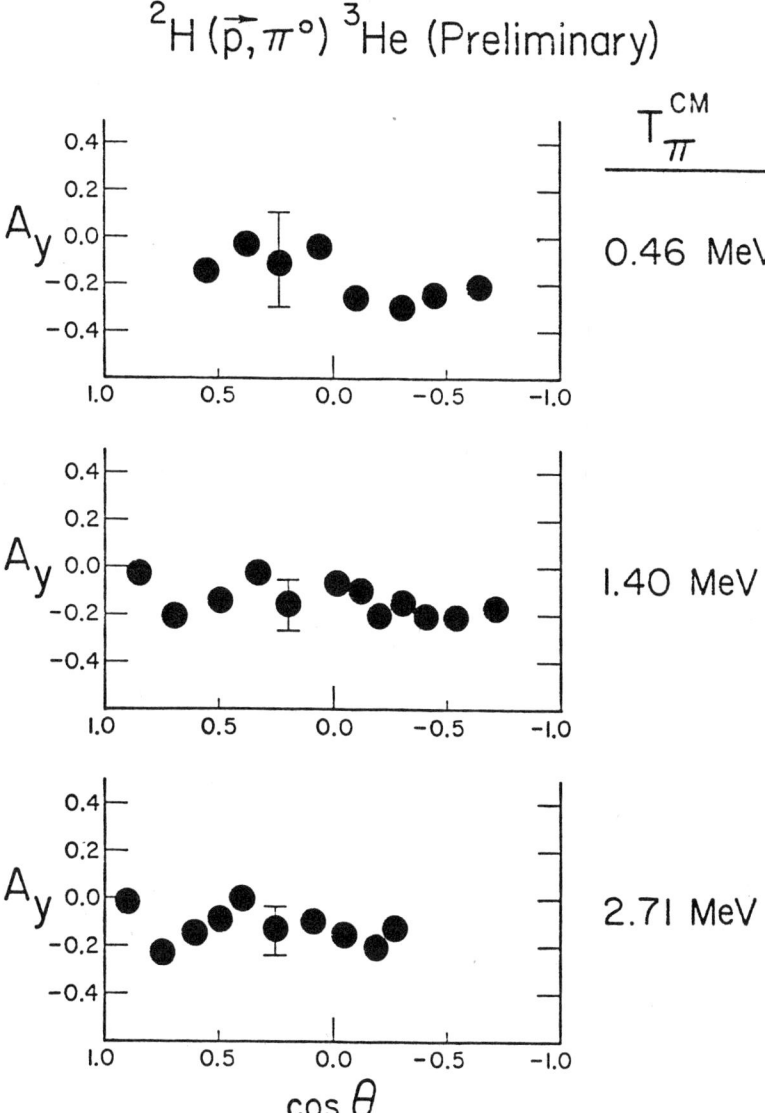

Fig. 4 Pion analyzing power as determined by a first order analysis of a small fraction of the data.

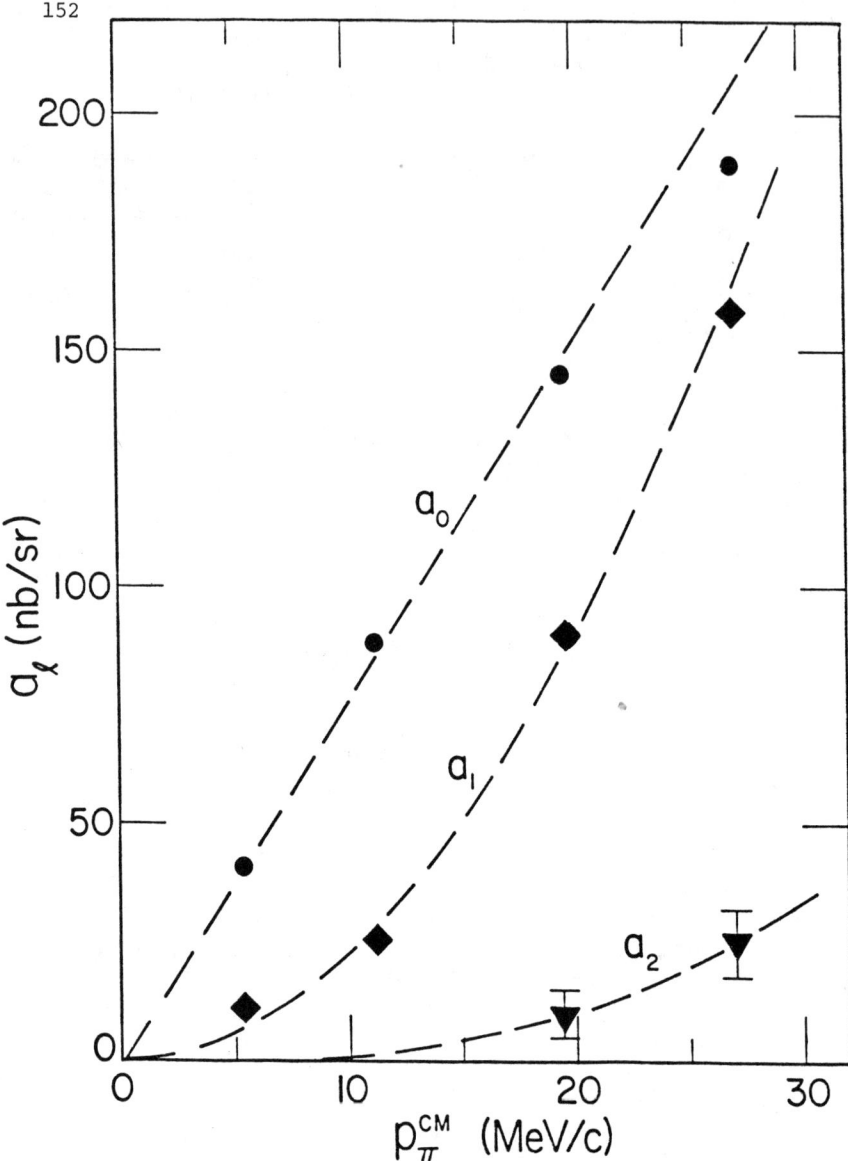

Fig. 5 Variation of the coefficients a_l describing the pion differential cross section with the pion momentum in the center of mass.

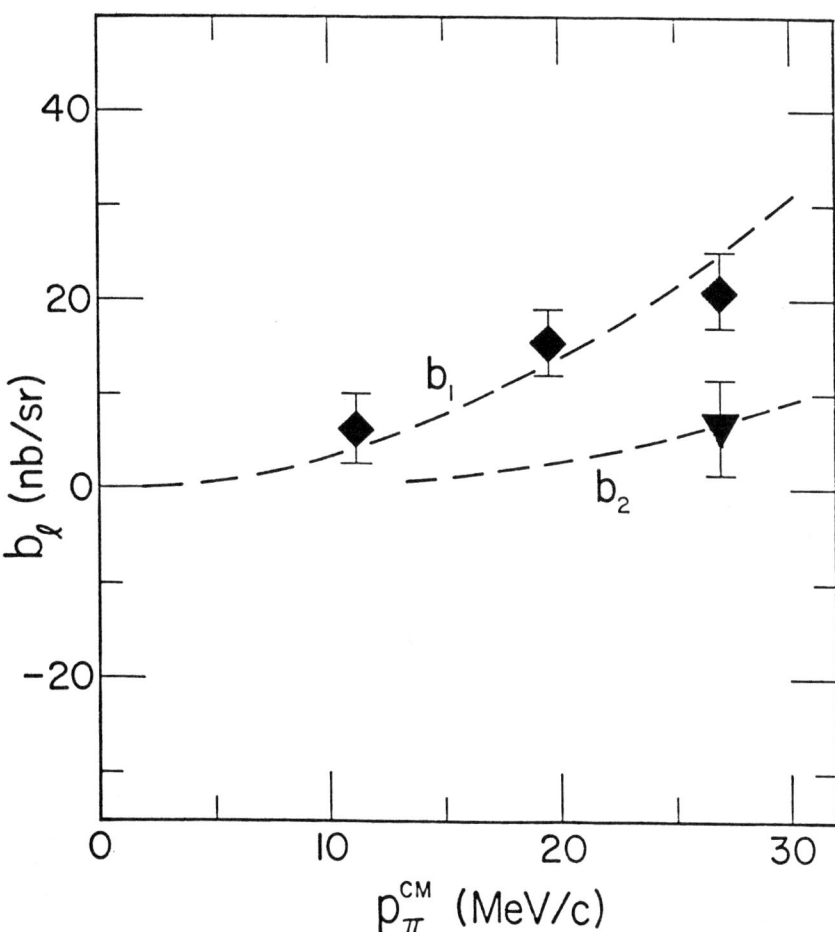

Fig. 6 Variation of the coefficients b_1 describing the product $A_y d\sigma/d\Omega$ with the pion momentum in the center of mass.

ORSAY AND SATURNE NEW RESULTS
ON (p,π) AND (ION,π) EXPERIMENTS

Y. Le Bornec and N. Willis
Institut de Physique Nucléaire, BP n°1
91406 Orsay, France.

ABSTRACT

New results of (p,π) reaction on light target nuclei (^{3}He, ^{4}He, ^{6}Li, ^{10}B) have been obtained at IPN Orsay. Data on (^{3}He,π$^+$) reaction on the same targets, in the exclusive kinematical region are presented together with data on ^{6}Li(d,π$^-$)^{8}B reaction obtained at Saturne.

INTRODUCTION

High momentum transfer processes such as coherent (p,π) or (p,d) reactions have been investigated intensively in medium energy nuclear physics over the past decade but there is still much controversy surrounding the basic reaction mechanisms. In order to disentangle the mechanism from the nuclear structure, we have performed (p,π) experiments [1] on light nuclei for which the wave functions are considered well-known. It could be that further valuable clues are also provided by reactions which involve the transfer of several nucleons and for that purpose we have started a program to study coherent pion production with composite projectiles. The hope was therefore to determine to what extent the different nucleons of the projectile and the target are collectively involved in the production of the pion.

In this paper experimental data on (^{3}He,π) [1] and (d,π) [2] reactions near and below the threshold for production in free NN → NNπ reactions will be presented.

I - EXPERIMENTAL PROCEDURE

1° The synchrocyclotron. After a shutdown for change over, the rebuilt Orsay synchrocyclotron has been operating since the end of 1978. Proton, deuteron, ^{3}He and ^{4}He external beams have been delivered. The main characteristics of the machine are presented in the table n°1.

Moreover, for one year, the energy has been continuously variable between the lower and the upper values given in the table. A scheme of the machine and of the experimental areas is presented in figure n°1.

A duty cycle up to 40 % can be obtained when using the slow extraction system. The energy dispersion of the extracted beam is about $\frac{\Delta E}{E} = \pm 3.5 \times 10^{-3}$. However, slits on the analysed beam line (used for the π production experiments) permit one to reduce this value down to $\pm 3 \times 10^{-4}$. An overall resolution of 40 keV has been achieved with 201 MeV protons by means of an energy loss spectrometer located on the other beam line.

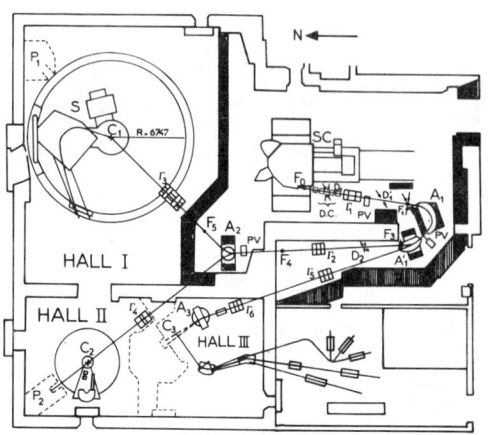

Fig. 1. Orsay synchrocyclotron : the machine and the experimental areas.

On line spectroscopic measurements are performed on the third line "Isocele".

2° **Experimental set-up for pions production experiments.**
The layout is shown in figure 2. The pions were focussed by a quadrupole doublet onto a scintillator A located in the object focal plane of a 180° spectrometer (radius R = 57.5 cm). This makes possible the analysis of particles with a maximum magnetic rigidity of .9Txm with a solid angle $\Delta\Omega = 6.2 \times 10^{-3}$ sr. This last value is for the studied pions instead of 6×10^{-5}Sr for particles of different momenta coming from the target. This factor of 100 is important for counting rates in detector A.

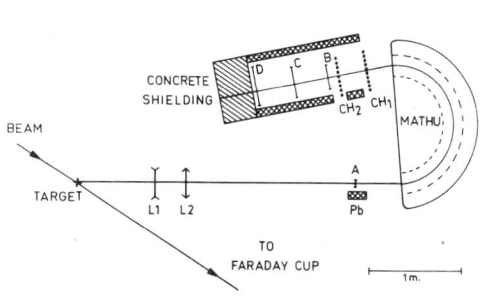

Fig. 2. Experimental set-up.

The momentum acceptance is about ± 2.5 %. The entire apparatus can be rotated around the target axis between 20° and 155°. The whole flight path of the pions is about 6.8 meters, the main part of which is in a continuous vacuum. It has to be remarked that all the elements of the spectrometer were salvaged from other equipment and that this whole apparatus was quite inexpensive.

The incident beam is in vacuum up to the Faraday cup in the beam dump. An additional relative monitoring is furnished by a 3 scintillators telescope which views the target.

The particles trajectories were determined by means of two multiwires chambers with cathode readout providing a spatial resolution of about .3 mm. These chambers which accept high counting rates are triggered by a four fold coincidence A.B.C.D.

Event-by-event calculation of the focal plane position and the trajectory angle relative to the optical axis provided the on-line monitoring of the experiment. Only those particles satisfying conditions on the time of flight between A & B and B & D and vertical position in the chambers were considered. However all events were kept for storage on magnetic tape to allow replay and optimal event selection. Dead times are measured with a pulse generator on each photomultiplier and chamber.

The reactions involving ^{3}He or ^{4}He targets were studied with the help of the Orsay Cryogenic helium target, the walls of which are very thin ($\sim$ 12 μm steel). Target-empty runs determined the background from interactions in the target walls.

II - (p,π) RESULTS

A - TYPICAL SPECTRA

1° ^{3}He(p,π$^+$)^{4}He and ^{4}He(p,π$^+$)^{5}He. Details on ^{3}He(p,π$^+$)^{4}He experiment can be found in ref. [3] and we just briefly summarize here some points.

On-line typical spectra (without background subtraction) are shown on figure 3 for T_p = 201 MeV, with an average intensity of 130 nA on a 110 mg/cm^2 ^{3}He target and 103 mg/cm^2 ^{4}He target.

The pion energy was typically 42 MeV in the case of ^{3}He(p,π$^+$)^{4}He reaction and 24 MeV for the ^{4}He(p,π$^+$)^{5}He.

The ^{5}He nucleus is unbound and not well known (position of the first excited state 4 ±1 MeV and width 4± 1 MeV) so that the peak corresponding to the ground state could not be precisely extracted. The dashed line represents a phase space calculation for the 3 body reactions subtraction (6).

2° ^{6}Li(p,π$^+$)^{7}Li and ^{10}B(p,π$^+$)^{11}B. These spectra (figure 4)were obtained in measurements of only 30 minutes with a 150 nA beam intensity for the 43.25 mg/cm^2 ^{6}Li target thickness. The energy resolution was about 300 keV. It can be noticed that the 7/2$^-$ (4.63 MeV) level is highly excited despite the fact that the one step process is strongly suppressed. This already was observed at 600 MeV in Saclay experiments [4] a few years ago.

It is well known that the high spin levels are generally the most excited states in the (p,π) reactions.

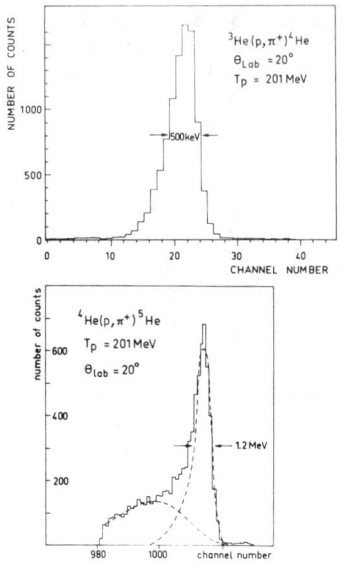 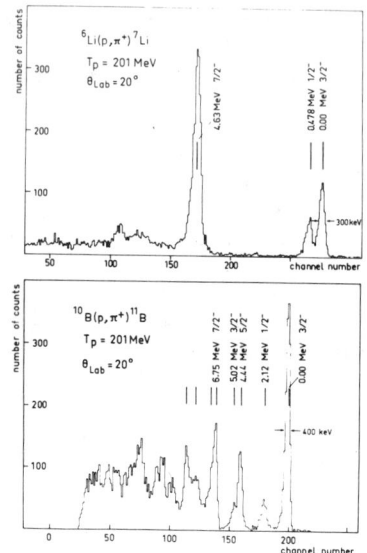

Fig. 3. ^{3}He(p,π$^+$)^{4}He and ^{4}He(p,π$^+$)^{5}He typical spectra at T_p = 201 MeV, θ_{lab} = 20°.

Fig. 4. ^{6}Li(p,π$^+$)^{7}Li and ^{10}B(p,π$^+$)^{11}B typical spectra at T_p = 201 MeV, θ_{lab} = 20°.

B - ANGULAR DISTRIBUTIONS

The angular distributions of the ^{3}He data at several energies are shown in the figure 5 where the error bars take into account statistical effects and uncertainties due to the overlapping of the three different magnetic fields which lead to the same peak. This effect becomes important for the lowest energy pions and hence for the largest angles. The absolute cross sections are obtained with an overall uncertainty of ± 20 %. The main uncertainties are due to the determination of the solid angle, the detection efficiency, the target thickness, and the beam monitoring.

It should be pointed out that the cross sections are high, of the order of magnitude of several μb/sr. The angular distributions are structureless and smooth with θ_{cm} variation. The transferred momentum at forward angles which is not very energy dependent is typically 2 fm^{-1}. The same type of comments can be done for the angular distribution on ^{4}He and the cross sections are of the same order of magnitude. Figure 6 represents the cross section variations versus the incident energy at an angle of 20°. The data from Saturne [5] are also presented. One can see the bump probably due to the influence of the (3,3) resonance and the strong decrease of the

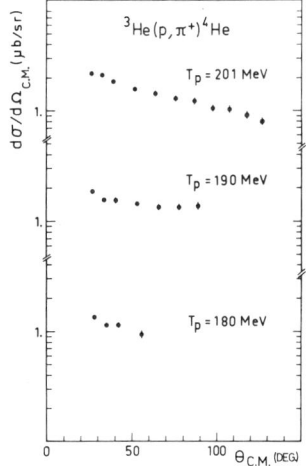
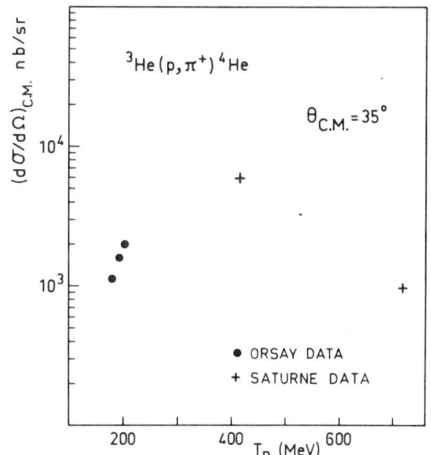

Fig.5. Angular distributions for ^{3}He(p,π^+)^{4}He reaction at different incident energies.

Fig.6. Differential ^{3}He(p,π^+)^{4}He cross-section versus incident energy. Crosses correspond to data from ref.(5) and circles correspond to our data.

cross sections at lower energies, partly due to the decrease of the phase space factor.

The range of pion energies covered allows us to make an extrapolation to zero energy so that a comparison with the results obtained from pionic atoms becomes possible.

In terms of a centre-of-mass amplitude f and momentum k^*, the unpolarised pion production cross section is

$$(d\sigma/d\Omega)^*_{p^3He \to \pi^4He} = (k^*_\pi/k^*_p)\overline{|f^2|} \qquad (1)$$

where the bar denotes averaging over the initial proton and ^{3}He spins. At threshold the only contributions to the imaginary part of the elastic $\pi\alpha \to \pi\alpha$ amplitude, calculated via the optical theorem, comes from the absorption cross section so that

$$k^*_\pi \sigma_{\pi\alpha \to abs} = 4\pi \, \text{Im}(f_{\pi\alpha \to \pi\alpha}) \qquad (2)$$

where the limit $k^*_\pi \to 0$ is understood. The right hand side may be estimated from the pionic atom shifts and widths [7] which give

$$\mathrm{Im}(f_{\pi\alpha \to \pi\alpha}) = 0.042 \pm 0.003 \text{ fm}$$

The branching ratio in the pionic atom to the particular nt channel is [8] [9] B = (19 ± 1) % and, assuming that the capture takes place from the s orbit, this enables us to calculate the nt production rate

$$k_\pi^* \sigma_{\pi\alpha \to nt} = 4\pi B \mathrm{Im}(f_{\pi\alpha \to \pi\alpha}) \qquad (3)$$

Since there is no angular dependence in the threshold cross sections we can then extract the nt → πα scattering amplitude, defined in equation (1), through the use of detailed balance:

$$\overline{|f^2|}_{k_\pi^* = 0} = (B/4\ k_p^*)\ \mathrm{Im}\ (f_{\pi\alpha \to \pi\alpha})$$

$$= (9.4 \pm 0.7) \times 10^{-4}\ \mathrm{fm}^2$$

This value can be compared to the values (included in the table n°2) deduced from our data.

Angular distributions on ^{6}Li et ^{10}B targets are plotted in the figure n° 7. Results from Indiana are in good [21] agreement with ours on the ^{10}B target.

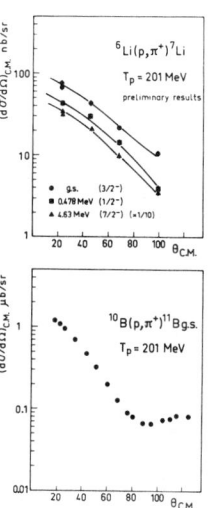

We have just to mention the strong decrease of cross - sections by about an order of magnitude between ^{3}He, ^{4}He and ^{6}Li.

Fig. 7. Angular distributions for ^{6}Li(p,π$^+$)^{7}Li and ^{10}B(p,π$^+$)^{11}B reactions at T_p = 201 MeV.

III - (^{3}He,π) AND (D,π) REACTIONS

Although a great deal of experiments (cross section and asymetry measurements) in proton induced pion production is now available, few results on exclusive production with heavier projectiles have been reported until now. The most recent ones are from LAMPF [10] with (π$^+$d) studies at T_π = 48 MeV on light nuclei and from CERN by Aslanides et al. [11] with (^{3}He,π$^-$) measurements done on ^{6}Li at 900 MeV incident energy. This last one shows evidence for exclusive final states with a cross section of about 10 pb/sr. The kinetic energy/nucleon of the projectiles for experiments that we have carried out in Orsay and at Saturne was below the threshold for the production on a free proton. Moreover the transferred momenta are important, so that very low cross sections were expected

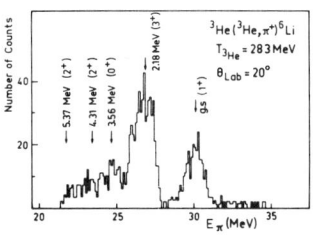

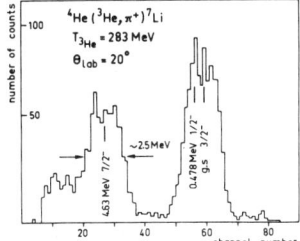

Fig. 8. Typical spectra for ^{3}He(^{3}He,π$^+$)^{6}Li and ^{4}He(^{3}He,π$^+$)^{7}Li at $T_{^3He}$ = 283 MeV.

A - (^{3}He,π). The first set of measurements we will present was performed at the synchrocyclotron in Orsay with ^{3}He projectiles. The incident energy was about 90 MeV/nucleon. The data were taken in the same experiments as the (p,π) data on the same targets.

Typical spectra for ^{3}He(^{3}He,π)^{6}Li and for ^{4}He(^{3}He,π$^+$)^{7}Li at $T_{^3He}$ = 283 MeV are shown on figure n° 8. For this type of experiment the different energy losses of the ^{3}He and π through the target was the main contribution to the experimental peak which was found to be ∿ 1.5 MeV. In the case of the ^{3}He target, the ground state (1$^+$), 2.18 MeV (3$^+$) and 3.56 MeV (0$^+$) levels are resolved. The relative excitation of the different levels will be discussed later on. The important excitation of the 3$^+$ level (2.18 MeV) should be noticed. For the ^{4}He(^{3}He,π$^+$)^{7}Li reaction it can be seen with the preliminary results that the ground state (3/2$^-$) and 0.478 MeV (1/2$^-$) level were not resolved. The 4.63 MeV (7/2$^-$) state is clearly seen. A measurement was also made for pions beyond the kinematical limit and as it can be seen the signal/background ratio is quite good.

Corresponding angular distributions are shown in figure n° 9. The error bars include statistical uncertainties. A systematic uncertainty of ± 20 % was found due to beam calibration, target thickness, solid angle and efficiency determinations.

Several features can be emphasized.

1° The cross sections have about the same order of magnitude (∼ a few tens of nb/sr) on ^{3}He and ^{4}He targets. This is quite high yield at such a low energy with transferred momenta of about 3 fm^{-1}. It must be kept in mind for comparison that (p,π) reactions cross sections are about two orders of magnitude higher with transferred momenta of about 2 fm^{-1}.

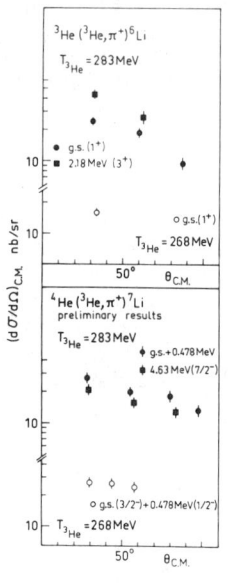

Fig. 9. ^{3}He(^{3}He,π$^+$)^{6}Li and ^{4}He(^{3}He,π$^+$)^{7}Li angular distributions at different energies.

The low pion energy in ^{3}He(^{3}He,π$^+$)^{6}Li reaction allows us to make a comparison with the results obtained from pionic atoms for the ^{6}Li ground state as we did earlier in the case ^{3}He(p,π$^+$)^{4}He reaction. With very simple approximations described in ref. [12] agreement with experiment is as good as could be expected i.e. the amplitudes in the centre of mass system are of the same order of magnitude.

2° The ratio R of the cross sections yielding the 2.18 MeV and ground states of ^{6}Li at the same laboratory angle is about 1.7. In low energy transfer reactions, where the mechanism could be completely different, values of the same order of magnitude are found [13]. In the case of ^{4}He(^{3}He,π$^+$)^{7}Li the 7/2$^-$ excited state and the two first levels doublet are equally excited. Any theoretical model has to reproduce these features.

The (^{3}He,π) reaction on ^{6}Li and ^{10}B targets has also been investigated at 260 MeV, 270 MeV and 283 MeV at θ$_{lab}$ = 20°. A typical spectrum on a ^{6}Li target is shown in figure 10. The ground state (3/2$^-$) and 2.43 MeV (5/2$^-$) levels are clearly resolved. The peak/background ratio is fairly good. This spectrum was obtained in 16 hours with an average intensity of 350 nA. A very preliminary analysis leads to cross sections of about 100 pb/sr for the ground state of ^{9}Be which is more than two orders of magnitude lower than cross sections on ^{3}He and ^{4}He targets. Nevertheless the transferred momentum is about 3.6 fm^{-1}.

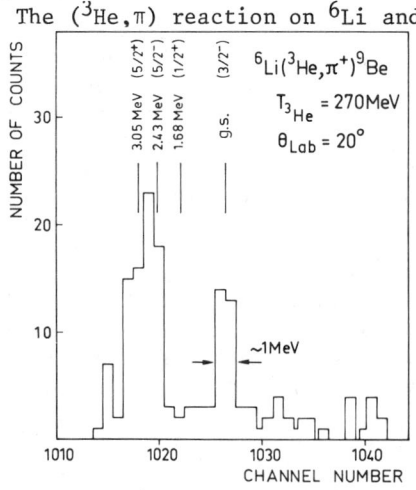

Fig. 10. Typical spectrum for ^{6}Li(^{3}He,π$^+$)^{9}Be reaction at T$_{3He}$ = 283 MeV.

B - (D,π) REACTION

Another experiment using deuteron beam was carried out at the Saturne National Laboratory (LNS) by a collaboration CRN Strasbourg, IPN Orsay and DPhNME/Saclay with the high resolution spectrometer SPES I. (d,π^-) reactions have been studied at 150 MeV/nucleon and 300 MeV/nucleon on ^{6}Li, ^{9}Be and ^{10}B targets [14]. The detection of π^- instead of π^+ minimizes the background due to the target through the spectrometer. Due to the very low counting rates, inclusive spectra only are measured on ^{9}Be and ^{10}B near the kinematical limit. Both exclusive and inclusive spectra were obtained on the ^{6}Li target.

We briefly describe the experimental set-up which will be covered in greater detail, by P. Couvert [15]. The basic detection system consisted of five planes of scintillation hodoscopes and three lucite Cerenkov counters. The particle trajectories were determined with 4 two fold drift chambers triggered by a coincidence of the plastic counters. In addition the time of flight was measured between the first and fifth plane of scintillation counters. All calibration and efficiencies were checked using the $p + p \rightarrow d + \pi^+$ reactions. The absolute cross sections were obtained with an overall uncertainty of 20 %. The measurements were performed at 15° (lab) which was a compromise to lower the background while keeping the pion rate measurable. The maximum intensity of the deuteron beam was $\sim 10^{11}$ deuterons/burst. ($\sim$ 15 nA). The data for the inclusive reactions are plotted in figure n° 11 in the form of Lorentz invariant cross sections versus the usual variable $x = \dfrac{k_{//}(CM)}{k_{//Max}(CM)}$, for the three targets. The previous data from Papp et al. [16] at 1.05 GeV/nucleon are also partially presented for comparison. One can summarize several features.

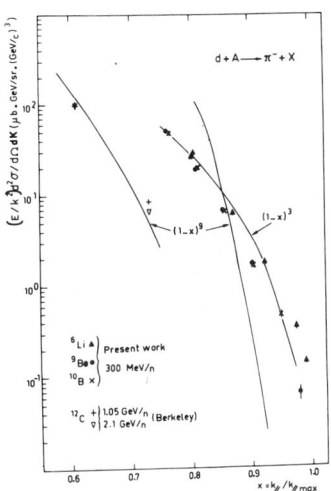

Fig. 11. Inclusive pion spectra induced by 600 MeV deuterons. Results of ref. [16] are also partially presented. The curves are the theoretical binomial shapes $(1-x)^n$.

1° The shape of the spectra is independent of the target as has previously been observed for small x, implying that the projectile structure dominates the pion spectra observed at forward angles.

2° For x = 0.75 the data from Berkeley [16] are lower than ours by about one order of magnitude. This difference cannot be explained even if the transverse momentum $k_\perp$ due to non zero experimental angles is removed. Hence the scaling behaviour which was one of the most striking feature at energies above

1 GeV/nucleon does not persist down to 300 MeV/Nucleon.

3° In the frame of recent theoretical models [17], the invariant cross sections can be parametrized as $(1-x)^n$ where the exponent n is related to the number of constituents and to the basic interactions of the model. For (d,π^-) experiment [16] at high energy, the value n = 9 is clearly favoured as can be seen in figure n° 11, in agreement with the theory [17]. Our data can be fitted by a function $(1-x)^n$ with n = 3. This exponent behaviour is not explained near $x \sim 1$ for such low incident energy.

Typical spectra for the two body reaction $^6Li(d,\pi^-)^8B$ are shown in figure n° 12 at 600 MeV and 300 MeV incident energies. The ground and first excited states are not well separated at 600 MeV due to the target thickness, but the second excited state is clearly seen. The spectrum at 300 MeV was obtained with a thinner target and the experimental resolution of 0.3 MeV (FWHM) permits a clear separation of the three levels.

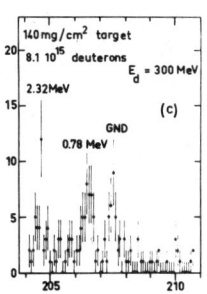
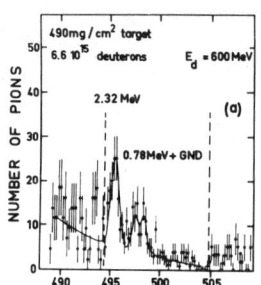

Fig. 12. Pion spectra of the reaction $^6Li(d,\pi^-)^8B$ at 15° (lab.) at 600 MeV and 300 MeV incident energy.

Differential cross sections are presented in table 3. They are found to be very low, and the most striking feature is that they are higher by about a factor of 5 to 8 at the lower energy of 150 MeV/nucleon, suggesting a strong influence of the transferred momentum (q = 4.6 fm^{-1} and 5.8 fm^{-1} at 300 MeV and 600 MeV respectively) eventhough the energy per nucleon is far below the NN → NNπ threshold.

C - SOME TRENDS OF THE (ION,π) REACTIONS

Despite the scarcity of the data, summarized in figures N° 13 and N° 14, we can try to see some trends in the (ion,π) reactions.

1° The cross sections obtained with an ^3He projectile at $T_{^3He}$ = 283 MeV are about the same for ^3He and ^4He targets, then decrease drastically (almost three orders of magnitude) when changing the target mass number from A = 4 to A = 6. Although much less pronounced, this decrease with A seems to be confirmed by a very preliminary result we have obtained with a ^{10}B target. Indeed this is just a rough comparison because of the difference between the transferred momenta involved in the different reactions.

2° For each composite projectile and a given target (^{6}Li) the cross sections first increase near threshold then strongly decrease when the incident energy goes up. It is clear that a maximum occurs at an incident energy higher than 283 MeV with an ^{3}He projectile. It is difficult to conclude definitively for (d,π) reaction because the data at 100 MeV/nucleon was obtained on a different target nucleus (reverse reaction ^{12}C(π$^+$,d)^{10}C at LAMPF [10]), however the same behaviour seems to occur. This effect has not been observed for (p,π) reaction on ^{6}Li considering results from Orsay (201 MeV) LAMPF [18] (equivalent energies T_p = 245 and 360 MeV) and Saturne (T_p = 600 MeV) [4].

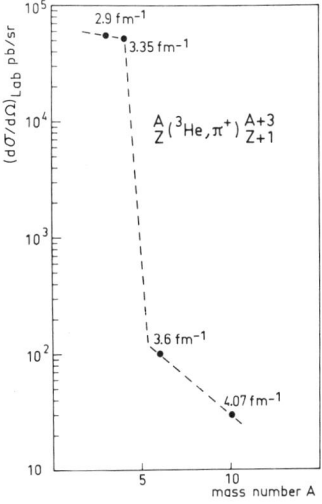

Fig. 13. Variation of the cross-sections for (^{3}He,π$^+$) on various targets versus the mass number A at 283 MeV incident energy.

3° The ratio of the pion production cross sections for p, d, ^{3}He incident projectiles on the same target are approximately 1 : 10^{-3} : 2.5 x10^{-5} respectively.

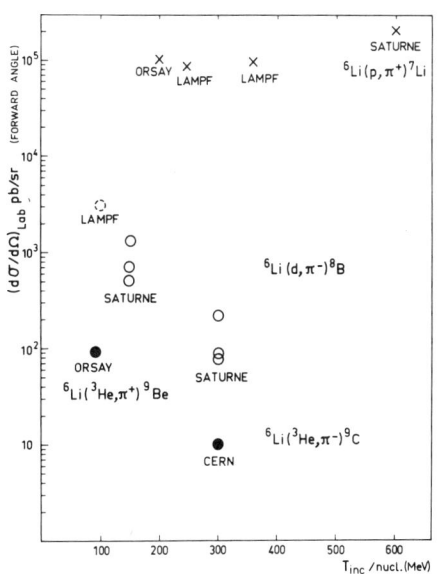

Fig. 14. Variation of the cross sections for the reaction A(a,π)B versus the incident energy/nucleon for different projectiles on ^{6}Li target.

IV - CONCLUSION AND PLANS FOR THE FUTURE

The (ion,π) reactions present very typical characteristics implying strong constraints for theoretical calculations. The present data has been useful as a starting point for theoretical works which will be presented at this workshop by their authors [19] [20].

Further studies of this type of reactions will be carried out at different energies near threshold on ^{6}Li and heavier targets. This is possible considering that high intensity ^{3}He beam is available in Orsay and that pions are clearly identified. Pion production with α particles will also be investigated.

Moreover these measurements should be extended at higher energies with Saturne II facilities.

Table I : Main characteristics of the Orsay synchrocyclotron external beams.

Particles	Energy (MeV)	Maximum extracted intensity (used for on line spectroscopy)
p	167 → 201	2∼3 µA
d	91 → 108	
^{3}He	238 → 283	
α	182 → 218	

Table II : Coulomb corrected average squared matrix element extracted from the present data. The overall normalization uncertainty of 20 % has not been included in the error bars.

T_p (MeV)	k_π^* (MeV/c)	$\overline{|f^2|}$ (10^{-4} fm^2)
180	59	10.9 ± 0.2
190	75	11.9 ± 0.2
201	90	11.9 ± 0.2

Table III : Values of ^{6}Li(d,π$^-$) differential cross section measured at 300 MeV and 600 MeV incident energy, leading to the ground state (2$^+$), and the two first excited states (0.78 MeV), (2.32, 3$^+$) of ^{8}B.

Levels	Incident energy E_d	
	300 MeV	600 MeV
0	521 ± 120 pb/sr	75 ± 27 pb/sr
0.78 MeV	707 ± 142 pb/sr	84 ± 27 pb/sr
2.32 MeV	1333 ± 163 pb/sr	237 ± 38 pb/sr

REFERENCES

1. This work in Orsay has been done by : L. Bimbot, M.P. Combes, J.C. Jourdain, N. Koori, Y. Le Bornec, F. Reide, A. Willis and N. Willis.

2. This work at Saturne national Laboratory has been done by E. Aslanides, A.M. Bergdolt, O. Bing, P. Fassnacht, F. Hibou (CRN Strasbourg), N. Willis, P. Kitching, Y. Le Bornec, B. Tatischeff (IPN Orsay), K. Baba, A. Boudard, G. Bruge, P. Couvert and B. Nefkens (DPhN/ME Saclay).

3. N. Willis et al., Journal of Physics G 7 (1981) 195.

4. T. Bauer et al. Phys. Lett. 69B (1977) 433.

5. B. Tatischeff et al. Phys. Lett. 63B (1976) 158.

6. K. Gabathuler et al. Nucl. Phys. B40 (1972) 32.

7. J. Hüfner et al., Nucl. Phys. A231 (1974) 455.

8. M. Bloch et al., Rev. Lett. 11 (1963) 301.

9. R. Bizzarri et al., Nuovo Cimento 33 (1964) 1497.

10. J.F. Amann et al., Phys. Rev. Lett. 40 (1978) 758.

11. E. Aslanides et al., Phys. Rev. Lett. 43 (1979) 1466; 45 (1980) 1738.

12. Y. Le Bornec et al., to be published.

13. J.S. Vincent and E. Boschitz, Nucl. Phys. A143 (1970) 121.

14. E. Aslanides et al., to be published in Phys. Lett.

15. P. Couvert, contribution to this workshop.

16. J. Papp et al., Phys. Rev. Lett. 34 (1975) 601.

17. I.A. Schmidt and R. Blankenbecler, Phys. Rev. D15 (1977) 3321.

18. J. Källne et al., Phys. Rev. C21 (1980) 2681.

19. J.F. Germond, contribution to this workshop.
    J.F. Germond and C. Wilkin, to be published in Phys. Lett.

20. M.G. Hüber, contribution to this workshop.
    M.G. Hüber, M. Dillig and K. Klingenbeck, contribution to 9th ICOHEPANS, Versailles (1981) 202.

21. F. Soga et al., Phys. Rev. C $\underline{22}$ (1980) 1348.

DISCUSSION

J. Noble (University of Virginia): I was interested by your ^{3}He on ^{3}He data going to ^{6}Li. Has anything been done with deuteron on ^{4}He going to ^{6}Li also? That would be a very interesting complimentary reaction, presumably at 180 MeV.
Willis: This would give a π°.
Noble: Well, then ^{6}Be, or ^{6}He which would give a π^+. Has that been done?
Willis: No, I don't think it has been done.
Noble: I'm suggesting that it ought to be done.

A.D. Bacher (IUCF): Julian, do you have a prediction for these reactions?
Noble: Not off the top of my head.

NEUTRON INDUCED PION PRODUCTION PROCESSES[+]

E. Rössle, W. Dutty, J. Franz, L. Lehmann[*], G. Nicklas
and H. Schmitt
Fakultät für Physik der Universität Freiburg,
D-7800 Freiburg, Fed.Rep.Germany

ABSTRACT

The unpolarized neutron beam at SIN has been used for measurements of elementary pion production processes. Recent results on differential cross sections of the following reactions will be reported: $np \to \pi^+ nn$ with the extraction of the non-resonant isoscalar cross section; $np \to d\pi^0$, including the threshold region; $nd \to t\pi^0$ and $nd \to {}^3He\pi^-$ mainly at backward angles of the pions.

INTRODUCTION

Most of the elementary pion production processes have been studied with proton beams. The informations on neutron induced pion production is still very scarce or even missing. After some pioneering work about 20 years ago little progress has been made until the meson factories started operation. But still the knowledge on these particular reactions is increasing very slowly. For a complete understanding of the pion production mechanism it is however necessary to have informations on both, p and n induced processes. This is not only for tests of symmetries or invariance principles. The pion production from an isoscalar initial state can only be measured in inelastic neutron proton scattering. A particular feature of the isoscalar channel is the non-resonant behaviour. The dominant resonant intermediate state of $N\Delta$ with its influence down to threshold can only be reached from the isovector initial state. Some details of the pion production mechanism could be obscured by this dominant resonant process. The neutron proton initial channel consists of equal amounts of isoscalar and isovector components. The isoscalar part is therefore not directly accessible but has to be extracted from the difference of two measurements. In addition technical difficulties have to be encountered with a neutron beam.

EXPERIMENTAL TECHNIQUES

All experiments which will be discussed have been performed at the neutron beam at SIN with the same experimental set up.

[+] Work supported by the German Bundesministerium für Forschung und Technologie

[*] Deceased by accident August 22, 1981

The layout of the neutron beam together with the experimental set up [1] is shown in figure 1. The primary proton beam of 590 MeV with a time structure of bunches of less than 1 ns width and a spacing of 20 ns (normal mode used parasitically) or 60 ns (main user mode) produces a continuous neutron spectrum at a 12 cm thick beryllium or carbon target. Neutrons emitted at an angle of 60 mrad are taylored by means of collimators of variable cross sections to a neutron beam of variable size with an area up to 5 x 5 cm² at 60 meters distance from the production target, where the reaction target is placed. With an overall time resolution of 0.8 ns the energy resolution is better than 1.5 %. A magnetic spectrometer of large angle and momentum acceptance together with a time of flight measurement is used for identification and momentum determination of the charged particles emitted from the reaction target of liquid hydrogen or deuterium respectively.

RESULTS AND DISCUSSION

A. $np \rightarrow \pi^+ nn$

An inclusive experiment of this process has been performed [1] in the energy region $470 < T_n < 590$ MeV by detecting the pion only for laboratory angles up to $20°$. The aim of this experiment was the extraction of the isoscalar part of the single pion production cross section. With the conventional decomposition into partial cross section $\sigma_{T_i T_f}$ according to the isospin of the two nucleons in the initial (T_i) and final (T_f) state, single pion production from nucleon-nucleon collisions can be described by four independent cross sections, namely $\sigma_{10}(d)$, $\sigma_{10}(np)$, σ_{11} and σ_{01}, where the brackets of the first two cross sections indicate the bound (d) or unbound (np) final isoscalar state of the two nucleons. From the expression

$$\sigma(np \rightarrow \pi^+ nn) = \frac{1}{2}(\sigma_{01} + \sigma_{11}) \quad (1)$$

the isoscalar part

$$\sigma_{01} = 2\sigma(np \rightarrow \pi^+ nn) - \sigma_{11} \quad (2)$$

can be evaluated if the cross section σ_{11} is known. It can be taken from other experiments, most directly from the process $pp \rightarrow \pi^0 pp$, which is a pure σ_{11} transition. The particular interest in the isoscalar cross section lies in the fact that it is non-resonant, because the resonant intermediate state involving the Δ_{33}-resonance cannot be reached from a T=0 state by isospin conservation. It is therefore of great value for testing models of pion production.

Two examples of pion energy spectra are shown in fig. 2 together with the predictions of the statistical model with a pure phase space distribution (curve a) and the isobar model (curve b) in which is assumed that the pion production is mediated by an NΔ intermediate state. The experimental spectra fall somewhere in between of the two predictions. Clearly, the isobar model which reproduces the π^0-spectra of the process pp → π^0pp fails to reproduce the π^+-spectra particularly at the lower energies.

The differential cross sections are commonly parametrized in the threshold region by

$$d\sigma/d\Omega = K(1/3 + b \cdot \cos^2\theta) \qquad (3)$$

We therefore plot the results as function of $\cos^2\theta$ in fig. 3 and give the resulting fit parameter b in fig. 4. A pronounced anisotropy can be observed contrasting the almost isotropic π^0 angular distribution of the process pp → π^0pp. This result again is a strong indication of an isoscalar contribution in π^+-production. The corresponding value of the fit parameter b_{11} for the process pp → π^0pp in the energy range 480 < T_p < 680 MeV is [2] b_{11} = 0.065 ± 0.056.

The cross section obtained by integration of the expression above is given in fig. 5 together with other results [3]. The extraction of σ_{01} from these data, however, can not be performed uniquely because of the mass differences of the particles involved in the different reactions. In other words, isospin invariance as the underlying assumption of (2) is disturbed by these mass differences. The subtraction procedure becomes model dependent. The two limits, statistical model and isobar model, are displayed in fig. 6a and compared to other measurements [3] in fig. 6b. In the statistical model assumption a strong energy dependence is observed indicating that at least s- and p-wave pion production together with a p-wave final state of the two neutrons have to be considered in order to reconcile with this slope of the excitation function.

From the pion spectra and the angular distribution supported by the integrated cross section in the upper limit a non-vanishing non-resonant isoscalar pion production can be discerned of comparable magnitude to the resonant cross section at the highest energy considered here.

B. np → dπ^0

First results obtained for this process [4] have been restricted to the energy region 470 < T_n < 590 MeV and the center of mass angle θ_d < (75° to 83°). The availability of the accelerator mode with 60 ns spacing allowed the extension to the full angular range and down to the very near threshold region. Neutron energies as low as 286 MeV corresponding to a pion kinetic energy of about 5 MeV or the equivalent value of $\eta = p_\pi^{cm}/m_\pi c = 0.26$ could be attained. The absolute values of the cross sections have been obtained by normalization at 0° to the process pp → dπ^+, which is related to our process by

$$d\sigma(np \to d\pi^0)/d\Omega = \frac{1}{2} d\sigma(pp \to d\pi^+)/d\Omega \tag{4}$$

assuming isospin invariance. For the reference cross section we included all available experimental data falling within the conventional confidence limits. The averaging and interpolation procedures were based on the parametrization of Spuller and Measday [5]. A normalization error of about 5 % is encountered by this procedure.

For the parametrization of the differential cross section of unpolarized beam and target still different presentations are customly used

$$d\sigma/d\Omega \begin{cases} = \frac{1}{4\pi} \Sigma\, a_k P_k(\cos\theta) & k \text{ even} \\ = \Sigma\, \gamma_n \cos^n\theta & n \text{ even} \\ = K \cdot (A + \cos^2\theta + B\cos^4\theta + \ldots) \end{cases} \tag{5}$$

where the latter is historical in origin. The limitation to even powers of the $\cos^n\theta$-series or Legendre polynomials also for the np cross section presumes isospin invariance. The differential cross sections are shown in fig. 7 and 8. Different energy bins have been applied in the presentation. The lowest energy bin at $T_n = 288$ MeV ranges from 286 MeV to 290 MeV. It is followed by 10 MeV intervals up to 360 MeV and 20 MeV intervals for the higher energies. Results of the fits with Legendre polynomials are shown in fig. 9, where relative coefficients are presented. Significant finite values of the Legendre coefficients a_4 and a_6 can be discerned above 450 MeV and 500 MeV respectively with corresponding noticeable d- and f-wave pion production. Below 350 MeV the angular distributions flatten off with a decreasing P_2-term and a relative decrease of p-wave production. According to Maxwell et al. [6] the coefficient γ_4 should be sensitive to the coupling constant $\alpha_\rho = f_{\rho N\Delta}/f_{\rho NN}$. The comparison is shown in fig. 10. The experiment favors the value of the static quark model of $\alpha_\rho \simeq 1.7$ to 2.0 over the value of $\alpha_\rho = 1.0$ proposed by Kisslinger [7].

C. $nd \to t\pi^0$ and $nd \to {}^3\text{He}\pi^-$

These two reactions together with their charge symmetric counterparts can be considered as the basic processes of coherent pion production on nuclei. Thus one would like to understand this production mechanism for its application to heavier nuclei. However in spite of increasing efforts, both experimentally and theoretically, it is fair to say that this problem is still far from being solved. The most critical part of understanding the reaction mechanism is the backward pion emission with its extremely high momentum transfer, and it is not surprising that the discrepancies are largest in this region. Most of the proton induced experiments pd → tπ^+ detect the π^+ and encounter difficulties at backward angles due to the decreasing pion energy, and are therefore limited to $\theta_\pi < 150°$. Our experiments are complementary to those measurements. By detecting the recoiling nucleus in forward direction with the magnet spectrometer the region of pion backward angles up to almost 180° are covered. This method allows furthermore the simultaneous measurement of both channels in a wide angular range, limited only by the ambiguity of the kinematics. In addition, the continuous neutron energy spectrum yields the excitation functions of the differential cross sections at the same time.

In the extension of a first experiment [8] of nd → ^{3}Heπ^- we measured both channels in the energy range $350 < T_n < 560$ MeV. The absolute differential cross sections plotted versus the pion center of mass angle are given in fig. 11. For a discussion of reaction mechanism it is more appropriate to present the data as a function of the momentum transfer $q = |\vec{p}_n - \vec{p}_\pi|$, which is done in fig. 12. In the figures, the data for the triton channel are multiplied by the "isospin factor" 2 (see below). At higher energies where the angular distributions could be extended to about 50° the known tendency of a decreasing cross section with increasing angle of roughly the same slope is observed as would be the expected from impulse approximation [9]. At backward angles all cross sections show a rather flat or slightly increasing behaviour with a minimum around 120°. A pion exchange mechanism has been proposed by several authors [10] to reconcile this effect. This is supported by recent results of πd scattering which also show a similar backward angular distribution. A comparison with theory [9,10] and with other experimental results [11] at two energies is given in fig. 13. The excitation functions for several angles are displayed in fig. 14. A maximum cross section at about 420 MeV can be observed, which is most pronounced at the extreme backward angles. It reflects the influence of the Δ_{33}-resonance in the reaction mechanism, which is shifted down in energy for kinematical reasons. A calculated excitation function of Barry [12] for 180° based on an exchange mechanism is included in the figure. The calculated curve has been scaled by a factor 0.35.

The simultaneous measurement of both reaction channels is an excellent opportunity to check on the ratio of the two channels. Assuming isospin invariance the two channels are related by

$$R = \sigma(nd \to {}^3He\pi^-):\sigma(nd \to t\pi^0) = 2 \qquad (6)$$

The experimental result on R as a function of the momentum transfer is given in fig. 15. There is almost no q-dependence and the average value of the ratio of (6) is $R = 1.76 \pm 0.09$. The deviation from the "isospin" value may be ascribed to several reasons, which all can be followed back to external or internal electromagnetic effects like simple Coulomb force, mass differences and the presumably important difference in the wave functions of the two final nuclei. From a calculation for the ratio of the charge symmetric processes ^{13} of $\sigma(pd \to t\pi^+) : \sigma(pd \to {}^3He\pi^0) = 2.2$ it can be inferred that the deviations are of the same origin.

REFERENCES

1. M. Kleinschmidt et al., Z.Physik A298, 253 (1980).
2. A. F. Dunaitsev et al., JETP 36, 1179 (1959).
3. R. Handler, Phys. Rev. B138, 1230 (1965).
   V. P. Dzhelepov et al., JETP 23, 993 (1966).
   Y. M. Kazarinov and Y. N. Simonov, Sov. J. Nucl. Phys. 4, 100 (1967).
   W. R. Thomas, Thesis, University of New Mexico (1977).
4. W. Hürster et al., Phys. Lett. 91B, 214 (1980).
5. J. Spuller and D. F. Measday, Phys. Rev. D12, 3550 (1975).
   L. Schmitt, private communication.
6. O. V. Maxwell et al., Nucl. Phys. A348, 388 (1980).
7. L. S. Kisslinger, in Theoretical methods in medium energy and heavy ion physics, ed. K. W. McVoy and W. A. Friedman (Plenum Press, 1978), 307.
8. J. Franz et al., Phys. Lett. 93B, 384 (1980).
9. H. W. Fearing, Phys. Rev. C11, 1210 (1975), Phys. Rev. C16, 313 (1977).
10. M. P. Locher and H. J. Weber, Nucl. Phys. B76, 400 (1974).
    V. S. Bhasin and I. M. Duck, Phys. Lett. 46B, 309 (1973).
    W. R. Gibbs and A. T. Hess, Phys. Lett. 68B, 205 (1977).
11. A. V. Crewe et al., Phys. Rev. 118, 1091 (1960).
    J. Carroll et al., Nucl. Phys. A305, 502 (1978).
    W. Dollhopf et al., Nucl. Phys. A217, 381 (1973).
    D. Harting et al., Phys. Rev. 119, 1716 (1960).
    J. Källne et al., Phys. Rev. C24, 1102 (1981).
12. G. W. Barry, Phys. Rev. D7, 1441 (1973).
13. H. S. Köhler, Phys. Rev. 118, 1345 (1960).

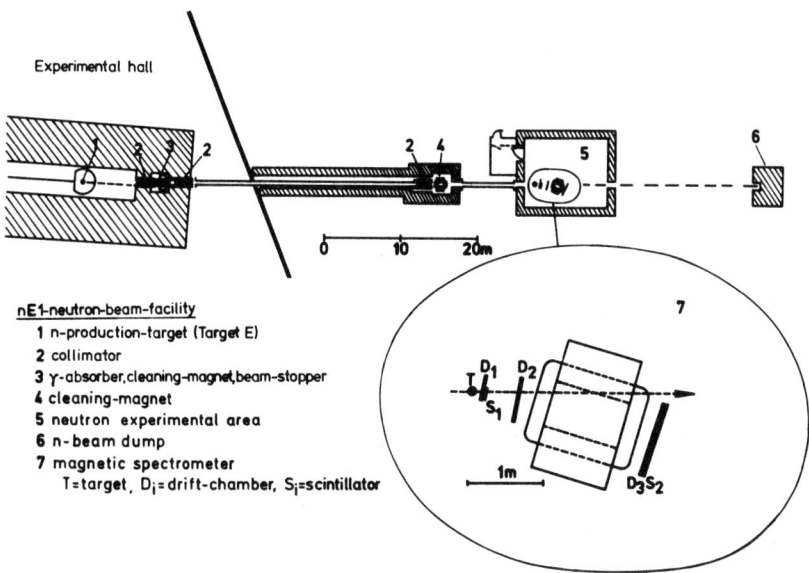

Fig. 1. Layout of the neutron beam at SIN and of the experimental set up.

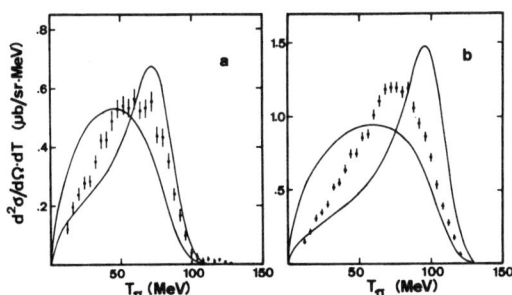

Fig. 2. Double differential cross section summed over the angular bins for the incident energies of 500 MeV and 560 MeV. The curves represent the predictions of the statistical model (curve a) and the isobar model (curve b).

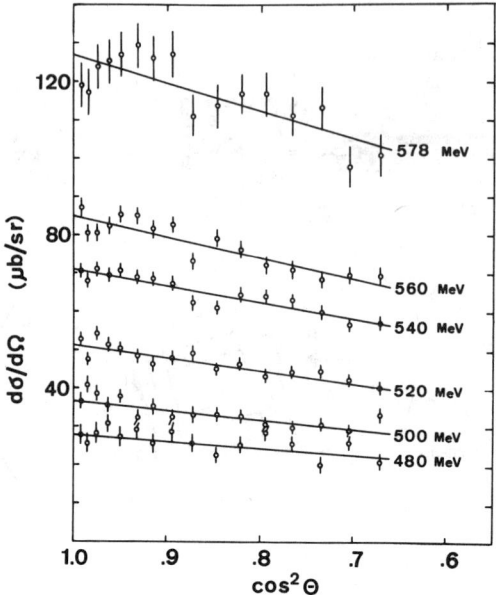

Fig. 3. Differential cross sections of the reaction np → π⁺nn. Solid lines are fits to the data according to eq. 3.

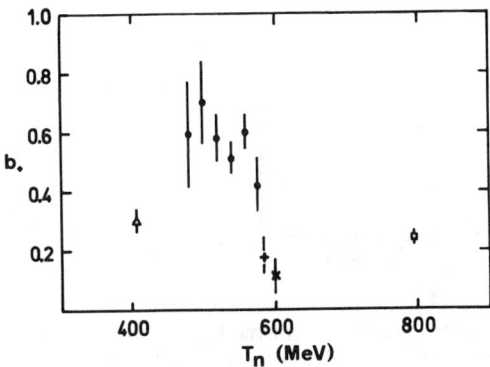

Fig. 4. Angular distribution coefficient of eq. 3. The experimental data are taken from reference /3/ Δ:a, +:b, *:c, ☐:d and · /1/.

Fig. 5. Energy dependence of the cross section for the reaction np → π⁺nn. The experimental data are taken from /1/ · and /3/ ∆, +, *, ▢.

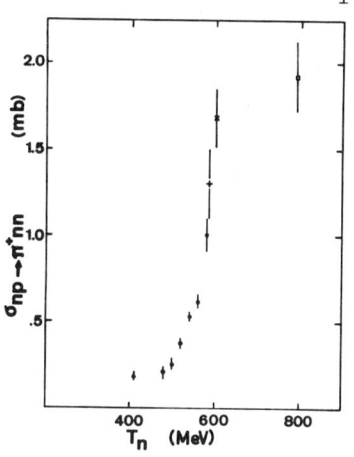

Fig. 6. The isoscalar part σ_{01} of the reaction np → π+nn according to (5). In the lower part (a) two different kinematics are compared (see text). The upper part (b) gives the comparison with other data, which are taken from ref. 3.

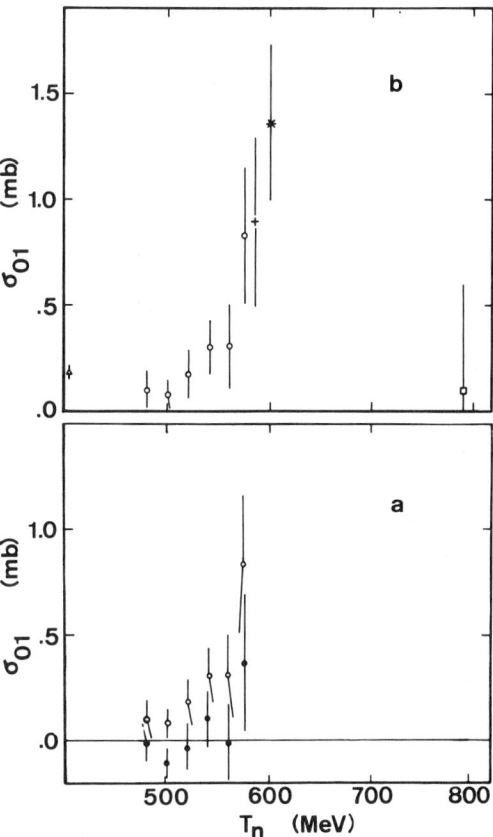

Fig. 7. Differential cross sections of the reaction $np \to d\pi^0$ at lower energies.

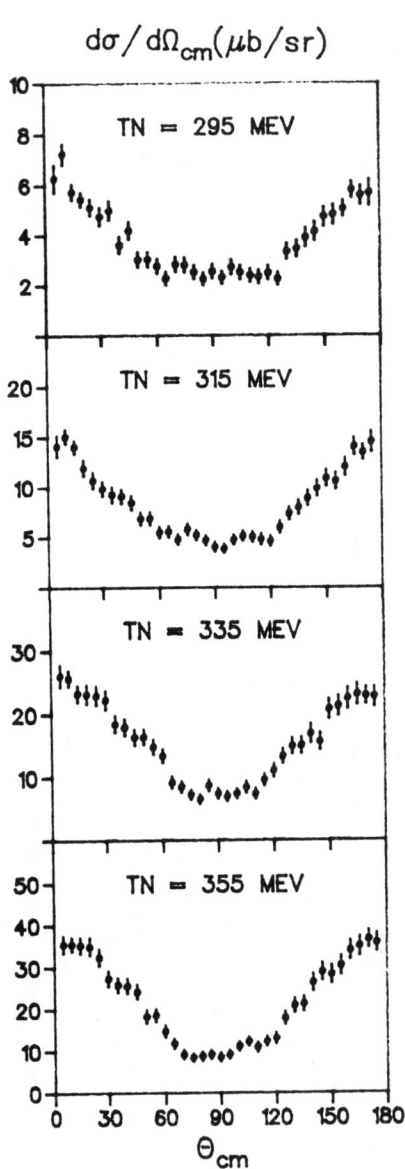
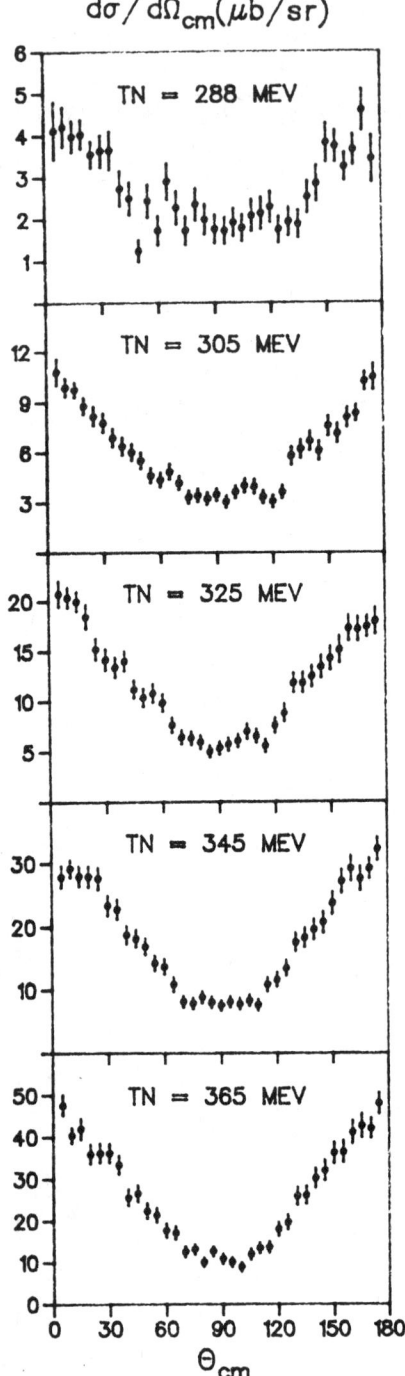

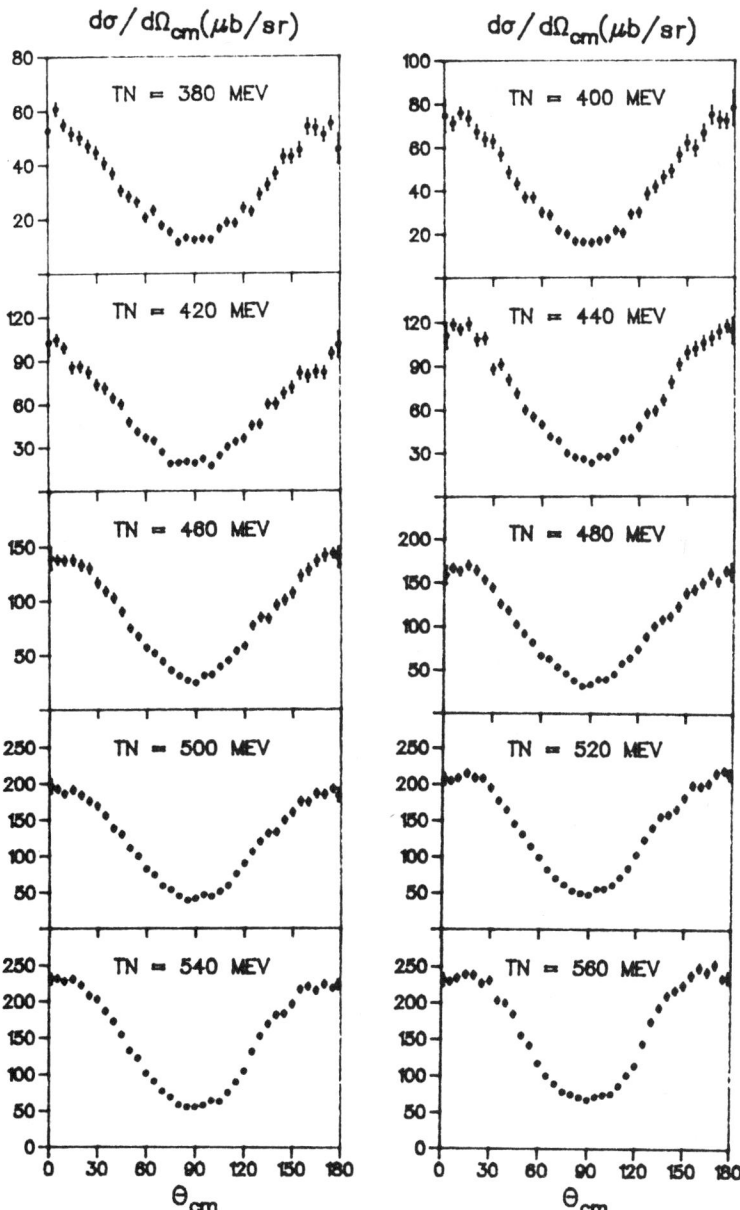

Fig. 8. Differential cross sections of the reaction np → dπ⁰ at higher energies.

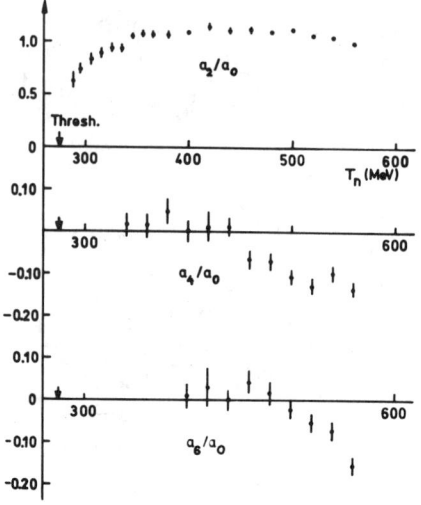

Fig. 9. Relative Legendre coefficients of the fits to the differential cross sections of the reaction $np \to d\pi^0$.

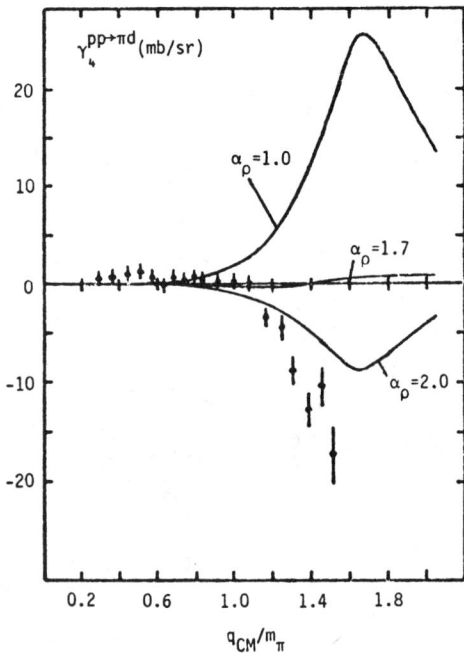

Fig. 10. The coefficient γ_4 of eq. 5 with predictions of Maxwell et al. (6) for different values of the coupling constant α_ρ.

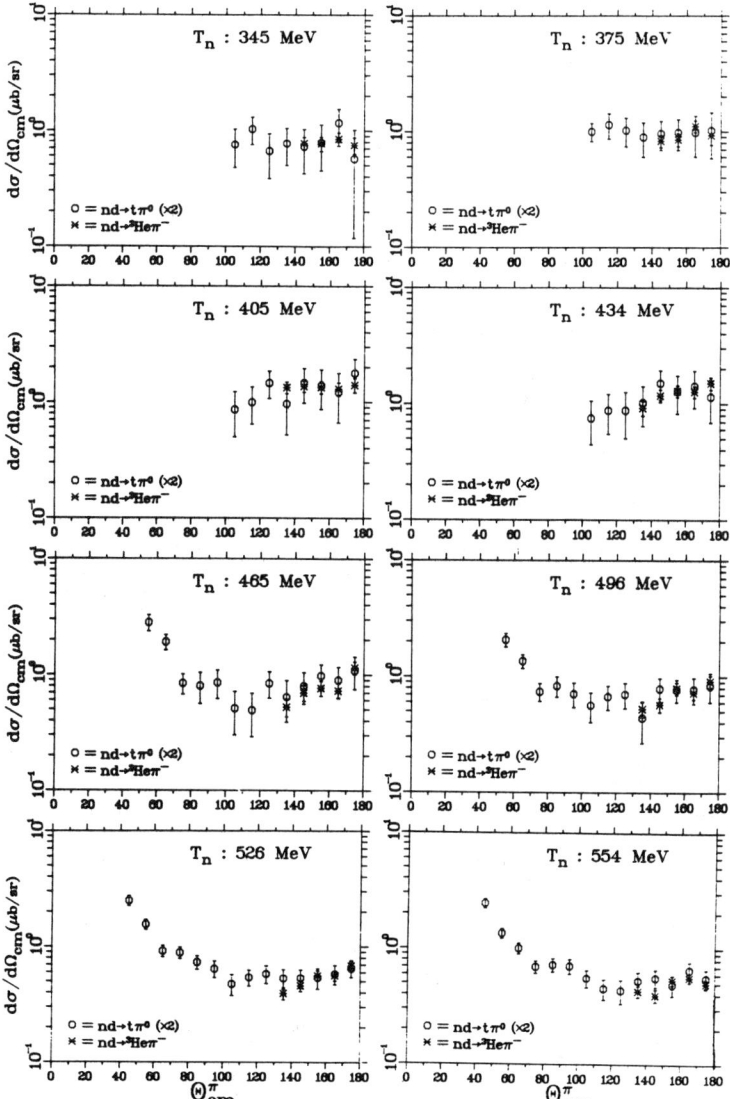

Fig. 11. Differential cross sections of the reactions nd → tπ⁰ (x2) and nd → ³Heπ⁻ as a function of θ_π.

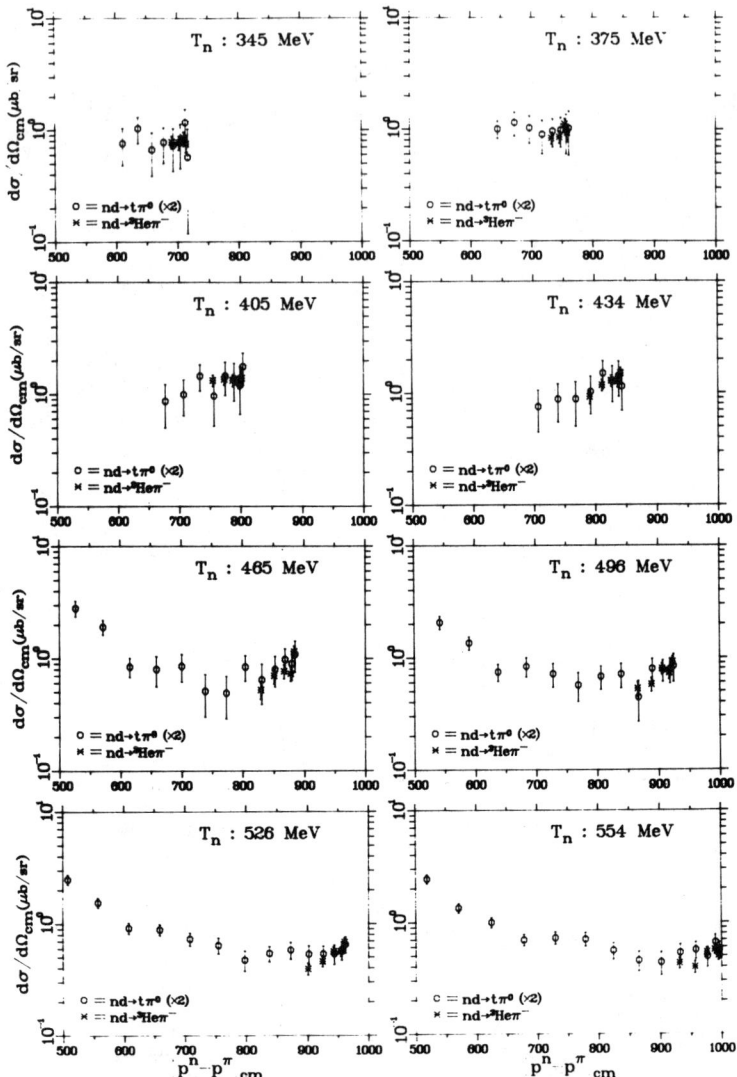

Fig. 12. Differential cross sections of the reactions $nd \to t\pi^0$ (x2) (o) and $nd \to {}^3He\pi^-$ (*) as a function of the momentum transfer.

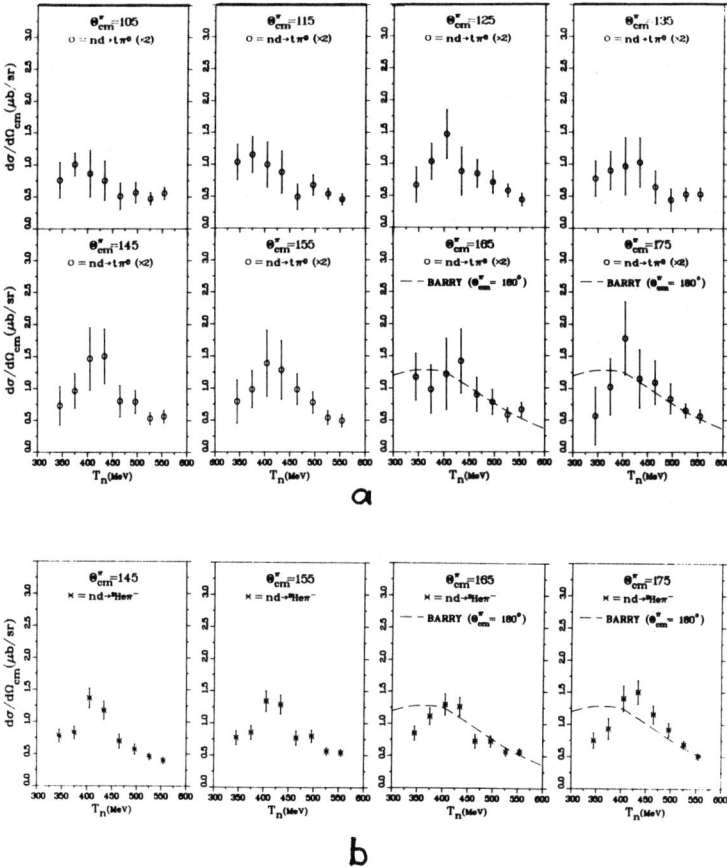

Fig. 14. Excitation functions for the two reactions
a) nd → tπ⁰ (x2) and b) nd → ³Heπ⁻ for several angles.
Dashed line: Theoretical prediction (12), which was
multiplied by a factor 0.35.

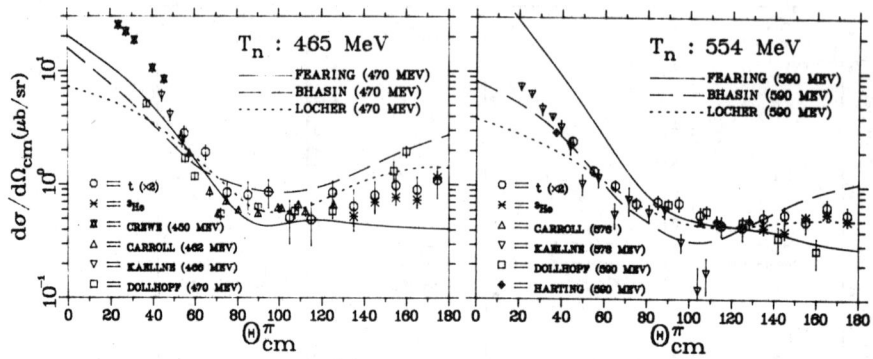

Fig. 13. Comparison with other experimental results (11) and theory (9,10) at 465 and 554 MeV.

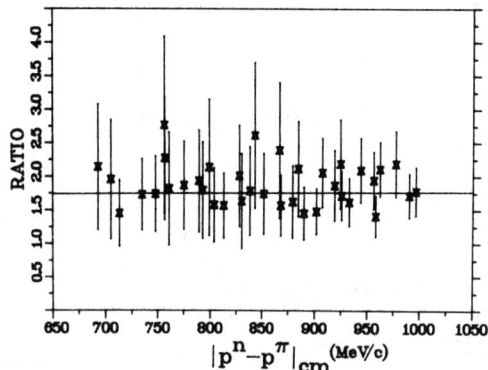

Fig. 15. Ratio $R = \sigma(nd \to {}^3He\pi^-):\sigma(nd \to t\pi^0)$, the average value is $R = 1.76 \pm 0.09$.

NEW EXPERIMENTAL RESULTS ON PION PRODUCTION AT LNS (SACLAY)

Pierre Couvert
DPh-N/ME, CEN Saclay, 91191 Gif-sur-Yvette Cedex, France

ABSTRACT

The present status of the experimental situation on pion production reactions at Saturne 2 is reviewed. Improved or new magnetic spectrometers are rapidly described and new experimental results obtained since the renovation are presented. As a conclusion, present and future possibilities of pion production physics at LNS are examined.

INTRODUCTION

Since the pioneering works of J.J. Domingo et al. at CERN and S. Dahlgren et al. at Uppsala[1] who showed the first evidences of exclusive pion production on complex nuclei, a large amount of experimental results on $A(p,\pi)A+1$ reactions (or charge equivalent related reactions) have been accumulated throughout the world.[2] In the meantime, important theoretical work has attempted to understand the reaction mechanism and to stress the possibilities and the promises of this kind of high momentum transfer reactions.

The need for a complete and coherent set of experimental data was pointed out early and important work was done from the beginning using the old synchrotron Saturne with incident energy protons far above the pion production threshold.[3] These high incident energy (p,π) reactions have been pursued at LAMPF and TRIUMF.[2]

In this talk, I will present results of some proton induced* pion production experiments which were done at LNS (Laboratoire National Saturne) since the new synchrotron Saturne 2 was built. These experimental data were obtained either with the high resolution magnetic spectrometer SPES 1 or the new high energy magnetic spectrometer SPES 4.

1. EXPERIMENTS AT SPES 1

The high resolution spectrometer SPES 1 is an energy-loss magnetic spectrometer. It has now been used for almost ten years alongside the synchrotron Saturne and has been described in many publications.[4] Thus, I will present only major modifications which were made on the spectrometer during the construction of the new accelerator Saturne 2 in 1978.

Due to the high intensity beam (several 10^{11} protons per burst) hitting the target, the background problems were crucial and have been considered very carefully. The shielding between the analyser, the

---

*Light ion induced pion production reactions done at Saturne 2 are reviewed in the N. Willis and Y. Le Bornec talk during this workshop.

beam transport line up to the target, and the spectrometer area itself was greatly improved (Fig. 1). In addition, a floor-to-ceiling, 9 meters long, concrete wall moves with the spectrometer and protects the detection from target background emissions. Furthermore, a vacuum chamber and two quadrupoles bring the focalised beam (for detection angles above 13° lab) into part of the distance between the target and a deep heavily shielded beam dump.

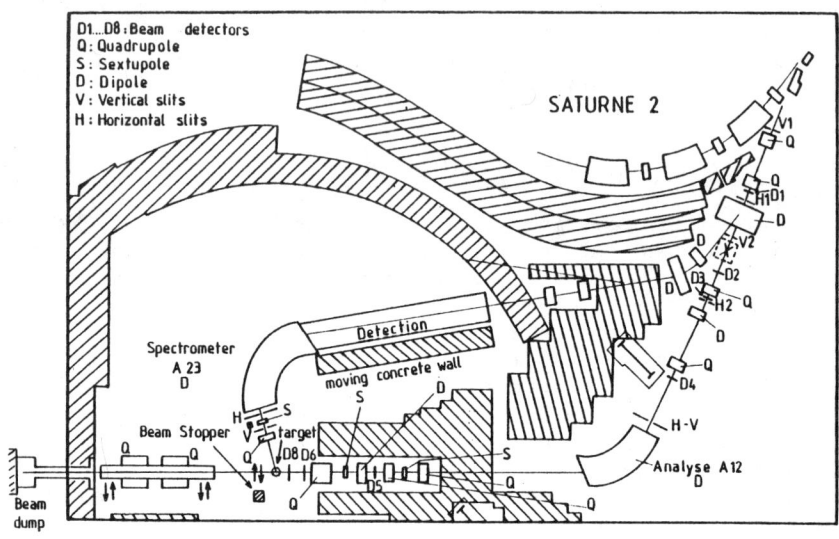

Fig. 1. Experimental set-up of the spectrometer SPES 1.

Concerning the detection located on the spectrometer's focal plane (Fig. 2), we still use the four double horizontal drift counters, the major modifications being the use of scintillator hodoscope planes to lessen the single counting rate and new high length Čerenkov counters in the trigger.

For the three experiments described below, the beam intensity was monitored by two six-fold scintillation-counter telescopes respectively viewing the target at angles of 140° in the reaction plane and 30° out of the reaction plane. The absolute normalization of the cross sections is determined by the calibration of a secondary emission monitor, located just upstream of the target, using a ^{11}C activation technique.

1.a. Excitation function of the $^{10}B(p,\pi^+)^{11}B$ reaction

The first pion production experiment realized at Saturne 2 was the achievement of the $^{10}B(p,\pi^+)^{11}B$ reaction excitation function started a few years ago[3] and it was undertaken by the same Saclay-Orsay-Strasbourg collaboration.[5]

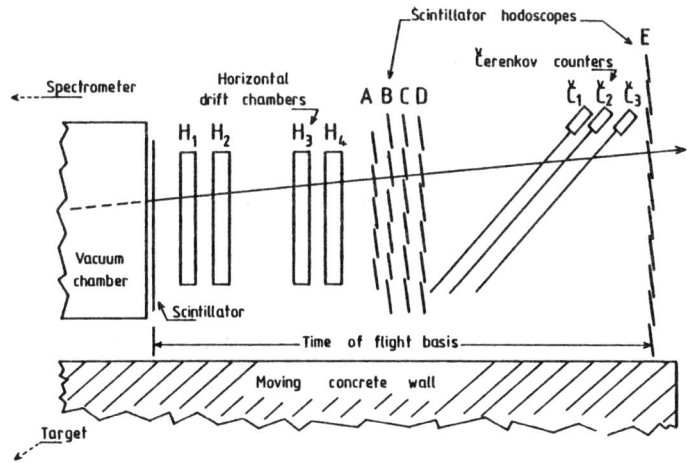

Fig. 2. Detection system on the SPES 1 spectrometer's focal plane.

The study of the variation of the cross section versus the incident proton energy, at constant transferred momentum, should be a useful tool to check the different reaction mechanism theories. Given such conditions, one hopes to reach, in a first approximation, only a small momentum range around a unique component of the nuclear wave function and so be sensitive only to well-known kinematical effects and to the reaction mechanism itself.

To complete the existing data, we studied the $^{10}B(p,\pi^+)^{11}B$ reaction at four new energies (250, 291, 465 and 800 MeV). Fig. 3 shows the angular distributions for the ^{11}B ground state (3/2$^-$) and for the first excited level (2.125 MeV, 1/2$^-$) at these four energies ; angular distribution of the second excited level of ^{11}B(4.445 MeV, 5/2$^-$) has also been measured at 465 MeV incident proton energy. Two more points have been redone at 320 MeV to check the consistency with the previous experiment. Taking into account the respective absolute normalization uncertainties, the agreement is excellent. An interesting feature of these angular distributions is the slight but significant difference in shape at forward angles between the ground state and the 1/2$^-$ first excited state which should distinguish the ^{11}B from the other nuclei already studied at high energy at SPES 1, such as ^7Li or ^{10}Be for example.[3]

The corresponding ^{11}B ground state excitation function for five different constant transferred momenta is shown in Fig. 4. The new data at 250 and 291 MeV give a more precise position of the maximum of the $\Delta(1232)$ resonance effects which can be located around 330 MeV incident proton energy. Concerning the 800 MeV measurements, they confirm the exponential decrease of the excitation function at high transferred momentum. Finally we must notice the change in shape of the two highest q_{cm} excitation functions which comes from the oscillation of the angular distributions observed at about 100° cm in the near threshold data. Excepting this last feature, the major trends of

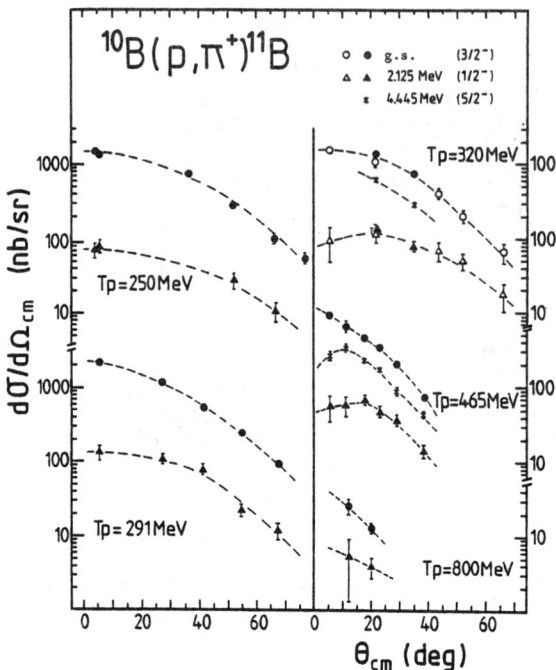

Fig. 3. Angular distributions of the $^{10}B(p,\pi^+)^{11}B$ differential cross section at 250, 291, 320, 465 and 800 MeV (full symbols). Open symbols are results of a previous experiment (M. Dillig et al).[3]

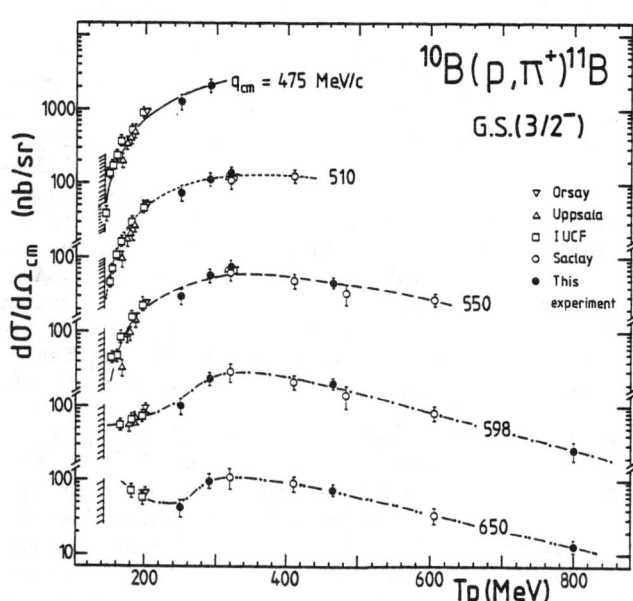

Fig. 4. Excitation function of the $^{10}B(p,\pi^+)^{11}B_{g.s.}$ differential cross section at constant transferred momentum q_{cm} ($q_{cm} = |\vec{k}_p - \vec{k}_\pi|_{cm}$).

these excitation functions are well described by a microscopic two
nucleon model involving a Δ(1232) intermediate excitation,[6] indicating that backward angle pion production should involve more complicated processes.

1.b. The $d(p,\pi°)^3He$ reaction around the (3,3) resonance

Despite the large amount of data accumulated on this reaction or on charge symmetric related reactions over the last few years,[2] some discrepancies and lack of coherence remain . Furthermore the (p,π) reaction on such a light nucleus should be an excellent tool to study the reaction mechanism and the influence of the (3,3) resonance.

The data presented below were obtained by a Saclay-UCLA collaboration,[7] and were done, under very similar experimental conditions, in connection with the p+d → ^{3}He+γ reaction. Using such a procedure the study of the p+d → ^{3}He+π°)/(p+d → ^{3}He+γ) ratio should give interesting information about Δ effects, independently of many experimental parameters (beam, target, etc.) or nuclear structure uncertainties (^{3}He form factor).

Several measurements of each experimental point have been done using a solid polyethylene CD_2 target at lower energies (≤ 450 MeV) and a liquid deuterium target at higher energies (≥ 450 MeV) with an overlapping control at 450 MeV. The recoil ^{3}He nuclei were detected in SPES 1. Differential cross sections of the p+d → ^{3}He+π° and p+d → ^{3}He+γ reactions have been measured at several angles and energies around the (3,3) resonance. In particular, the p+d → ^{3}He+π° experiment has been studied at 49° and 98° center-of-mass angles at eight different energies in the Δ region (300, 350, 400, 425, 450, 470, 500 and 550 MeV).

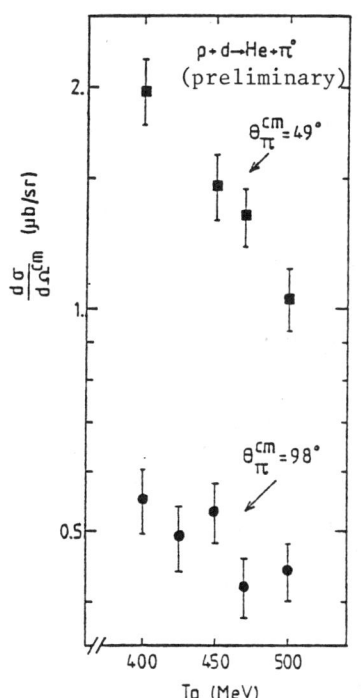

Partial and very preliminary results are given in Fig. 5. To locate the maximum of the cross section due to the Δ resonance influence, one must still wait for the analysis of the lower energy measurements. However, one can already notice that these effects are more pronounced at 49° than at 98°. This trend suggests, as for the ^{11}B case, that Δ resonance could no longer be the dominant feature of the reaction mechanism at backward angles.

Fig. 5. Excitation function of the p+d → ^{3}He+π° differential cross section at 49° and 98° c.m.

1.c. Polarized proton induced p+p → d+π⁺ reaction around 800 MeV

I will now briefly present the results of one of the very first experiments using the new polarized proton beam of Saturne 2. These measurements have been realized by a Saclay DPh-N - LNS collaboration.[8]

In addition to its important contribution to nucleon-nucleon interaction studies and to pion production theories as an elementary process, the p+p → d+π⁺ reaction could also be able to give interesting information about eventual dibaryonic resonances.

For that last purpose, angular distributions and analyzing powers of the $\vec{p}$+p → d+π⁺ reaction have been measured in a wide angular range at three incident proton energies around 800 MeV, which corresponds to the region of an eventual $^3F_3$ (2.24 GeV) resonance.[9] During this experiment, the average intensity of the polarized proton beam was around 3×10^8 $\vec{p}$ per burst with about 80 % polarization, the polarization being flipped at each beam burst. The spectrometer SPES 1 is used in a standard way to detect either the pions at forward angles (up to about 100°cm) or the recoil deuterons at backward angles. The target is a 14 mm thick liquid hydrogen target. The pion selection is done by Čerenkov counters and the identification of deuterons uses the time-of-flight procedure. The beam polarization is measured at each angle shift by shunting the incident protons in a secondary beam line where the polarimeter is located. Using a CH_2 target, the asymmetry for proton-proton scattering at 17° (where many absolute polarization data exist) is measured by detecting the scattered and the recoil protons in coincidence. Furthermore, an asymmetry measurement of the p-p scattering at 17° is done at each energy with the spectrometer itself. Finally, a symmetric two-armed monitor viewing the target controls the relative beam polarization during the data acquisition.

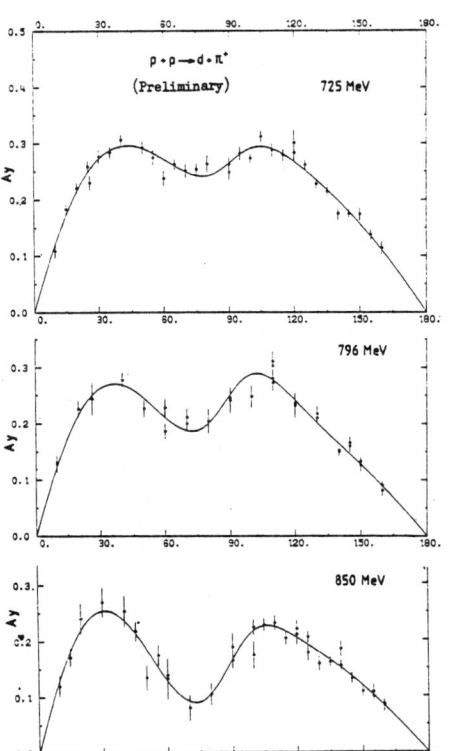

Preliminary asymmetry measurements at 725, 796 and 850 MeV are given in Fig. 6 and show a rapid change in shape for rather small incident energy steps. A parametrization of these asymmetries together with the differential cross sections, in terms of Legendre polynomials, is under analysis.

Fig. 6. Analyzing power of the $\vec{p}$+p → d+π⁺ reaction at 725, 796 and 850 MeV incident proton energies. Curves are Legendre polynomial fits to the data.

2. EXPERIMENTS AT SPES 4

SPES 4 is a new high energy magnetic spectrometer of medium resolution which has been working alongside Saturne 2 since the end of 1979.[10] From the target up to the focal plane, the scattered particle analyzing system is composed of 4 dipoles, 6 quadrupoles and 2 sextupoles (Fig. 7). The dipoles are built from sections of the old Saturne 1 synchrotron yokes. This spectrometer can be used in different analyzing configurations. Two extreme cases have been studied. The first configuration, with wide angular acceptance (2.4×10^{-3} sr solid angle), allows only a reduced analyzed momentum range (± 0.5 %). The second one is a large accepted momentum range ($\pm 4,7$ %), smaller solid angle (3.5×10^{-4} sr) version. Both configurations give an energy resolution of about 7×10^{-4}. The spectrometer itself is motionless but the beam line, right before the target, can be moved to obtain a small range of forward or backward angular distribution (-9° to +39° depending on configuration and incident beam momentum).

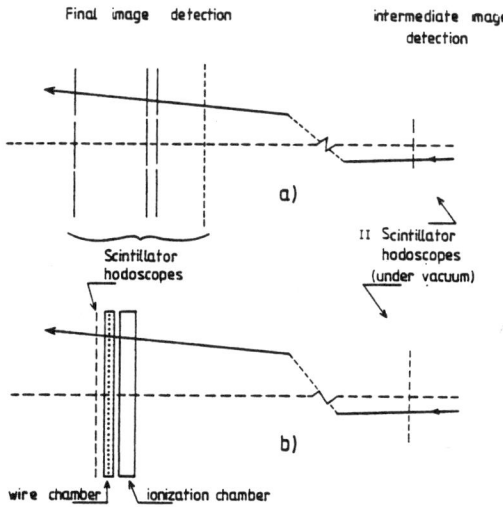

The detection system on the focal plane depends on the experiment. Up to now there are two different types of detection : one is very simple (Fig. 8.a) and uses only scintillator hodoscopes ; the second is more elaborate (Fig. 8.b) with large wire and ionization chambers and a scintillator hodoscope, giving better resolution measurements of the different parameters. Both detections use a 16.2 meters long basis time-of-flight system between the horizontal intermediate image and the final double image on the focal plane.

Fig. 8. Experimental arrangement for two different detection systems on the SPES 4 focal plane. The II scintillator hodoscopes are located on the intermediate horizontal image of the spectrometer.

2.a. High energy p+d → t+π⁺ reaction excitation function at 180°

If, as I pointed out above, an important effort was made on the experimental investigation on the exclusive pion production on deuterium, most of it has been done at incident proton energies under 1 GeV. Recently an Orsay-Saclay-Frascati collaboration[11] has used the SPES 4 spectrometer to study the p+d → t+π⁺ reaction at θ_π = 180° between 700 MeV and 1.7 GeV incident proton energies.

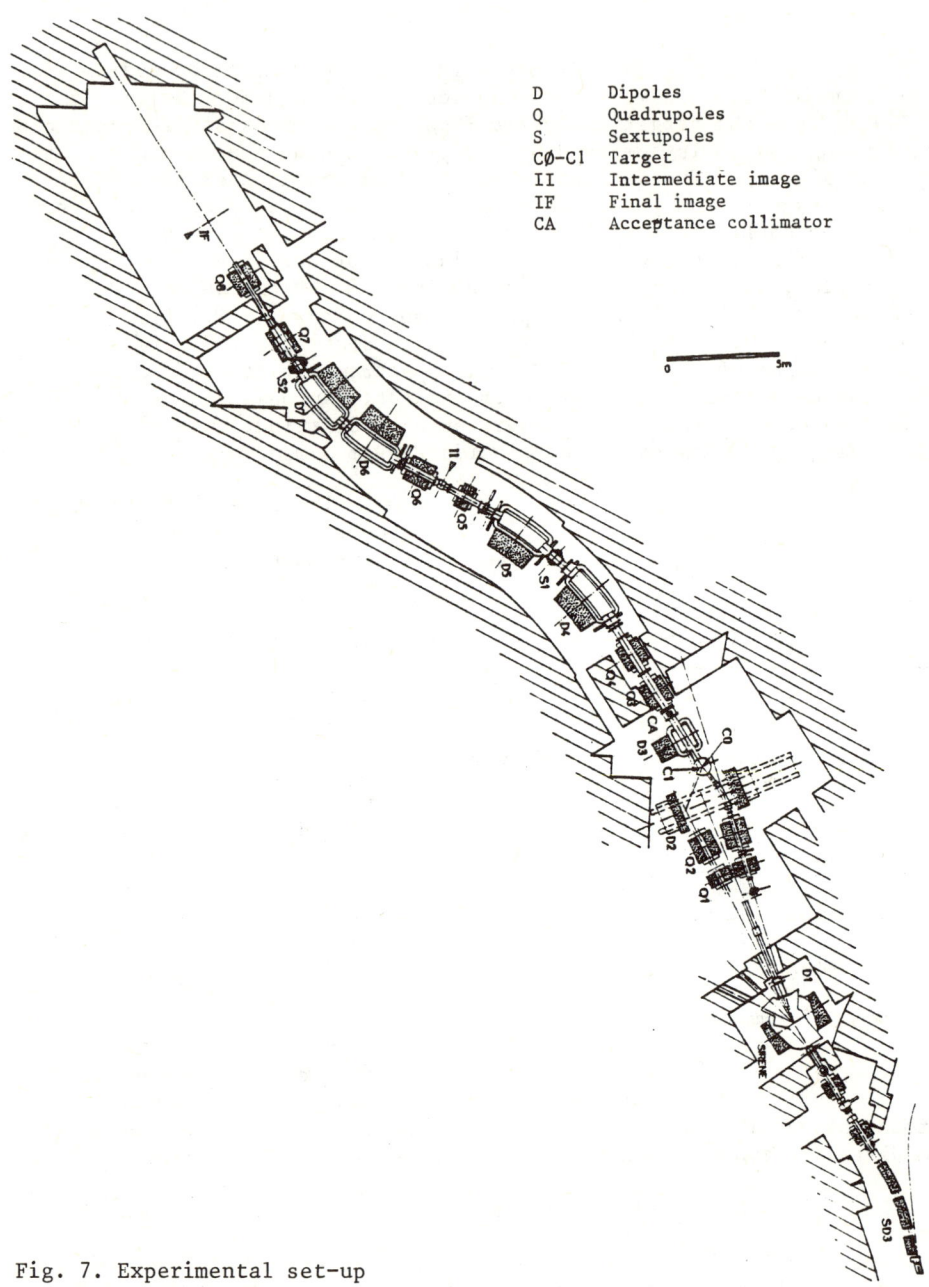

Fig. 7. Experimental set-up of the SPES 4 spectrometer.

Using a 500 mg liquid deuterium target and the simple hodoscope detection, the identification of the recoil triton nuclei by energy loss and time-of-flight procedures is very clear. The absolute normalization of the cross section is determined by monitor calibration, using the $^{12}C(p,X)^{11}C$ activation method.

Fig. 9 presents the differential cross section at $\theta_\pi = 180°$ for 10 different energies between 600 MeV and 1.5 GeV obtained during this experiment, together with previous lower energy data.[2]

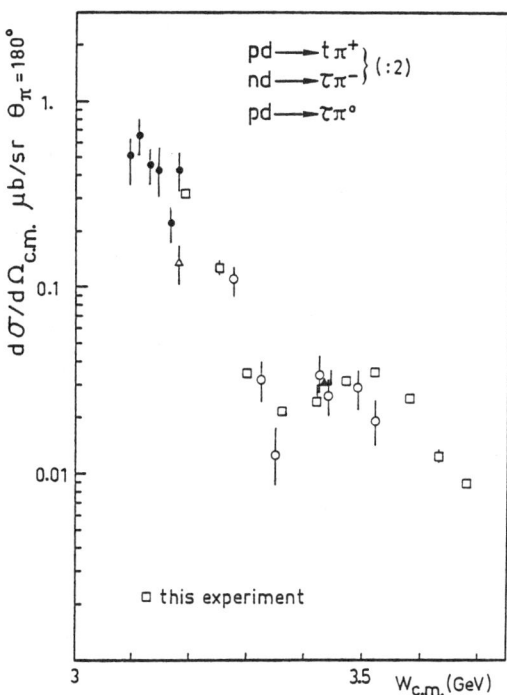

In addition to the well known first maximum at 3.1 GeV total center-of-mass energy due to the $\Delta(1232)$ resonance excitation, a clear broad second maximum is observed at about 3.5 GeV. Simple kinematical arguments, considering different intermediate systems in the reaction mechanism, give indications that such a structure in the excitation function could be interpreted in terms of a heavier Δ resonance excitation, as $\Delta(1650)$, or, more probably, of a two $\Delta(1232)$ propagation in the nucleus. Theoretical analysis of these data is necessary to investigate these kinds of assumptions.

Fig. 9. Excitation function of the $p+d \rightarrow t+\pi^+$ differential cross section at 180° measured during this experiment (open squares) together with previous data.[2]

2.b. Experimental test of the $^9Be(p,\pi)^{10}X$ at 2 GeV

Search for eventual nucleon-antinucleon $N\overline{N}$ bound states is going to be undertaken at SPES 4 by an Orsay-Lyon collaboration.[12] As a test of the experimental set-up and detection, pion production by 2 GeV incident protons on ^9Be target was studied at the two-body kinematics upper limit, corresponding to the three exclusive reactions:

$$p + {}^9Be \rightarrow {}^{10}Be + \pi^+$$
$$\rightarrow {}^{10}B + \pi^0$$
$$\rightarrow {}^{10}C + \pi^-.$$

The spectrometer was used to look for the recoil mass-10 nuclei at $\theta_{lab} \sim 6°$ which corresponds to pion production at around $\theta_{lab} = 160°$. High accuracy identification of the detected nuclei is possible with very low background level (momentum, mass and charge of the recoil nucleus are given respectively by the wire chamber, the time-of-flight and the ionization chamber). During a twelve hour run with a high intensity beam, not a single mass-10 was identified, leading to an upper limit of the differential cross section of the (p,π) reaction on ^{9}Be at 2 GeV of the order of 100 pb/sr. If this limit is very low, it might not be significant enough given the tremendous transferred momentum (about 18 fm^{-1}). However, such an experimental test is important and must be pursued to study the feasibility of high energy exclusive pion production at SPES 4.

SUMMARY AND OUTLOOKS

In this talk I reviewed the experimental situation on pion production physics at LNS over the last three years and I tried to stress some of the interesting features underlying these data. A small amount of results have been presented here which nonetheless have pointed out some of Saturne's possibilities. In fact, the outlooks concerning pion production experimentals at LNS are far more promising.

Within the next few months, two experiments on (p,π) reaction are planned to be done. They are the continuation of two of the experiments described above : analyzing powers and cross sections of the $\vec{p}+p \rightarrow d+\pi^+$ reaction are going to be measured at energies above 850 MeV ; furthermore, the backward angle $p+d \rightarrow t+\pi^+$ excitation function will be studied at energies up to 2.7 GeV.

However, the possibilities of LNS facilities are very large and many experiments could be done during the next few years to complete the experimental work on (p,π) reactions. SPES 1, for example, is a perfect tool to achieve some kind of systematic study of (p,π) reactions, on heavier nuclei in particular. Furthermore, and despite the very low cross sections, π^- experimental production is easy to do at SPES 1 because of the very low background level due to the spectrometer's inverse polarity. From a theoretical point of view, (p,π^-) reactions are important to the understanding of the reaction mechanism and are also an almost unique spectroscopic tool.

Even more promising is the opportunity to now measure analyzing powers at high energies. The present situation of Saturne's polarized beam should be improved in the near future in a way to reach one or a few 10^9 $\vec{p}$ per burst with around 80 % polarization, even above 1 GeV where some depolarizing resonances have to be crossed.

As I mentioned in the second part of my talk, the spectrometer SPES 4 is an excellent experimental facility to study pion production reactions up to 2.7 GeV proton kinetic energy, especially for few-body problem physics. A further step on LNS's experimental possibilities will be reached next year when SPES 3 will be ready to work. It is a wide solid angle ($\sim 10^{-2}$ sr), large momentum acceptance ($p_0 \pm 40\%$), good resolution (better than 10^{-3}) magnetic spectrometer. These characteristics and its multi-particle detection should be particularly well adapted to study three-body reactions as $(p,p\,\pi)$ and Δ or heavy meson production.

Finally I must point out the ability of Saturne 2 to give high intensity beams of variable energy light ions (d,^{3}He and α) and, above all, polarized deuterons in the near future.

ACKNOWLEDGMENT - REMERCIEMENTS

I wish to thank Dr R.D. Bent and the organizing committee for their invitation at this successful workshop, which was supported in part by the U.S. National Science Foundation and the U.S. Department of Energy.

Je remercie R. Bertini, R. Frascaria, T. Hennino et B.H. Silverman d'avoir accepté de me communiquer des résultats expérimentaux encore préliminaires ou non publiés et de m'apporter les précisions nécessaires sur des expériences qui m'étaient parfois peu familières. Je tiens enfin à remercier tout particulièrement G. Bruge pour la lecture critique qu'il a faite de ce manuscrit ainsi que Elizabeth Devlin-Couvert qui a corrigé et clarifié la rédaction en anglais du texte présenté ici.

REFERENCES

1. J.J. Domingo et al., Phys. Lett. 32B, 309 (1970).
   S. Dahlgren et al., Phys. Lett. 35B, 219 (1971).
2. For a complete review of experimental results on (p,π) reactions, see for example : H.W. Fearing, Prog. Part. Nucl. Phys.(1981) under print and H.W. Fearing, "a bibliography and summary of data for the (p,π) reaction", TRIUMF internal report n° TRI-80-3.
3. B. Tatischeff et al., Phys. Lett. 63B, 158 (1976).
   T. Bauer et al., Phys. Lett. 69B, 433 (1977).
   E. Aslanides et al., Phys. Rev. Lett. 39, 1654 (1977).
   P. Couvert et al., Phys. Rev. Lett. 41, 530 (1978).
   M. Dillig et al., Nucl. Phys. A333, 477 (1980).
4. J. Thirion, J. Saudinos and P. Birien, Note CEA-N-1248 (1970) and R. Beurtey, Brentwood Summer School Lectures, B.C. Canada, June 1975.
5. E. Aslanides, K. Baba, A.M. Bergdolt, O. Bing, A. Boudard, G. Bruge, P. Couvert, J.L. Escudié, F. Hibou, P. Kitching, Y. Le Bornec, B.M.K. Nefkens, B. Tatischeff and N. Willis - DPh-N/ME Saclay - IPN Orsay - CRN Strasbourg collaboration.
6. P. Couvert and M. Dillig, Abst. Contrib. papers, 9-ICOHEPANS, Versailles (1981) p. 192.
7. A. Boudard, W.J. Briscoe, G. Bruge, L. Farvacque, C. Glashausser, J.C. Lugol, B.M.K. Nefkens and B.H. Silverman, DPh-N/ME Saclay - UCLA collaboration, private communication.
8. J. Arvieux, R. Bertini, G. Bruge, Ph. Catillon, H. Catz, J.M. Durand, L. Farvacque, C. Glashausser, B. Mayer, G. Smith, A. Yavin and C. Whitten - DPh-N/ME - DPh-N/HE - LNS (Saclay) collaboration ; poster session at 9-ICOHEPANS (Versailles) July 1981 and to be published.
9. See review article by H. Spinka, Proceed. Workshop Nucl. Part. Phys. at Energies up to 31 GeV, Los Alamos, January 1981.
10. J. Thirion et P. Birien, Projet d'analyseur à 3,8 GeV/c, DPh-N/ME internal report, March 1975, unpublished.

11. J. Banaigs, J. Berger, J. Duflo, L. Goldzahl, F. Plouin, R. Frascaria, P. Berthet, B. Tatischeff, F. Fabbri, P. Picozza, L. Satta and M. Boivin ; ER54 Saclay, IPN Orsay, Frascati, LNS Saclay collaboration, Abst. Contrib. papers, 9-ICOHEPANS, Versailles, July 1981, p. 186 and private communication.
12. D. Bachelier, J.L. Boyard, T. Hennino, J.C. Jourdain, M. Roy-Stephan, P. Radvanyi, M. Gusakow, J.R. Pizzi, J.Y. Grossiord, A. Guichard, M. Bedjidian, R. Haroutunian, E. Descroix and P. Foessel, IPN Orsay - IPN Lyon - LNS Saclay collaboration, private communication.

DISCUSSION

G. Jones (TRIUMF-UBC): Purely a technical question. What is the time structure of the beam and the burst rate? Can you extract the beam over a long spill?
Couvert: The time structure of the beam is much better on the new accelerator than on the old one. Most of it is suppressed by the use of a feedback system but some still remains. Concerning the beam time repetition it was about one burst per second for 1 GeV protons with a 400 milliseconds spill on the tartet. More recently an even better duty cycle was obtained with a 850 milliseconds spill for 1.4 second cycles.

L.S. Kisslinger (Carnegie-Mellon Univ.): I have a question about that second bump in $pd \rightarrow t\pi^+$. You said that you felt it was two deltas instead of a single resonance. Is that only from the energy, or did you have some other criterion.
Couvert: That is from the total energy in the center of mass. I only made the calculation corresponding to the prediction of one nucleon and two deltas. It gives 3.4 GeV in the center of mass.
Kisslinger: Do you know where the isospin zero dibaryon would fit on that?
Couvert: No.
M. Huber (Univ. or Erlangen-Nürnberg): The various dibaryons should be around 3.0 GeV within ± 100 MeV.
Kisslinger: Could it be down at that second shoulder?
Couvert: In the sense of your question I can just mention that the the Saclay group who reported on that experiment at Versailles speculated that it could be a tribaryon effect. But I know they gave up on that assumption and prefer the two deltas interpretation.
Kisslinger: There is still a surviving dibaryon. The question is whether it might be showing up in your experiment.
What are the solid lines on this graph?
Couvert: I do not have precise information about these calculations. I wanted only to show the experimental results.

A. Bacher (IUCF): What is your interest in looking for rho production?
Couvert: The rho meson has an important role in intermediate energy physics, but from an experimental point of view it is very difficult to detect because of its 160 MeV width. For the moment, the possibility to do such a rho production experiment on deuterium, for example, or on very simple nuclei without any excited states (because of the width of the rho) is just a speculation. We don't know if we can or will do it. If there are some good arguments for doing such an experiment, we would be glad to hear them.

THE (p,π) PROGRAM AT TRIUMF: PAST, PRESENT AND FUTURE

G.J. Lolos
University of British Columbia, Vancouver, B.C., Canada V6T 2A6

1. INTRODUCTION (THE PAST)

The (p,π) program was established at TRIUMF as early as 1976 with the development of a dedicated pion spectrometer. The Browne-Buechner spectrometer employed was coupled to a hodoscope array of plastic scintillator counters along the focal plane for pion momentum definition. The overall resolution of the system was typically ~1.5 MeV FWHM at ~65 MeV of pion kinetic energy. The maximum central ray energy for the pions was ~65 MeV which meant that for exclusive reactions on nuclear targets the maximum incident proton energy was of the order of 200 MeV. The solid angle was of the order of ~4 msr.[1]

Although the spectrometer was of limited use as a pion spectrometer because of the above restraints, it produced some very important results especially in exploring polarization effects on (p,π) reactions. One defines the analyzing power A_y as

$$A_y = \frac{d\sigma/d\Omega(\uparrow) - d\sigma/d\Omega(\downarrow)}{P(\downarrow)d\sigma/d\Omega(\uparrow) + P(\uparrow)d\sigma/d\Omega(\downarrow)},$$

where ($\uparrow$) and ($\downarrow$) indicate the spin configuration of the incident polarized proton as up and down, respectively (according to the Madison convention[2]) and P is the magnitude of the polarization.

The pioneering work at TRIUMF measuring analyzing power for exclusive nuclear (p,π) reactions revealed some surprising results. For the two nuclei studied, ^{9}Be and ^{12}C, the analyzing power angular distributions for all states observed in ^{10}Be or ^{13}C were very similar.[3] In addition the observed $A_y(\theta)$ was very similar to the analyzing power for the $\vec{p}p \to d\pi^+$ reaction also measured at TRIUMF with the same spectrometer.[4] These results seemed to indicate a strong reaction mechanism domination, rather than nuclear structure effects as one would expect from a simple single-nucleon model (SNM).

With this spectrometer showing the way and with these early A_y results whetting their appetite, the (p,π) group at TRIUMF decided to move on to bigger and better things, in this case a new Browne-Buechner magnetic spectrograph employing helically wound delay line multiwire proportional chambers (MWPC)[5] for pion momentum definition. This spectrograph is in use up to the present time, and it enabled the useful pion energy range to be extended from ~30 MeV to ~120 MeV. That in its turn enables one to extend the incident proton energy to ~260 MeV for exclusive nuclear (p,π) reactions. It has improved resolution of the order of ~800 keV, and it can handle beam intensities in excess of 30 nA, in contrast to ~1-5 nA useful current with the old spectrometer. Limiting factors are the small solid angle, ~3 msr, and the restricted angular range of 46° to 135°.

The new spectrometer and the use of MWPC's has greatly enhanced not only the intrinsic resolution but the background rejection as well. The event definition is provided by a three-scintillator

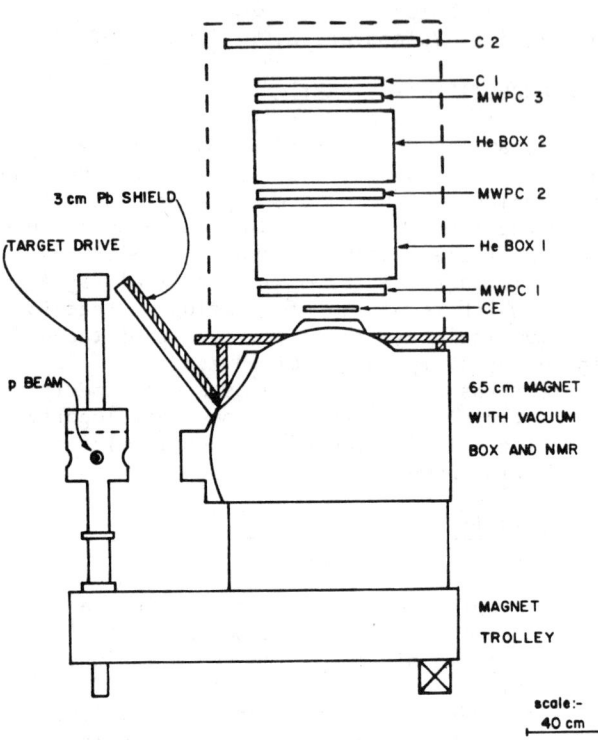

H.M.S. RESOLUTION

Fig. 1. The present pion spectrometer for the study of (p,π) reactions at TRIUMF.

counter telescope, one covering the exit of the spectrometer and optimized for timing and the other two larger ones providing timing as well as dE/dx information. The spectrometer layout is shown in Fig. 1. The He boxes are placed between the MWPC's in an attempt to reduce the multiple scattering of the pions in their flight path through the three chambers. Particle identification is based on time of flight and energy loss in the scintillators. The effects of pole-face scattering, multiple scattering as well as pion decay are greatly reduced by track reconstruction, restrictions on both the x and y plane pion trajectories as well as extrapolation to the target. An example of the background rejection efficiency is shown in Fig. 2 for "worst case scenario" of (p,π^-) reactions.

2. EXPERIMENTAL RESULTS (THE PRESENT)

A. Reactions of the type $A(\vec{p},\pi^{\pm})A+1$

The first set of results completed in the 200 to 250 MeV incident proton energy range was a study of the $^9\text{Be}(\vec{p},\pi^-)^{10}\text{C}^*$ reaction. This reaction has also been studied at IUCF[6] for 200 MeV. At TRIUMF the energy dependence of the analyzing power $A_y(\theta)$ for the ground state, as well as for the first three excited states, was explored.[7-9] While the ground state showed very little energy dependence in the 200-250 MeV region, both the 3.35 and 6.60 MeV showed significant variation. In addition the states examined displayed markedly different $A_y(\theta)$ depending on state. The energy variation of $A_y(\theta)$ for the 3.35 and 6.60 MeV $^{10}\text{C}^*$ states is shown in Fig. 3(a,b).

The normalized yield for the $^{10}\text{C}_{g.s.}$ transition exhibits a more forward peaked structure at 200 MeV than suggested by the essentially

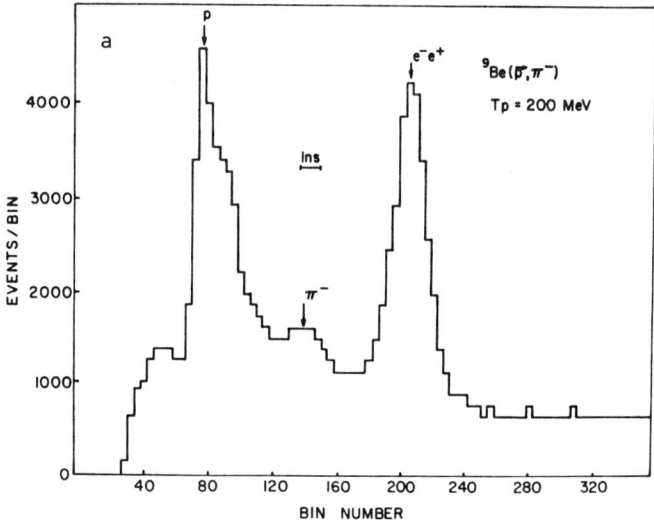
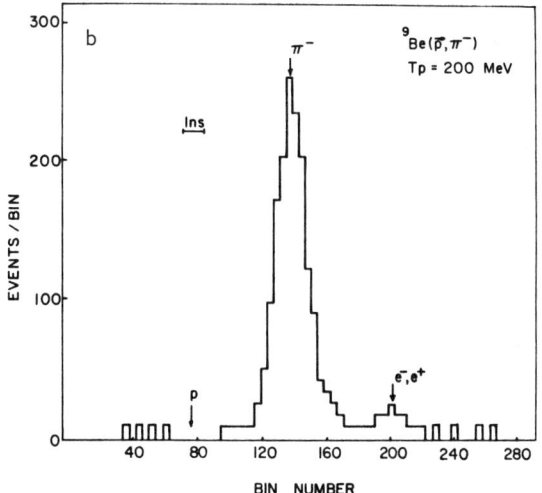

Fig. 2. Time-of-flight spectra for the $^9\text{Be}(\vec{p},\pi^-)$ reaction at T_p = 200 MeV (a) before any background subtraction, and (b) after event-by-event analysis and data reduction.

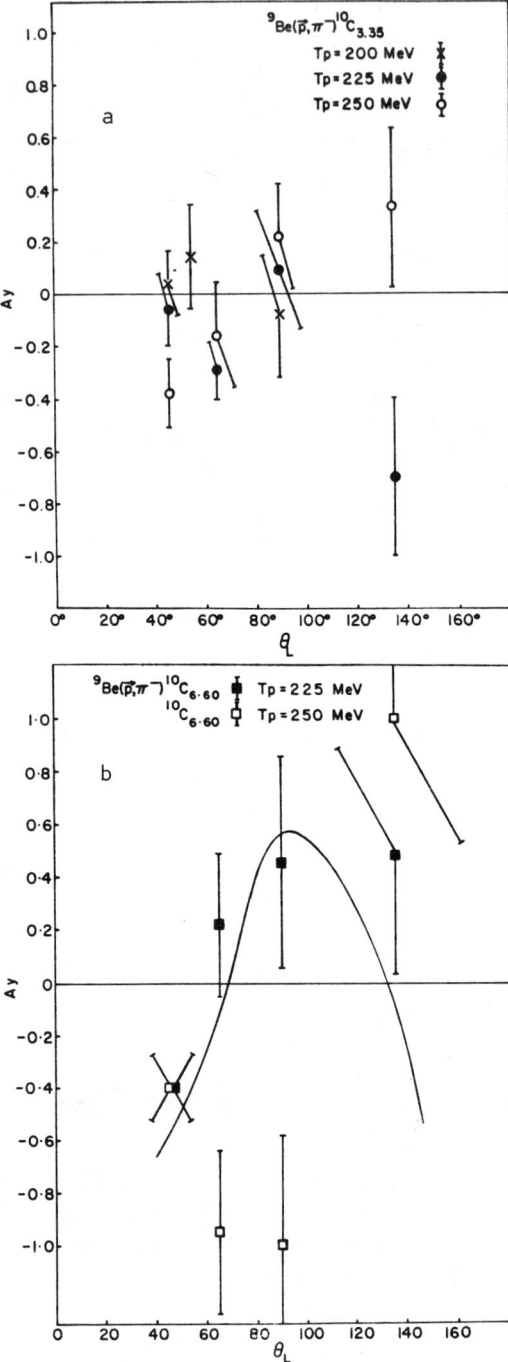

Fig. 3. The energy dependence of $A_y(\theta)$ for (a) the $^{10}C^*_{3.35}$ and (b) $^{10}C^*_{6.60}$ states. The solid line serves as a guide to the eye for the shape of $A_y(\theta)$ exhibited for the transition to the $^{10}C_{g.s.}$

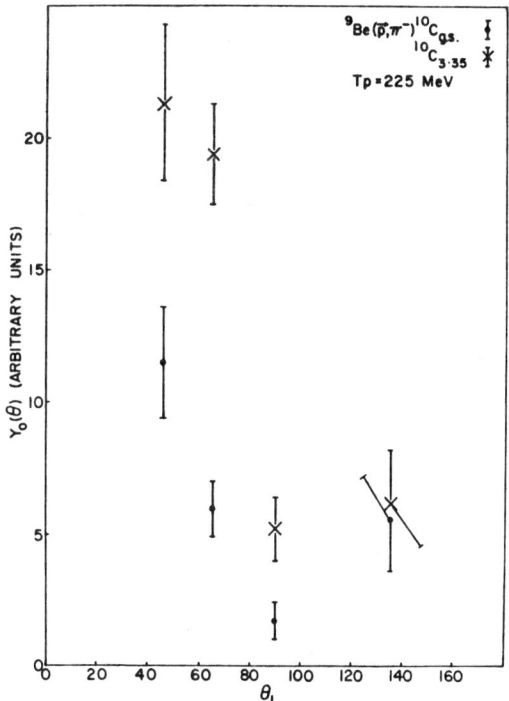

Fig. 4. The yield angular distribution for the $^{10}C_{g.s.}$ (solid points) and the $^{10}C^*_{3.35}$ (crosses) states from the $^9Be(\vec{p},\pi^-)^{10}C$ reaction, for $T_p = 225$ MeV.

flat IUCF results.[6] At 225 MeV there is a strong forward peaking, Fig. 4, which becomes even more pronounced at 250 MeV. There is also evidence for a back angle increase in the yield, which at 225 and 250 MeV is in good qualitative agreement with the backward rise in the cross section in the 90°-135° region reported by the IUCF group.

Turning to the more prolific $(\vec{p},\pi^+)$ reactions, two nuclei have been examined in detail recently. The $^9Be(\vec{p},\pi^+)^{10}Be$ reaction studied at 225 and 250 MeV, with some points taken also at 200 MeV, investigates the energy dependence of the $A_y(\theta)$ in the above energy range. To my knowledge no data using polarized protons have been collected from ^9Be below 200 MeV; within the 200-250 MeV range no significant overall energy dependence in $A_y(\theta)$ has been seen.[10,11] The characteristic dip with large negative values seen by Auld et al.[3] is still prominent for both the $^{10}Be_{g.s.}$ and the $^{10}Be^*_{3.37}$ states, as can be seen in Fig. 5. In the back angles, however, the picture is quite different introducing another minimum at ~120° for the $^{10}Be_{g.s.}$ transition at 250 MeV. Within the limited angular range, this large angle oscillation is as prominent at 250 MeV as was the 60° dip at all the energies examined here. In addition the $A_y(\theta)$ for the two states is not as similar at 250 MeV as it appeared at 200 MeV. Although, then, some energy dependence is seen in the large angle analyzing power, the characteristic

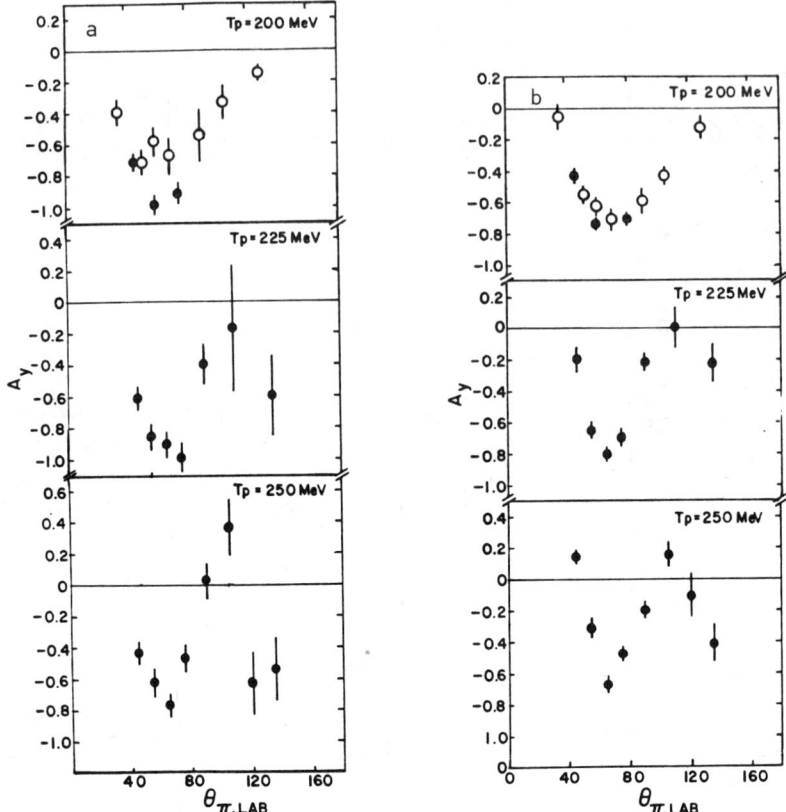

Fig. 5. The analyzing power dependence on incident proton energy for (a) the $^{10}Be_{g.s.}$ transition and (b) the $^{10}Be^*_{3.37}$ transition. The solid circles are results obtained with the present spectrometer, the open circles are from Ref. 3.

features of the "pp → dπ⁺" like channel observed at 200 MeV[3] is still very much present.

The other nucleus studied at TRIUMF was ^{12}C; analyzing power measurements existed for this nucleus below 200 MeV taken by the IUCF at 159 MeV.[12] Their results indicated that no variation with energy, to any significant degree, occurs between 159 and 200 MeV. Our results at 225 and 250 MeV, however, indicated a very dramatic energy dependence.[13] The ground state transition (Fig. 6) shows the energy dependence that takes place primarily in the 200-225 MeV region. The solid circles represent the latest set of data while the crosses represent earlier attempts on the $^{12}C(\vec{p},\pi^+)^{13}C$ reaction. The open circles are taken from the Auld et al.[3] results with the old spectrometer.

In light of the ^9Be results as well as the energy independence observed for the ^{12}C from threshold to 200 MeV region, this sudden change in analyzing power distribution came as a surprise, to say the least.

One hopes that this new element in the $(\vec{p}, \pi^+)$ subject will shed light on the reaction mechanism instead of complicating our limited understanding even further. The analyzing power for the $^{13}C_{3-4}$ MeV group of transitions exhibits the same effect as observed for the $^{13}C_{g.s.}$ transition; the only transition not observed to show any significant variation between 225 and 250 MeV (no 200 MeV data exist) is the $^{13}C^*_{9.50}$ as seen in Fig. 7. The $^{13}C^*_{9.50}$ configuration is a probable 2p-1h in the shell model while the $^{13}C_{g.s.}$ is best described by a single particle (sp) configuration. Whether, of course, this differentiation in the $A_y(\theta)$ observed in Figs. 6 and 7 indicates a reaction sensitivity to these particular nuclear structure configurations remains to be verified.

The differential cross sections for the $^{13}C_{g.s.}$ transition at 225 and 250 MeV indicate a backward peaking with a cross section minimum at ~105° (Fig. 8). This is the first time a back angle increase has been observed in the $^{12}C(p,\pi^+)^{13}C_{g.s.}$ reaction. The backward peaking is more pronounced at 250 MeV than it is at 225 MeV.

In an attempt to look in more detail at the energy dependence seen in the $^{12}C(\vec{p},\pi^+)^{13}C$ reaction, measurements were taken at 216 MeV as well as 237 MeV of incident proton energies. The experiment has just been completed and some of the data at 216 MeV have been analyzed. The

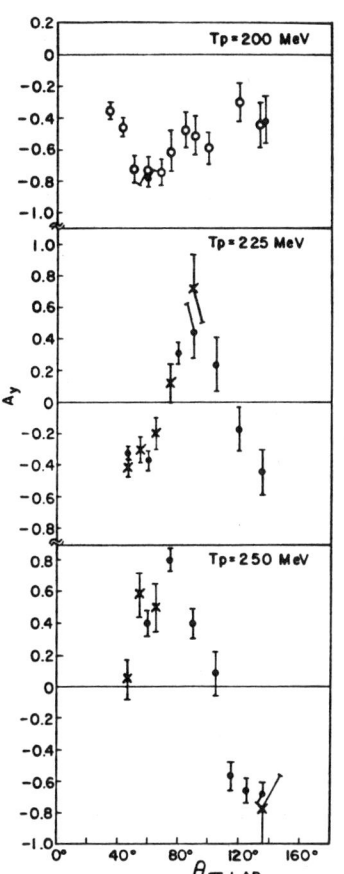

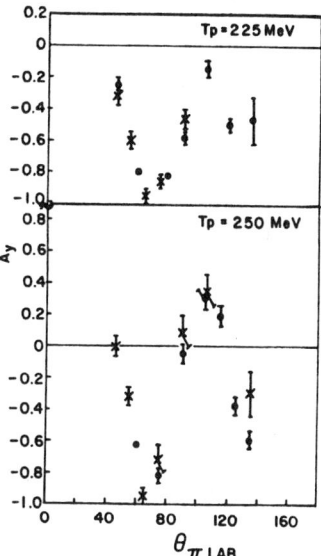

Fig. 6. The energy dependence of $A_y(\theta)$ for the $^{13}C_{g.s.}$ transition in the 200-250 MeV incident proton energy range.

Fig. 7. The analyzing power for the $^{13}C^*_{9.50}$ transition at 225 and 250 MeV incident proton energy.

The results are only preliminary but they indicate a smooth transition between the 200 MeV and 225 MeV results; a spot check on the "critical angle" of 60° for 210 MeV protons also indicates a smooth transition from 200 to 225 MeV at least for the $^{13}C_{g.s.}$ transition analyzing power. It is hoped that the analysis will be completed shortly and the data published in the future.

Some initial measurements were also obtained from the $^{10}B(\vec{p},\pi^+)^{11}B$ reaction and the analyzing power angular distribution was extracted for a limited number of angles. Our preliminary results do not indicate any energy dependence between 225 MeV and the IUCF results at 154.5 MeV. There is, however, an indication of a strong energy dependence between 225 and 250 MeV as can be deduced from Fig. 9(a,b). Again these are preliminary results and we are

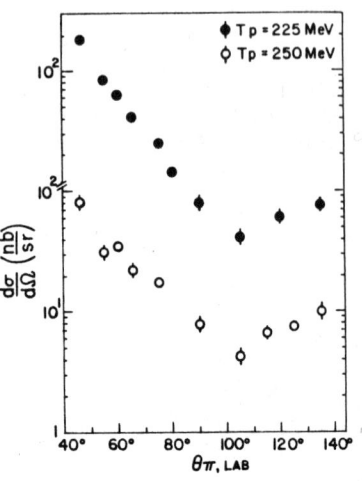

Fig. 8. The differential cross section for the $^{12}C(\vec{p},\pi^+)^{13}C_{g.s.}$ reaction at 225 and 250 MeV incident proton energy.

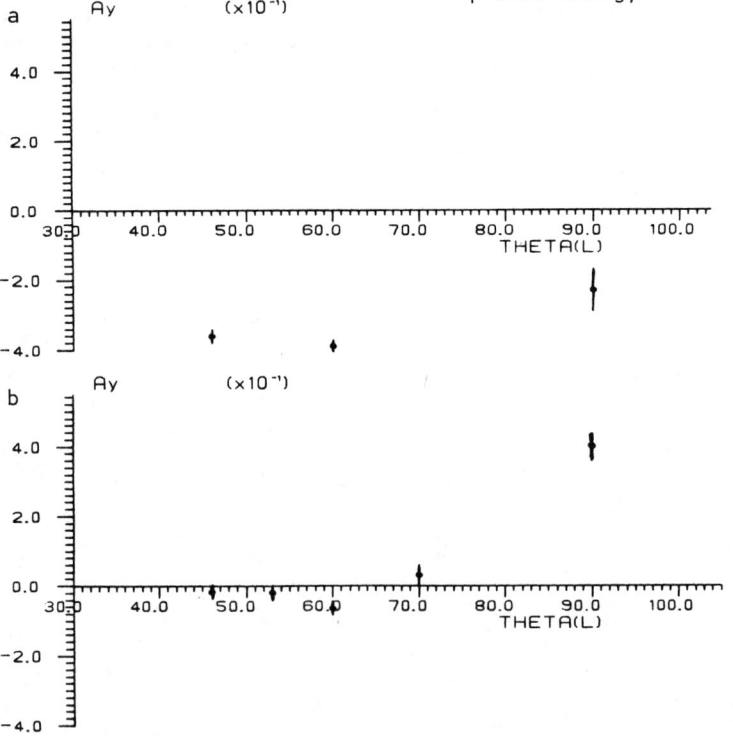

Fig. 9. Preliminary results of the analyzing power $A_y(\theta)$ for the $^{10}B(\vec{p},\pi^+)^{11}B_{g.s.}$ reaction at (a) T_p = 225 MeV and (b) T_p = 250 MeV.

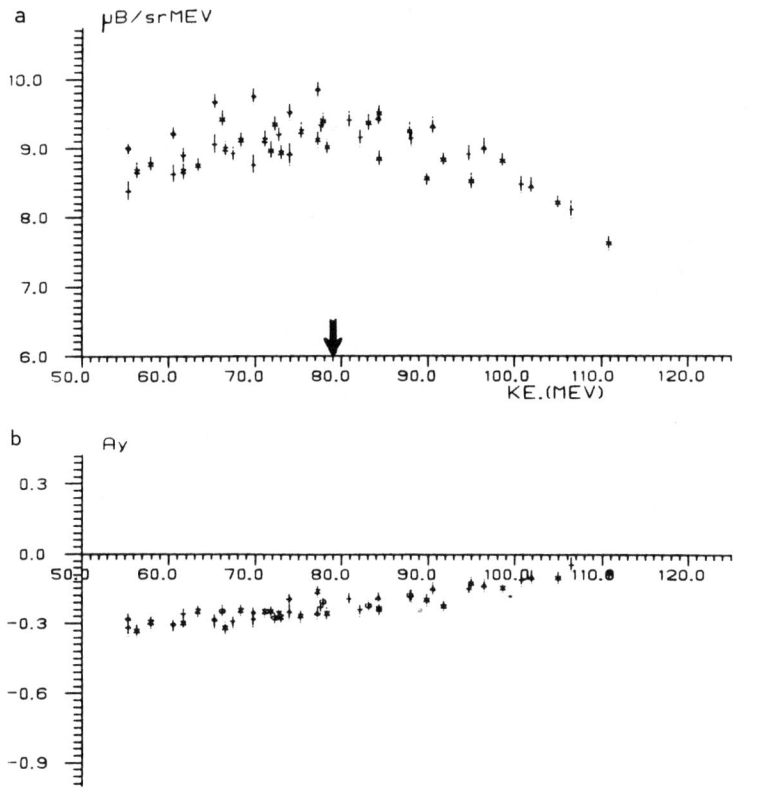

Fig. 10. Results from the $^{12}C(\vec{p},\pi^+)X$ reaction at $\theta_{lab} = 46°$ and $T_p = 400$ MeV for (a) the differential cross section (the arrow indicates the equivalent π^+ energy from the $\vec{p}p \to d\pi^+$ reaction) and (b) the analyzing power.

awaiting the delivery of new enriched ^{10}B targets to complete our measurements.

B. Reactions of the type $A(\vec{p},\pi^+)X$

The inclusive reaction $^{12}C(\vec{p},\pi^+)X$ results presented here are a byproduct of the spectrometer calibration via the $\vec{p}p \to d\pi^+$ reaction at 400 and 450 MeV. The hydrogen target was a CH_2 polymer, and as a result a ^{12}C background subtraction is necessary. As a result of the inclusive reaction analysis results we see evidence of a cross section enhancement in the quasi-free deuteron pion kinematic range. This is the first indication of a quasi-free deuteron process from (p,π) reactions. The data are still under analysis and we hope to have all the final results soon. The cross section and analyzing power results for one angle at 400 MeV incident proton energy are shown in Fig. 10(a,b).

C. Reactions of the type NN → πd and NN → NNπ⁺

The reaction $\vec{p}p \to d\pi^+$ has been used extensively to calibrate our spectrometer for extraction of absolute cross sections as well as for investigations of line-shape measurements, effects of pole-face scattering on resolution and solid-angle acceptance as a function of beam spot location on the target. For all these measurements a polyethylene type of target (CH_2 polymer) was used. Included in the interest of performing these measurements is the physics, in addition to spectrometer characteristics, that one can extract. We have obtained the analyzing power for the $\vec{p}p \to d\pi^+$ reaction at 400 and 450 MeV. The 400 MeV data were obtained as a check of consistency with earlier results[4] while the 450 MeV $A_y(\theta)$ are first time results.[14] The results are shown in Fig. 11. There is good agreement with the old 400 MeV results as well as with the theoretical values of Niskanen.[15]

One more interesting piece of information extracted out of the $\vec{p}p \to d\pi^+$ measurements with a CH_2 target is the evidence of a singlet state in the pn system. The deuteron breakout threshold occurs at 2.12 MeV excitation of the deuteron. In the past attempts at TRIUMF to observe a singlet state in the pp → pnπ⁺ reaction had been foiled by the pressure of a long, low-energy pion tail in the pion spectrum. This low-energy pion tail was mainly the result of magnet-poleface scattering in the spectrometer. Monte-Carlo simulation revealed that the effect due to scattering was so severe that even breakout cross section could not be extracted reliably for deuteron excitation less than ~15 MeV.[1]

The present spectrometer is equipped with antiscattering baffles but the biggest advantage with the present system originates from the pion track reconstruction that the three MWPC's made possible. Tight

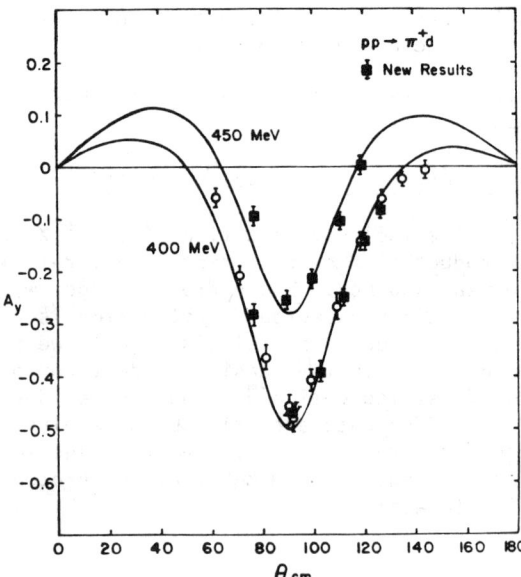

Fig. 11. The analyzing power for the $\vec{p}p \to d\pi^+$ reaction for T_p=400 MeV and T_p=450 MeV. The solid line is the calculated $A_y(\theta)$ from Ref. 15. The open circles are from Ref. 4.

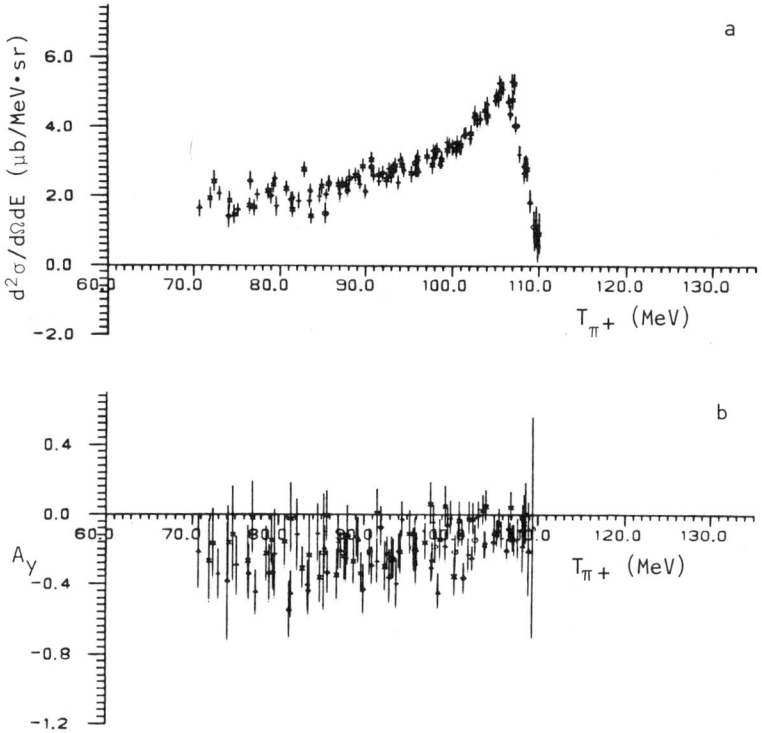

Fig. 12. The deuteron breakup reaction at $\theta_{lab} = 46°$ and $T_p = 450$ MeV (a) the double differential cross section and (b) the analyzing power. The different symbols represent different sets of experimental data.

constraints on the pion track exit angles in the non-bend plane of the spectrometer have reduced the pole-face scattering significantly, to the extent that pions due to real deuteron events can be subtracted from the total pion spectrum. The results, still of preliminary nature, strongly suggest an enhancement of the $pp \to pn\pi^+$ differential cross section corresponding to a singlet state in the pn system. A representative set of the cross section and analyzing power for the deuteron breakup reaction is shown in Fig. 12. The data obtained in this series of measurements are in the last stages of analysis. Again we hope to have the final results in the near future.

One assumption that is very critical in the analysis of the $\vec{p}p \to pn\pi^+$ data is concerning the shape of the low-energy tail of the pion spectra due to pole-face scattering; a pronounced shoulder on the low-energy side of the $\vec{p}p \to d\pi^+$ pion spectrum could artificially create the cross-section enhancement seen in Fig. 12. In order to gain an insight to the true shape of the pion spectrum a number of experiments were performed using a silicon surface barrier detector to pick up the deuteron. The results were encouraging and a purposely designed beam line horn was installed to accommodate a plastic scintillator counter

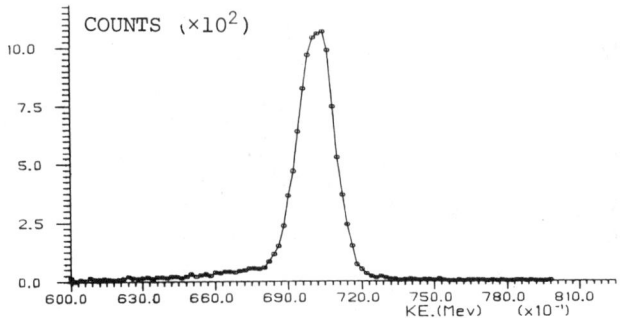

Fig. 13. Pion spectrum from the pp → dπ⁺ reaction when both the π⁺ and the deuteron are detected.

to pick up the recoil deuterons. The results, just being completed and in a very preliminary stage of analysis, indicate a pion spectrum very similar to the spectrum shape used to subtract the pions from the $\vec{p}p \to d\pi^+$ contribution. The pole-face scattering is small and at this stage of the analysis does not appear to contribute to the shape of the cross section in Fig. 12. A characteristic pion spectrum in the $\vec{p}p \to d\pi^+$ reaction in a double-arm experiment is shown in Fig. 13.

The study of the $\vec{p}p \to d\pi^+$ reaction described above has a modest primary justification, that of spectrometer "calibration". At TRIUMF there is an ongoing study of the pp → dπ⁺ reaction, on the other hand, that has more solid physics justifications. The two-arm experiment aims to measure the pion production cross section to ~1% accuracy in the 325 to 500 MeV incident proton energy range. The pion differential cross section can be given by the expression in terms of associated Legendre polynomials:

$$\frac{d\sigma}{d\Omega} = \frac{1}{4\pi}\left(a_0 + a_2 P_2(\cos\theta) + a_4 P_4(\cos\theta) + \ldots\right).$$

In the TRIUMF energy range the a_0, a_2 and a_4 parameters' energy dependence is not all that well known. A precision measurement in the energy region up to 500 MeV will complement the data taken at SIN (513-583 MeV)[16] and LAMPF (500-800 MeV).[17] It may also help determine whether a nonzero a_6 term is necessary as might be suggested from the SIN results. This experiment is in progress and although data have been obtained for T_p = 500 MeV the analysis is at the early stages yet.

3. FUTURE DEVELOPMENTS

A. Technical developments

The future course of action for the $(\vec{p},\pi^\pm)$ program at TRIUMF depends greatly on the equipment that will be available for $(\vec{p},\pi^\pm)$ work. A new spectrometer is urgently needed that is built on the particular requirements associated with (p,π) reactions. A design has been completed, on such a spectrometer, and certain items such as iron for the magnets and copper for the coils have been purchased. The design aims for a solid angle of Ω = 14.8 msr, a good momentum acceptance, $\Delta P/P = \pm 22\%$, and a central ray length $\ell = 5.04$ m. Two most important characteristics of the new spectrometer are an extended angular range of 20° to 160° and a maximum momentum of 600 MeV/c.

Such a spectrometer would allow TRIUMF to make use of the full 200-500 MeV incident proton energy range for a complete energy dependence investigation of pion production on nuclei. The improved resolution, from future machine and beam line improvements, in combination with the larger solid angle, will allow experiments, like $(\vec{p},\pi^-)$ reactions, to be completed in a realistic length of time with good state definition. We believe the above spectrometer could be operational by 1983 if it gets funded soon on a high priority basis.

In the event of delays in deciding to allocate funds for the new spectrometer, there are certain short term improvements that can be made on the present spectrometer. Highest in the shopping list would be new coils that will increase the useful beam energy for exclusive (p,π) reaction to ~350 MeV; this would allow a wider energy dependence study of the $A_y(\theta)$ as well as the cross section. The angular range of the present spectrometer can also be extended with the addition of a quad to compensate for the increased target-to-spectrometer length. The solid angle should not be reduced from the present value.

There is no doubt decisions will have to be made soon if the (p,π) program at TRIUMF is to remain viable. All the above options are under active consideration at present, but decisions cannot be further postponed. Below is a list of the future experimental plans that do not depend on spectrometer design at this stage.

B. $A(\vec{p},\pi^{\pm})A+1$ and $A(\vec{p},\pi^{\pm})X$ reactions

Although some measurements have been taken at 225 and 250 MeV from the $^{10}B(\vec{p},\pi^+)^{11}B$ reaction, the results were limited. We plan to complete the ^{10}B measurements in the 200-275 MeV energy range; since data already exist for ^{10}B below 200 MeV[12] we will have for the first time a thorough investigation of the energy dependence of both analyzing power and differential cross section. Two new ^{10}B targets enriched to ~92% have been ordered and should be available soon.

The reaction we plan to investigate next is the $^{16}O(\vec{p},\pi^+)^{17}O$. For this reaction there are also IUCF results at threshold; in addition the $^{17}O^*_{0.87}$ state is the only state below 200 MeV ever observed to display $A_y(\theta)$ different from the "$\vec{p}p \to d\pi^+$"-like structure observed in all the other light nuclei.[12] It would be of interest to trace the energy dependence of this state, as well as the $^{17}O_{g.s.}$, in the 200-275 MeV range.

The $A(\vec{p},\pi^-)A+1$ reaction studies have barely scratched the surface of the physics hidden underneath. With the recently displayed capability of sustained operations with 30-40 nA of beam intensity, it would be worth while to look at selected $(\vec{p},\pi^-)$ reactions; the $^{12}C(\vec{p},\pi^-)^{13}O$ is a likely candidate. This reaction has been studied at 200 MeV[18] as well as 613 MeV[19]; at 613 MeV the observed first excited state is at ~2.82 MeV excitation and therefore the separation is more than adequate for our resolution capabilities to resolve them.

C. Reactions of the NN → πd and NN → NNπ⁺ type

We plan to continue the experiment already in progress (see preceding section) of the accurate determination of the pp → dπ⁺ reaction cross section. Using the same experimental equipment we plan to measure the pion production asymmetry in the $\vec{p}p \to d\pi^+$ reaction in the 400-520 MeV incident proton energy range. The A_{y_0} parameter has not been measured in detail between 425 and 513 MeV; the experiment to be performed will investigate the region where the pion analyzing power becomes from strongly negative at 425 MeV[4] to positive at 513 MeV.[20] The accuracy will also be in the ~1% region.

D. Pion absorption experiments

Three different pion absorption experiments have been proposed. All three are to use light nuclei as targets:

(i) <u>Proton polarization in the $\pi^+ + d \to \vec{p} + p$ reaction</u>. Incident pion energy will be initially 145 MeV; at this incident pion energy the c.m. energy will be the same as that of the inverse reaction $\vec{p} + p \to d\pi^+$ at $T_p = 578$ MeV studied at SIN.[16] The test of such equivalence should provide a test at time reversal invariance via the P-A measurements. The pion energy will be subsequently raised to 165, 185 and 205 MeV. Such measurements are necessary to define the contribution of pion s-wave rescattering in this energy region.

The above experiment is viewed as the stepping stone for the $\pi^+ \vec{d} \to 2\vec{p}$ reaction to follow as a part of a long program of precision measurements of the elementary channel $(\vec{p})(\vec{p}) \to d\pi^+$ and its inverse.

(ii) <u>π± absorption on ³He</u>. Incident pion energies will be 30, 50 and 85 MeV. The reactions will be ³He(π⁺,2p) and ³He(π⁻,pn), and both outgoing nucleons will be detected. These experiments should provide information on pion absorption in a ³S₁ T=0 nucleon pair at normal nuclear density; the deuteron density is lower than nuclear density. They will provide, in addition, information on absorption on three nucleons (³He+π⁺ → 3p) as well as information on absorption in a ¹S₀ T=1 nucleon pair ³He(π⁻,pn).

(iii) <u>Study of the π⁺+⁶Li → ³He+³He reaction</u>. Incident pion energies will be 30, 70 and 90 MeV. The scientific value of this experiment lies in its capability to investigate coherence effects in the reaction mechanism. The inverse reaction has been studied earlier close to threshold,[21] but the present experiment will extend the energy dependence of the cross-section measurements. It will be of interest to determine whether the ³He production is due mainly to a two-nucleon emission (π⁺,2p) followed by deuteron pick-up (p,³He) or it is due to absorption on a ³H cluster in ⁶Li.

It is hoped this will be a first leg of pionic fission studies extending later to π⁻+⁶Li → ³H+³H and π⁺+⁷Li → ³He+⁴He reactions.

E. Conclusion

There is a wealth of pion production and pion absorption experiments either under way, or proposed for the near future, at TRIUMF.

The nuclear $(\vec{p},\pi)$ program has produced some notable firsts in analyzing power measurements that have introduced severe constraints in the theoretical treatment of $(\vec{p},\pi)$ exclusive reactions. We have seen strong evidence of quasi-free deuteron process in inclusive $(\vec{p},\pi^+)$ reaction(s) as well as evidence of a singlet state in the pn system. The pion absorption experiments, and the $\pi^+\vec{d} \rightarrow 2\vec{p}$ in particular, will be at the forefront of target and detector development. We are looking forward to interesting and productive next few years.

REFERENCES

1. E.L. Mathie, Ph.D thesis, University of British Columbia 1980 (unpublished).
2. Madison Convention, in Proc. Third Int. Symposium on Polarization Phenomena in Nuclear Reactions, Madison, Wisconsin, 1970, eds. H.H. Barschall and W. Haeberli (Univ.Wisconsin Press, Madison, 1971).
3. E.G. Auld et al., Phys. Rev. Lett. 41, 462 (1978).
4. P.L. Walden et al., Phys. Lett. 81B, 156 (1979).
5. D.M. Lee et al., Nucl. Instrum. Methods 120, 153 (1974).
6. T.P. Sjoreen et al., Phys. Rev. Lett. 45, 1769 (1980).
7. G.J. Lolos et al., 5th Int. Symp. on Polarization Phenomena in Nuclear Physics, Santa Fe, 1980, AIP Conference Proceedings No. 69, Part 1 (AIP, New York, 1981), p. 550.
8. G.J. Lolos et al., 9-ICOHEPANS, Versailles, 1981, contributed paper.
9. G.J. Lolos et al., Phys. Rev. C (to be published).
10. E.G. Auld et al., 9-ICOHEPANS, Versailles, 1981, contributed paper.
11. E.G. Auld et al., Phys. Rev. C (to be published).
12. T.P. Sjoreen et al., IUCF preprint No. 142 (1981).
13. G.J. Lolos et al., Phys. Rev. C (to be published).
14. W.R. Falk et al. (to be published).
15. J.A. Niskanen, Nucl. Phys. A298, 417 (1978).
16. NESIKA collaboration, A. Green "The Few Body Problem", August 1980, 113.
17. H.S. Nann et al., Phys. Lett. 88B, 257 (1979).
18. B. Höistad et al., Phys. Lett. 94B, 315 (1980).
19. P. Couvert et al., Phys. Rev. Lett. 41, 530 (1978).
20. P. Chatelain et al., SIN Newsletter No. 30, November 1980.
21. N. Willis et al., in Abstracts of Int. Conf. on Nuclear Physics, Berkeley, August, LBL report LBL-1118, p. 914 (1980).

DISCUSSION

<u>G.A. Miller</u> (Univ. of Washington): In pp → dπ, why do you want to measure the angular distribution to one percent accuracy?
<u>Lolos</u>: If you look at the state of affairs in the TRIUMF energy range, the pp → dπ amplitudes are not known to any precision. I wasn't aware of the Rössle results. One of the reasons also is, one would like to have such accuracy to see if there is any a_6 component. SIN results have suggested there may be.

<u>P. Couvert</u> (Saclay): About the time reversal experiments you are going to do: πd → pp and pp → dπ ... .
<u>Lolos</u>: We don't do it specifically for time reversal reasons. It's just that we want to have a test.
<u>Couvert</u>: With what accuracy do you plan to do that experiment?
<u>Lolos</u>: The SIN results are about 5%, and we hope to do the experiment at the same 5% level.

<u>A.D. Bacher</u> (IUCF): Do you have a prediction for what the size of the time reversal violating effect might be?
<u>Lolos</u>: Probably smaller than 1%, if there is any.
<u>Bacher</u>: There are some calculations in the literature. It is expected to be larger in the pn system than it would be in the pp system.

<u>T. Londergan</u> (Indiana Univ.): I would like to point out that the time-reversal-violating effects calculated (by Bryan and Gersten) using Sudarshan's model should not be considered very seriously. The model of Sudarshan was constructed to have no weak neutral currents. One consequence of Sudarshan's model was a small time-reversal-violating amplitude. However, since weak neutral currents have now been discovered (confirming the Weinberg-Salam theory), there is no reason to use this theory as a guide to searching for time-reversal violations.

<u>J. Noble</u> (Univ. of Virginia): Do you have any measurements of the asymmetry in pp → npπ as well as pp → dπ?
<u>Lolos</u>: Yes. I showed them. We have done four angle measurements at 450 MeV.

<u>Noble</u>: You mentioned a calculation of the asymmetry in the continuum of ^{12}C. Did the person who did the asymmetry calculation also try and do the spectrum calculation?
<u>W. Falk</u> (Univ. of Manitoba): No, we didn't do any calculations of the energy spectrum. Its a very simple model, and we just concentrated on the analyzing powers.

<u>R.E. Pollock</u> (Indiana Univ.): I believe you have some analyzing power data for $^9Be(p,\pi^-)$. When the Indiana group measured that reaction at a lower energy about a year and a half ago, we were somewhat surprised to find that the angular distribution of the

cross section was backward-peaked. That behavior is really a strange beast in this game. I know that you were taking your data at a time when your Faraday cup was not yet calibrated, but do you have any indication of whether the angular distribution at somewhat higher energies remains backward peaked?

Lolos: It does. At 200 MeV we've got only some forward angles for the ^{9}Be(p,π^-)^{10}C(g.s.) reaction, and we saw a little bit more structure than the Indiana results would suggest. For the ground state we see a forward peak at 200 MeV, but if you look between 90 and 135° at higher energies, there is an indication of a backward rise, and if one takes the relative ratio of the yields between 90 and 135° from Indiana at 200 MeV and our yields at 225 MeV, we are in good agreement. So there is an indication of a backward peaking that is there as well at 250 MeV.

The $(p,\pi^{\pm})$ Reactions at 800 MeV

H. Nann

Physics Department, Indiana University, Bloomington, IN 47405

ABSTRACT

Angular distributions of the (p,π^+) reaction on ^{6}Li, ^{9}Be, ^{10}B, ^{12}C, ^{16}O and ^{40}Ca and of the (p,π^-) reaction on ^{9}Be have been measured at 800 MeV. Most of the differential cross sections show a straight exponential slope over the momentum transfer interval between 600 and 800 MeV/c which does not seem to be characteristic of the properties of the final states. High-spin states are excited with larger cross sections than low spin states.

INTRODUCTION

The understanding of the reaction mechanism of pion production has greatly hindered the interpretation of experimental results from the (p,π) reaction at all energies. Various experimental attempts to isolate specific attributes of the reaction mechanism and/or the nuclear structure involved have been made in the past with negligible success. So far no systematic behavior in the experimental differential cross sections has been found.

Pion production experiments began at the Los Alamos Meson Physics Facility about three years ago[1]. Some results from these studies are already published[2,3] and more data are still being analyzed[4,5]. In this report, results of (p,π) reaction at 800 MeV incident energy on target nuclei ranging from ^{6}Li to ^{16}O are discussed with the aim of extracting some systematic behavior in the observed differential cross sections.

EXPERIMENTAL PROCEDURE

The $(p,\pi^{\pm})$ experiments were performed with the 800 MeV proton beam from the Clinton P. Anderson Meson Physics Facility (LAMPF) using the high-resolution spectrometer (HRS). A detailed description of the spectrometer and focal plane detection system is given elsewhere[6]. Particle identification was achieved by measuring the time of flight in a 2 m flight path in the detector system, the energy loss in the scintillation counters and by a lucite Cherenkov counter. In general, the discrimination of pions against all other particles was quite good, resulting in a very small background level. The energy resolution was typically about 350 keV FWHM, where the main contribution came from the energy straggling in the targets.

Sample spectra from the (p,π^+) reaction on ^{6}Li, ^{10}B and ^{12}C are shown in Fig. 1. A large number of states with both

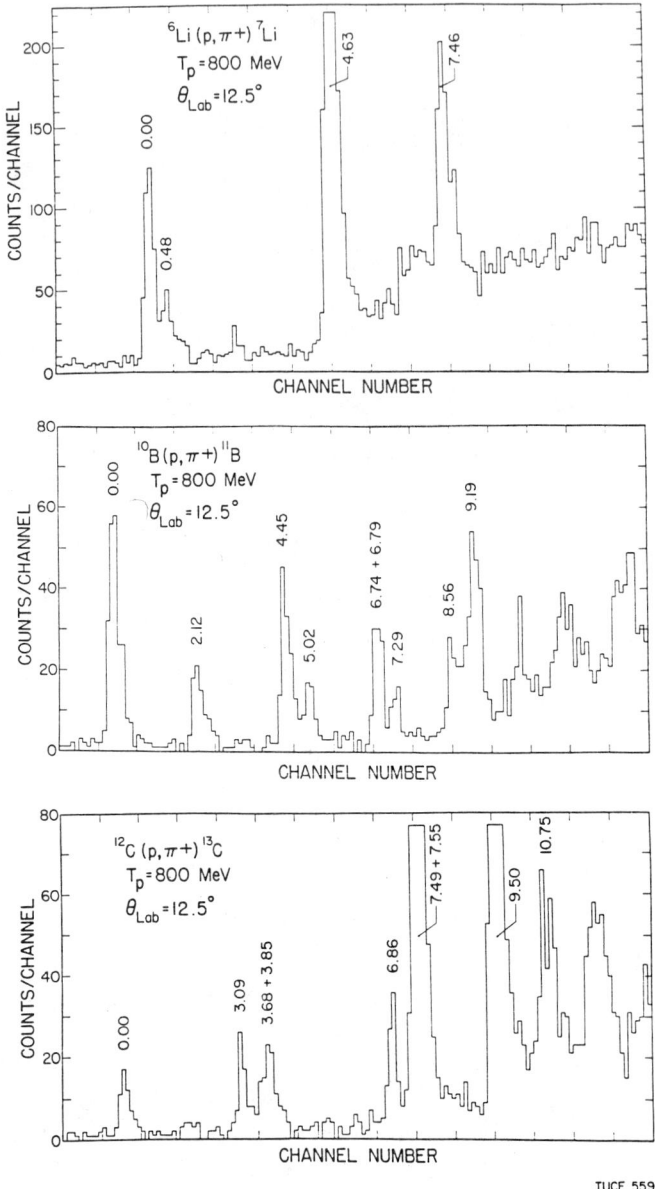

Fig. 1. Pion spectra from the (p,π+) reaction on ⁶Li, ¹⁰B and ¹²C at 800 MeV incident energy.

single-particle and more complex structure are populated.
High-spin states seem to yield larger cross sections than low-spin states, but no other selection rule could be deduced.

RESULTS AND DISCUSSION

The angular distributions from the π^+ production on ^{6}Li, ^{10}B, ^{12}C and ^{16}O are presented in Figs. 2-5 as a function of the momentum transfer, $q_{c.m.}$. The conspicuous feature of these angular distributions is the exponential behavior of the differential cross section with the momentum transfer. There are three different

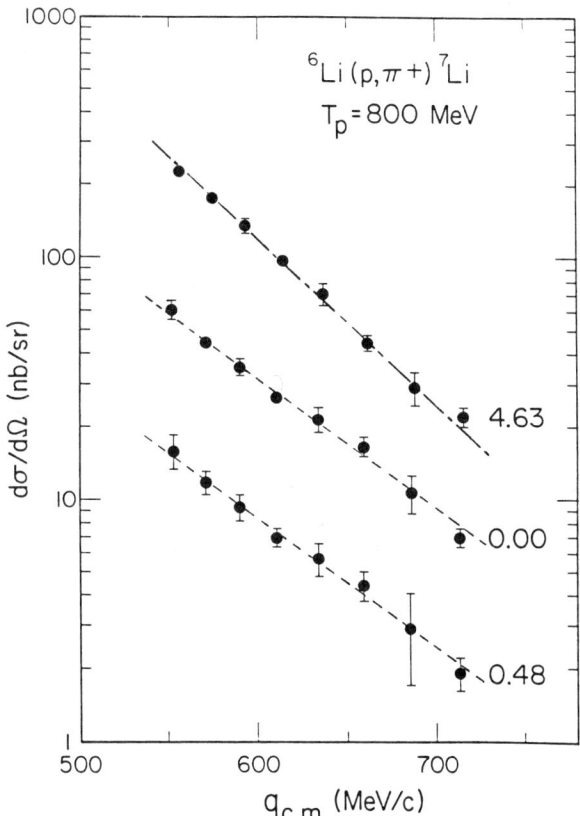

Fig. 2. Angular distributions of the ^{6}Li(p,π^+)^{7}Li reaction.

straight slopes which can be extracted to describe most of the
data. One slope fits the 0.00 and 0.48 MeV transition in ^{7}Li, the
2.12 and 5.02 MeV transition in ^{11}B, and the 0.00 MeV transition in
^{13}C. A second slope was observed for the 4.63 MeV transition in
^{7}Li, for the 0.00, 4.45 and 6.74 + 6.79 MeV transitions in ^{11}B, for
the 7.49 + 7.55 MeV transition in ^{13}C, and for the 3.06 and 3.84
MeV transitions in ^{17}O. A third slope fits the 7.29 MeV transition
in ^{11}B, the 9.50 MeV transition in ^{13}C and the 7.75 MeV transition
in ^{17}O.

Figure 2 shows angular distributions of the ^{6}Li(p,π^+)^{7}Li
reaction leading to the ground and first two excited states in ^{7}Li.
The 0.00 and 0.48 MeV states in ^{7}Li are known to be predominant
$1p_{3/2}$ single particle states from low energy (d,p) results[7] with
spectroscopic factors of S = 0.90 and 1.15, respectively. The
^{6}Li(p,π^+)^{7}Li angular distributions for these two transitions show
the same exponentially straight slope; their relative intensities,
however, resemble by no means these spectroscopic factors. The
transition to the 0.48 MeV state should be only about 1.6 times
stronger than the ground state transition on the basis of (2J+1)S
instead of about 3.8 as the data show. The transition to the 4.63
MeV, J^π = 7/2$^-$ state shows a steeper slope. This state is strongly
excited as a $2F_{7/2}$ resonance in the ^{4}He + t channel[7]. Here the
^{6}Li(p,π^+) reaction excites quite strongly a predominant cluster
state.

Angular distributions of the ^{10}B(p,π^+)^{11}B reaction to states
up to 7.3 MeV of excitation are shown in Fig. 3. From low energy
^{10}B(d,p)^{11}B studies it is known that the 0.00, 4.44 and 6.74 MeV
states have a predominant $1p_{3/2}$ single particle configuration[8]. In
the ^{10}B(p,π^+)^{11}B reaction, these three transitions exhibit
identical exponential slopes, which are, however, different from
the 0.00 and 0.48 MeV transitions in the ^{6}Li(p,π^+)^{7}Li reaction.
Thus, $1p_{3/2}$ single particle states in ^{7}Li and ^{11}B are excited in
the (p,π^+) reaction at 800 MeV incident energy by angular
distributions which show different straight slopes. The transition
strengths observed in the (p,π^+) reaction are quite different from
what is expected on the basis of their spectroscopic factors. The
other transitions lead to more complicated states with cross
sections comparable to the "single-particle" transitions.

A variety of patterns is exhibited by the angular distributions of the ^{12}C(p,π^+)^{13}C reaction displayed in Fig. 4. The
transitions to the ground state, to the doublet at 7.49 ± 7.55 MeV
and to the 9.50 MeV state can be described by the three different
exponentially straight slopes already established above.
Considering the large momentum transfer, it is remarkable that the
transition to the 1/2$^+$ state at 3.09 MeV, which has a predominant
$2s_{1/2}$ single particle configuration, shows an almost isotropic
angular distribution.

A similar pattern is observed for the transition to the 3.68 +

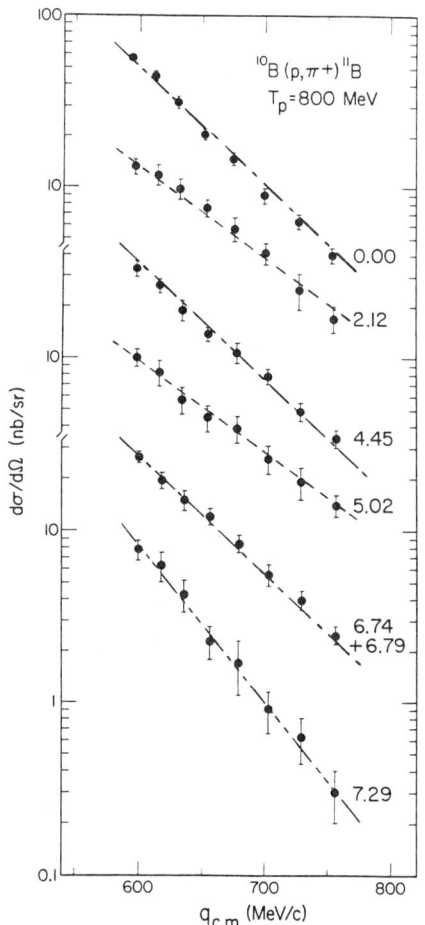

Fig. 3. Angular distributions of the $^{10}B(p,\pi^+)^{11}B$ reaction.

3.85 MeV doublet which consists of a predominant single particle and a predominant 2p-1h state. The transition to the stretched 2p-1h configuration $J^\pi = 9/2^+$ state at 9.50 MeV is one of the strongest observed in the (p,π) reaction at 800 MeV.

In Fig. 5 angular distributions from various transitions in the $^{16}O(p,\pi^+)^{17}O$ reaction are displayed. Only the transition to the ground and 0.87 states, which have predominant $1d_{5/2}$ and $2s_{1/2}$ single particle character, respectively, do not show a straight slope at forward angles. The angular distribution of the transition to the $1/2^+$ state at 0.87 MeV is almost isotropic at the forward angles as was observed for the transition to the $1/2^+$ state at 3.09 MeV in the $^{12}C(p,\pi^+)^{13}C$ reaction, but the exponential drop off occurs at much lower momentum transfer. The strongest transition again belongs to the excitation of the stretched configuration 2p-1h state at 7.75 MeV with $J^\pi = 11/2^-$. It is interesting to note that the ground state, $J^\pi = 5/2^+$, and the $5/2^-$ state at 3.84 MeV which have completely different configurations are about equally excited. This indicates again that the (p,π^+) reaction at 800 MeV does not distinguish between the excitation of simple and more complex states.

Angular distributions of the ground and first excited state transitions of the $^9Be(p,\pi^+)^{10}B$ and $^9Be(p,\pi^-)^{10}C$ reactions are presented in Fig. 6 as a function of the momentum transfer, $q_{c.m.}$. Again the same patterns are found as before. The (p,π^+) angular distributions have the same straight slope as, for example, the $^{10}B(p,\pi^+)^{11}B$ ground state transition, whereas the (p,π^-) angular

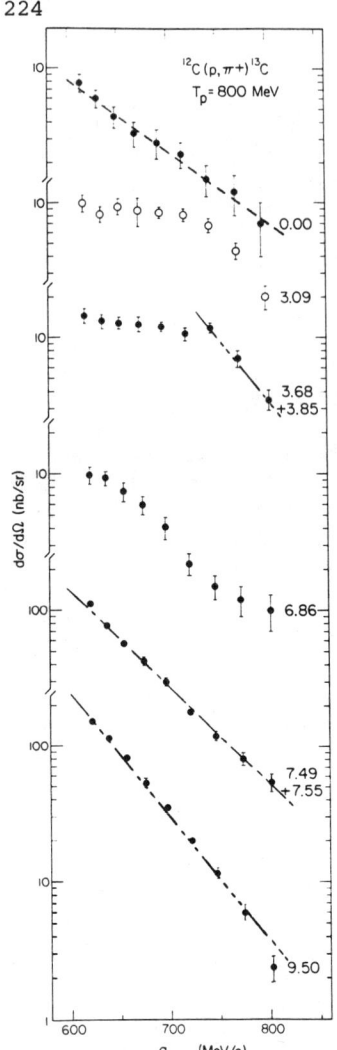

Fig. 4. Angular distributions of the $^{12}C(p,\pi^+)^{13}C$ reaction.

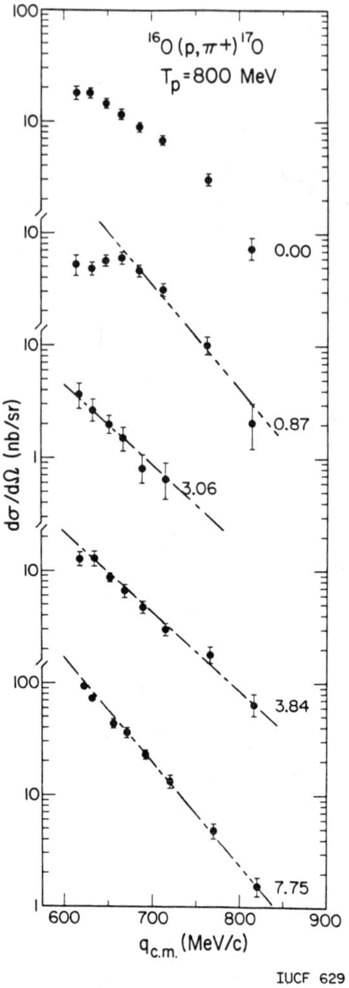

Fig. 5. Angular distributions of the $^{16}O(p,\pi^+)^{17}O$ reaction.

distributions follow the slightly less steep slope of the $^{12}C(p,\pi^+)^{13}C$ ground state transition.

It is interesting to compare the ground state transitions of the $^{9}Be(p,\pi^+)^{10}Be$ and $^{10}B(p,\pi^+)^{11}B$ reactions. Neutron stripping results at low energy show that the two transition strengths are about equal[9] in agreement with shell-model predictions[10]. At 800 MeV incident energy, the two (p,π^+) cross sections are about equal,

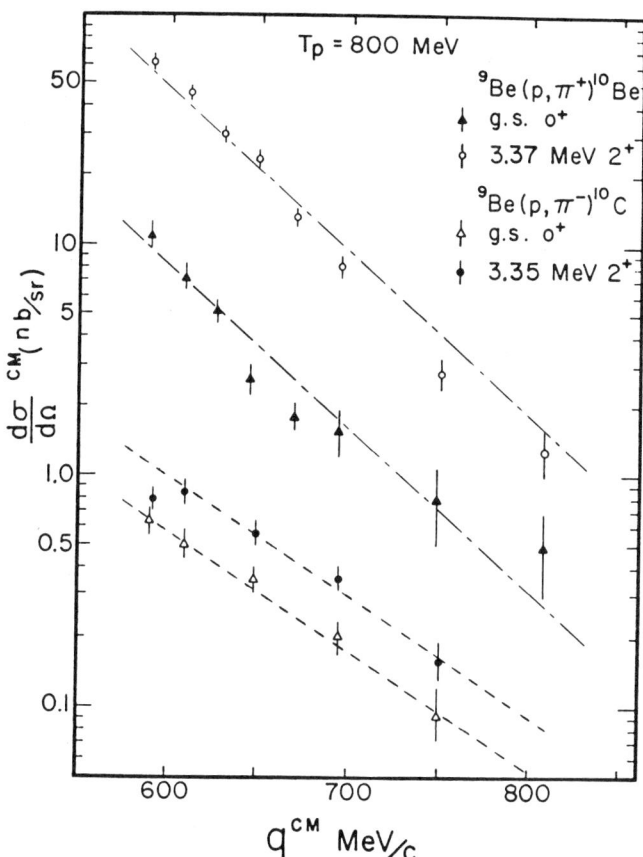

Fig. 6. Angular distributions of the ^{9}Be(p,π^+)^{10}Be and ^{9}Be(p,π^-)^{10}C reactions.

thus agreeing with the neutron stripping results, but disagreeing with the (p,π^+) results at 185 MeV incident energy[11], where the measured differential cross sections differ by a factor of about 10.

Another way to extract a possible systematic behavior is to compare the (p,π^+) angular distributions from different energies. Figures 7 and 8 show such a comparison for the ground state transitions of the ^{6}Li(p,π^+)^{7}Li and ^{10}B(p,π^+)^{11}B reactions. The

data at the lower incident energies are from Saclay[12]. For the
^{6}Li(p,π^+)^{7}Li ground state transition, shown in Fig. 7, the 800 MeV

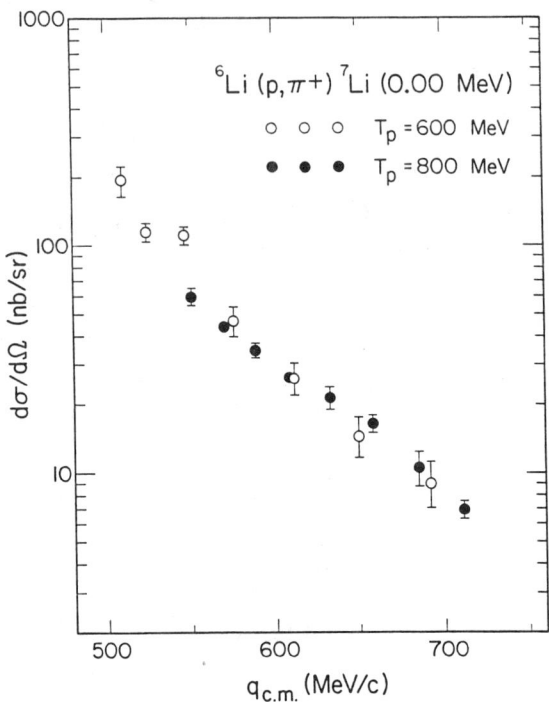

Fig. 7. Angular distributions of the ^{6}Li(p,π^+)^{7}Li (0.00 MeV) reaction at different incident energies.

data seem to be just a continuation of the 600 MeV data of Aslanides et al. (see Ref. 12). But for the ^{10}B(p,π^+)^{11}B ground state transition in Fig. 8, no simple connection between the three different incident energies is prevalent. Many other cases can be found which belong to either of the two above categories and no state dependence has emerged from this comparison.

CONCLUSION

The angular distributions of the (p,π^+) and (p,π^-) reactions at 800 MeV show little individual behavior. The only

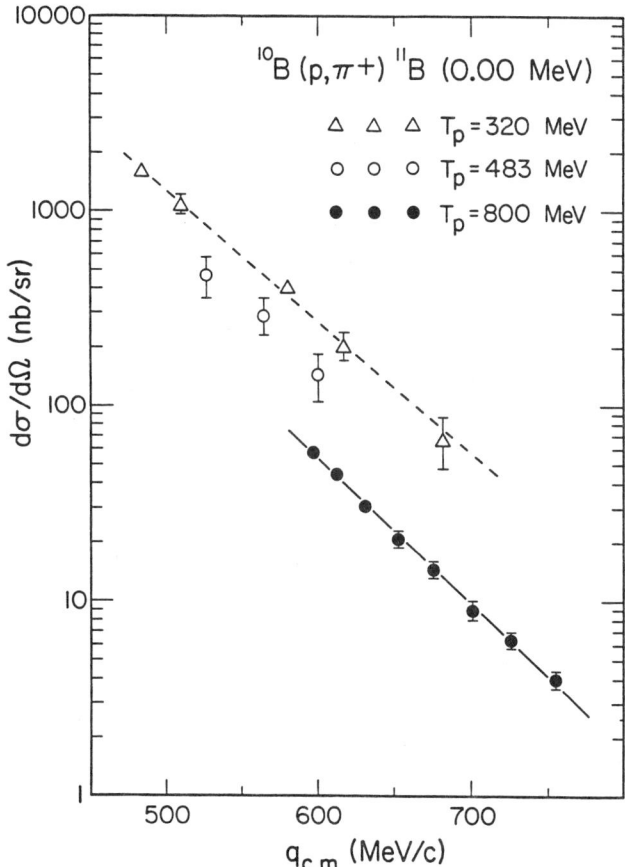

Fig. 8. Angular distributions of the ^{10}Be(p,π+)^{11}B (0.00 MeV) reaction at different incident energies.

characteristic features which could be extracted were three different exponentially straight slopes describing most of the data. Unfortunately, no connection between these slopes and the properties of individual transitions could be made. What this means in terms of details in the underlying nuclear structure at high momentum transfer and/or in the reaction mechanism is, of course, not clear at present.

REFERENCES

1. This work was done in collaboration with: S. Iversen and K.K. Seth, Northwestern University; B. Hoistad, Los Alamos Scientific Laboratory; G.S. Adams, M. Gazzaly, G. Igo and F. Irom, University of California, Los Angeles.
2. B. Hoistad et al., Phys. Rev. Lett. $\underline{43}$, 487 (1979).
3. H. Nann et al., Phys. Lett. $\underline{88B}$, 257 (1979).
4. H. Nann, S. Iversen, K.K. Seth, B. Hoistad and G. Kyle, to be published.
5. B. Hoistad et al., to be published.
6. G.S. Blanpied et al., Phys. Rev. Lett. $\underline{39}$, 1447 (1977); also G.W. Hoffmann et al., Phys. Rev. Lett. $\underline{40}$, 1256 (1978).
7. F. Ajzenberg-Selove, Nucl. Phys. $\underline{A320}$, 1 (1979).
8. F. Ajzenberg-Selove, Nucl. Phys. $\underline{A336}$, 1 (1980).
9. R.J. Slobodrian, Phys. Rev. $\underline{126}$, 1059 (1962).
10. S. Cohen and D. Kurath, Nucl. Phys. $\underline{A101}$, 1 (1967).
11. B. Hoistad, Advances in Nuclear Physics, Vol. 11, 1978, Plenum Press, New York; and references therein.
12. G. Bruge, CEN Saclay, internal report DPh-N/ME/78-1, 1978.

DISCUSSION

<u>P. Couvert</u> (Saclay): What is the magnitude of the (p,π^+) cross section on ^{40}Ca?
<u>Nann</u>: It is about 0.5 nb/sr.

<u>A.D. Bacher</u> (IUCF): I realize one is groping for some systematics, but it appears that the highest spin states that you excite are stronger at 800 MeV relative to the low lying states than they are at 200 MeV? Is that correct?
<u>Nann</u>: That's not always true. We have a $11/2^-$ state in ^{17}O which has about the same strength as the $7/2^-$ state in ^7Li.

III Current Models: Calculations and Comparisons with Data

Top: J.V. Noble, E.D. Cooper, B.D. Keister
Middle: L.S. Kisslinger, M. Dillig, F. Soga, W.R. Gibbs
Bottom: J. Vary, G.E. Walker, H. McManus, H.W. Fearing
Photos by Kent Berglund and Chuck Foster

THE RELATIVISTIC ONE NUCLEON MODEL

E.D. Cooper* and H.S. Sherif
Nuclear Research Centre, University of Alberta
Edmonton, Alberta, CANADA T6G 2N5

ABSTRACT

A one nucleon mechanism description of the $(\vec{p},\pi^+)$ reaction is given where all explicit nucleon wavefunctions are represented by Dirac Spinors. Great sensitivity to distortion effects in both incident and final channels is found. However, providing the distorting potentials used give good fits to elastic scattering data, then the $(\vec{p},\pi^+)$ predictions do not appear to be critically sensitive to any remaining ambiguity in the potentials. It is found that using pseudoscalar coupling for the πNN vertex renders the theory incapable of describing the data, whereas using pseudovector coupling gives very good agreement with the data.

INTRODUCTION

The one nucleon mechanism (ONM) for the $(\vec{p},\pi^+)$ reaction is one of the oldest models[1], and is the model which has received the most theoretical attention over the last two decades[1-22].

The model has become unpopular in the last few years. One reason for this is the increased interest in microscopic two nucleon mechanism (TNM) calculations. With faster computers and more sophisticated techniques, such microscopic calculations are now becoming feasible. Another reason for the ONM's fall from grace is that the model has demonstrated extreme sensitivity to its inputs; this is especially true for the off-shell parts of the pion distorting potentials used[9], and also to an ambiguity in the non-relativistic πNN vertex[22].

The latter ambiguity in the vertex can be neatly side-stepped by employing Dirac Spinors for the nucleon wavefunctions, and using directly a Dirac operator for the vertex. Such an approach has been considered before[23,24,25]. However, the incoming and outgoing wavefunctions were either taken as plane waves, or else the distortion effects were handled in a crude manner. We have recently extended this relativistic approach to properly treat the effects of proton and pion distortions[26,27], and report on our results so far.

*Present Address: Indiana University Cyclotron Facility, Milo B. Sampson Lane, Bloomington, Indiana USA 47405

FORMALISM

We wish to represent the interaction of a nucleon (described by a Dirac Spinor) and a nucleus using some time-independent potential. In general such a potential will be 4 x 4 matrix; however, it has been shown[28] by Hartree-Fock calculations that the largest components are an attractive potential which transforms as a Lorentz scalar, and a repulsive potential which transforms as the time-like component of a 4-vector potential (hereinafter referred to as a vector potential). We assume only these potentials are important.

For the continuum case both vector and scalar potentials acquire imaginary parts[32]. These imaginary parts are determined by fitting elastic scattering data, and at present there are two distinct sets of Optical model parameters which afford good fits to the elastic scattering data, which we refer to as sets A & B[29]. The essential difference between these sets of potentials is that in set A the potentials have imaginary parts of opposite sign and of order 100 MeV; whilst in set B we have potentials whose imaginary parts are smaller and both absorptive. Rather than go into the merits of the two sets here, we will calculate the $(\vec{p},\pi^+)$ observables by using members of both sets to generate the proton distorted waves.

The Dirac equation describing the nucleon motion with scalar and vector potentials is:

$$H_0\psi = [-i\underline{\alpha}\cdot\underline{\nabla} + \beta mc^2 + \beta U_s(r) + U_v(r)]\psi = E\psi \qquad (1)$$

We solve equation (1) for the bound state wavefunction and the proton distorted wave. These spinors are then put into the expression for the T-Matrix.

$$T_{\mu_i M_B} = i\sqrt{2}g_\pi \int \bar{\psi}_n^{M_B}(x)\Gamma(x)\psi_p^{\mu_i(-)}(x)\phi_\pi^{*(-)}(x)d^4x \qquad (2)$$

where $\Gamma(x)$ is the πNN vertex function, taken to be either of pseudoscalar (γ_5) or pseudovector $[-i(2M)^{-1}\gamma_5\partial_\pi]$ form. $\phi_\pi^{(-)*}(x)$ is a pion distorted wave generated from a suitable pion Optical potential. Just as for the proton distorted waves, we demand that our pion potential give the correct elastic scattering. This we do by taking an "off the shelf" pion potential[30], which gives not only a reasonable account of the elastic scattering data but also the pionic atom shifts and widths. This potential includes the Lorentz-Lorentz effect, which goes some way to reducing its large off-shell components.

DISCUSSION

Since in the plane wave Born approximation the T-matrix is given by the Fourier transform of the bound-state wavefunction, hopes were held long ago that the $(\vec{p},\pi^+)$ reaction would give a direct probe of the high momentum components of single particle nuclear wavefunctions. We show in figure 1 the momentum space single particle bound state wavefunction arising from solving equation (1) for the $1f_{7/2}$ state in ^{41}Ca. Shown by the arrows in figure 2 is the region corresponding to the momentum transfers involved in the reaction ^{40}Ca$(\vec{p},\pi^+)^{41}$Ca(g.s.) at a proton (lab.) kinetic energy of 160 MeV. One striking feature of the curves is that, at the momentum transfers shown by the arrow, there are regions where the lower component is larger than the upper component. This gives us grounds for believing that any attempt at producing a non-relativistic formalism is going to run into problems. In particular, a reduction scheme based on identifying the upper component as the Schrodinger equivalent wavefunction and transforming the operator so that it is "even"[37] to some order, is going to run into convergence problems when used for processes which involve high momentum transfer.

One other point to notice is that in equation (2) the operator $\Gamma(x)$ is (or can be expressed as) a purely odd operator. The T matrix therefore depends <u>linearly</u> on the lower components of the nucleons, making these components <u>vital</u> to the (p,π^+) calculations.

Our programme then is as follows:

(i) Obtain the bound-state wavefunction by some suitable assumption about the bound state vector and scalar potentials.

(ii) Take Woods-Saxon forms for the proton Optical potentials and vary their parameters so as to fit the elastic scattering data. Then obtain the wavefunctions corresponding to these potentials.

(iii) Put the bound state wavefunction and the proton distorted wave generated as above, along with a pion distorted wave, into equation (2), and then obtain the $(\vec{p},\pi^+)$ cross-sections and analyzing powers.

RESULTS

We show in figure 2 the results of our calculations for the reaction ^{40}Ca$(\vec{p},\pi^+)^{41}$Ca(g.s.) at 148, 160 and 185 MeV using a proton distorting potential from set A. For the 148 MeV calculations we have taken the proton distortion from the fit to the 160 MeV elastic scattering data. The dashed curves correspond to using pseudoscalar coupling for the vertex in equation (2), while the solid curves correspond to using pseudovector coupling.

Clearly the curves qualitatively agree with the data for pseudovector coupling, but not for pseudoscalar coupling. By adjusting the parameters of the potentials within reasonable limits it turns out to be impossible to make the curves generated using pseudoscalar coupling agree with the data. Here we have a clear experimental preference for the πNN vertex to be of pseudovector, not pseudoscalar, form.

We have (by hand) made the bound state potential's diffuseness parameters smaller than those values obtained from the Dirac Optical Model fit (from 0.65 fm to 0.5 fm) to lift the curves up to the data at 160 MeV. This results in a factor of 1.8 increase in the cross-sections.

Thus, in figure 3, we show the predictions for $^{16}O(\vec{p},\pi^+)^{17}O(g.s.)$ using bound state potentials with radial and diffuseness parameters 1.0 fm and 0.4 fm respectively.

From figure 3 we see confirmation of the trends seen in figure 2 namely that the data show a marked preference for pseudovector coupling over pseudoscalar coupling. The disagreement between the predicted cross-sections and the data at back angles may be due to uncertainties in the proton Optical potentials, or perhaps to some other mechanism becoming important at threshold.

The analysing power for $^{16}O(\vec{p},\pi^+)^{17}O(g.s.)$ has been measured at 157 MeV. We showed[26] previously that the analyzing power prediction using pseudovector coupling is in reasonable agreement with the data, while that using pseudoscalar coupling is not.

As it stands at present, the model probably should not be applied to the $(\vec{p},\pi^+)$ reaction on ^{12}C. However, since most of the analyzing power measurements have been done on this nucleus, we show in figure 4 the results of our calculations in which we have had to assume that the nuclear structure of ^{13}C can be represented as a single neutron orbiting a spherical closed ^{12}C core. In our model, the analyzing power does not depend strongly upon the final nuclear state quantum numbers, and so hopefully any configuration mixing effects will not be too important. This does not apply to the cross-sections where configuration mixing is very important[9].

The agreement with the data at 159 MeV is fairly good, whereas at back angles at 200 MeV the theoretical curve goes positive in contrast to the data, which remain negative. This is disappointing; however, we are encouraged to see that recent IUCF analyzing power data[34] at 170 and 183 MeV do indeed show this positive region at back angles.

CONCLUSION

We have seen that, when a $(\vec{p},\pi^+)$ reaction leads to a final state which is reasonably described as a single particle outside of a spherically symmetric closed core, the one nucleon model is quite capable of describing the process.

In order to get agreement with the data it is necessary to have proton elastic scattering data at the same energy as the $(\vec{p},\pi^+)$ data so as to constrain sufficiently the proton Optical model parameters.

We have also seen that the $(\vec{p},\pi^+)$ reaction, in this model, provides a sensitive probe of the πNN vertex, and that we are able to rule out pure pseudoscalar coupling as a suitable candidate for this vertex. In this regard we note that Noble[35] has shown how using pseudovector coupling for the πNN vertex is equivalent to a model where sigma mesons are explicitly included, and pseudoscalar coupling is used for the πNN vertex.

REFERENCES

1. J. LeTourneux & J.M. Eisenberg, Nucl. Phys. 87 (1966) p. 331.
2. J.M. Eisenberg et al., Phys. Lett. 43B (1973) p. 20.
3. W.B. Jones & J.M. Eisenberg, Nucl. Phys. A154 (1970) p. 49.
4. E. Rost & P.D. Kunz, Phys. Lett. 43B (1973) p. 17.
5. M.P. Keating & J.G. Willis, Phys. Rev. C7 (1973) p. 1336.
6. J.M. Eisenberg, Phys. Lett. 45B (1973) p. 93.
7. S. Dahlgren et al., Nucl. Phys. A211 (1973) p. 243.
8. B. Hoistad et al., Phys. Scripta 9 (1974) p. 201.
9. G.A. Miller, Nucl. Phys. A224 (1974) p. 269.
10. Y. LeBornec et al., Phys. Lett. 49B (1974) p. 434.
11. S. Dahlgren et al., Nucl. Phys. A227 (1974) p. 245.
12. Z. Grossman et al., Ann. Phys. 84 (1974) p. 348.
13. A. Reitan, Nucl. Phys., A237 (1975) p. 465.
14. J.V. Noble, Nucl. Phys. A224 (1975) p. 526.
15. T.S.H. Lee & S. Pittel, Nucl. Phys. A256 (1976) p. 509.
16. Y. LeBornec et al., Phys. Lett. 61B (1976) p. 47.
17. J.V. Noble, Phys. Rev. Lett. 37 (1976) p. 123.
18. B. Tatischeff et al., Phys. Lett. 63B (1976) p. 158.
19. W.R. Gibbs & S. Young, Phys. Rev. C17 (1978) p. 837.
20. B. Hoistad et al., Phys. Lett. 79B (1978) p. 384.
21. H.J. Weber & J.M. Eisenberg, Nucl. Phys. A312 (1979) p. 201.
22. M. Tsangarides, Ph.D. Thesis, Indiana University (1980).
23. R. Brockmann & M. Dillig, Phys. Rev. C15 (1977) p. 361.
24. L.D. Miller & H.J. Weber, Phys. Lett. 64B (1976) p. 279.
25. L.D. Miller & H.J. Weber, Phys. Rev. C17 (1978) p. 43.
26. E.D. Cooper & H.S. Sherif, Phys. Rev. Lett. 47 (1981) p. 818.
27. E.D. Cooper & H.S. Sherif, in preparation.
28. M. Jaminon et al., Nucl. Phys. A365 (1981) p. 371.
29. E.D. Cooper & H.S. Sherif, in preparation.
30. K. Stricker et al., Phys. Rev. C19 (1979) p. 929.
31. L.L. Foldy & S.A. Wouthuysen, Phys. Rev. 78 (1950) p. 29.
32. L.G. Arnold & B.C. Clark, Phys. Rev. C19 (1979) p. 912.
33. M. Jaminon et al., Phys. Rev. C24 (1981) p. 353.
34. M.C. Green et al., contribution to this conference.
35. J.V. Noble, Phys. Rev. C20 (1980) 225.
36. S. Dahlgren et al., Nucl. Phys. A227 (1974) p. 245.
37. P.H. Pile et al., Phys. Rev. Lett. 42 (1979) p. 1461.
38. T.P. Sjoreen et al., Phys. Rev. C24 (1981) p. 1135.
39. E.G. Auld et al., Phys. Rev. Lett. 41 (1978) p. 462.

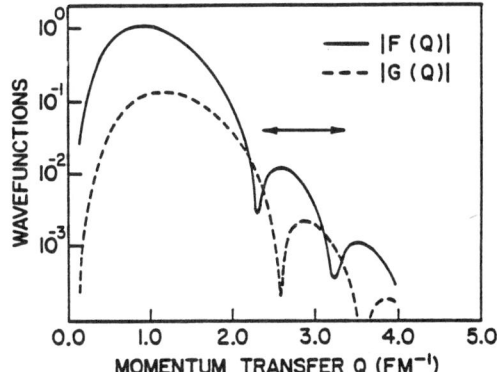

Figure 1. The bound state wavefunction for the $1f_{7/2}$ state in ^{41}Ca, given in momentum space. The arrows show momentum transfers corresponding to those of the reaction ^{40}Ca$(\vec{p},\pi^+)^{41}$Ca at a proton lab kinetic energy of 160 MeV.

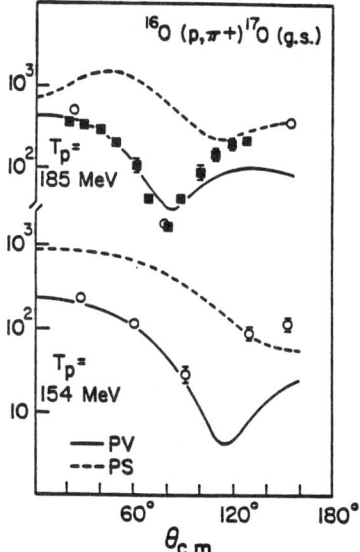

Figure 3. The calculated ^{16}O$(\vec{p},\pi^+)^{17}$O(g.s.) cross-sections at proton (lab.) kinetic energies of 154 MeV and 185 MeV. The solid curves were calculated using pseudovector coupling for the πNN vertex, the dashed curves were calculated used pseudoscalar coupling. The open data points are from IUCF[38], the closed points at 185 MeV are from Uppsala[36].

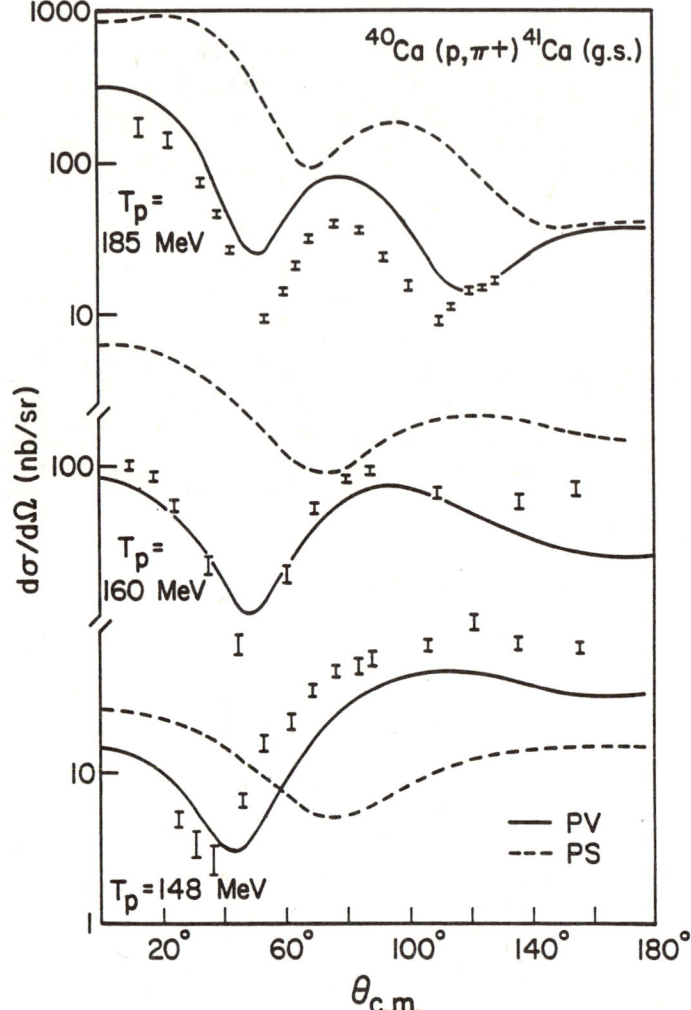

Figure 2. The calculated $^{40}Ca(\vec{p},\pi^+)^{41}Ca(g.s.)$ cross-sections with incident protons of energy 148 MeV, 160 MeV and 185 MeV. The solid lines are calculated using pseudovector coupling, the dashed lines are calculated using pseudoscalar coupling. In each case a proton distorting potential from set A is used. The 185 MeV data are from Uppsala, the rest from Indiana[37]

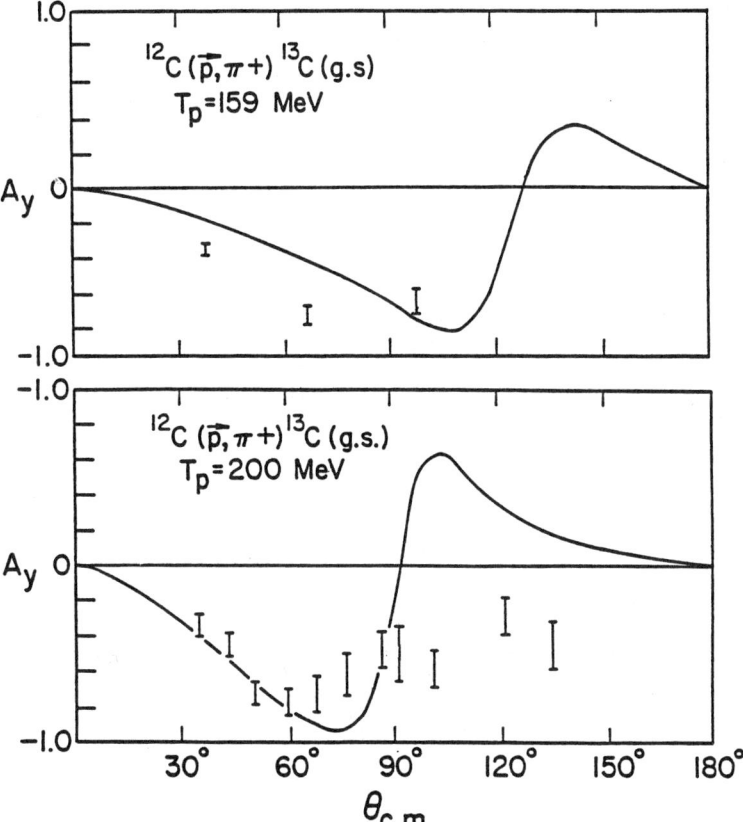

Figure 4. The analysing power predictions for the $(\vec{p},\pi^+)$ reaction in ^{12}C. Pseudovector coupling is used for all the calculations shown. The data at 200 MeV are from TRIUMF[39], those at 159 MeV are from IUCF[38].

DISCUSSION

M. Banerjee (Univ. of Maryland): Did you use a ^{13}C wave function to see if you get the right ft-value and magnetic moment?
Cooper: No. We have to explore these bound states in some other way.

J. Eisenberg (Tel Aviv): This is neither a question nor a comment, nor is it intended to annoy the speaker, but rather to encourage him to pursue vigorously and quickly what he calls extension A, because the results he shows have all the classic symptoms of pionic wave catastrophies. The cross sections go up by an order of magnitude or more when you turn on the pionic distortion, the asymmetries flip sign and become very large, and this is in the face of pionic phase shifts on the full nucleus that are not terribly large. This suggests that something in the uncontrolled part of the off-shell behavior may be producing this, and I would be very hesitant to draw firm conclusions, say, about which coupling to prefer until that point has been pursued in depth.
Cooper: When it comes to the question of coupling, we also address the question of the reaction mechanism--but you're right, I have to look at that.

P. Hwang (Indiana Univ.): The only conclusion you can draw here is based on the one-nucleon mechanism. It is imperative to study two-body mechanisms, since you may resolve the disagreements between the pseudoscalar coupling and the data by pursuing further reactions in this process.
Cooper: That reminds me of Banerjee's talk this morning, in which he said something to the effect that if you have a pseudovector coupling in a vertex, you can actually get the same answer through evaluating this graph if you write it as a pseudoscalar at the vertex plus, interesting enough, a term that corresponds to target emission. In other words, pseudovector coupling here has a piece in it which collapses the propagator and gives you a contact term. So, in some sense pseudovector coupling does have a two-nucleon mechanism within it. So it's not that clear what's going on. Distortion effects bring in some of the effects of the other nucleons in an average sort of way.

J. Noble (Univ. of Virginia): I would like to point out a paper that I wrote some time ago, in which I pointed out that pseudoscalar and pseudovector coupling for pion emission or absorption are indeed equivalent as long as you insist that the pseudoscalar coupling is within the framework of a theory that respects PCAC. What happens is that the pion will scatter off the sigma field of a nucleus if you have the pseudoscalar coupling theory like the sigma model, and that will cause all sorts of cancellations so that the end result of the two effective vertex functions will look exactly the same to all leading orders of perturbation theory that one cares to check. So, I don't believe that one can make a statement like this, that such a reaction is telling you anything about the nature of the coupling of a pion and a nucleon. Unfortunately it doesn't, and the fact that your fits look better in one case than in another case I think is an irrelevant statement.

Cooper: The statement is that you shouldn't use pure pseudoscalar coupling without other pieces.

Noble: Yes, but one has known that for 25 years.

M. Banerjee (Univ. of Maryland): Allow me to make two comments. The first one is a continuation of the beautiful point that Eisenberg made. The Adler consistency condition which is part of chiral symmetry demands that s-waves should have a very strong momentum dependence. This will also help cut down the catastrophy that Eisenberg talked about. That is something you could take into account.

I think I should also make a summary comment on what Walker and I did, which is very simple. We have a logical point. You have to examine what we have said--we may be right or we may be wrong. If you find that we are right, a theory for (p,π^+) surely must meet our criteria before it can be compared with experiment. The point we are making cannot be tested by comparing with experiments. It is a logical point.

M. Dillig (Univ. or Erlangen-Nurnburg): In the calculations so far no exchange terms were included in calculating the wave function. Would you expect that, for example, exchange terms for the γ_5 for the pseudovector interaction would in part cancel that strong sensitivity and come back to a kind of insensitive result? As far as I remember, it is known from other calculations using exchange terms (in a Fock sense), that these terms were large and had a strong influence. Couldn't one expect internal cancellations if the calculation of the relativistic wave function is done consistently, including both Hartree and Fock terms?

Cooper: The Fock terms give you the energy dependence of the Dirac optical model parameters, and so they are certainly important in that sense. They aren't going to change things. To some extent, by treating the Dirac potentials phenomenologically you're mocking them up. They're in there because you fit the elastic scattering data which sees the same Fock terms. So, to some extent they're there already.

MODELS OF PION PRODUCTION AND A QUARK MODEL OF THE
PHYSICS AT SHORT DISTANCE

Leonard S. Kisslinger
Carnegie-Mellon University, Pittsburgh, PA 15213

and

University of Washington, Seattle, WA 98195

ABSTRACT

An isobar doorway treatment of the two-baryon plus core model is discussed, with sample results. Short distance properties of interacting hadrons are involved. A six-quark model for the short-range two-baryon current is presented, with application to electromagnetic interactions with deuteron targets and pion absorption.

INTRODUCTION

Pions interact with nucleons and with nuclei mainly by being absorbed. On the scale of distances usually involved in nuclear physics, the absorption of a 140 MeV meson is a short-range phenomenon. When only one nucleon is directly involved, high momentum components of the nuclear wave function must be involved. This has led both to studies of distortion with optical potentials and to multinucleon models for momentum-sharing mechanisms. However, one cannot avoid considerations of interactions at short-distance.

In the present brief discussion I shall review the main theoretical questions which are involved, and discuss the Isobar-Doorway treatment of the two-nucleon model of the (p,π) reaction on complex nuclei which Brad Keister and I have been developing. It will be pointed out that the ranges of the interactions involved are so short that regions of overlap of baryons play a large role. For this reason, a six-quark model allowing full color transfer among the quarks in the two baryons involved in the reaction is natural. Such a six-quark model will be discussed with application to 1) the deuteron magnetic form factor, 2) the electrodisintegration of the deuteron below pion threshold, 3) the possible dibaryon seen in $d(e,e'p)n$ polarization, and 4) to $\pi^+D \leftrightarrow pp$. In other parts of these proceedings, more detailed results of our (p,π) calculations are discussed by Brad Keister, and Jerry Miller gives a detailed discussion of our work on $\pi D \leftrightarrow pp$ in the $\Delta(1232)$ region.

THE PHYSICS OF PION ABSORPTION/PRODUCTION: $pp \leftrightarrow \pi^+ D$

Let us briefly review the work of the past quarter century on the $pp \leftrightarrow \pi^+ D$ reaction. Because of the extensive literature on this subject I shall not attempt a review of the past or recent research, but shall concentrate on the physical processes involved. The main purpose is to explicitly introduce the physics which must be treated

0094-243X/82/79243-21 $3.00 Copyright 1982 American Institute of Physics

in the (p,π) process discussed in the next section. Results of recent research on pp↔π⁺D is given in other parts of these proceedings.

Nucleon Pole Processes

The lowest order process consists of a single pion interaction with production (absorption) on a single nucleon, illustrated in Fig. 1a. This involves high momentum transfer at the deuteron vertex, i.e., high momentum components of the deuteron wave function. For example, for a pion of kinetic energy T_π = 188 MeV in the laboratory system, corresponding to an invariant mass of the Δ(1232) resonance for a target nucleon at rest, the relative momentum of the two nucleons is

$$|\vec{q}|_{Pole} = \frac{|\vec{N}_2 - \vec{N}_2'|}{2} \approx 0.5 \text{ GeV/c}. \tag{1}$$

Since the Fourier components of the deuteron wave function are very small at large relative momenta, this process is quite small.

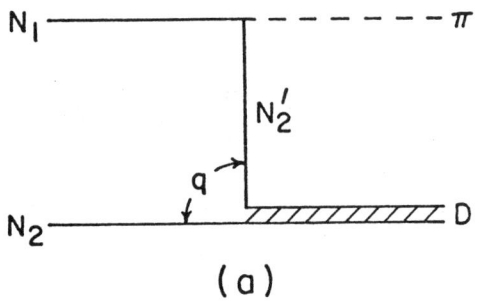

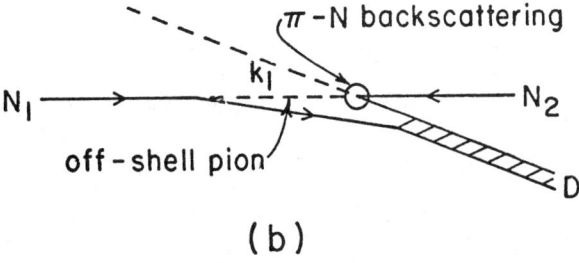

Fig. 1. Processes for πD→NN. a) Nucleon pole diagram; b) backscattering (Mandelstam process).

Mandelstam Mechanism: Two Organizations

As was pointed out more than two decades ago,[1] if a pion first backscatters on one nucleon and then is captured on the second nucleon, the final two nucleons in the $\pi D \to NN$ reaction can achieve the necessary "back-to-back" configuration without any relative momentum in the initial deuteron. Since the deuteron is a weakly-bound system, it is expected (and found) that this mechanism, illustrated in Fig. 1b, dominates the nucleon pole process if Fig. 1a at low-medium energies by an order of magnitude. However, one must not be led into believing that this mechanism involves only long ranges. Taking the pion with $T_\pi = 188$ MeV as above, for the desired small momentum transfer at the deuteron vertex

$$\vec{q} \simeq 0 \tag{2a}$$

one finds that the intermediate pion is very far off mass shell, and that

$$|\vec{k}_1| \simeq 3m_\pi. \tag{2b}$$

In other words, the range of the $\Delta N \to NN$ interaction is of the order of $3m_\pi$ rather than m_π

$$r_{\Delta N \to NN} \simeq \frac{1}{3m_\pi}.$$

At pion threshold the kinematics no longer favor Δ formation. However, the dynamics of π-N',n interactions leads to important contributions from intermediate Δ's even here. In this case the target nucleon must supply momentum and energy, along with the pion mass, for the formation of the Δ. In this case

$$|\vec{q}|_{threshold} \simeq 500 \text{ MeV/c} \tag{3}$$

as in the case of the pole term.

With these things in mind there are [at least] two different ways which one can organize the calculation of the process, illustrated in Fig. 1b, in order to emphasize different aspects of the physics. The first organization which we consider is illustrated in Fig. 2a. In words, the pion absorption takes place on the second nucleon after an off-shell scattering from the first nucleon. In the language of diagrams, the amplitude for the process is given by the loop integral

$$T_{\pi D \to NN} = \int d^4 q \, G_N \Psi_D G_N t_{\pi N} G_\pi \gamma_{\pi NN} G_N, \tag{4}$$

where G_N, G_π are nucleon, pion propagators, $\gamma_{\pi NN}$ is the π-N vertex function and $t_{\pi N}$ is the off-shell π-N scattering amplitude. This organization was mainly used in the early days. Thus the physics

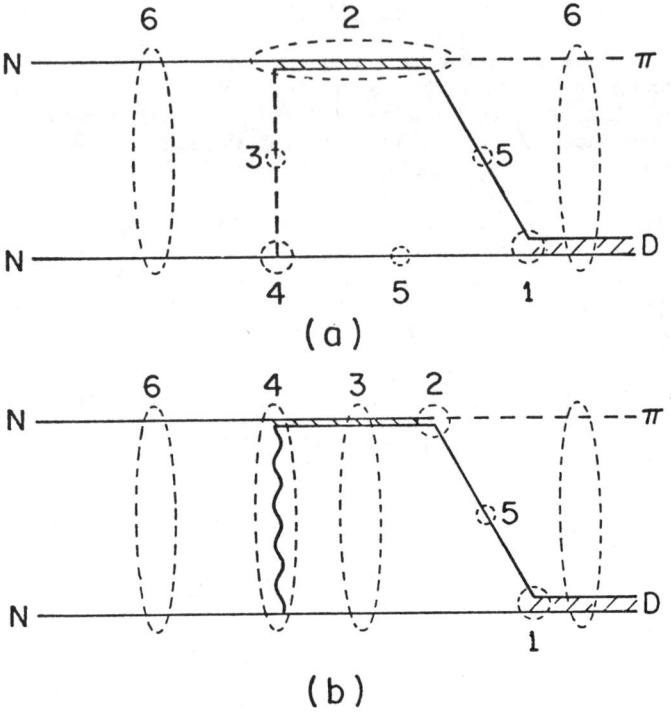

Fig. 2. Mandelstam mechanism organized a) for off-shell πN scattering, and b) for ΔN→NN interaction.

corresponding to the areas marked in Fig. 2a is
1. The D→pn vertex function (relativistic or nonrelativistic) wave functions
2. The off-shell π-N scattering amplitude
3. The off-shell meson propagator
4. Meson-nucleon vertex functions
5. Off-shell nucleon propagators
6. Initial and final distortions (not included otherwise).

It is in this organization which is utilized in the two nucleon-model for the (p,π) reaction discussed in the next section. Recently, the relativistic aspects of this organization have been studied, and shown to be important near threshold and at energies above the Δ. A nonrelativistic model should be adequate near the Δ.

The second organization is depicted in Fig. 2b. In words, the pion interacts with a nucleon in the deuteron to form a Δ-N doorway state, which converts to a N-N system via the ΔN→NN interaction. As a perturbation diagram the amplitude is

$$T_{\pi \to NN} = <\pi D|V_{\Delta \to N\pi}|\Delta N> G_{\Delta N} <\Delta N|V_{NN \to \Delta N}|NN>, \qquad (5)$$

where $G_{\Delta N}$ is the ΔN Green's Function. This organization emphasizes the $\Delta N \rightarrow NN$ interaction. It has been used, e.g., in the extensive work of Niskanen.[3] It is also most useful for studying the short-range Δ-N system, which is treated in our six-quark model below. The physics involved is
1. The D→pn vertex function
2. The $\pi N\Delta$ vertex
3. The interaction and propagation of the Δ-N system
4. The $\Delta N \leftrightarrow NN$ interaction
5. Off-shell nucleon propagation
6. Distortions (not included otherwise).

Let us now turn our attention to absorption in complex nuclei.

THE (p,π) REACTION ON COMPLEX NUCLEI

The Two-Baryon Model: Direct and Exchange Processes

As is evident from the previous section, in treating pion absorption it is important to treat the coordinates of at least two nucleons explicitly. This leads to the two-baryon model of the (p,π) reaction, which was first proposed by Grossman et al.[4] The process

$$A(p,\pi)B$$

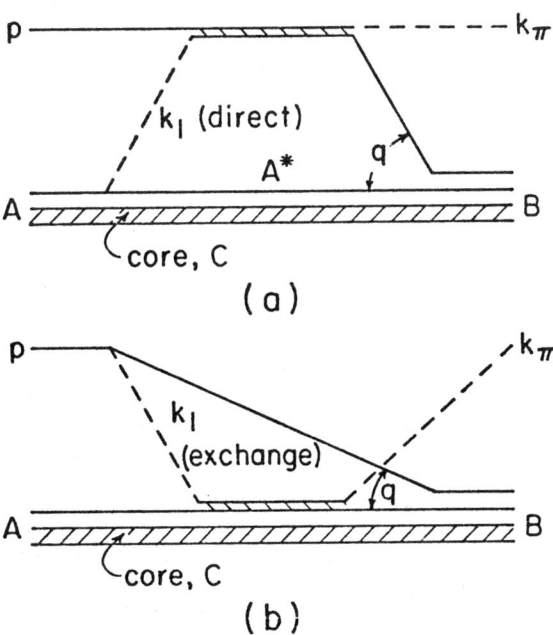

Fig. 3. Two-baryon plus core model for the A(p,π)B reaction with a) Direct (target emission) and b) Exchange (projectile emission) processes.

is illustrated in Fig. 3. In the mechanism shown in Fig. 3a a virtual pion emitted from the target nucleus, A, is scattered by the proton projectile and emitted, mainly, by a Δ in the resonance region. We refer to this as the direct (or target emission) process. Obviously the mechanism resembles the Mandelstam mechanism discussed above (Fig. 2). Note, however, that there are important differences. Since the pion emission vertex

$$<A|j_\pi|A^*>$$

involves a bound nucleon initially and finally, there is a tendency for the virtual meson to have low momentum

$$k_1(\text{direct}) \sim 0. \qquad (6)$$

This results in very different kinematics for the $A(p,\pi)B$ in comparison to the $d(p,\pi)n$ reaction. For the situation in which Eq. (6) holds, it is quite obvious that the relative momentum, q, at the (A,pB) vertex is comparable with the pole term

$$q^{\text{Direct}}(k_1 \sim 0) \sim 500 \text{ MeV/c}. \qquad (7)$$

The exchange (or projectile emission) process, which must be present by antisymmetry of the nuclear wave functions, is illustrated in Fig. 3b. This process is closely related to "pionic stripping" model (reviewed by Measday and Miller[5]). In this case the intermediate pion is close to its mass shell, and

$$k_1(\text{exchange}) \sim k_\pi, \qquad (8)$$

so that the scattering by the nucleus of the pion emitted by the projectile proton at first seems to be almost identical to the optical model treatment of the pionic stripping model. However, there are important differences. In the two-nucleon exchange process:
 1. The pion propagation is treated explicitly, with finite range effects.
 2. The πNN and πNΔ form factors are introduced.
 3. The scattering of the pion by the bound nucleon can lead to off-shell effects, and nuclear structure effects are included which are not present in the optical potential.

<u>The Isobar-Doorway Model for $A(p,\pi)B$</u>

Since the pion final-state interactions in the two-nucleon model must be included, there can be serious problems of double-counting. One way to avoid this and to include the bound-state π-N T-matrix is to make use of the projection operator formalism of the Isobar-Doorway Model[6] with parameters given by fits to elastic scattering.[7] This was first studied by Hirata.[8] I shall briefly review the formalism here as applied to the (p,π) reaction.[9] This work is being carried out jointly with Brad Keister.

The $\pi N \Delta$ interaction is introduced as a basic interaction, H_Δ. This gives the Δ resonance in π-N scattering by

$$t^\Delta_{\pi N} = \frac{<\pi N|H_\Delta|\Delta><\Delta|H_\Delta|\pi N>}{E - M_\Delta + i\,\Gamma_\Delta/2} .$$

Utilizing the projection operators onto Δ-nucleus doorway states,[6] $|D_i>$, one can separate the reaction into nonresonant parts and parts mediated by the Δ

$$T_{(p,\pi)} = T^{nonresonant}_{(p,\pi)} + T^\Delta_{(p,\pi)} .$$

We expect that the nonresonant part can be treated by the stripping model with the Δ contributions removed from the pion optical potential. The resonant part is obtained by

$$T^\Delta_{(p,\pi)} = \sum_i \frac{<\pi \eta_A|H_\Delta|D_i><D_i|H_\Delta|p \eta_{A-1}>}{E - E_i + i\,\Gamma_i/2} . \qquad (9)$$

The bound-state energies E_i and widths Γ_i of the Δ-propagator are determined by elastic scattering.[7] The pion in Eq. (9) is distorted only by the non-resonant potential. This avoids the problem of double-counting.

In our present calculations the loop integrals implied in Fig. 3 are explicitly carried out. Shell model wave functions are used, with spectroscopic factors for the nucleons and core states obtained from the Argonne computer program. Details will be published elsewhere.[10]

Some results for the cross sections are shown in Fig. 4 for $^{12}C(p,\pi)^{13}C_{g.s.}$ at the proton energy of 200 MeV. Note that exchange term >> direct term for most of the angular distribution. Therefore, one might expect that the results would be similar to the stripping calculations shown in Fig. 5. Since hard form factors are used for the results of Fig. 4, this should be a good case for comparison. For the distorting potential of the CMU experimental group at low π-energies the results are similar, as can be seen from comparison of the solid curves of Figs. 4 and 5. Note that both the two-nucleon and stripping calculations are sensitive to the distorting potentials, as can be seen by the curves marked II, which are obtained using the Miller Potential II[5] for the elastic pion scattering. However, the exchange process cannot be treated as stripping.

It is hoped that the polarization can help sort out the mechanisms for this process. A few of our results with comparison to the stripping model are shown in Fig. 6. One of the important new experimental results is that there is a strong energy dependence of the polarization for $^{12}C(p,\pi)^{13}C_{g.s.}$ as one goes from T_p = 200 MeV to 250 MeV.[11] Although the present theoretical results are not very good in comparison with the polarization data, rapid changes in the

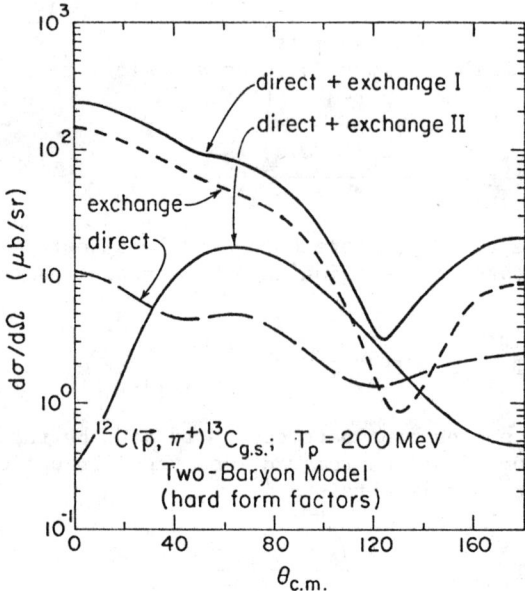

Fig. 4. The $^{12}C(p,\pi)^{13}C_{g.s.}$ reaction in the two-baryon + core model. The curves I and II are results using pion optical potentials I and II of Ref. 5.

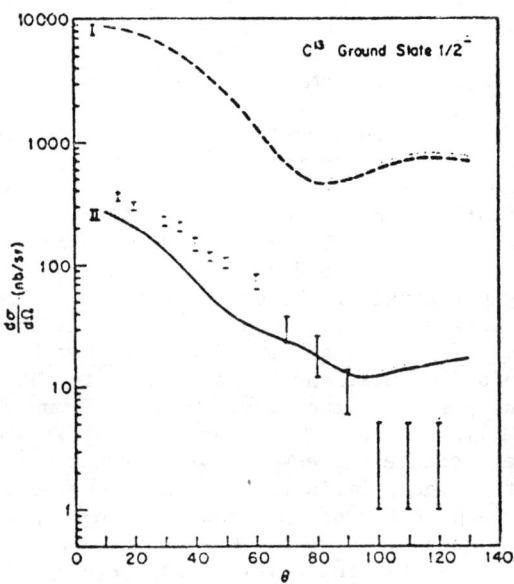

Fig. 5. Same as Fig. 4 in the pionic stripping model. Taken from Ref. 5.

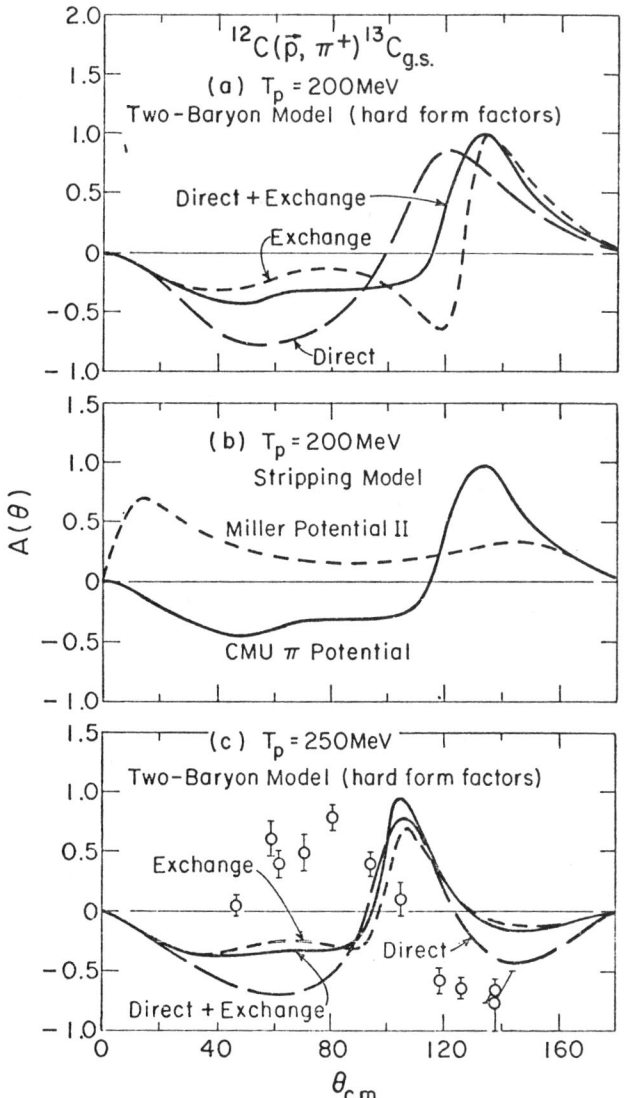

Fig. 6. The asymmetry in $^{12}C(p,\pi)^{13}C_{g.s.}$ in the two-baryon plus core model a) for 200 MeV protons; b) with the two potentials of Fig. 4; and c) for 250 MeV protons.

theoretical results also can be obtained in our model. The interference between the direct and exchange terms leads to the results seen in Figs. 6a and 6c. It is also important to notice that the theoretical results for the asymmetry are strongly dependent on the distortions, as illustrated in Fig. 6b. See Brad Keister's contribution for further results.

We conclude that there are two major processes, the direct and exchange mechanism, and systematic comparison with experimental data is necessary. The threshold region gives additional problems, since the factorization of the π-N amplitude is not a satisfactory approximation. However, with detailed comparison between theory and experiment, and further technical improvements in the calculations we should begin to obtain the rich information contained in the (p,π) process.

SIX-QUARK MODEL OF THE TWO BARYON CURRENT: SHORT-RANGE PHENOMENA

As has been emphasized in earlier sections, pionic absorption involves short-range nucleon-nucleon and other baryon-baryon interactions. Since nucleons are of the size of about 1 fm, for all processes which involve internucleon distances of $r \gtrsim 1.5$ fm, one can expect the quarks in the two nucleons to communicate with no confinement barrier. The purpose of this section is to explore the possibility of treating all short-range two-nucleon effects in a six-quark shell model. The model proposed is that only pion exchanges, and perhaps baryon excitation with pion exchange, be considered explicitly, and all heavier boson exchanges be treated by quark processes. In this section I shall give recent results for electromagnetic interactions, and briefly discuss the $\pi D \leftrightarrow pp$ reaction.

Six-Quark Model of the Two-Baryon System

The model proposed is based on the MIT Bag Model.[12] Let me briefly review that model. The quarks are confined to a spherical region (the bag)

$$r < r_o,$$

and in the lowest order are free within the bag. Thus they satisfy the Dirac equation within the bag and thus can be represented by linear combinations of the Dirac spinors

$$\psi_\kappa^j = \eta \begin{pmatrix} j_\ell(\omega_\alpha r)[Y_\ell \chi]^j \\ \\ j_{\ell-|\kappa|/\kappa}(\omega_\alpha r)[Y_{\ell\pm 1}\chi]^j \end{pmatrix}. \quad (10)$$

The color confinement is achieved by the vanishing of the color quark current at the bag radius

$$-i\,\hat{r}\cdot\vec{\gamma}\psi_\kappa^{\,j}\Big|^{r_o} = \psi_\kappa^{\,j}\Big|^{r_o}. \tag{11a}$$

This gives the ω_α eigenvalues. Finally, the condition of no energy flux across the bag surface

$$\hat{r}\cdot\nabla(\bar\psi\psi)\Big|^{r_o} = -2B, \tag{11b}$$

where B is a constant, gives the energy eigenvalue of the hadrons, i.e., the hadronic masses. Only $j = 1/2$ solutions satisfy the boundary condition 11b. In addition, there are one-gluon potentials[13] which give hyperfine splitting, resulting in further mass splitting between the hadrons. Detailed fits to the hadron spectroscopy have been studied by Isgur and Karl, and Cutkosky and Forsyth with considerable success.[14]

Let us now consider two-baryon systems in a state LSJI. The model which I shall use is defined as follows. The wave functions of the system are:

$$\psi^{LSJI} = \begin{cases} \psi_{NN}^{LSJI}(r) & r > r'_o \\ \psi_{6q}^{LSJI} & \text{otherwise} \end{cases}, \tag{12a}$$

with

$$\psi_{6q}^{\alpha} = \sum_i c_i^{\,\alpha}\,\psi_{6q}^{i}, \tag{12b}$$

where ψ_{6q}^{i} are six-quark configurations in the free Dirac representation (Eq. 10). The boundary condition of Eq. (11a) is satisfied to confine the color, but that of Eq. (11b) is not relevant, since we are not looking for hadronic energy eigenvalues in this quark model. The quark configurations are simply giving the short-distance behavior in terms of the phenomenological spectroscopic factors C_i.

For short-distance behavior QED and QCD dynamics can be used, since this is defined for the quarks. Thus the parameters of the theory are r_o and the spectroscopic factors, C_i. The dynamics are given by fundamental theory, where possible. The job is thus to explore a variety of phenomena to see if a consistent picture emerges.

The Electromagnetic Form Factor of the Deuteron

The electromagnetic form factors of the deuteron are quite favorable for a test of the six-quark model of the two nucleon current. Since the pion current contribution vanishes, only short range mesonic current and isobar mechanisms are possible. Processes

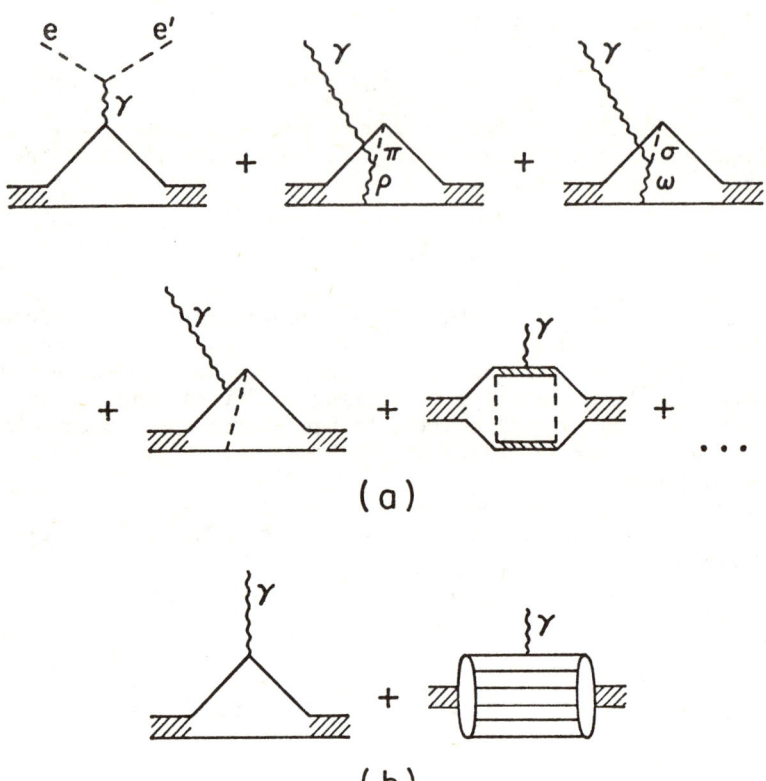

Fig. 7. Deuteron electromagnetic form factor using a) one nucleon (impulse) term plus meson currents, and b) using the six-quark model for the two-baryon currents.

which have been considered[15] are illustrated in Fig. 7a. The alternative picture being proposed here is illustrated in Fig. 7b. That is, we should like to see if quark electromagnetic transitions can successfully model all of the meson current contributions for the form factors. Only the magnetic form factor is discussed here, as this is the most favorable case for comparison with experiment.

The six-quark model for the deuteron is

$$\Psi_{6q}^{Deut} = \Psi_{D_0} + \Psi_{D_1} + \Psi_{D_2} \;.$$

The large components of these three configurations are

$$|D_0\rangle = |[S_{1/2} \times \text{core}(\tfrac{1}{2}+)]^1\rangle$$

$$|D_1\rangle = |[1p_{1/2} \times \text{core}(\tfrac{1}{2}-)]^1\rangle$$

$$|D_2\rangle = |[1d_{3/2} \times \text{core}(\tfrac{1}{2}+)]^1\rangle.$$

For the present calculation all spectroscopic factors are set equal, so there are two parameters: 1) r_0, and 2) quark probability. The small components can be obtained from Eq. (10), with obvious consideration of the quantum numbers. The results are shown in Fig. 8. They are quite sensitive to the parameter r_0, and suggest that $r_0 \approx$ 0.8 and that the probability of the quarks is $\approx 5\%$.

It is also interesting to note that the theoretical results show that the two-baryon effects which have been calculated over the past thirty years for this problem essentially are that there is a region of about 1 fm with a probability of a few % involved in the two-baryon process. Of course, other phenomena must be considered for more information, and pion currents must be included when they contribute.

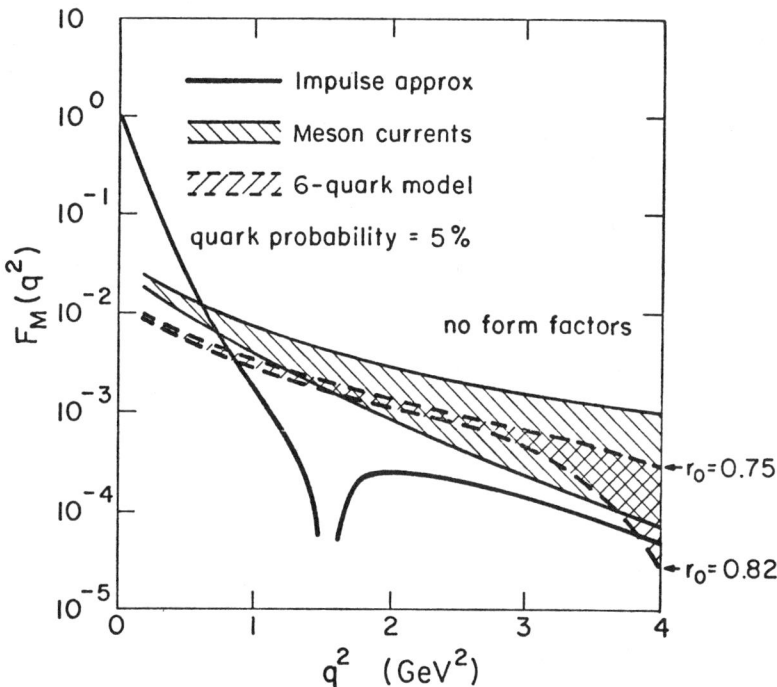

Fig. 8. Magnetic form factor of deuteron.

The d(e,e'p)n Reaction

The electrodisintegration of the deuteron has long been known to be favorable for showing effects of meson currents and related phenomena. The very large discrepancy between the impulse approximation and the data even for rather small q values is well established. The ability of theorists[16] to obtain fits to the data using meson currents is often cited as evidence for correct treatment of the latter.

This process is considerably more involved than the deuteron form factor. One of the most significant differences is that pion currents can contribute, and thus the six-quark part of the two-baryon current cannot be expected to be adequate to represent all two-baryon effects. For this reason we consider the process illustrated in Fig. 9. The terms illustrated in Figs. 9a,b, and c are the impulse term and the long-to-medium range pion and pair contributions to the two-nucleon current. The six-quark current of Fig. 9d gives the short range part of the two-baryon current. The model is

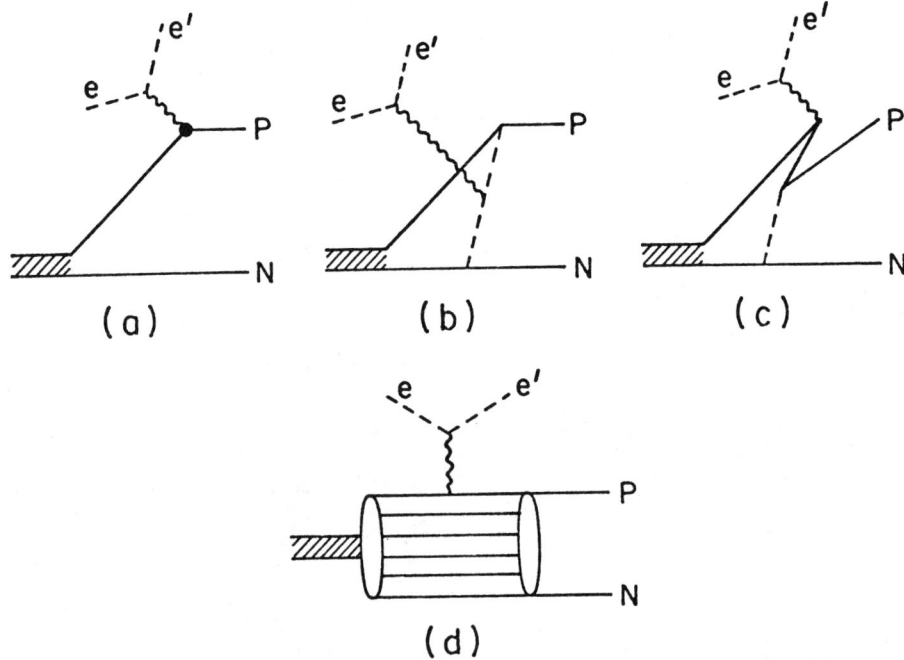

Fig. 9. Processes used in the present work for the electrodisintegration of the deuteron. a) Impulse term, b) pion current, c) pair current, d) short range two-baryon current given by six-quark model.

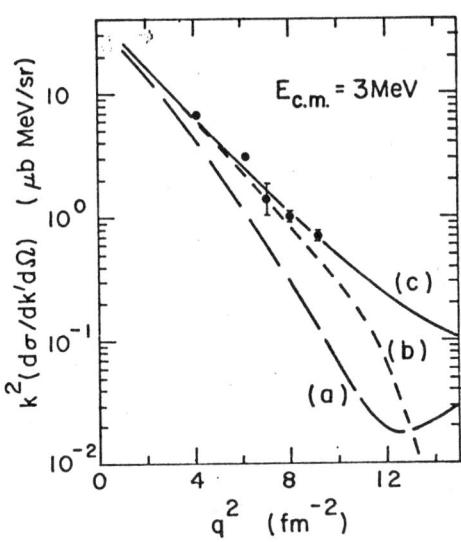

Fig. 10. Cross section for D(e,e'p)n with a) impulse approximation, b) impulse approximation plus pion and pair currents, and c) all the processes of Fig. 9.

very similar to that of the preceding subsection, with the same quark states being used and configurations chosen to represent the singlet and triplet P-N states with due consideration of quantum numbers.

The results are shown in Fig. 10. Note that now the pion-pair currents are necessary to represent the low to medium-range two-nucleon phenomena, while the 6q contribution is able to represent the high-q aspects. The results are compatible with those of the form factor of the deuteron (preceding subsection), and the model seems quite satisfactory.

The I = 0 "Dibaryon"

One of the most significant phenomena for the study of quark representations of two-baryon states is that of dibaryons. We define a dibaryon as a resonance in a two-baryon system which cannot be explained by interactions arising solely from the exchange of color singlets (hadrons). There is now an interesting and extensive debate on whether such objects have been seen in p-p experiments. I shall not review this subject.

The first possible dibaryon which has been detected[18] is probably an isospin zero object. Some experimental results are shown in Fig. 11. In extensive calculations,[19] including a recent extensive study of relativistic effects,[20] it has not been possible to obtain the large, negative, resonant-like polarization peak. The solid and

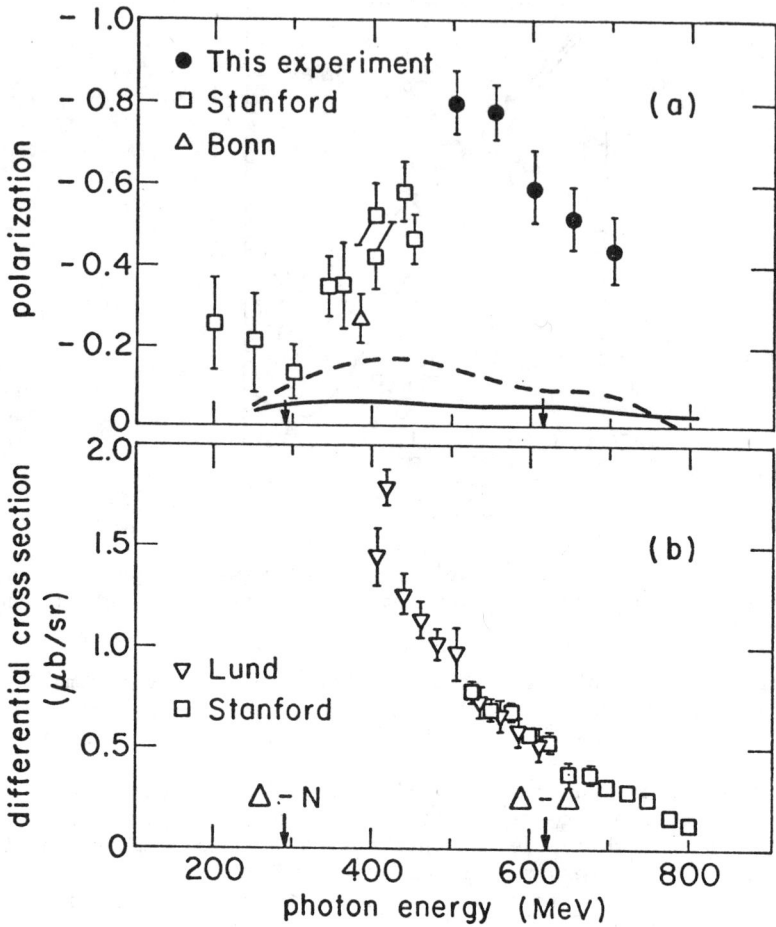

Fig. 11. The polarization and differential cross section showing the possible I = 0 dibaryon at $\sqrt{s}$ = 2.35 GeV, from Ref. 18.

dashed curve in Fig. 11 shows attempts by the Japanese group[18] to fit the data with conventional models. In Fig. 12 the dashed curve shows he results of Chi-Yee Cheung and myself of extending the conventional physics by including relative energy (or time) in the deuteron using results of Bethe-Salpeter calculations. All parameters have been chosen to give the largest possible polarization, and we fail to reproduce the experiment. The solid curve in Fig. 12 is the result of our[20] adding a six-quark dibaryon.

The electrodynamics and six-quark model used to get this result are the same as in the previous section, but the dibaryon is a state of the system. Although from these results one cannot necessarily conclude that there is no other explanation for the experimental results of Kamae et al., it seems likely that there is a six-quark resonance at 2.35 GeV.

Pion Absorption

Let us now return to the subject of pion absorption. Under the conditions for pion absorption with no relative momentum at the deuteron vertex (Eqs. 2a,b), one would expect a peak in the energy dependence of the cross section arising from almost free $\Delta(1232)$ production at an energy of $E_{c.m.}$ = 2.18 GeV. Experimentally the peak is at 2.14 GeV. This suggests that in this energy region one can apply the quark model being discussed in this section, without treating the difficult problem of pion absorption within the quark model. The idea is as follows.

At $E_{c.m.} \stackrel{\sim}{\sim} 2.15$ GeV, the mechanism of π absorption is almost entirely $\pi N \to \Delta$. The deuteron simply provides the nucleon at long range. The interaction $V_{\pi N \to \Delta}$ is treated in the Isobar-Doorway theory as described above. The short-range $\Delta N \to NN$ transition is treated in the six-quark bag using configurations with 1s, 1p, and 1d (large component) single-quark wave functions, as in the electromagnetic processes. These configurations have two "active" quarks and a four-quark core, so the spectroscopic factors are two-quark spectroscopic factors, which can be related to the one-quark spectroscopic factors, c_i, of the previous sections in the usual way. The $\Delta N \to NN$ interaction is taken as the Isgur-Karl potential,[14] so there are no free parameters, except that the spectroscopic factors and r_o are adjusted as phenomenological parameters.

The resulting model is a version of Eq. (5) which corresponds to Fig. 13a. The amplitude is

$$T_{\pi D \to pp} = \sum <PN|6q(NN)_j><6q_j|H'|6q_i>$$
$$\times <6q(\Delta N)_i|\Delta N> G_{\Delta N} <\Delta N|V_{N\Delta}|\pi D> \quad (13)$$
$$= \sum_{ij} c_i^{NN} c_j^{\Delta N} <i|H'|j> b_{\Delta N} <\Delta N|V_{N\Delta}|\pi D>.$$

The quark-transition matrix elements $<i|H'|j>$ are calculated from the Isgur-Karl potential,[14] and $<\Delta N|V_{N\Delta}|\pi D>$ is determined by the $\pi N \Delta$

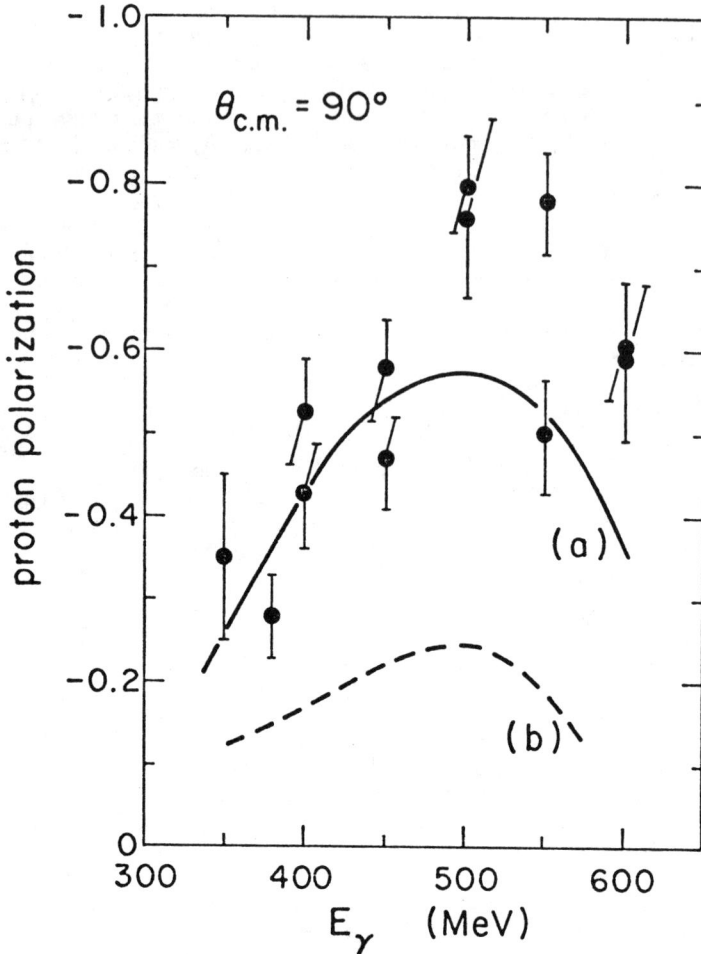

Fig. 12. a) The polarization in D(e,e'p)n calculated with a dibaryon treated in the six-quark model (see text) plus background (see text), and b) without the dibaryon.

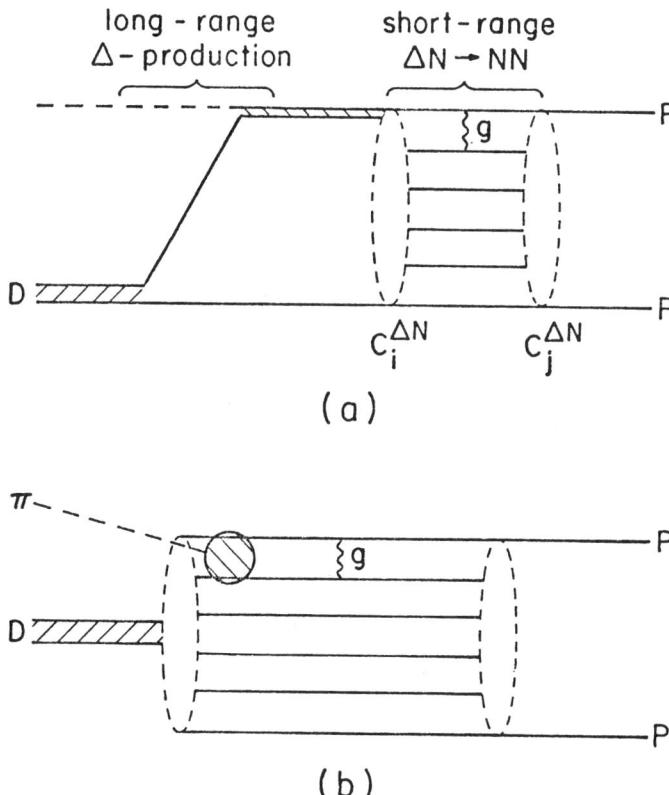

Fig. 13. The six-quark model used to calculate the short range processes in the pp↔πD reaction a) near the Δ resonance, and b) away from the Δ resonance.

form factor and the deuteron and ΔN wave functions. The results of our calculation are given in Jerry Miller's paper elsewhere in these Proceedings. The model is quite satisfactory, but contributions from the long-range $\Delta N \rightarrow NN$ via pions must be calculated, and polarization studies must still be carried out.

For energies above or below the Δ-resonance, the model of Fig. 13a will not be adequate, as short-range interactions will be important for the pion absorption. This can be seen in the figures in Miller's paper. Thus contributions illustrated in Fig. 13b must be included. This is a subject of research for the near future. We conclude that progress in understanding short-range phenomena in the quark models with QCD leads to important clarification of the complicated pion absorption processes. This is an encouraging new aspect for this area.

CONCLUSION

In conclusion, the (p,π) reaction in complex nuclei requires large momentum and energy transfer, and there are a variety of quantum mechanical effects involved in momentum sharing. Therefore, theoretical models must treat carefully a number of processes, some of them quite novel. From theoretical considerations, models with two active nucleons and a core might provide an adequate framework for including the necessary effects, particularly within Isobar Doorway models. However, present theoretical calculations have not yet reached a level of accuracy to provide firm conclusions. A great deal of continuing theoretical effort is needed in order to make use of the systematic experimental data now becoming available.

In the light of developments in particle physics it is desirable to attempt to treat the short-range aspects of absorption in quark/QCD models. If QCD is the correct theory of strong interactions, it is the proper basis for treating interactions at short distance. We have seen that near the Δ resonance a quantitative treatment of the $pp \leftrightarrow \pi D$ reaction can be obtained from a six-quark bag model, with parameters largely determined by deuteron form factors and electrodisintegration. I believe that this will provide a unified theory of pionic absorption and other short-distance nuclear phenomena in the future.

This work is supported in part by NSF Grant #PHY 78-19757, and in part by the U.S. Department of Energy.

The author would like to acknowledge hospitality and useful discussion with the Theory Groups at SIN during the Summer, 1981, and at the University of Washington during the Fall, 1981.

REFERENCES

1. S. Mandelstam, Proc. Royal Soc. $\underline{A244}$, 491 (1958).
2. B.D. Keister and L.S. Kisslinger, Nucl. Phys. $\underline{A326}$, 445 (1979).
3. J.A. Niskanen, Nucl. Phys. $\underline{A298}$, 417 (1978).
4. Z. Grossman, F. Lenz, and M.P. Locher, Ann. Phys. (N.Y.) $\underline{84}$, 348 (1979).
5. G.A. Miller, Nucl. Phys. $\underline{A224}$, 269 (1974); D. Measday and G.A. Miller, Ann. Rev. Nucl. and Part. Phys. $\underline{29}$, 121 (1979).
6. L.S. Kisslinger and W.L. Wang, Phys. Rev. Lett. $\underline{30}$, 1071 (1973); Ann. Phys. (N.Y.) $\underline{99}$, 374 (1976).
7. A.N. Saharia, R.M. Woloshyn, and L.S. Kisslinger, Phys. Rev. $\underline{C23}$, 2140 (1981).
8. M. Hirata, Phys. Rev. Lett. $\underline{40}$, 704 (1978).
9. L.S. Kisslinger, Proceedings of Los Alamos Conference, January, 1980, LA-8303-C.
10. B.D. Keister and L.S. Kisslinger, to be published.
11. G.J. Lolos et al., TRIUMF preprint TRI-pp-81-21 (1981).
12. A. Chodos et al., Phys. Rev. $\underline{D10}$, 2599 (1974); T.A. DeGrand and R.L. Jaffe, Ann. Phys. $\underline{100}$, 425 (1976).
13. T. DeGrand et al., Phys. Rev. $\underline{D12}$, 2060 (1975); A. DeRujula et al., Phys. Rev. $\underline{D12}$, 147 (1975).
14. N. Isgur and G. Karl, Phys. Rev. $\underline{D20}$, 1191 (1979); C.P. Forsyth and R.E. Cutkosky, Phys. Rev. Lett. $\underline{46}$, 576 (1981).
15. R.J. Adler, Phys. Rev. $\underline{141}$, 1499 (1966); M. Chemtob, E.J. Moniz, and M. Rho, Phys. Rev. $\underline{C10}$, 344 (1974).
16. W. Fabian and H. Arenhövel, Nucl. Phys. $\underline{A258}$, 461 (1976).
17. J. Hockert et al., Nucl. Phys. $\underline{A217}$, 14 (1973).
18. T. Kamae et al., Phys. Rev. Letters $\underline{38}$, 468 (1977); $\underline{42}$, 132 (1979).
19. K. Ogawa et al., Nucl. Phys. $\underline{A340}$, 451 (1980); M. Anastasio and M. Chemtob, Saclay Preprint DPh-T/80/153.
20. C.-Y. Cheung and L.S. Kisslinger, submitted for publication.

RESULTS OF MICROSCOPIC CALCULATIONS

B. D. Keister[*]
Carnegie-Mellon University, Pittsburgh, PA 15213

ABSTRACT

This section contains an outline of recent microscopic calculations for the $(\vec{p},\pi)$ reaction using the isobar-doorway approach of Keister and Kisslinger. These include various model comparisons and tests, as well as comparisons to experiment for ^{12}C and ^{3}He targets.

INTRODUCTION

While the primary goal of our calculations has been ultimately to make contact with the wealth of experimental data now available, we are also in a position to test other models and approximations, of both one-nucleon (ONM) and two-nucleon (TMM) variety. The results of these model tests are presented first, followed by comparisons to experiment.

MODEL COMPARISONS

The "direct" and "exchange" diagrams which form the basis for our calculations allow us to make at least formal contact with previous ONM and TNM calculations. Our "exchange" diagram is formally similar to an ONM, or "stripping" amplitude, though practical calculations may differ. The "direct" and "exchange" diagrams together are also similar in spirit to the TNM of Grossmann, Lenz and Locher,[1] but without certain approximations found in many TNM calculations. These connections will now be considered in more detail, using the reaction $^{12}C(\vec{p},\pi^+)^{13}C_{g.s.}$ at 200 MeV.

Almost all TNM calculations to date employ some sort of zero-range approximation, in which the two participating nucleons lie at the same point. In our theory, this amounts to letting the pion propagator become a delta function: $G_\pi(\vec{r}_1-\vec{r}_2) \to -\delta^3(\vec{r}_1-\vec{r}_2)/(\vec{k}^2+m_\pi^2)$, where $\vec{k}$ is some average momentum, and the πNN and $\pi N\Delta$ form factors are set to unity. Figures 1 and 2 show the results of zero-range calculations. The cross sections are large (> 1μb/sr); for the case of a $^{13}C_{g.s.}$ final state, the exchange term is clearly dominant, and the two terms have opposite asymmetries. The effect of distortions is not small: the distorted wave (DW) and plane wave (PW) cross sections differ by a factor of 10.

The effect of restoring the finite range of the pion is shown in Fig. 3. The cross section is 100 times smaller than the zero-range case, the roles of direct and exchange terms are reversed, and the asymmetry is very different. From this analysis, we conclude that the zero-range approximation is not very useful for microscopic calculations, since it leads to very different sizes and shapes of the differential cross section, as well as different asymmetries.

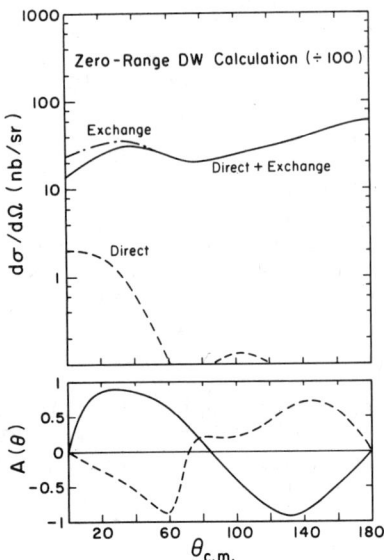

Fig. 1

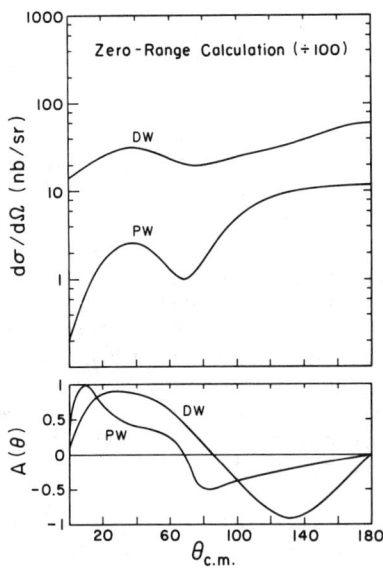

Fig. 2

We have also examined other elements of these calculations, from which we observe the following:

(1) Varying the πNN and πNΔ cutoff masses changes only the scale of the cross section, while the angular distribution and asymmetry are essentially unaffected. For example, if $\Lambda_{\pi NN} = 5$ fm^{-1}, and $\Lambda_{\pi N\Delta}$ is increased from 1.5 fm^{-1} to 4 fm^{-1}, the cross section increases by a factor of four.

(2) The use of Woods-Saxon in place of harmonic-oscillator wave functions also changes mostly the scale (x3) of the cross section, with little effect on the shape or asymmetry (although the roles of "direct" vs. "exchange" are again reversed). These two points, along with the large effect of the range of the pion propagator, can be understood through the following simple analysis. If one ignores the effects of distortions and considers the nucleon wave functions to be s-shell Gaussians, then the (p,π) amplitude depends upon a three-dimensional loop integral which can be reduced to the approximate form

$$I \sim \int_0^\infty p^2 dp\, [p^2/(p^2+m_\pi^2)] v(p) \qquad (1)$$
$$\exp(-3p^2/4 + pq)c^2 ,$$

where the term in brackets is the pion propagator with p-wave coupling factors, $v(p)$ represents the πNN and πNΔ form factors, c is the oscillator length, and q is the momentum transfer to the nucleus. For the zero-range calculation, one simply sets the pion denominator to $(k^2+m_\pi^2)$ using the on-shell momentum k, $v(k)=1$, and then performs the

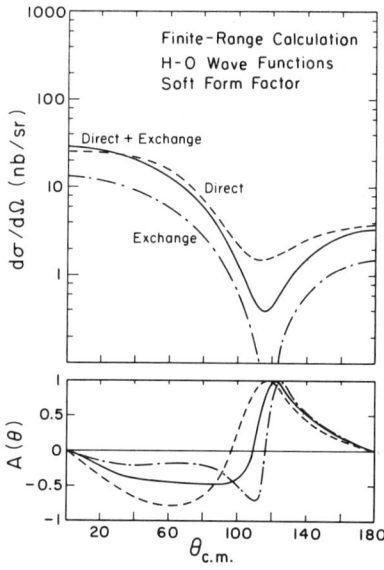

Fig. 3

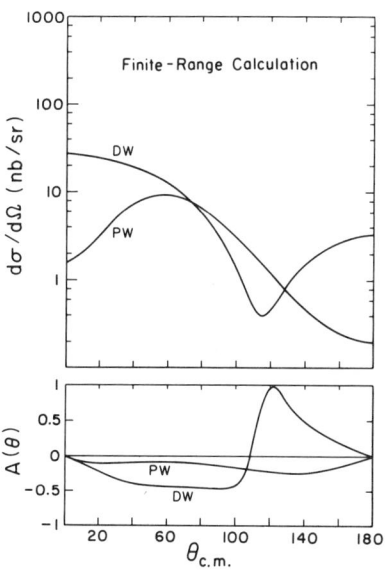

Fig. 4

integral. However, the exponential functions in the integrand are sharply peaked at $p \simeq 2/3\, q$. For typical momentum transitions to the nucleus, say 450 MeV/c, the virtual pion momentum is of order 300 MeV/c; i.e., momentum is shared between nucleons at the expense of making the pion momentum large. The propagator is then reduced from its zero-range value by the ratio $(\bar{k}^2+m_\pi^2)/(p^2+m_\pi^2) \sim 0.2$, which gives a factor of 25 reductions in the cross section. The cross section is further reduced when the πNN and $\pi N\Delta$ form factors are included. Since the from factors are very smooth functions of p compared to the remainder of the integrand, they can be factored outside the integral at the peak value. If one substitutes Woods-Saxon for harmonic oscillator wave functions, the analysis is basically the same, as above, in the sense that the integrand peaks at approximately the same virtual pion momentum, but the amplitude of the integrand is larger at its peak value.

Our calculations also indicate the following:

(3) Intermediate coupling to higher spin and isospin states is an important ingredient, as expected from the fact that two-particle excitations are enhanced in this approach.

(4) The effect of distortions and the sensitivity to optical potentials are still substantial: evidently the sharing of momentum by two nucleons is not enough to eliminate the sensitivity to rather far-off-shell values of the external pion and proton momenta. Fig. 4 shows the difference between plane-wave and distorted-wave calculations - still rather large, even when the pion propagator has finite range.

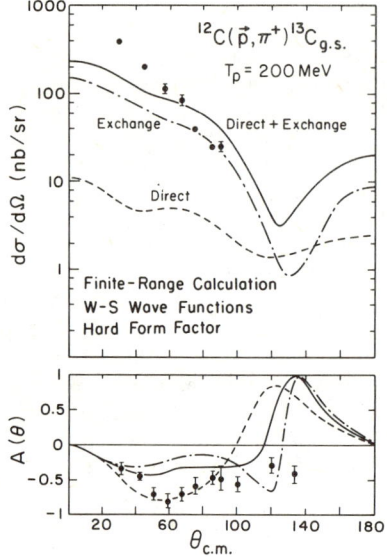

Fig. 5

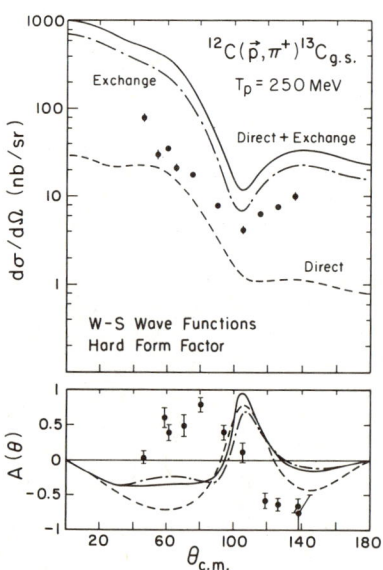

Fig. 6

The proton distortions alone (generally assumed to be a small effect) can change the cross section by a factor of two. One also finds similar dramatic sensitivities to the pion optical potential as those seen by Miller in a standard one-nucleon calculation.[2]

COMPARISON TO EXPERIMENT

Figure 5 shows our best calculation to date against TRIUMF[3] data for $^{12}C(\vec{p},\pi^+)^{13}C_{g.s.}$ at 200 MeV proton laboratory energy. Considering the uncertainties and sensitivities still inherent in the calculation, the agreement is satisfactory.

At 225 and 250 MeV, the measured[4] asymmetry oscillates between negative and positive values in contrast to the relatively flat negative asymmetry seen at lower energies for a variety of targets. Our calculation at 250 MeV (Fig. 6) reproduces approximately the shape of the differential cross section, and the predicted asymmetry changes sign, but at the wrong angles.

Finally, we compare to data for $^3He(\vec{p},\pi^+)^4He$ at 201 MeV (Fig. 7).[5] Our calculations assume ^4He to be an antisymmetrized s-shell with harmonic oscillator wave functions which reproduce[6] the ^4He charge form factor up to $Q^2 \sim 6$ fm^{-2}. The measured flat angular distribution is not reproduced. In this particular case, the direct and exchange contributions are comparable, and their sum is similar to the result of a one-nucleon (DWBA) calculation.

SUMMARY

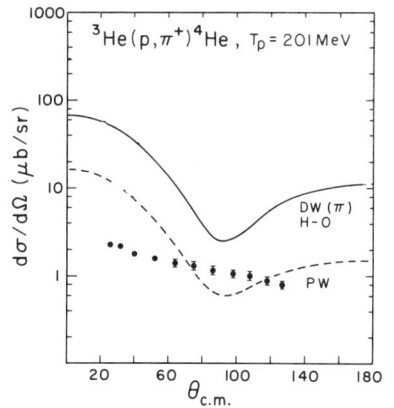

Fig. 7

Our approach to the (p,π) reaction provides us with the following opportunities:
(i) to provide a unified picture of the (p,π) reaction and to make contact with prevailing ONM and TNM calculations;
(ii) to test quantitatively certain approximations used in other calculations;
(iii) to reproduce experimental data.

A number of questions and refinements still remain for our calculations, which include:
(i) the effect of energy dependence, or nonlocality of the Δ in intermediate pion rescattering;
(ii) the actual inclusion of Δ propagators and pion wave functions which are consistent within the isobar-doorway model.

On the basis of calculations to date, our main conclusions are:

(1) The approximation of zero separation between the two nucleons leads to substantially different results than when the intermediate pion is allowed to propagate with its appropriate range.

(2) While the idea that two-nucleon momentum sharing reduces parameter sensitivity is attractive in principle, in practice the momentum carried by each nucleon can still be rather large, and one's choice of cutoff masses, nucleon wave functions and optical potentials can affect the results considerably.

(3) The agreement with experiment is qualitatively satisfactory, but quantitative differences still remain.

REFERENCES

1. Z. Grossman, F. Lenz and M. P. Locher, Ann. Phys. (N.Y.) 84, 348 (1979).
2. G. A. Miller, Nucl. Phys. A224, 269 (1974).
3. F. G. Auld, et al., Phys. Rev. Lett. 41, 462 (1978).
4. G. J. Lolos, et al., TRIUMF preprint TRI-PP-81-21 (1981).
5. N. Willis, et al., Orsay preprint IPNO-PhN-81-07 (1981).
6. R. F. Frosch, et al., Phys. Rev. 160, 874 (1967).
* Supported by the N.S.F.

DISCUSSION

G.E. Walker (Indiana Univ.): If the sensitivity of these results holds up it's kind of a depressing result. I want to ask a question. One would not expect, if not much is going on with pion distorted waves at certain energies, large differences to come from distorted waves. Then the only way I can understand that you can have so much sensitivity to distorted waves and form factors, is that it really is an incredibly short-range interaction; that short range behavior is crucially important. On the other hand, yesterday we heard from both Thomas and Niskanen that maybe the situation is not so dependent on short-range physics, because sometimes the intermediate pion can be almost a real pion. Where's my confusion?

Keister: In $\pi d \to 2p$, both of the protons are free. For (π,p) in a larger nucleus, then the second nucleon remains in the nucleus, and there is a price for that. You introduce extra sensitivities because there are extra wave functions.

Walker: So, if you were to apply your model to (p,π) in the quasi-elastic continuum, where the final proton didn't have to be bound, it wouldn't be quite the same situation?

Keister: That's right.

Walker: That might be something to try.

Keister: Yes.

M.K. Banerjee (Univ. of Maryland): The pion must lose half its energy. It cannot go on the mass shell.

M. Hynes (Los Alamos Scientific Lab.): Kisslinger mentioned some Japanese (γ,p) work having to do with dibaryons. What was that experiment?

L.S. Kisslinger (Carnegie-Mellon Univ.): The "dibaryons" that one usually talks about are these pp "dibaryons", and they have been studied by Kloet and Silbar and some other people. One can reproduce those things by colorless exchanges -ordinary nucleons, isobars and pions. Not as much attention has been paid to the first dibaryon that was seen: the I=0 dibaryon at 2.35 GeV. In the 6-quark calculations, this is the dibaryon that shows up to be most isolated from the other dibaryons. The cross section doesn't show anything, but in $\gamma d \to pn$ the polarization shows a very sharp peak. The Japanese group has tried for many years to do the kind of calculations that people do in the standard models, without getting anything. Chi-yee Cheung and I threw in some extra relativistic effects which we knew could help, and it didn't do the job. So our conclusion is that we can't do it in conventional physics, but the dibaryon was able to do it. The 6-quark model that I mentioned is just another use of the 6-quark model, so it doesn't really prove anything, but it is at least consistent.

T. Londergan (Indiana Univ.): I just wanted to mentioned that I will be briefly discussing that experiment - the deuteron photodisintegration - in my talk on related reactions tomorrow, and I'll discuss other non-quark attempts to fit the observed polarization spectrum.

J. Eisenberg (Tel Aviv): Do I understand correctly that in the way you calculate the (p,π), your pion always suffers at least one rescattering?
Keister: Yes.
Eisenberg: Both speakers agree on this?
Kisslinger: Do you mean in the nucleus? It suffers many?
Eisenberg: But at least one rescattering off of one nucleon. Then I would like to correct a misapprehension of one of the earlier speakers, namely the first. He suggested that what you've done effectively sums what's conventionally called the single-nucleon mechanism and the two-nucleon mechanism, with the interpretation that Noble gave at the time, which is one that I think we all more or less accept. This is because there surely is a Feynman diagram in this story where the pion is not rescattered at all, and that should be added in. It is known from tons of other calculations to be of the same order of magnitude in its contribution to the amplitude as these other effects.
Kisslinger: One has a pion optical potential. It's a little like saying that in elastic scattering there is no scattering. There is a pion optical potential.
J.V. Noble (Univ. of Virginia): There is a term where it doesn't scatter.
Eisenberg: No, there isn't, because in the amplitude itself there is always at least one rescattering.
Keister: Our distorted-wave matrix element of the pion-rescattering two-body operator undercounts by the plane wave Born approximation.
Kisslinger: That's the plane wave part of the elastic scattering.
Eisenberg: No, I believe that's incorrect. Your basic mechanism insists that the pion always rescatters at least once. It may then rescatter more through the optical potential. If I'm wrong I'd like to get that straight. That's quite crucial. It undercounts by one diagram -- a diagram which is known to be of the same magnitude as the ones you calculated. I'm sorry to put it so bluntly, but there it is.
Kisslinger: In the Isobar-Doorway Model which we have been discussing, there is no such term, as one always interacts with the nucleons through a Λ. There is such a term in the nonresonant background. The latter could be important near threshold.

P. Hwang (Indiana Univ.): Regarding the quark model presented by Kisslinger, my question is: Once we start playing with the quark degree of freedom, simultaneously with the nucleon degree of freedom, do you think we already introduce a superfluous degree of freedom into the model?

At one junction you have to decide, for instance, whether the two components of the deuteron interfere--or if they don't. Could you explain the ground rule for this?
Kisslinger: Here I gave a wave function interpretation. But actually you can write this in terms of projection operators, of course. Maybe that would make you feel better. We're projecting onto the two-nucleon space and onto the quark space. What we have not done, is to connect these two spaces, since we don't have a Hamiltonian theory and we don't know the boundary condition which allows us to calculate, for example, the nucleon pressure that would balance the quark part. In the bag model what's done is to introduce an artificial bag pressure. This is something that we are now looking at - to try to put those two parts together.

Banerjee: Do the two parts coexist in the same part of the space ?
Kisslinger: No. One of our next steps is to derive the spectroscopic factors. That's one of the things we've talked about, and we hope that we can make some progress soon on that.
R. Silbar (Los Alamos Scientific Lab): You're talking about the C's in your formula?
Kisslinger: Yes.

G.A. Miller (Univ. of Washington): Just a technical question about the (p,π) calculation. What did you use for the relative wave function of the two nucleons ? Was there any Jastrow correlation function in it?
Keister: No.
Miller: Well, in the current calculation you did, this comes in exactly the same way as the form factor --in fact mathematically instead of having V of P you have V of P times the Fourier transform of the correlation. So, in my opinion if you want to compare to data, you should have both of them in there.
Kisslinger: In a way, the off-shell pion goes between those two nucleons. It's a little bit like the final state interaction in $pp \to \pi d$.
Miller: These types of correlations in this kind of calculation are very important. The anticorrelation with the nucleons, for example, cuts off a lot of the stuff from the virtual pion.

K.F. Liu (Univ. of Kentucky): When you talk about six quark states, what are the quantum numbers of these six quark states?
Kisslinger: It's a shell model, so there are an infinite number of possible single particle states. The same as the nuclear shell model. The three configurations that we've used in all the calculations are 1s, 1p, and 1d. For all the results that I have shown, including the dibaryon, the form factor, the electrodisintegration, and $\pi d \to pp$, all used those three configurations.
Liu: Then is there a particular reason why the polarization would peak at such an energy?

Kisslinger: Are we talking about the dibaryon?
Liu: Yes
Kisslinger: The dibaryon is a separate question. The dibaryon is essentially a six quark object. It is not the same as that short range part of the nucleon-nucleon system. So its probability is essentially unity in the 6 quarks. It's the deuteron that has the 5% probability. The dibaryon in this model is a six quark object.

J. Igbal (Indiana Univ.): You showed the exchange and the direct term, and you showed that the exchange term dominates. How much does that depend on the distortions you put in?
Keister: It certainly depends upon whether distortions are included, as well as many other ingredients. I don't know about the case of distorted wave sensitivity.

Igbal: One more question. Don't you think if you put in the nonlocality of Λ, and if you treat the Λ isobar in the proper way as you are going to do, it's going to make the distortions less important?
Keister: It would certainly help, but how much I can't say.

D. Lichtenberg (Indiana Univ.): When you have six quarks in a bag, you have quite a few color configurations which can give a result of zero. Do you add them all up statistically ?
Kisslinger: Yes. The color moves freely through the system. At the end we have a color zero system for the whole thing. There are no color zero clumpings in the model. But when we use the Isgur-Karl potentials, we're doing it in such a way that $\lambda_i \lambda_j$ pieces add up properly.
Lichtenberg: Why should you exclude 3 of the quarks having color zero? That would have some statistical weight.
Kisslinger: Oh yes, there is some of that.

Proton-Induced Pion Production in the Rescattering Model*

M. Dillig
Institute for Theoretical Physics
University of Erlangen-Nurnberg
Erlangen, W. Germany

F. Soga
Institute for Nuclear Study, The University of Tokyo,
Tokyo 188, Japan

J. Conte
Indiana University Cyclotron Facility
Bloomington, Indiana 47405, U.S.A.

1. Introduction

Within the last years an impressive amount of experimental information on single-nucleon transfer reactions at intermediate energies and at large momentum transfers (large compared to the Fermi momentum of a bound nucleon) has been obtained. This holds particularly for the (p,π) reaction[1], where at projectile energies $T_p \lesssim 200$ MeV a variety of new systematic data exist for light (^{9}Be, ^{10}B, ^{11}B, ^{12}C, ^{13}C, ^{16}O)[2,3] and medium heavy (^{28}Si, ^{40}Ca)[3] nuclei.

This spectacular experimental progress was not matched by the theoretical developments in the same period. Current theoretical approaches show two obvious shortcomings: at the moment there exists

- no systematic analysis in a
- sufficiently detailed microscopic model.

(We consider the stripping model, used in the only extensive analysis done so far[4], as too restricted for a realistic description.) Recent theoretical reviews[5] reflect this situation in focusing more on the shortcomings of existing models, than presenting new information obtained from the (p,π) reaction itself.

Opposite to this somewhat disappointing theoretical status, it is clear that the (p,π) process is potentially a rich source of interesting information. A systematic analysis should provide information on

(i) details of the reaction mechanism,
(ii) the interplay between the reaction mechanism and nuclear structure and, ultimately,
(iii) new degrees of freedom and exotic phenomena, such as the nuclear π-field or the Λ-dynamics (to mention only two examples).

These goals, combined with the experience so far from the study of the (p,π) reaction itself as well as from other high momentum transfer reactions and processes at medium energies, impose strong conditions on the formulation and the application of any microscopic model. Such a model has to account for:

    (i) the different aspects of the elementary production mechanism;
    (ii) the conventional distortions of the proton and the pion;
    (iii) the complexities of the conventional nuclear structure.

Finally, as an important condition for any calculation
    (iv) physical and especially numerical approximations, which could imply misleading conclusions, have to be avoided.

It is clear that the two conflicting conditions--theoretical completeness and numerical feasibility--exclude at the beginning the formulation of a model which is adequate in detail for all existing (p,π) data. This is especially true for proton-induced pion production near threshold. Here the characteristics of the high energy data (forward scattering combined with very strong absorption, appropriate conditions for the application of the eikonal approximation[6]) or of pion production around the 33-resonance (Δ dominance and large medium corrections, suggesting the application of the Isobar Doorway Model[7]) are much less pronounced; additional background contributions, such as the S-wave and the nonresonant P-wave interaction, antisymmetrization effects, etc., make the whole picture more messy. The formulation of a microscopic model, both adequate and practical, necessarily involves physical intuition, practical experiences and personal prejudices. Our model, born in this spirit, is presented and discussed in the next section.

2. The Model.

a. Details

In formulating our model, we start out from the assumption that for a large momentum transfer reaction like (p,π), with typical recoil momenta ≃ 500 MeV/c (as compared to the average momentum of a bound nucleon in light nuclei of ≃ 100 MeV/c), momentum sharing has to be build in explicitly in the elementary interaction. A rather natural starting point is then - as shown in Fig. 1(a) - a two-nucleon mechanism, based on the NN → NNπ amplitude, supplemented by the direct production piece (Fig. 1(b)), where an on-shell pion is directly produced from the projectile (from our philosophy, we expect that the direct term, frequently

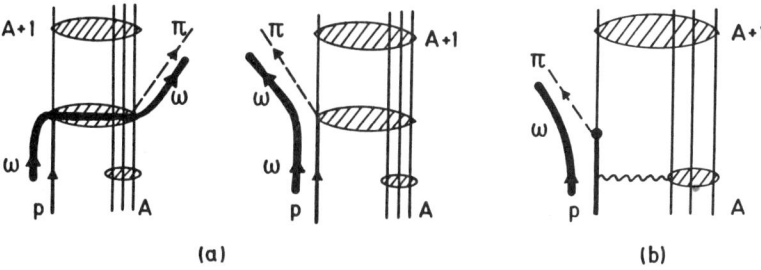

Fig. 1. a) Schematic representation of the rescattering term in (p,π). b) Direct (one-step) production of a pion from the projectile. For all diagrams the flow of the scattering energy ω is schematically indicated.

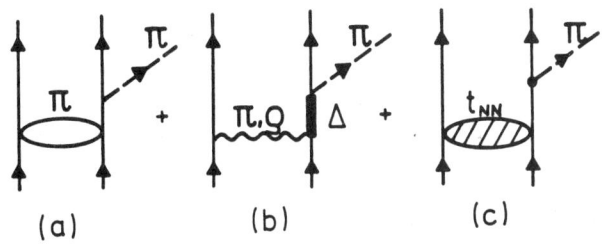

Fig. 2. Structure of the elementary NN → NNπ amplitude: (a) π-induced s-wave rescattering; (b) Δ-dominated, π and ρ induced, p-wave rescattering; (c) nonresonant p-wave rescattering via an intermediate nucleon; t_{NN} is the NN t-matrix.

called the "one-nucleon mechanism", is a small correction to the "two-nucleon" contribution). The various diagrams building up the elementary NN → NNπ subamplitude are shown in Fig. 2; being close to the pion threshold, we include both S-and P-wave rescattering. Our explicit formulation is thereby guided by recent investigations of the π-absorption on the deuteron.[8] Based on an effective Lagrangian, we include S-wave π-rescattering within an extended Koltun-Reitan ansatz[9] (including σ and ρ exchange together with an effective hard core contribution), while the P-wave rescattering is mediated by the whole NN t-matrix for the (nonresonant) N-pole term and by π and ρ exchange for the (resonant) Δ-pole term, which is treated in the closure approximation. Furthermore, off-shell corrections and short range correlations are included via

phenomenological monopole form factors (at each vertex) and a
Jastrow correlation function (with an effective correlation length
$\lambda \simeq m_\omega^{-1}$; m_ω being the mass of the ω-meson[10]). Actually, the
various diagrams are evaluated using old-fashioned perturbation
theory (see the example in Fig. 3(a)), in order to allow for a

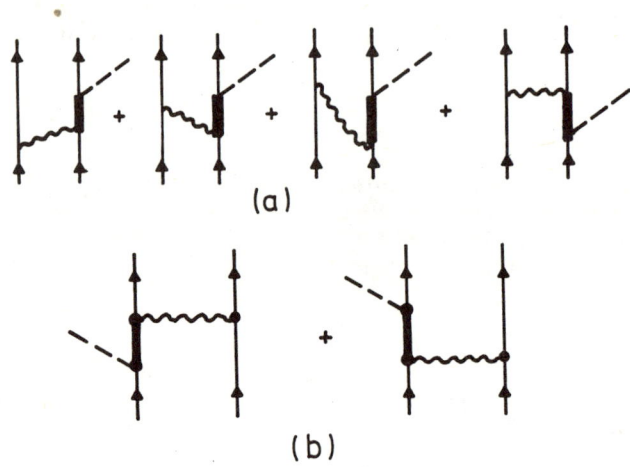

Fig. 3. Schematic representation of the crossed and noncrossed
projectile (a) and target emission (b) diagrams. In (a) the
various time ordered pieces for the noncrossed diagram are shown
explicitly.

refined treatment of nonstatic effects for the intermediate mesons
(in refs. 8 they are included in an ad hoc prescription for the
energy transferred by the exchanged meson). Further details of the
evaluation are discussed in the following paper by Soga et al.,[11];
a comprehensive presentation of the model will be given in a
forthcoming article. In spite of the momentum sharing in our
model, corrections beyond the first rescattering contribution are
by no means negligible. Such multiple scattering effects might be
included either in the framework of the Isobar Doorway Model
(IDM)[7,12] or within the Distorted Wave Born Approximation,
employing a (off shell corrected) π-nucleus optical potential
(DWBA).[13] Though being in principle completely equivalent, in
practice the two approaches differ substantially in the underlying
physics and, especially, in their technical implementation. From
recent progress in the IDM, it seems clear that, given Δ-dominance
in the rescattering mechanism (compare Fig. 4(a)), it is necessary
to treat this additional degree of feedom on the same footing as

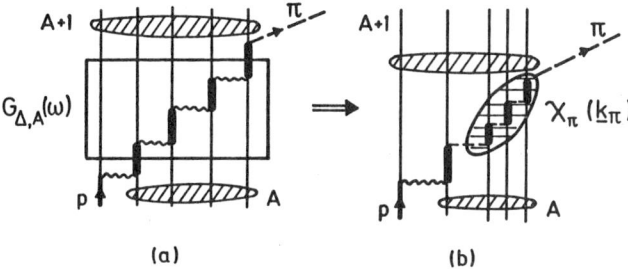

Fig. 4. Schematic relation of the Greens-function of the generalized (A+1)-baryon system $G_{\Delta,A}(\omega)$ (a) and the distorted wave Born approximation (b); $\chi_\pi(\underline{k}_\pi)$ denotes the distorted pion wave.

the nucleon and to solve the generalized Δ-baryon problem; such a procedure would allow medium corrections for the Δ-isobar to be handled in a systematic way. Unfortunately, for scattering energies close to the π-threshold the mechanism, as already stressed above, is much more complex and the Δ-doorway hypothesis seems much less realized. Technically we arrive at a substantial simplification by introducing the closure approximation for the Δ-propagator, i.e., by replacing the Greens function of the generalized Δ-baryon system by its closure limit[12]

$$G_{\Delta,A-1}(\omega) = \sum_{J^\pi} \frac{|(\Delta,A-1)^{J^\pi}\rangle\langle(\Delta,A-1)^{J^\pi}|}{\varepsilon_{J^\pi}-\omega} \qquad (1)$$

$$\rightarrow G^{CL}_{\Delta,A-1}(\omega) = \frac{1}{M_\Delta - M_N - i\Gamma_\Delta/2 - \Sigma_\Delta - \omega} \qquad (2)$$

(where Σ_Δ denotes an appropriate (complex) closure energy) and by including higher order corrections in a distorted pion wave $\chi_\pi(\underline{k}_r)$ (see Fig. 4(b)). The actual calculation is supplemented by the initial state interactions of the proton, which are also included in a p-nucleus optical potential, and by Coulomb corrections both for the pion and the proton.

b. General Features

From the diagrammatical structure of our model it is clear that the reaction mechanism involves dynamically very different pieces. This is indicated in Fig. 1, where the flow of the scattering energy ω (which is the kinetic energy of the proton) through the nucleus is shown schematically. It is evident that the projectile emission diagram (the first diagram in Fig. 1(a)) and the target emission (as well as the one-nucleon) diagram (see Fig. 1) probe rather different aspects of the reaction mechanism and the nuclear structure. In the projectile emission the total scattering is transferred onto the nucleus, exciting the nucleus high into the continuum. In kinematical situations where the projectile emission dominates the reaction mechanism, the experimental data predominately reflect information on the "hot" nucleus at energies above the pion production threshold.

If target emission dominates, the information from the (p,π) process is rather different. There, except for very central collisions (i.e., for backward scattering), the scattering energy ω is immediately converted into the pionic degree of freedom, leaving the nucleus practically "cold" in or close to its conventional ground state. Consequently, this part of the reaction mechanism reflects and probes mainly static properties of the nucleus. Without entering into a detailed discussion, we just present two interesting examples: the nuclear pion field and the virtual Δ-admixture in nuclei.

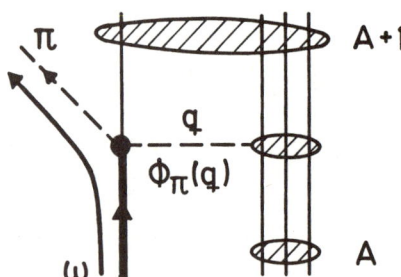

Fig. 5. Relation between the (p,π) cross section and the pion field $\phi_\pi(q)$ of the nucleus A.

For the target emission term only, the (p,π) cross section is easily cast into the form (schematically)

$$\frac{d\sigma}{d\Omega}(\theta_\pi,\omega) \simeq \int \Psi_\alpha(Q-q) t_{\pi N}(Q,q,\underline{k}_\pi) \phi_\pi(\underline{q}) d\underline{q} \qquad (3)$$

where $\psi_\alpha(Q)$ is the bound state wave function of the projectile after the π-emission, and the recoil momentum $Q = K_p = k_\pi$, while $t_{\pi N}(Q,q,k_\pi)$ is the πN (off shell) t-matrix. The most interesting quantity is, however, the pion field $\phi_\pi(q)$ at the momentum component q, which carries the signature of the specific nuclear transition involved; for the excitation of single particle states, $\phi_\pi(q)$ is just the pion field in the nuclear gound state

$$\phi_\pi(q) \sim \frac{f_\pi}{m_\pi} \frac{1}{q^2+m_\pi^2} \int e^{i\underline{q}\underline{r}} \langle \underline{\sigma}\underline{r}\vec{\tau}_\pi \rangle \rho_A(\underline{r}) d\underline{r} \qquad (4)$$

($\rho_A(r)$ is the corresponding nuclear density averaged over the spin and isospin operators included above). It seems that a systematic study of the (p,π) reaction should provide detailed information on the nuclear pion field and help to clarify controversial issues like precritical phenomena, reflecting a nearby pion condensate[14]. For further elaboration of this and related aspects, we refer to the contribution to this workshop of Gibbs, who focusses in his (p,π) model exclusively on the target emission contribution[15].

A quantity of similar interest as the nuclear pion field is the virtual Δ admixture in low lying nuclear states[16]. Diagrammatically the corresponding piece is given in Fig. 6,

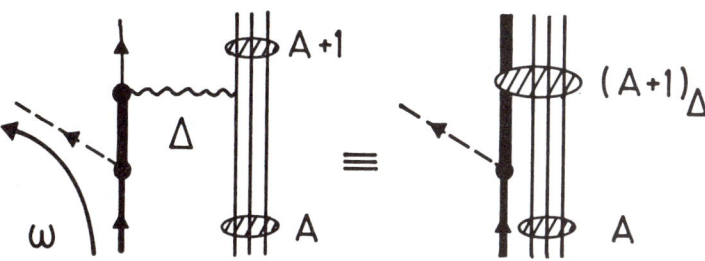

Fig. 6. (a) Contribution of the virtual Δ-excitation to the (p,π) cross section; (b) relation to the virtual Δ-admixture in the residual nucleus

yielding a differential cross section

$$\frac{d\sigma}{d\Omega}(\theta_\pi,\omega) \propto |\psi_\Delta(\underline{Q})|^2 \qquad (5)$$

Thus, similarly as in the one-nucleon model for the conventional
nuclear wave function, in this approach the cross section directly
maps out the wave function of the Δ-isobar at the recoil momentum
Q[17]. Actually, however, the situation is more complicated.
Comparison with Fig. 3(b) shows that the piece we identified as
Δ-admixture in the nuclear ground state, is, at least in a
perturbative calculation of the virtual Δ wave function, only part
of the target emission contribution; unfortunately, the sum of the
other diagrams in general dominates the reaction mechanism. As the
information on the Δ-components is even more indirect than that on
the nuclear pion field, only a careful and systematic anaysis,
supplemented by experimental and theoretical information from
related fields, in particular, on the quenching of magnetic and
Gamow-Teller transitions[18], will provide specific and reliable
information.

3. Preliminary results

Already from the rather fragmentary survey of our model it
becomes clear that its actual application on a large scale requires
extensive numerical work. Two codes, with slightly different
emphasis on various model assumptions, are under construction:
first results from the code of F. Soga are presented and
discussed in the following contribution; the code of J. Conte will
run in the near future.

We present here preliminary results on the very light nuclei
^{3}He and ^{4}He, which were obtained from a simple analytical
calculation as a test of the complete program. We used the model
as presented in the previous section, except that the proton and
pion distortions were included in the eikonal approximation. For
the nuclear structure of ^{3}He and ^{4}He only s-waves were kept in an
harmonic oscillator basis.

As the calculations will be reported in more detail in a
separate publication[19], we only present a few typical results
together with some qualitative details. In Fig. 7 the calculated
differential cross section is compared with experimental data from
IUCF and Orsay.[20] We find that the model accounts for the
magnitude and qualitatively for the shape of the angular
distribution; however, the slope of the theoretical curve is too
steep, reflecting the minimum around $\theta_\pi \sim 100°$. To clarify the
origin of the discrepancy, further and more realistic calculations
are needed, which should be complemented by large angle data to
look for a possible backward peak. The asymmetry prediction in
Fig. 8 is of similar quality, though it has to be kept in mind that
it is rather sensitive to a variation of various input parameters
(in obtaining the dashed curve the π-optical potential, the π and
ρ cutoffs as well as the ρNN coupling constant were varied by 20%),
and in addition the spin-orbit interaction of the incoming proton
has been neglected completely.

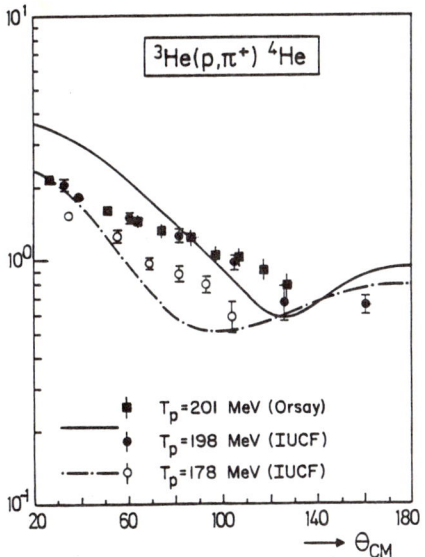

Fig. 7. Comparison of the calculated differential cross section for the reaction ^{3}He(p,π^+)^{4}He with data from Ref. 20 at two different bombarding energies.

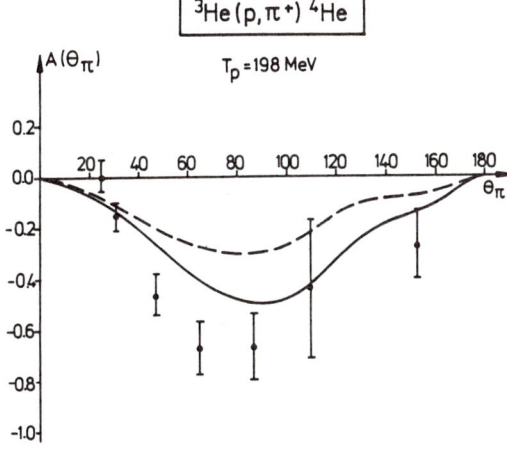

Fig. 8. Angular dependence of the analyzing power of ^{3}He(p,π^+)^{4}He at 198 MeV. Compared are predictions for two parameter sets (as explained in the text) with experimental results from IUCF (Ref. 20).

Extending our model to higher energies, a similar qualitative agreement exists for the inverse reaction ^{4}He(π^-,n)^{3}H at pion kinetic energies T_π between 50 MeV and 295 MeV (see. Fig. 9). The reason for such finding might be that, though our closure approximation for the intermediate Λ-isobar gets worse with increasing energy, the eikonal approximation for initial and final state interactions improves for larger T_π. Again, to trace the origin of the quantitative disagreement, especially in the structure of the various angular distributions, more detailed calculations are required.

We close our brief discussion, by pointing out an interesting particular feature of the ^{3}He(p,π^+) cross section, which is its dependence on a variation of the cut off mass Λ_π at the $\pi N\Lambda$ vertex and the ρNN coupling constant (note the relation $f_{\rho N\Lambda} \sim f_{\rho NN}$ in the static quark model). We find (see Fig. 10), with all the other parameters fixed, that at low energies the cross section at forward angles increases for an increasing ρ-meson coupling strength. Such a trend is opposite to conventional findings (see for example ref. 8), that due to the dominance of the tensor term in the $N\Lambda$ interaction a stronger ρ-coupling cuts down the cross section. Exploiting this feature in more detail, it should be possible to disentangle off shell effects and the influence of the ρ-meson exchange.

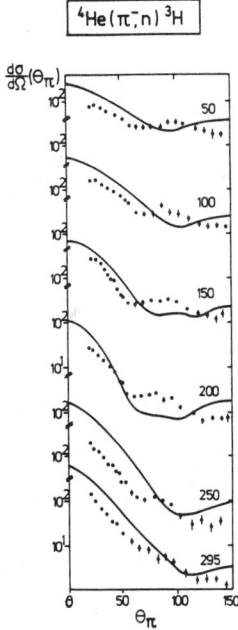

Fig. 9. Comparison of calculated differential cross sections for the reaction ^{4}He(π^-,n)^{3}H with data at several energies as indicated (T_π in MeV). The data are from Ref. 21.

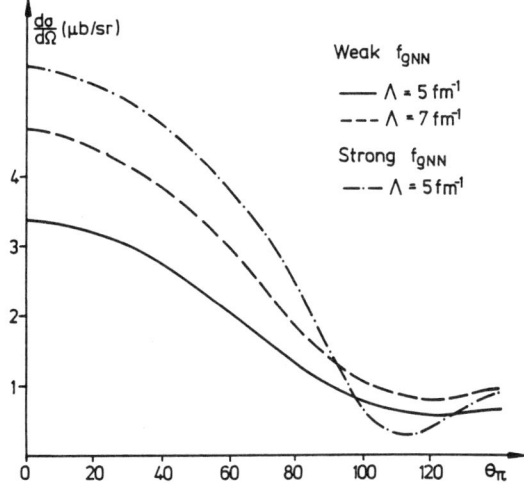

Fig. 10. Variation of the differential cross section for different cut off masses $\Lambda = \Lambda_\pi = \Lambda_\rho$ and two different ρNN (and $\rho N\Delta$) coupling constants.

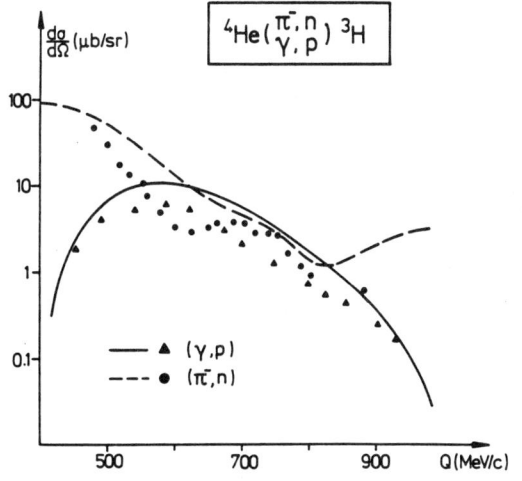

Fig. 11. Comparison of the differential cross section for $^4\text{He}(\pi^-,n)^3\text{H}$ at $E_\pi = 339$ MeV and $^4\text{He}(\gamma,p)^3\text{H}$ at $E_\gamma = 334$ MeV. The data are taken from Refs. 21, and 23, respectively.

4. Summary and conclusions

Summarizing and extending our discussion in the preceding sections, we would like to stress the following points:

(i) We feel, as already demonstrated in detail from the experimental side, that the (p,π) reaction is potentially a rich source of new and interesting information on various aspects of reaction mechanisms, conventional nuclear structure, new degrees of freedom, etc.

(ii) In order to uncover this richness, we need a detailed model. For energies close to the pion threshold our starting point is a microscopic rescattering model. Its main justification and advantages are its experimental and—especially in the study of light nuclear systems—threoretical support, as well as its unifying aspect both with respect to other single nucleon transfer reactions at medium energies such as (γ,p) or (d,p)[22] (as an example we present in Fig. 11 a comparison of the reaction $^4He(\pi,n)^3H$ and $^4He(\gamma,p)^3H$, calculated in a similar rescattering model) and to coherent pion and photon production with complex projectiles ("pionic fusion")[24]. Finally, such a model is naturally extended into the isobar doorway model at higher energies[7].

(iii) In order to pin down microscopic details of the model, extensive use has been made from findings and recent progress in related fields, such as from the study of the NN interaction and the π absorption on the deuteron for the two-nucleon subamplitude, from pion scattering up to the 33 resonance to study the Δ-isobar propagation in the medium, from proton scattering at high momentum transfers to pin down many-body corrections for the π-propagator and the proton nucleus optical potential, etc.

(iv) Finally, codes coming into operation are flexible enough to study and test different ingredients of the model over a wide range of experimental conditions without invoking serious numerical assumptions and approximations.

It seems to us that especially the last point is crucial for the success of the whole (p,π) project. We are convinced that a comprehensive and systematic analysis of all the beautiful existing (and forthcoming) experimental data will ultimately lead us to a more profound understanding of an important part of medium energy and nuclear physics.

One of us (M.D.) would like to acknowledge the warm hospitality and the support from the IUCF, where the main part of his work was done.

REFERENCES

*Supported in part by the U.S. National Science Foundation, by the Deutsche Forschungsgemeinschaft (Az. Di 148/2-1) and by the NATO (No. 23381).

1. H.W. Fearing - Bibliography of (p,π) data, TRIUMF Rep. TRI-80-3 (1980).
2. F. Soga, R.D. Bent, P.H. Pile, T.P. Sjoreen and M.C. Green, Phys. Rev. C22, 1348 (1980); B. Hoistad, P.H. Pile, T.P. Sjoreen, R.D. Bent, M.C. Green and F. Soga, Phys. Lett. 94B, 315 (1980); T.P. Sjoreen, M.C. Green, W.W. Jacobs, R.E. Pollock, F. Soga, R.D. Bent and T.E. Ward, Phys. Rev. Lett. 45, 1769 (1980); F. Soga, P.H. Pile, R.D. Bent, M.C. Green, W.W. Jacobs, T.P. Sjoreen, T.E. Ward and A.G. Drentje, Phys. Rev. C24, 570 (1981); T.P. Sjoreen, P.H. Pile, R.E. Pollock, W.W. Jacobs, H.O. Meyer, R.D. Bent, M.C. Green and F. Soga, Phys. Rev. C24, 1135 (1981).
3. P.H. Pile, R.D. Bent, R.E. Pollock, P.T. Debevec, R.E. Marrs, M.C. Green, T.P. Sjoreen and F. Soga, Phys. Rev. Lett. 42, 1461 (1979); T.P. Sjoreen, P.H. Pile, R.D. Bent, M.C. Green, J.J. Kehayias, R.E. Pollock, F. Soga, M.C. Tsangarides and J.G. Wills, Phys. Rev. C24, 2569 (1981).
4. M. Tsangarides, Ph.D. thesis, Indiana University, 1979 (unpublished).
5. B. Hoistad - Advances in Nucl. Physics, Vol. 11 (1979) 135, (Ed. J.W. Negele and T.W. Vogt, Plenum Press, N.Y.), D.F. Measday and G.A. Miller, Ann. Rev. Nucl. Part. Sci. 29 (1979) 121; H.W. Fearing, Prog. Part. Nucl. Phys. edited by D.H. Wilkinson (Pergammon, Oxford, 1981) Vol 7, p. 113.
6. J.M. Eisenberg, invited talk at this workshop; J.M. Eisenberg and D.S. Koltun, in preparation; P. Couvert and H. Dillig, Contrib. paper to the 9 ICOHEPANS, E14 (1981) 195 and in preparation.
7. B.D. Keister and L.S. Kisslinger, Contrib. paper to the 9 ICOHEPANS, E16 (1981) 197 and invited talk at this workshop; J. Iqbal and G. Walker--in preparation.
8. J.A. Niskanen, Nucl. Phys. A298 (1978) 417; Phys. Lett. 82B (1979) 187; J. Chai and D.O. Riska, Nucl. Phys. A338 (1980) 349; O.V. Maxwell, W. Weise and M. Brack, Nucl. Phys. A348 (1980) 388, 429; A.S. Rinat, Y. Starkand and E. Hammel, Nucl. Phys. A364 (1981) 486.
9. D. Koltun and A. Reitan, Phys. Rev. 141 (1966) 1413; J. Hüfner and F. Iachello, Nucl. Phys. A247 (1975) 441.
10. G.E. Brown, S.O. Bäckmann, E. Oset and W. Weise, Nucl. Phys. A286 (1977) 191.
11. F. Soga and M. Dillig, invited talk at the workshop (following paper).

12. M. Hirata, F. Lenz and K. Yazaki, Ann. Phys. 108 (1977) 116; K. Klingenbeck, M. Dillig and M.G. Huber, Phys. Lett. 41 (1978) 387.
13. G.A. Miller and S.C. Phatak, Phys. Rev. Lett. 51B (1974) 129; J.T. Londergan and E.J. Moniz, Phys. Lett. 45B (1973) 195
14. W. Weise, Invited contribution to the 9 ICOHEPANS (1981).
15. W.R. Gibbs, Los Alamos Rep. LA-8303-C (1980) 233, and invited talk at this workshop.
16. H.J. Weber and H. Arenhövel, Phys. Rep. 36C, (1978) 277.
17. L.S. Kisslinger and G.A. Miller, Nucl. Phys. A254 (1978) 493.
18. A. Richter, Invited Contribution to the 9 ICOHEPANS (1981); C. Goodman, Invited Contribution to the 9 ICOHEPANS (1981).
19. J. Conte and M. Dillig - in preparation.
20. N. Willis, L. Bimbot, N. Koori, Y. Le Bornec, F. Reide, A. Willis and C. Wilkin, J. Nucl. Phys. G7 (1981) L 195; J. Kehayias et al. (IUCF private communication).
21. J. Källne, J.E. Bolger, M.J. Devereaux, and S.L. Verbeck, Phys. Rev. C24 (1981) 1102.
22. J.T. Londergan, invited talk at this workshop; P. Couvert and M. Dillig, contributed paper to the 9 ICOHEPANS, C30 (1981) 119; A. Boudard, Y. Terrien, B. Beurtey, L. Bimbot, G. Bruge, A. Chaumeaux, P. Couvert, J.M. Fontaine, M. Garçon, Y. Le Bornec, D. Legrand, L. Schecter, J.P. Tabet and M. Dillig, Phys. Rev. Lett. 46 (1981) 218.
23. J. Arends, J. Eyink, A. Hegerath, H. Hartmann, B. Mecking, G. Nöldeke and H. Rost, Nucl. Phys. A322 (1979) 253.
24. M.G. Huber, invited talk to this workshop; J.-F. Germond, invited talk to this workshop; Klingenback, M. Dillig and M.G. Huber, Phys. Rev. Lett. 47 (1981) 1654; J.-F. Germond and C. Wilkin, Phys. Lett. 106B (1981) 449; Y. Le Bornec, L. Bimbot, N. Koori, A. Reide, A. Willis, N. Willis and C. Wilkin, Phys. Rev. Lett. 47 (1981) 1870.

STUDY OF THE (p,π) REACTION IN THE TWO NUCLEON MODEL*

F. Soga[†]
Institute for Nuclear Study, the University of Tokyo,
Tokyo 188, Japan

M. Dillig
Institute for Theoretical Physics,
University of Erlangen-Nurnberg, Erlangen, W. Germany

INTRODUCTION

Recently acquired high quality data on the (p,π) reaction,[1] especially in the near threshold region, has stimulated new theoretical efforts to understanding the reaction mechanism and associated nuclear structure. It is evident, in some cases, that both single-particle states and 2p-1h states in the residual nucleus are strongly excited in the (p,π+) reaction. Significant differences are observed, however, in the energy dependence of the reaction leading to final states of different structures.[2]

We have carried out preliminary calculations of the (p,π+) reaction based on the two-nucleon model investigating the importance of S-wave pion rescattering in the near threshold region.

FORMALISM

We assume that the basic mechanism of coherent pion production from nuclei is a two-nucleon process in which the essential ingredient is the NN → NNπ amplitude.

We use non-relativistic time ordered perturbation theory. At the pion production vertex, the operator form used is $\sigma \cdot q$, which comes from the non-relativistic reduction of the covariant form of the pseudo-scalar coupling γ_5.

It is well known that πN scattering is dominated by the P-wave interaction, and an operator form $(\sigma \cdot q)(S \cdot q)$ is used for the NN → NΔ transition potential[3] (q is the momentum of the virtual exchanged pion; σ and S are the spin matrices for the nucleon and Δ, respectively). The interaction Hamiltonian for the P-wave pion rescattering term is written in momentum space as:

$$H_P = \frac{1}{(2\pi)^3} \int d\vec{q}\, e^{i\vec{q}\cdot\vec{r}_{12}} e^{-i\vec{k}_\pi\cdot\vec{r}_2} \frac{f_\pi}{m_\pi} (\vec{\sigma}_1 \cdot \vec{q}) \{ \overline{V}_N(1,2;\vec{q}) + \overline{V}_\Delta(1,2;\vec{q}) \}$$

*Supported in part by the U.S. National Science Foundation, by the Deutsche Forchungsgemeinschaft (Az. Di 148/2-1) and by the NATO (No. 23381).
[†]Part of this work was carried out at the Indiana University Cyclotron Facility, Bloomington, IN, 47405, U.S.A.

$$V_N(1,2;q) = N_{post}\, \underline{\sigma}_2 \cdot (\underline{k}_\pi + \lambda \underline{p}')\, \underline{\sigma}_2 \cdot \underline{q}\, (\underline{\tau}_2 \underline{\phi}_\pi)(\underline{\tau}_1 \underline{\tau}_2)$$
$$+ N_{pre}\, \underline{\sigma}_2 \cdot \underline{q} \cdot \underline{\sigma}_2 (\underline{k}_\pi + \lambda \underline{p})\, (\underline{\tau}_1 \underline{\tau}_2)(\underline{\tau}_2 \underline{\phi}_\pi)$$

$$V_\Delta(1,2;q) = D_{post}\, \underline{S}_2 \cdot (\underline{k}_\pi + \xi \underline{p}')\, \underline{S}_2^\dagger \cdot \underline{q}\, (\underline{T}_2 \underline{\phi}_\pi)(\underline{\tau}_1 \underline{T}_2^\dagger)$$
$$+ D_{pre}\, \underline{S}_2 \cdot \underline{q}\, \underline{S}_2^\dagger (\underline{k}_\pi + \xi \underline{p})\, (\underline{\tau}_1 \underline{T}_2)(\underline{T}_2^\dagger \underline{\phi}_\pi)$$

where the symbols $\underline{\tau}_1$, $\underline{\tau}_2$, $\underline{\phi}_\pi$ and $\underline{T}_2$ denote the isospins for nucleon 1, 2, π and Δ. The outgoing pion has a momentum $\underline{k}_\pi$. The momenta of the recoil nucleons are $\underline{p}$ and $\underline{p}'$, and λ and ξ are the associated coefficients.

The terms N_{post}, N_{pre}, and D_{post}, D_{pre} are the postemission and preemission terms for the intermediate nucleon pole and Δ pole, respectively. For example, in case of the nucleon pole and the post emission term, the propagator N_{post} is expressed as:

$$N_{post} = \frac{F_N^2(q)}{2\omega(q)} \left\{ \frac{1}{E_N + E_1' - (E_1 + E_2)} \left(\frac{1}{E_1' - E_1 + \omega(q)} + \frac{1}{E_N - E_2 + \omega(q)} \right) \right.$$
$$\left. + \frac{1}{E_2' - E_2 + \omega(q) + E_\pi} \cdot \frac{1}{E_N + \omega(q) - E_2} \right\}$$

where $\omega(q)$ is the total energy of the virtual pion. E_1, E_2, E_1', and E_2' are the total energies of the initial and final states of the nucleons, and E_N is that of the intermediate propagating nucleon. $F_N(\underline{q})$ is the NNπ vertex form factor with the appropriate cut off parameter.

From the symmetrization of the Hamiltonian and the antisymmetrization of the two-nucleon final state configuration, there are four different diagrams for each operator (see Fig. 1).

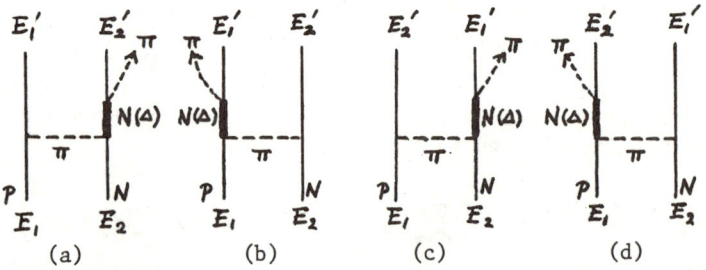

Fig. 1. Four different diagrams (a) Direct (b) Operator Exchange (c) Final state Exchange (d) Operator and Final state Exchange

As we are interested in the near threshold region, the S-wave pion scattering amplitude is also incorporated together with the P-wave term. For the S-wave rescattering term, the parametrization of the scattering lengths for σ meson and ρ meson exchange and the hard core contribution are included in the amplitude.[4]

The distorted waves for the incident proton and the outgoing pion are calculated using the the pionic stripping model code of Tsangarides.[5] Several options for the pion optical potential are available (Kisslinger potential and Laplacian potential with and without off-shell damping).

The bound state wave functions are caluclated using either a harmonic oscillator or a Woods-Saxon potential well. Our initial calculations are for a closed shell target nucleus and either single particle or 2p-1h final states.

RESULTS FOR THE (p,π⁺) REACTION

To illustrate the capabilities of the code, we show preliminary results for the $^{12}C(p,\pi^+)$ reaction leading to the ground state of ^{13}C (1p1/2 single-particle state) and the 6.86 MeV state (2p-1h state assumed to have a $1d5/2(1p3/2^{-1},p1/2)^{2+}$ configuration).

In Fig. 2, calculations for the single-particle final state in

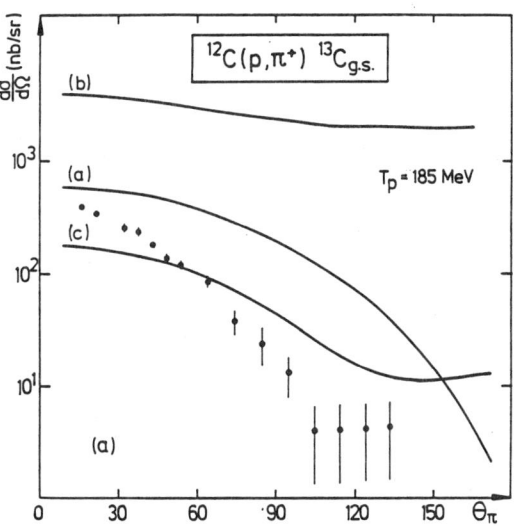

Fig. 2. Angular distributions of the differential cross section for the reaction $^{12}C(p,\pi^+)^{13}C(g.s.)$ at T_p=185 MeV:
(a) Plane waves
(b) Kisslinger potential without off-shell damping
(c) Damped Laplacian potential

^{13}C using three different pion optical potentials are shown together with the experimental data at E_p = 185 MeV.[6] Differential cross sections calculated with plane waves for protons and pions show a monotonic decrease toward the backward angles. When the Kisslinger potential is used for pions, the calculated cross sections are much too large, which is suggestive of the wrong off-shell behavior of the πN t-matrix. Using the Laplacian potential with off-shell damping, the overall trend of the calculated angular distribution of the differential cross section is similar to the data, though the calculated forward to backward cross section ratio is much smaller than the data indicate. In both cases of pion distorted waves, the parameters used for the pion optical potentials were those deduced from πN phase shift analysis.[7]

The cross section in the above example can be decomposed into two parts involving pion rescattering from 1s1/2 nucleons and from 1p3/2 nucleons. An example of this is shown in Fig. 3. The contribution from 1s1/2 rescattering dominates at forward

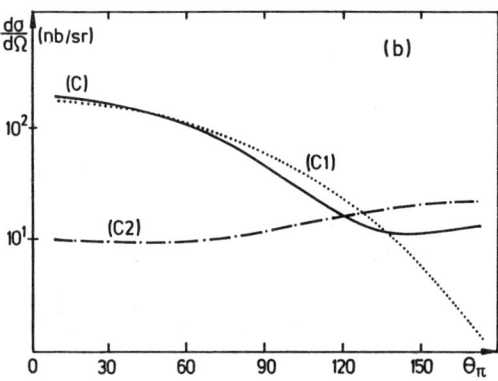

Fig. 3. Decomposition of the differential cross section (c) in Fig. 2 into a 1s1/2 rescattering part (C1) and a 1p3/2 rescattering part (C2).

angles, whereas 1p3/2 rescattering dominates at large angles (high momentum transfer).

An example of an asymmetry calculation is shown in Fig. 4. Here a calculation employing the Kisslinger potential is compared with the corresponding experimental data at E_p = 200 MeV.[8]

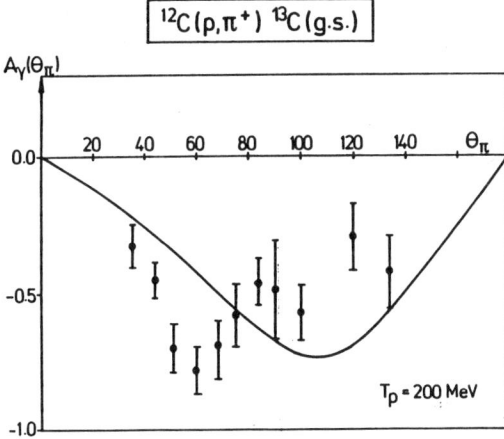

Fig. 4. Analyzing power of the reaction $^{12}C(p,\pi^+)^{13}C(g.s.)$ at Tp=200 MeV. The calculation uses the Kisslinger potential

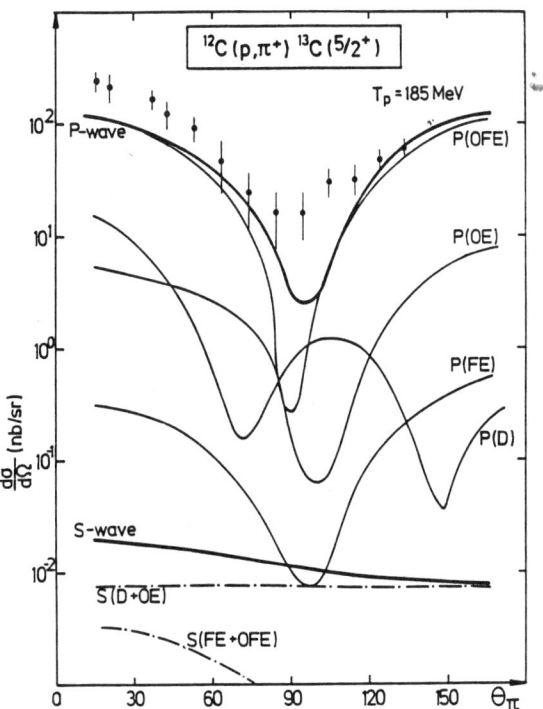

Fig. 5. Angular distribution of the differential cross section for the 6.86 MeV state in ^{13}C at Tp=185 MeV. The calculation uses the Laplacian potential with off-shell damping. The cross sections are decomposed into different components.

Calculations for the 6.86 MeV 2p-1h state in ^{13}C are shown in Fig. 5. The differential cross section is decomposed into several components from the different diagrams, which are added coherently for the cross section. At E_p = 185 MeV, the P-wave contribution overwhelms the S-wave contribution. Among the four different P-wave diagrams, the operator and final state exchange terms determine the overall shape of the angular distribution, while the direct term tends to fill in the deep minimum around 90 degree.

We would like to stress that these results are preliminary and represent only a first step toward understanding the complicated interplay between the (p,π) reaction mechanism and the associated nuclear structure.

REFERENCES

1. See report by B. Hoistad at this conference.
2. J.F. Amann, P.D. Barnes, K.G.R. Doss, S.A. Dytman, R.A. Eisenstein, J.D. Sherman and W.R. Wharton, Phys. Rev. Lett. 40, 758 (1978)
   F. Soga, P.H. Pile, R.D. Bent, M.C. Green, W.W. Jacobs, T.P. Sjoreen, T.E. Ward and A.G. Drentje, Phys. Rev. C24, 570 (1981)
3. H. Sugawara and F. von Hippel, Phys. Rev. 172, 1764 (1968)
4. D.S. Koltun and A. Reitan, Phys. Rev. 141, 1413 (1966); O.V. Maxwell, W. Weise and M. Brack, Nucl. Phys. A348 (1980) 388
5. M. Tsangarides, Ph. D. thesis, Indiana University, 1979 (unpublished)
6. S. Dahlgren, P. Grafstrom, B. Hoistad and A. Asberg, Nucl. Phys. A211, 243 (1973)
7. G. Rowe, M. Salomon and R.H. Landau, Phys. Rev. C18, 584 (1978)
8. E.G. Auld, A. Haynes, R.R. Johnson, G. Jones, T. Masterson, E.L. Mathie, D. Ottewell, P. Walden and B. Tatischeff, Phys. Rev. Lett. 41, 462 (1978)

DISCUSSION

G. Jones (TRIUMF-UBC): I'm getting confused. Sometimes one heard that pre-emission was important and other times one heard it wasn't. I would like clarification. If it was a vote, I obtain 2 out of 3 saying it isn't. Kisslinger doesn't have it, Dillig says it's not very important, but over here we heard that it was very important. It was stated that this arises when there is no rescattering but where you have proton distortion, which in some sense is a pion rescattering. So is one saying that this pre-emission in some sense accounts for the distortion without explicit pions, or does it rather refer to pion rescattering without Δ formation (like the p_{11} pion nucleon scattering), in which case it would be Dillig's one nucleon mechanism? I think a lot of us probably aren't really sure of what you people mean when you talk about one-nucleon or two-nucleon, and I would appreciate some clarification.

L.S. Kisslinger (Carnegie-Mellon Univ.): I'll stick to concentrating on the isobar doorway aspect through the Λ. There is no such term, because one enters into a doorway state and one then gets out a doorway state. Especially at threshold, when the Λ is not the mechanism for the attraction, then one has this term and it can be quite large at threshold. I think that's what the difference in that discussion was. That should be in a nonresonance part. It's not in the isobar doorway part.

One thing should make you feel better: there are many, many things to do in this program. When you finally get to doing the isobar doorway part, then you won't have to worry about the double counting because the theory is very clear there. Once you do that, then there is a very clear prescription.

Dillig: I agree that with the isobar doorway we have no problems with double counting, because the Λ-resonance is treated explicitly, whereas all the rest (s-wave interaction, nonresonant p-wave interaction) is included in the distortions.

J. Noble (Univ. of Virginia): Just let me remind you that there are energies, afterall, in the 33 resonance region where you see that the data actually go up through a 33 resonance. It's clear that if you're in the 33 resonance region, rescattering in some sense through a delta must be important compared to the single emission term, otherwise you wouldn't see that bump with such strength. But in the threshold region I think that Eisenberg is absolutely correct and that single emission is important.
Dillig: I guess we can get a feeling for the importance of the single nucleon emission if we look at different nuclei. Especially for very light nuclei such as ^{3}He or ^{4}He, where the wave function is known reasonably well, the single nucleon emission can be estimated with reasonable accuracy.

J. Niskanen (Univ. of Helsinki): I might have a word of warning. I hope that the closure approximation for the $N\Lambda$ component contribution that you mentioned in your talk is not very essential, because that approximation smells very much like the one that is done by Brack, Riska & Weise, and the descendents of that group. Essentially, that closure amounted to destroying the angular momentum structure of $N\Lambda$ states. This happens through omission of the angular momentum barrier in $N\Lambda$ states, so the higher $N\Lambda$ states are not suppressed compared to lower values.
Dillig: I didn't elaborate on that point in detail. However, I could imagine that away from the resonance - below 200 MeV - you could correct for the structure of the intermmediate Λ-AN states in the Λ propagator in a simple way, by including in the closure propagator in addition a density dependent Λ-self energy, thus weighting the different ℓ structures in a different way. I espect that close to the π threshold such an inclusion of medium corrections for the Λ-isobar should be a good approximation, but I definitely agree that in obtaining the correct selection rules the isobar doorway model has to be applied in a rigorous way.

Niskanen: Well below the resonance and especially in s-wave
NΛ's, you have a very good possibility of having good results
in that way too.

Silbar (Los Alamos Scientific Lab.): I'd like to give a warning to
the last questioner and to people in general, which we haven't been
hearing much about. There are complexities in the NN to NΛ
ingredient that are coming in to all of these calculations. A
transition amplitude that goes like $\vec{\sigma}\cdot\vec{q}$, $\vec{S}\cdot\vec{q}$ is only one of
sixteen different transition amplitudes. In a model calculation
done by Lombard, Kloet, and myself, we have seen that, at least at
800 MeV and at 500 MeV, that that is not necessarily the dominant
spin-dependent transition amplitude. What the situation would be
for the kinds of calculations you were just talking about now I
don't know. It is likely to be a very complicated thing. We have
found that the dependence upon the invariant mass in the π-nucleon
system that is resonating is severe. The dependence on the beam
energy is extreme. This reaction has probably been oversimplified
in the discussions that we have been hearing so far.

Dillig: I think one can make two comments. On the one side, if
it turns out that other invariants in the transition potential are
important, they can be easily included in the code. The way the
code is constructed, it's easy to add a new invariant (which
involves the spin and the momentum operator). I think such an
extension is not a serious problem.

On the other side, experience tells us that even with a rather
simple effective elementary interaction in a nucleus one can get a
reasonable description. In most cases the many-body effects are
dominant and only moderately influenced by details of the
elementary interaction. It may be that the many body nature of the
problem determines more or less the whole spectrum, much more than
details of the ingredients. However, I admit this reflects my own
prejudice; a convincing answer can only come out from a detailed
comparison.

PION PRODUCTION BY TARGET EMISSION
by
W. R. Gibbs
Los Alamos National Laboratory
Los Alamos, New Mexico 87545

ABSTRACT

The coherent production of pions in a proton nucleus collision is calculated. The energy region considered is below the nucleon-nucleon threshold so that the entire nucleus must take part in the process. The amplitude for this reaction is broken into two parts, the first part being characterized by initial emission of the pion from the projectile and the second part by initial emission from the target nucleus. It is the second part which is emphasised in this talk. The target nuclei ^{12}C and ^{209}Bi are considered. Total production cross sections as a function of energy and some angular distributions are calculated.

INTRODUCTION

Let me begin by trying to put my talk in perspective. To do this I will consider only the one pion rescattering graph. It is known [1] that a reasonable representation of the p+p →π^++d reaction can be achieved with this type of calculation. This is not to deny the possible importance of ρ- exchange [2], interacting Δ's [3], relativistic effects [4] etc. but to try to take the simplest graph for use in the nuclear calculation. Taking a static pion approximation immediately we may write

$$M = -2 \sum_{q,i,j} \frac{(\bar{q}'|H_s^i|\bar{q})(\bar{q}|H_p^j|0)}{\omega(q)} . \qquad (1)$$

Here i and j label the nucleons involved in the reaction, $\bar{q}'$ is the final pion momentum and $\bar{q}$ the intermediate pion momentum. H^j_p is the production vertex operator and H^i_s is the corresponding rescattering vertex operator. The nuclear states have been suppressed but are implicit [Eg. |0) means a proton incident on ^{12}C and no pions etc.]. The rescattering operator can be related to the pion nucleon t-matrix and the production vertex is related to γ_5.

To obtain the two nucleon model [5] one calculates first a pion production "potential" by doing the $\bar{q}$ sum (integral):

$$V_{ij} = V(\bar{r}_i - \bar{r}_j) \sim \int d\bar{q}\; \frac{t(\bar{q},\bar{q}')\, \bar{\sigma}\cdot\bar{q}\; e^{i\bar{q}\cdot(\bar{r}_i-\bar{r}_j)}}{\omega^2(q)} \quad . \tag{2}$$

One can then calculate the expectation value in the nucleus of this potential

$$M \sim (\bar{q}'|\Sigma V_{ij}|0) \quad . \tag{3}$$

This is a perfectly reasonable way to proceed but I shall consider another organization of the work. We may note that in the i,j sum either i or j must be the projectile. If j is the projectile then i just labels the target nucleons. We may recognize

$$\sum_{i=1}^{A} (\bar{q}'|H_s^i|\bar{q}) \tag{4}$$

as the single scattering approximation to the pion nucleus t-matrix. If we improve the calculation by using the complete multiple scattering t-matrix and add in the plane wave part we recognize this result as

$$M \sim (\phi_\pi(\bar{q}')|H_p^p|0) \tag{5}$$

which is the usual one nucleon DWIA matrix element.

What about the other part? Let us consider the case where i labels the projectile. In this case the shell structure is more complicated. I will use the ^{12}C target as a specific example. The final state will be some state in ^{13}C and we can write it as an expansion of the final captured nucleon multiplied by core states of ^{12}C

$$|^{13}C\rangle = \Sigma\; c_{\ell j}^{J\,T\,E}\, |J\,T\,E\rangle\, |\ell j\rangle \quad . \tag{6}$$

This part of M will now be

$$\sum_{i\, E\, ^{12}C} \int d\bar{q}\; \frac{(\bar{q}'\ell j|H_s^p|\bar{q}\Psi_p)\, (\bar{q}\, J\,T\,E|H_p^i|0)\, c_{\ell j}^{J\,T\,E}}{\omega(q)} \tag{7}$$

Since $H^i \sim \bar{\sigma}_i \cdot \bar{q}^i \tau^i_\lambda$ the core states must be unnatural parity ($0^-, 1^+, 2^-, \ldots$) and have isospin one. Note that the matrix element on the left refers only to the projectile and final nucleon and the matrix element on the right contains only target coordinates. In order to make further progress let us expand the ^{12}C core states in particle-hole promotions from the ^{12}C ground state i.e.

$$|J\,T\,E\rangle = \Sigma\; B_{\ell_h j_h \ell_p j_p}^{J\,T\,E}\, \left\{|\ell_h j_h\rangle|\ell_p j_p\rangle\right\}^{JT} \quad . \tag{8}$$

Now the production matrix element can be written as

$$(\bar{q}\ J\ T\ E\ |H_p^j|0) \sim \sum B_{\ell_h j_h \ell_p j_p}^{J\ T\ E} \frac{1}{\sqrt{\omega(q)}} Y_J(\hat{q}) Q_{ph}^{JL}(q)\, a_{JL} \qquad (9)$$

$$Q_{ph}^{JL}(q) = \alpha f(q^2) \left(\frac{q}{m_\pi}\right) \hat{C}_{ph}^{JL} \int_0^\infty dr\, r^2 j_L(qr)\, R_p(r)\, R_h(r) \qquad (10)$$

where $\alpha=2$ for nucleons, $\alpha = 8/3$ for Δ-isobars. This last factor of α takes account of the fact that we could also excite a Δ-hole component. The reason for writing the function in this way is to make connection with the work of Toki and Weise, [5] and the Lyon group [7] as well. Toki and Weise show how the Q function may be corrected by means of RPA to include the effects of multiple particle-hole and delta-hole states. If we represent the simplest form of Q as defined above as

then, taking into account the possiblity of exciting virtual particle-hole states by either pion exchange or heavy meson exchange, the Q gets replaced by $\tilde{Q}$ where $\tilde{Q}$ can be represented by a sum of diagrams.

Note that, depending on one's point of view, this correction can be regarded as a correction to the basic particle-hole wave function or a modification to the pion propagator. The main effect in any case is to give more support to the high momentum components. The rest of the nucleons in the nucleus "back up" the simple particle-hole wave function so that the pion-nucleus vertex is given enhancement from the coherent many-body effects. This actually provides a more-than-two nucleon theory of pion production.

Leaving off the spin and isospin indicies one can write a typical element in the sum for M as the deceptively simple expression

$$\int d\bar{r} \; \phi_\pi^{(-)*}(\bar{r}) \; \Psi_f(\bar{r}) \; f(\bar{q},\bar{q}',\bar{p},\bar{p}') \; A(\bar{r}) \; \Psi_i(\bar{r}) \tag{11}$$

where

$\Psi_i(\bar{r})$ is the initial proton wave function,

$A(\bar{r})$ is the function $Q_{ph}^{JL}(q)$ in r-space,

$f(\bar{q},\bar{q}',\bar{p},\bar{p}')$ is a function derived from the pion-nucleon t-matrix as an operator in all four momenta,

$\Psi_f(\bar{r})$ is the final bound nucleon wave function,

$\phi_\pi^{(-)}(\bar{r})$ is the final pion wave function.

This has the form of a DWIA matrix element for knock out of a pion followed by a capture of the incident nucleon. The reason that the computation of this integral is not simple is the presence of the operators $\bar{q},\bar{q}',\bar{p},\bar{p}'$ in f. Here $\bar{q}$ is the gradient operator on $A(\bar{r})$, $\bar{q}'$ is the gradient operator on $\phi_\pi^{(-)}(\bar{r})$, $\bar{p}$ is the gradient operator on $\Psi_i(\bar{r})$ and $\bar{p}'$ is the gradient operator on $\Psi_f(\bar{r})$. It is only the assumption of a fairly simply form for f that makes computation possible. The expression used for f is

$$f(\bar{t},\bar{t}') = (\lambda_0 + \lambda_1 \bar{t}\cdot\bar{t}' + \tilde{\lambda}_1 \bar{\sigma}\cdot\bar{t}\times\bar{t}') \; v(t)v(t') \tag{12}$$

where

$$\bar{t} \equiv \bar{q} - \frac{\mu}{m}\bar{p} \quad \text{and} \quad \bar{t}' \equiv \bar{q}' - \frac{\mu}{m}\bar{p}' \; .$$

A schematic representation of this integral is given in fig. 1. The initial proton interaction is a standard (non-relativisitic) proton nucleus optical potential.[8] The pion "wave function" was calculated[9] with the methods of Toki and Weise. The needed shell model coefficients were taken from the Los Alamos version of Glasgow code[10] as implemented by Haxton and Dubach.[11] The final nucleon bound state wave functions were solutions in a Saxon-Woods potential. The shell model coefficients were obtained as above. The final pion wave function was taken from the coordinate space non-local optical model code developed in T-5 at Los Alamos.[12]

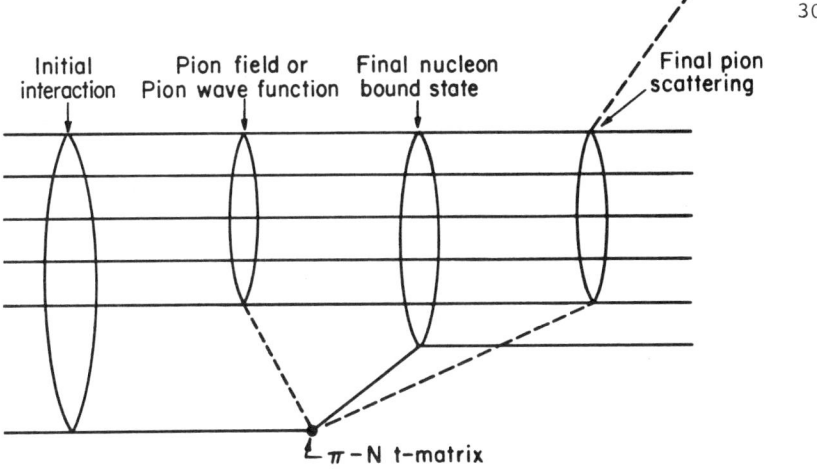

Figure 1. A schematic representation of expression 11.

I am going to consider only the target emission in most of what follows. The reasons for this are several. The first motivation is to separate the problem into manageable pieces and to study them independently. Since π^- production may well be dominated by target emission there is a chance to look at it (almost) alone. Another reason is the hope that the sensitivity to the final optical potential will be less than in the projectile emission since more vertices are present in the target emission. Another sensitivity in the projectile emission is due to the treatment of true pion absorption with an incident off-shell pion. Presumably the possiblity of reabsorption is much greater since the momentum of the pion is much nearer the pole for true absorption in the optical potential. [13] This would affect the projectile emission more than the target emission because in the latter the pion is more likely to be on-shell when it leaves the π-nucleon t-matrix.

II. ENERGY DEPENDENCE FOR $^{12}C(p,\pi^+)^{13}C$

There are many possible calculations that could be done with this technique. I will try to focus primarily on a single simple point, namely the energy dependence of the total pion production cross section. This data is now available on a large number of nuclei but for the first comparison with experiment I shall restrict myself to several states of ^{13}C. It was noted by the Indiana group [14] that there seemed to be two types of energy dependence of the total cross section; one in which there was an initial rapid rise with a "bend" to give a much more slowly rising function and a second type which simply increased from zero with no abrupt change in slope. The first type of behavior seemed to be associated with single particle states and the second type with 2p1h states. An exception to this rule is the $9/2^+$ state at 9.5 MeV.

Since simple projectile emission populates single particle states only, while target emission populates single particle and 2p1h states we might conjecture that the separation of reaction mechanisms might be made in this way.

The first step in this investigation was the calculation of a total cross section with projectile emission, figure 2 shows the results of a calculation using the Keating-Wills-Tsangarides [15] code.

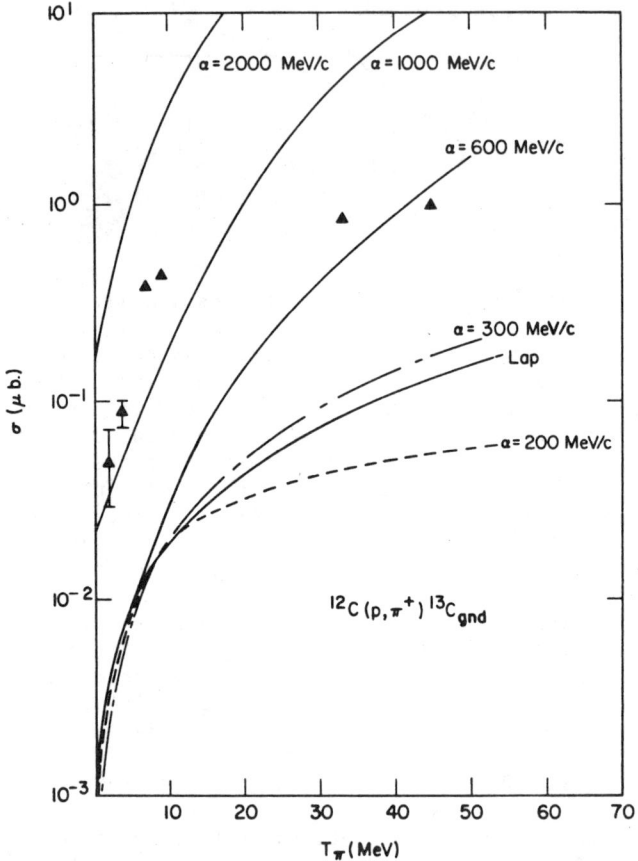

Fig. 2. One-nucleon-model calculations showing the dependence on the non-locality of the optical potential Note that this non-locality is a total effective value inlcuding e.g. the motion of the Δ.

The curve labeled "Lap" was calculated using the damped Laplacian form. The other curves were calculated using the T-5 non-local code with the off-shell range denoted as α. One might expect a similar variation with other optical model parameters, such as p^2 terms.

We may see that (within limits) the very low energy part is little affected by the optical potential and the variation is largely controlled by phase space and Coulomb penetrability. We note that a change in slope can be obtained at about the right energy. Thus the hoped-for separation hypotheses is still alive although it is somewhat disturbing that the calculated cross section showing the desired shape is so far below the data. The uncertainties in the optical potential may allow us to hope for the correction of this feature.

The next step is to see if the target emission can explain pion production to the 2p1h states. Let us first look at some characteristics of the target production mechanism. The effect of each of the terms in Eq. 12 is shown individually in Fig. 3.

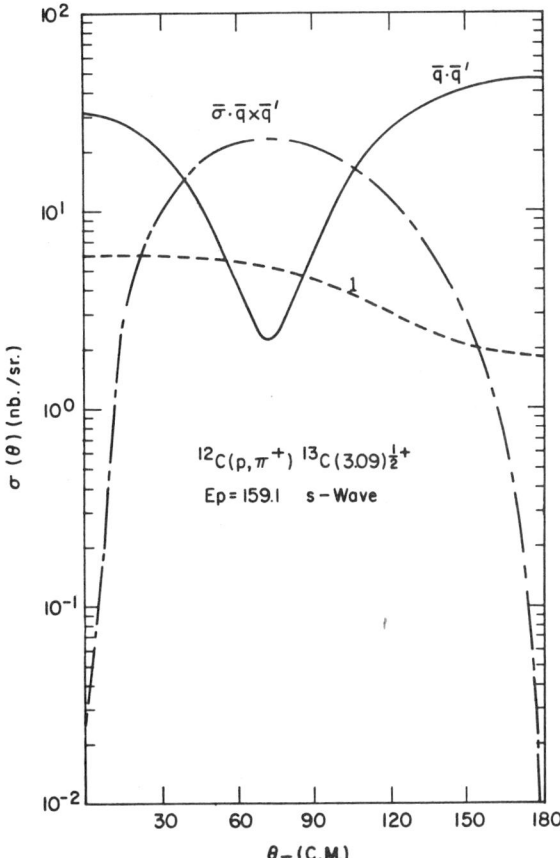

Fig. 3. Characteristic shapes of a target emission calculation showing the dependence on the various parts of the π-nucleon t-matrix.

The curve labled "1" contains only the S-wave pion-nucleon scattering and thus shows the form factor and distortion effects only. The other two curves show the $\cos^2\theta$ and $\sin^2\theta$ dependences as well. These features are very clear in the case presented, but not quite so clear in others. The spin dependent term is too large for the ^{13}C states. It is not known whether this is due to the fact that the pion-nucleon t-matrix should be modified in the nuclear medium to reduce the spin dependent piece, whether some feature of the calculation should be modified to reduce this term or if we are simply barking up the wrong graph. In any case, for the remainder of the comparisons with ^{13}C I shall set this term to zero.

The calculations are classified to some extent by the core excited state. For p-wave capture in the final state the positive parity states will be described by 0^- or 2^- (or higher in some cases) core states.

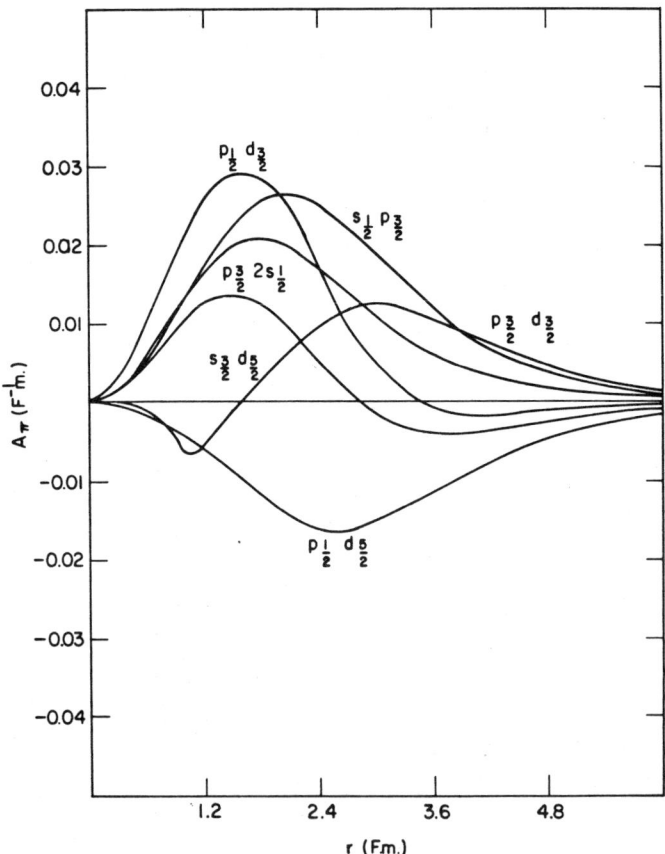

Fig. 4. The hole particle components entering into the structure of the 2^- core state.

The hole-particle pairs allowed for 0^- are
$1s_{\frac{1}{2}}^{-1} 1p_{\frac{1}{2}}$, $1p_{\frac{1}{2}}^{-1} 2s_{\frac{1}{2}}$ and $1p_{\frac{3}{2}}^{-1} 1d_{\frac{3}{2}}$ For 2^- the available states

are $1s_{\frac{1}{2}}^{-1} 1p_{\frac{3}{2}}$, $1p_{\frac{3}{2}}^{-1} s_{\frac{1}{2}}$, $1p_{\frac{1}{2}}^{-1} 1d_{\frac{3}{2}}$, $1p_{\frac{1}{2}}^{-1} 1d_{\frac{5}{2}}$, $1p_{\frac{3}{2}}^{-1} 1d_{\frac{3}{2}}$, $1p_{\frac{3}{2}}^{-1} 1d_{\frac{5}{2}}$

The hole-particle A functions for the 2^- case are shown in figure 4.

The actual $A(r)$ used in Eq. 11 is a linear combination of these functions with coefficients given by the shell model. It is common for one or two of these basis functions to dominate. The shape of $A(r)$ has a large effect on modification of the angular distribution.

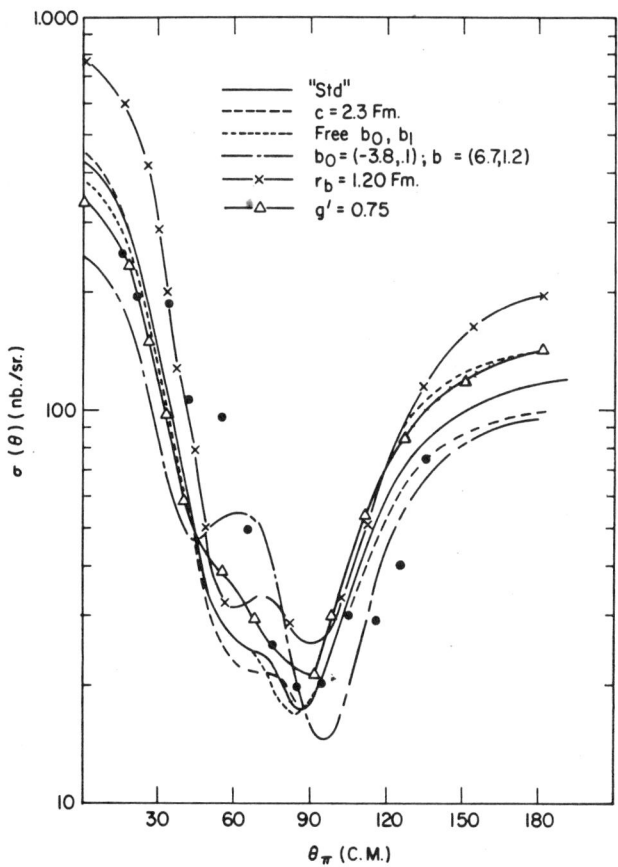

Fig. 5. The effect of variation of some of the input parameters on the angular distribution. The solid ("std") curve is calculated for c (the W-S radius for ^{12}C in the optical potential) = 2.41fm, b_0 = (2.7,.9) b_1 = (7.8,.5), r_b (radius of the W-S well for computing the final bound state) = 1.31 fm, and no enhancement from the RPA corrections.

The first state considered is the 6.86 $\frac{5+}{2}$ state in ^{13}C. This state is dominated by components

$$\left\{\begin{bmatrix} 1p^{-1}_{\frac{3}{2}} & 2S_{\frac{1}{2}} \end{bmatrix}_2 \times 1p_{\frac{1}{2}} \right\}_{\frac{5}{2}} \qquad \left\{\begin{bmatrix} 1p^{-1}_{\frac{3}{2}} & 1d_{\frac{5}{2}} \end{bmatrix}_2 \times 1p_{\frac{1}{2}} \right\}_{\frac{5}{2}}$$

and

$$\left\{\begin{bmatrix} 1p^{-1}_{\frac{1}{2}} & 1d_{\frac{5}{2}} \end{bmatrix}_2 \times 1p_{\frac{3}{2}} \right\}_{\frac{5}{2}}.$$

The 1p1/2 capture dominates and I show only these results in Fig. 5, i.e. all six components are included in the 2^- core but only the p1/2 coupling is calculated. Figure 5 shows some angular distributions compared with data at 185 MeV. As can be seen the qualitative agreement is quite good with only moderate variations in shape corresponding to rather substantial modifications in the model.

Figure 6 shows the angular distribution as a function of energy. Detailed agreement leaves a great deal to be desired but the dependence of magnitude on energy is good.

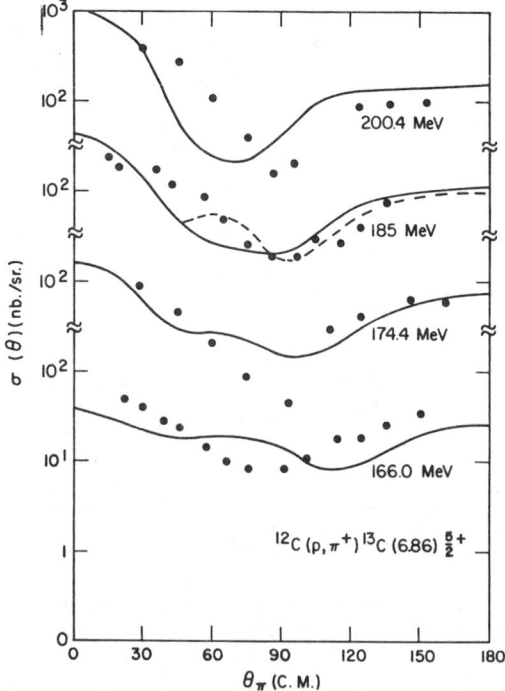

Fig. 6. Energy dependence of the angular distributions for the 6.86 MeV state corresponding to the "std" curve in Figure 5. The dash-dot curve is the same as in Figure 5.

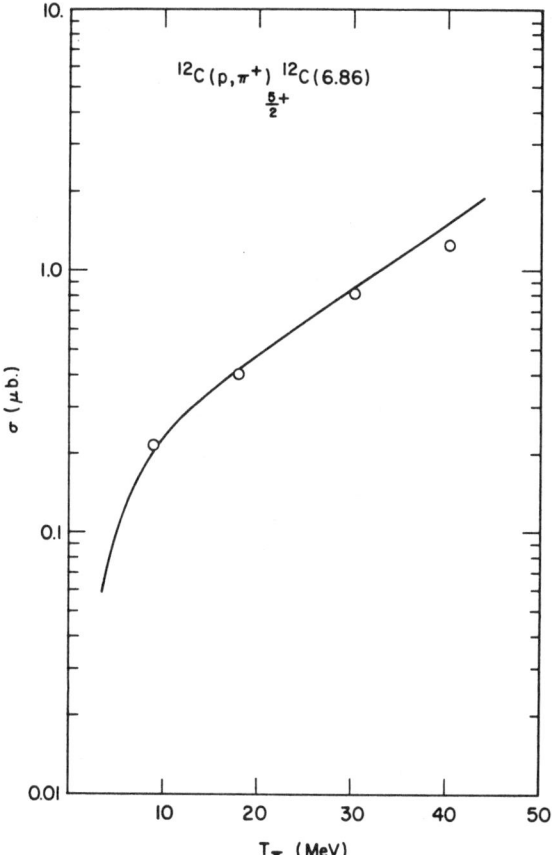

Figure 7. The energy dependence of the integrated cross section for the 6.86 MeV state.

This is reflected in Fig. 7 where the integrated cross section is plotted. Thus we see that the gradual energy dependence is correctly given by this calculation.

The third step in the study of the separation hypothesis requires that the target emission calculation fail to give the energy dependence of a single particle state. For this case I have chosen the 3.09 $1/2^+$ final state in ^{13}C. This state is well represented by ^{12}C$_{gs} \times 2s\frac{1}{2}$. The target emission must destroy the core, however, so to restore the state the captured particle must fill the hole. Thus since $\frac{1}{2}^+$ can be obtained by $0^- \times 1p\frac{1}{2}$ or $2^- \times 1p\frac{3}{2}$ the dominant components are

$$\left\{ \left[1p_{\frac{1}{2}}^{-1} \, 2s_{\frac{1}{2}} \right]_0 \times 1p_{\frac{1}{2}} \right\}_{\frac{1}{2}}$$

and
$$\left\{ \left[1p_{\frac{3}{2}}^{-1} \; 2s_{\frac{1}{2}} \right]_2 \times 1p_{\frac{3}{2}} \right\}_{\frac{1}{2}}$$

Figure 8 shows the result of the calculation from s- and d-core states (all hole-particle states included).

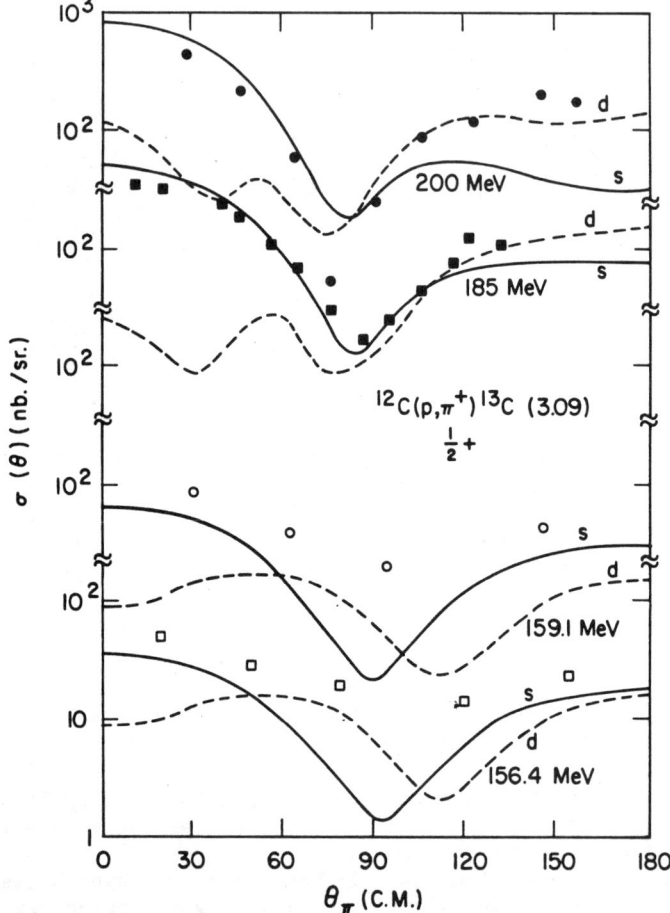

Figure 8. Separate angular distributions for the 3.09 MeV state.

Note that at high energy where the experimental minimum is deep the minima of the two pieces nearly coincide. At low energy where the experimental angular distribution becomes much flatter then the two theoretical minima do not agree. It looks almost accidental that the d-wave comes up in the back angles but recall that the relative strengths will be given by angular momentum coefficients obtained by the decomposition of the single particle state.

In figure 9 the cross sections from the coherent sum of the s- and d- state core are shown. The general agreement with the angular distributions, in particular the energy dependence of the forward hemisphere cross sections, is largely satisfactory. In fact the agreement seems rather better for this single particle state than for the 2p1h state.

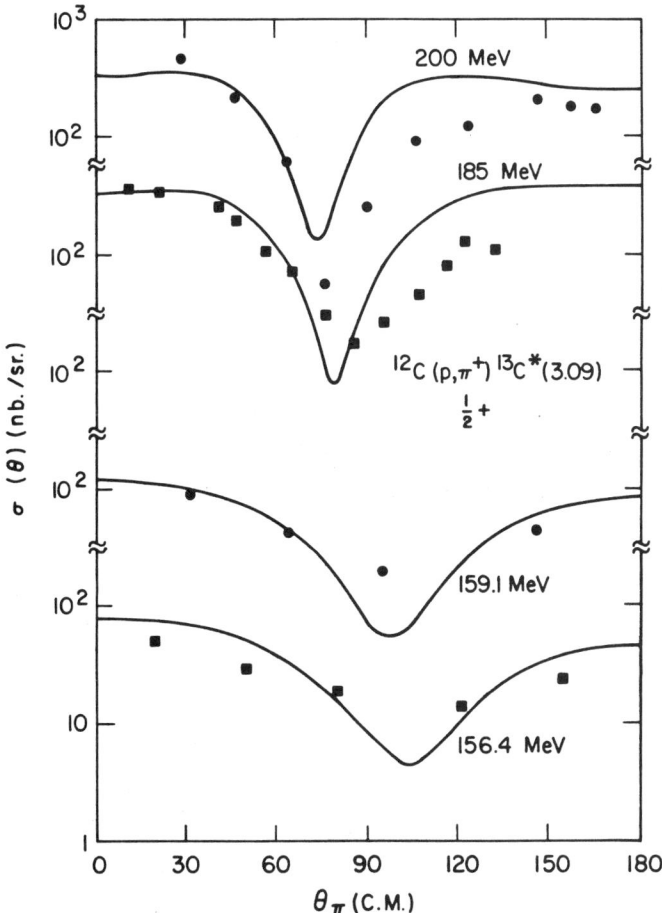

Figure 9. Combined angular distributons for the 3.09 MeV state.

The corresponding integrated cross sections are shown in figure 10.

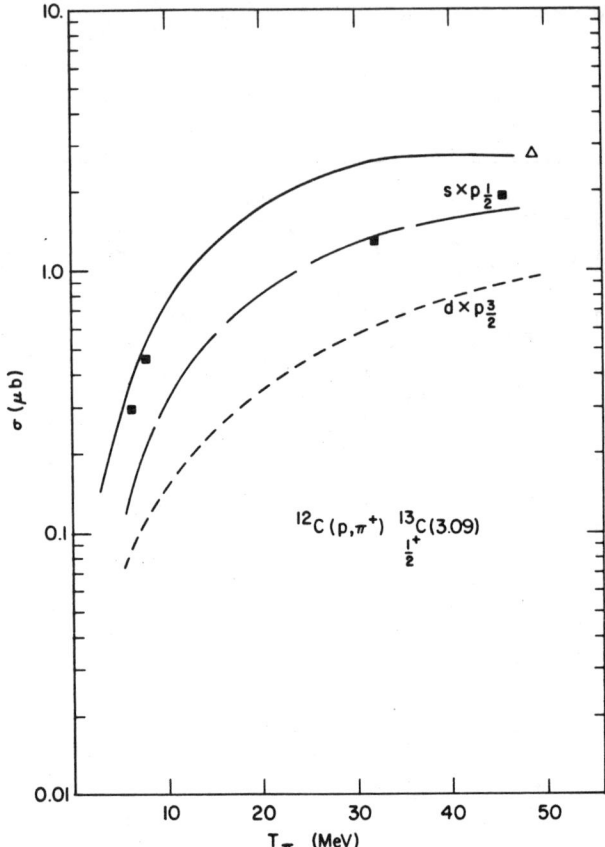

Figure 10. Separate and combined integrated cross sections for the 3.09 MeV state.

We note that the s- and d- state energy dependences are different with the s- state showing a dependence like the single particle shape. The s- and d- states tend to match better at low energy and the coherence holds up the cross section there.

Thus we see that a separation of diagrams by the energy dependence is not possible. This does not mean that a separation by structure cannot be made, just that the association of a dominance of target emission with 2p1h states is not true.

A fairly typical asymmetry is shown in fig. 11. This is in reasonable qualitative agreement with recent data.[16]

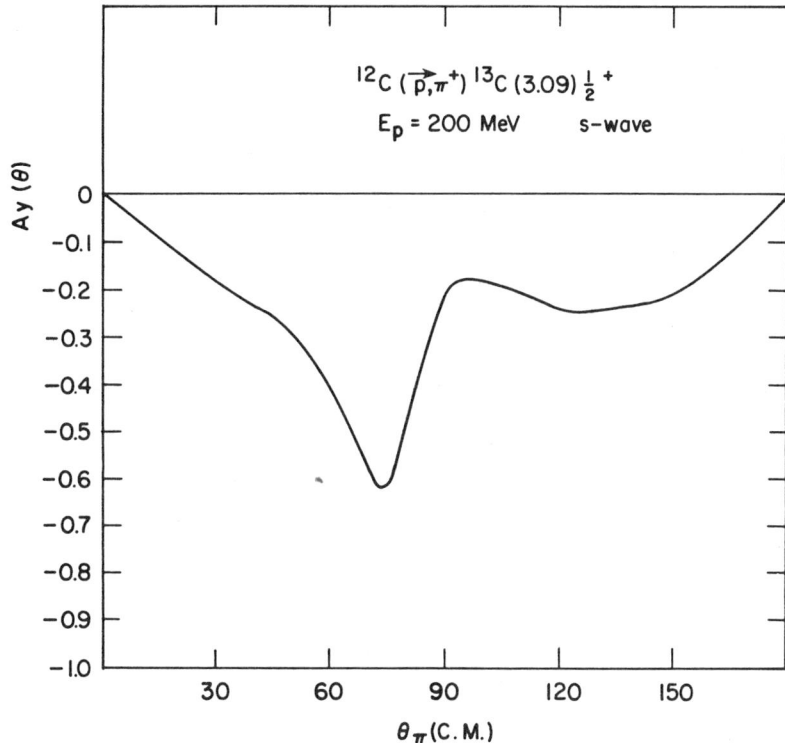

Figure 11. An example of asymmetry for the 3.09 MeV state. Only a 0^- core is inlcuded.

I shall briefly consider one other state, the 3.85 state in ^{13}C. Since this is also a $\frac{5}{2}^+$ state like the 6.86 one might expect the pion production to be similar. However, the particle-hole composition is rather different. For this state the dominant (predicted) contributors are

$$\left\{ \left[1p_{\frac{3}{2}}^{-1} \; 1d_{\frac{5}{2}} \right]_2 \times 1p_{\frac{3}{2}} \right\} \; \frac{5}{2}$$

and the $p_{\frac{1}{2}}$ couplings

$$\left\{\left[1p_{\frac{1}{2}}^{-1}\ 1d_{\frac{5}{2}}\right]_2 \times 1p_{\frac{1}{2}}\right\}_{\frac{5}{2}} \qquad \left\{\left[1p_{\frac{3}{2}}^{-1}\ 1d_{\frac{5}{2}}\right]_2 \times 1p_{\frac{1}{2}}\right\}_{\frac{5}{2}}$$

$$\left\{\left[1p_{\frac{3}{2}}^{-1}\ 2s_{\frac{1}{2}}\right]_2 \times 1p_{\frac{1}{2}}\right\}_{\frac{5}{2}} \qquad \left\{\left[1p_{\frac{3}{2}}^{-1}\ 1d_{\frac{3}{2}}\right]_2 \times 1p_{\frac{1}{2}}\right\}_{\frac{5}{2}}$$

The characteristics of the measured angular distributions are rather similar to the other $\frac{5}{2}^+$ state at 6.86. According to the shell model the underlying structure is different. This difference is reflected in the angular distributions of figure 12. I only give the separate pieces but it seems clear that, while the general magnitude may not be too bad there is a great deal lacking in further agreement. This leads us to think that a fair sensitivity to details of nuclear stucture exists.

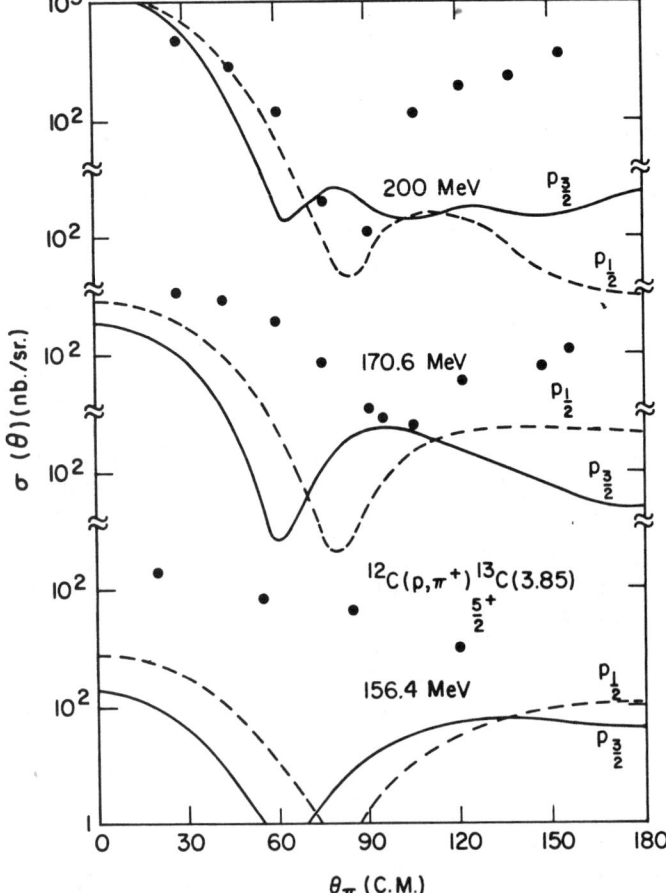

Figure. 12. Separate angular distributions for the 3.85 MeV state.

III. ENERGY DEPENDENCE IN ^{209}Bi$(p,\pi^-)^{210}$At

Let me now consider a second study of the energy dependence of the target emission. A large part of negative pion production may be reasonably thought of as coming from target emission. The projectile emission may produce π^- by π^0 production followed by charge exchange and more exotic processes (such as Δ^{++} mechanisms) can exist as well. I shall neglect these possibilities and assume that the negative pion is produced from one of the neutrons in the target so that target emission is the total mechanism for negative pion production, admittedly an extreme point of view.

The reaction considered is ^{209}Bi$(p,\pi^-)^{210}$At followed by the evaporation of neutrons. From the residual isotopic distribution, information about the original energy spectrum can be infered.
To make this connection some additonal assumptions will be needed. First we shall assume that the cross section to a state at a given energy can be represented (on average) by the cross section to a "typical" state.

For simplicity this state is taken to be a pure configuration.

$$\left\{ \left[1h_{\frac{9}{2}}^{-1} \; 2g_{\frac{9}{2}} \right]_0 \times 2g_{\frac{9}{2}} \right\}_{\frac{9}{2}}$$

The cross sections calculated under this assumption are shown in figure 13. There are several interesting features in these curves.

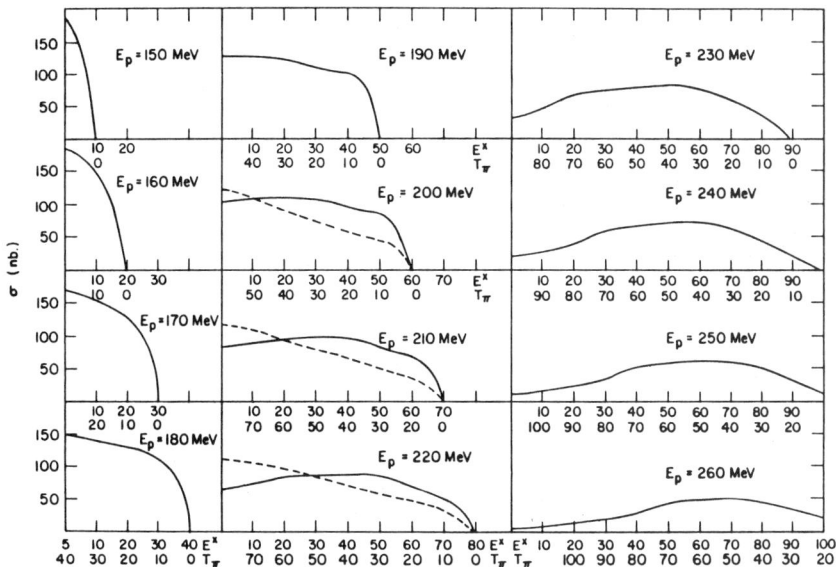

Figure 13. Calculated energy dependence to the "typical" state in the ^{209}Bi(p,π^-) ^{210}At reaction.

A rough idea of the isotopic abundance of ^{210}At to be left after this process can be estimated by integrating the first 10 MeV of this function (it should be multiplied by a level density, and will be shortly). We note that this integral rises from threshold and then drops quickly, peaking at the surprisingly low energy of 10-20 MeV above threshold. This feature has already been noted both experimentally and theoretically.[17] Another feature of interest is the importance of the energy dependence of the pion-nucleus optical potential. The dotted curves show the energy variation of these basic cross sections fixing the optical parameters at their value at 50 MeV. The solid curves were calculated with all parameters kept fixed except for the imaginary part of b_1 which was taken as a linear combination of a constant plus the "free" value such that the elastic scattering is well represented at 50 MeV.

To calculate the spectral strength we must know the density of our "typical" states. In a simple model of equal spacing of single particle states, for high energies where Pauli exclusion effects should be unimportant, the level density should vary as the energy raised to the number of "particles" minus one. Thus, for these 2p1h states, the density should vary at E^2. To make the density a little more realistic at low energies the form

$$\rho(E) = .004 \, (E+15)^2$$

(E in MeV) was used. The spectrum calculated using this function is shown in figure 14.

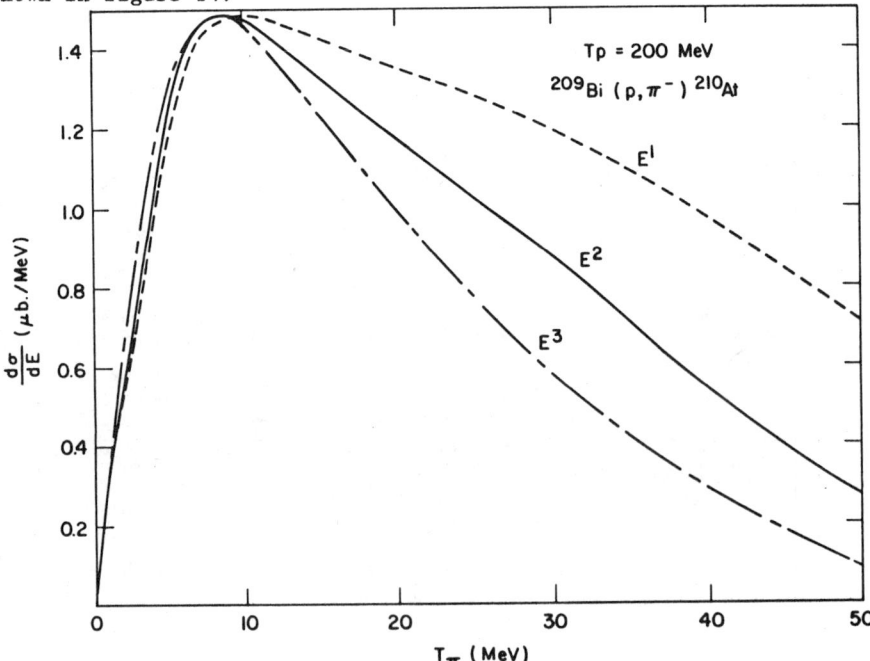

Figure 14. Calculated spectrum for negative pion production from protons on ^{209}Bi. The curves labeled E^1 and E^3 are shown to estimate sensitivity to the level density assumption.

Also shown is the shape of the spectrum for two neighboring powers (normalized to the same peak value). Since the normalization of the level density and the spectroscopic factors are very crude the results can be considered as normalized to the total integrated data to all observed isotopes.

Following an evaporation of successive neutrons, the original ^{210}At spectrum can be converted to a cross section for production of various isotopes. The comparison with data [18] is shown in figure 15.

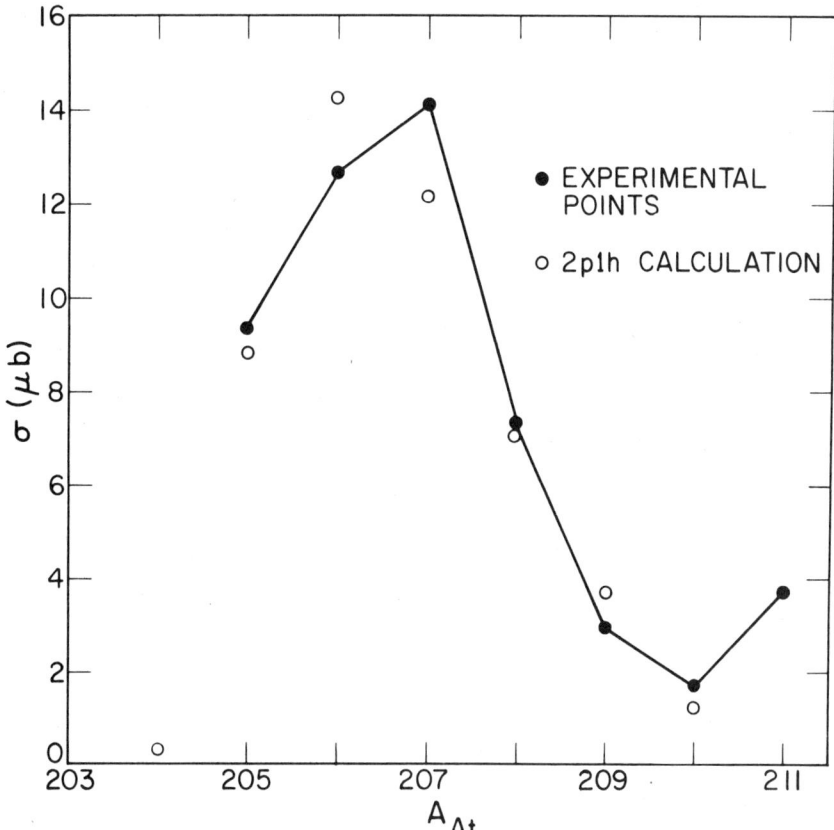

Figure 15. Comparison of the calculated relative isotope production cross section (open circles) with the data of Ref. 18.

The agreement is very good. The production of the various isotopes is extremely sensitive to the bombarding energy, especially near threshold. These cross sections are shown in figure 16.

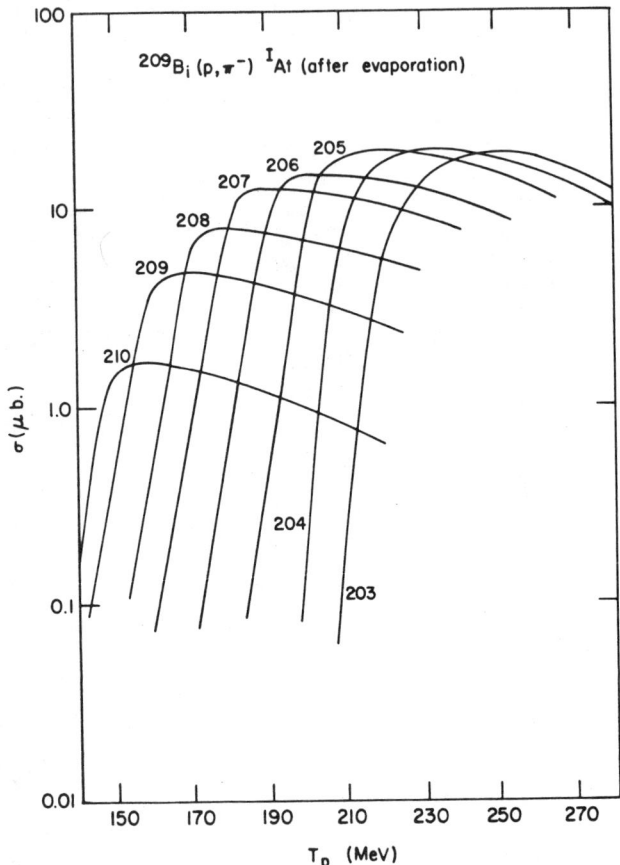

Figure 16. The dependence on proton bombarding energy of the cross section for production of At isotopes from a Bi target.

Clearly this technique provides an alternative to the use of Fermi motion for inclusive sub-threshold production of mesons. More detailed comparison with data is needed to select between these two models.

References
1) B. Goplen, W. R. Gibbs and E. L. Lomon, Phys. Rev. Lett. $\underline{32}$, 1012 (1974) B. M. Preedom et al. Phys. Lett. $\underline{65B}$, 31 (1976).
2) D. O. Riska, M. Brack and W. Weise, Phys. Lett. $\underline{61B}$, 41 (1976).
3) J. Niskanan (these proceedings).
4) Bruce J. Verwest, Phys. Lett. $\underline{83B}$, 161 (1979)
5) M. Dillig (these proceedings).
6) H. Toki and W. Weise, Z. Physik A$\underline{295}$, 187 (1980), Phys. Rev. Lett. $\underline{42}$, 1034 (1979).
7) M. Ericson and J. Delorme, Phys. Lett. $\underline{76B}$, 182 (1978).
8) G. R. Satchler and R. M. Haybron, Phys. Lett. $\underline{11}$, 313, (1964).
9) J. R. Comfort and W. G. Love, Phys. Rev. Lett. $\underline{44}$, 1656, (1980).
10) R. R. Whitehead, A. Watt, B. J. Cole and I. Morrison, Advances in Nuclear Physics Vol. 9, (Ed. by Baranger and Vogt) Page 123.
11) J. Dubach and W. C. Haxton, Phys. Rev. Lett. $\underline{41}$, 1453, (1978).
12) W. R. Gibbs, B. F. Gibson and G. J. Stephenson, Jr., Phys. Rev. Lett. $\underline{39}$, 1316 (1977). W. R. Gibbs, "Common Problems in Low-and Medium Energy Physics (Plenum Press New York and London 1979) p. 595.
13) H. Garcilazo and W. R. Gibbs, Nucl. Phys. A356, 248 (1981).
14) F. Soga et al., Phys. Rev. C24, 570 (1981).
15) M. P. Keating and J. G. Wills, Phys. Rev. $\underline{C7}$, 1336 (1973), M. Tsangarides - Thesis - Univ. of Indiana (1978).
16) M. C. Green (these proceedings)
17) T. E. Ward et al. Phys. Rev. C24, 588 (1981).
18) James L. Clark - Thesis - Carneigie-Mellon University, Sept. 1980.

DISCUSSION

G.A. Miller (Univ. of Washington): If I understood you correctly, when you compared to data you only included the target emission term, is that correct?
Gibbs: I only included the target emission term, which is fairly well justified for the 6.86 MeV state. It's not clear whether it is justified at all for the 3.09 MeV state. For Bismuth it's π^-production, and there isn't any projectile emission. There it is justified.

B. Keister (Carnegie-Mellon Univ.): In all cases do you drop the spin-flip term?
Gibbs: Not in the Bismuth calculation, but in all of the angular distributions I showed you I dropped it. I don't understand why I have to drop it. That's something that has to be looked at.
J. Noble (Univ. of Virginia): It's equivalent to saying that you have both $p_{3/2}$ and $p_{1/2}$ in the final nucleon scattering with equal weights.
Gibbs: Something like that. It doesn't make much sense, I agree. The orignal numbers that I put in there before, and that I showed, were taken from scattering lengths and scattering volumes as known.

T. Cooper (IUCF): Just a comment on the 3.09 MeV state. Basically you can represent it as a good $2s_{1/2}$ single-particle state. But in the pionic stripping case the cross-section is a whole order of magnitude lower than the data. It's interesting to see that the target emission piece appears to have the correct normalization.
Gibbs: It seems to be right on. I was shocked.
Cooper: And in lieu of the fact that the direct term is small, I'm not surprised.
Gibbs: I tried calculating the 3.09 state expecting it to fail and wanting it to fail, and I was shocked when it didn't fail. Not only it didn't fail, it did better on that than it did on the other one.

MODELS FOR (p,π) REACTIONS

Harold W. Fearing
TRIUMF, 4004 Wesbrook Mall, Vancouver, B.C., Canada V6T 2A3

INTRODUCTION

It has been several years since the first interest in pion production reactions arose, stimulated by the pioneering experiments of the Uppsala group.[1] There have now been a variety of models and permutations of models advanced to describe the production process p+A → π^++(A+1) where the nucleus is left in a definite state.[2] Here we want to make some comparisons among these models and some general observations applicable to most all models. We do this by discussing first some of the ways various models or classes of models differ and second some of the general unsolved problems characteristic of many of the models. Our hope is that this will lead into a more detailed discussion of the various approaches which have been used and eventually to a consensus on the best approach and on the most important outstanding problems.

There are at least three different ways in which model calculations can differ, which we plan to discuss in turn. First and most significant is in the basic physics which is included. Ideally one hopes it will be possible to choose among models on this basis, as it is the fundamental physical understanding of the process that we are after. A second difference can be the way in which the same basic physics is approximated. Different models, while agreeing on the most important components, can approximate those dominant components in quite different ways, leading to quite different numerical results. Finally a third difference, which overlaps the others somewhat, can be the emphasis given to different aspects of the calculations. That is, models can differ on the relative importance to be given to the particular components included.

MODELS WITH DIFFERENT PHYSICS

At some level, e.g. Fig. 1, all models are the same. In fact most practitioners would even admit to a description of their model as in Fig. 2, at least as long as the intermediate states are not specified. Most would also agree that there are four major ingredients in any calculation: the primary pion production interaction, pion distortion or re-scattering, proton distortion, and nuclear structure, as reflected in the nuclear wave functions. The

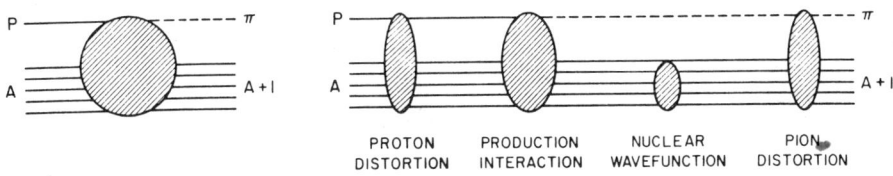

Fig. 1. Universal model of pion production.

Fig. 2. Basic ingredients of most pion production calculations.

0094-243/82/79319-14 $3.00 Copyright 1982 American Institute of Physics differences come when we try to detail each of these ingredients, i.e. to look inside the boxes in Fig. 2.

Traditionally, models have been separated into two classes, the single-nucleon models (SNM) and the two-nucleon models (TNM). The terminology is somewhat misleading, as all models except plane wave Born approximation are multi-nucleon models. However, one can make this distinction in a somewhat useful way if it is understood to depend only on the one- or two-nucleon character of the basic production operator used, since it is the restrictions imposed by matrix elements of this operator which lead to the different basic physics included in a typical SNM as compared to a TNM. Furthermore, by SNM one essentially always means a single nucleon operator taken in distorted wave Born approximation (SNM-DWBA). The plane wave Born approximation contribution (SNM-PWBA) is of course included as a part, but not all, of the SNM.

Once this operator distinction is made there are at least three ways in which the basic physics of these classes of models differs. The first of these is in the projectile versus target emission character of the process. The standard SNM treated in distorted wave Born approximation, since it involves a matrix element of a single-nucleon operator in the projectile co-ordinates and simple optical potentials also in these coordinates, leads to emission of pions from the projectile only (Fig. 3). In contrast the standard microscopic TNM, based on an operator in both projectile and target nucleon coordinates as derived from diagrams such as those of Fig. 4, allows both target and projectile emission. The importance of both contributions within these models was originally discussed by Grossman et al.[3] and later by Dillig and Huber.[4,5] Kisslinger and Keister[6] calculated both target and projectile emission contributions in their two-nucleon model and find in some cases strong interferences, some of which lead to interesting energy dependences in the analyzing power. Unfortunately in this model the relative importance of the two contributions depends very much on the details of the calculation, e.g. zero versus finite range or Woods-Saxon versus harmonic oscillator wave functions and also of course on the structure of the nuclear state considered. For the best calculation for $^{12}C(p,\pi^+)^{13}C_{g.s.}$ at 200 MeV, Kisslinger

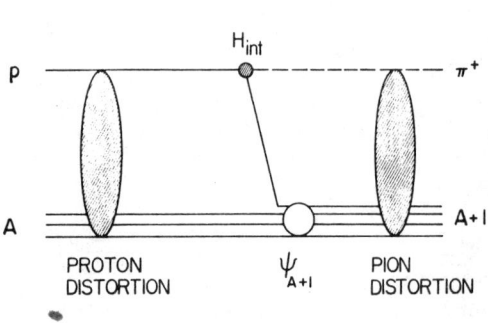

Fig. 3. Single nucleon model in distorted wave Born approximation.

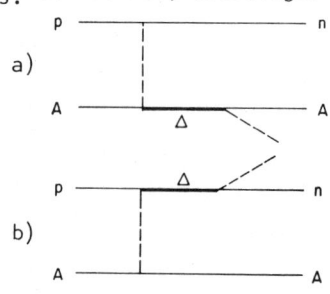

Fig. 4. a) Projectile emission and b) target emission contributions to the standard microscopic two-nucleon model. Spectator nucleons have not been shown explicitly.

and Keister find the projectile emission dominant. An alternative approach of Gibbs[7] goes toward the opposite extreme to the SNM, as it considers target emission only and hence describes contributions complementary to the SNM. For the states considered, which are different than those considered by Kisslinger and Keister, order of magnitude agreement with data is obtained, thus indicating that for these cases the target emission contribution is important.

For π^- emission both target and projectile (via charge exchange) emission graphs contribute in TNM while π^- cannot be obtained from the SNM in a single step. Experimentally π^- production is suppressed in the forward direction—one of the original arguments for the SNM—but is often comparable to π^+ production at more backward angles. Furthermore the most obvious way of getting π^- in a SNM-DWBA, i.e. via an additional step involving charge exchange of the π or p, does not seem to work[8] while at least one TNM including both the target and projectile emission contributions does give the right π^+/π^- ratio and cross sections.[5] Thus one would guess that again the target emission contribution is necessary, though how model-dependent its contribution is obviously needs further study.

A second difference in the basic physics included is the difference in the final nuclear states which can be excited. It arises also from the difference in the production operator in the two approaches. For the SNM the operator involves only co-ordinates of the projectile and thus can excite only single-particle states built on the ground state of the target nucleus as a core. To get any more complicated state requires some sort of second-order process or inelastic rescattering which is not present in the usual optical model distortion. Such inelastic effects were once estimated by Miller[9] and found to be non-negligible. They are not included in a normal SNM-DWBA calculation, however.

In the TNM, however, the production operator involves co-ordinates of both nucleons and so can excite one of the target nucleons to some state and then put the incident nucleon into the resulting hole, or into another excited state. Thus both single-particle and two-particle/one-hole states are accessible. The particular states may depend on the details of the model operator, e.g. in the Gibbs approach, with target emission only, the isovector-pseudoscalar nature of the operator $\sigma \cdot \nabla \phi_\pi$ at the pion-target nucleon vertex leads to a transition of the core, for example from a closed shell T=0 ground state to an unnatural parity T=1 state. The accessible final states are then those made by coupling the incoming nucleon to this excited core state.

There have been several attempts to compare experimentally the probability of excitation of single-particle states versus more complicated core excited states. So far results have been somewhat ambiguous and show no clear trend, as discussed by Hoistad[10] earlier in this conference. For ^{12}C targets single-particle and two-particle/one-hole states seem to be excited with similar strength[11] and for ^{28}Si angular distributions to both kinds of states are qualitatively the same.[12] However, for ^{13}C, ^{10}B and ^{11}B there seems to be a preference for single-particle final states.[11] It may be that other aspects of the nuclear structure are important also and hide effects of the basic operator, or different mechanisms may dominate for different states. Note, however, that any excitation of a two-particle/one-hole state indicates that for that state something more than the standard SNM-DWBA model is required. On the other hand in two-nucleon models both kinds of states can be excited so that in such models the relative populations of final states can, a priori, be anything. Obviously they will depend on the details of the particular model and a TNM will be successful only if it predicts correctly these relative populations in each specific case.

A third difference in the physics of the two classes of models originates in the way the $\Delta(1232)$ is put into the calculation. The difference is somewhat one of philosophy, or emphasis, with the SNM emphasizing the πNN vertex production operator and the TNM emphasizing the pion rescattering via the Δ. Since the Δ dominates much of the physics in the medium-energy range, however, this difference becomes really one of physics content, and rather more important than the differences in approximations which will be discussed in the next section.

In the SNM the Δ appears only in an average way via the distorting potential used to calculate the pion distorted wave. If a phenomenological form for the potential is used, e.g. square well or cut-off Laplacian, with parameters fit to elastic scattering, as has been done in many calculations, e.g. Ref. 13, then the Δ contribution is really hidden and one might expect to see little explicit indication of its presence. On the other hand, if a more fundamentally based multiple scattering type potential, e.g. that of Landau-Tabakin, is used, one with leading term proportional to the π-nucleon scattering amplitude, then effects of the Δ might appear somewhat more explicitly.

In contrast essentially all of the TNM start with the dominance of the Δ rescattering diagram and try to construct this contribution exactly. To the extent this is done completely and correctly one would expect to see effects of the Δ, reflected most likely as a bump in the cross section as a function of energy. In practice, however, many calculations use a crude closure approximation for the Δ intermediate state. Such approximations in the denominator of the Δ propagator may distort the energy dependence, especially near the resonance where the denominator is small. Furthermore in most TNM, results have not been given as a function of energy so it is not really known how dramatic an effect one should expect.

On the experimental side we have heard at this workshop about new data[14] from Saclay for $^{10}B(p,\pi^+)^{11}B$ which fills in some of the missing energies and gives for the first time a fairly complete energy dependence from threshold up to 800 MeV. The results seem to indicate a rather broad bump in the delta region, but nothing nearly as pronounced as is seen for pp→πd. New data[15] for nd→^{3}Heπ$^-$ together with older pd→tπ data[2] also show a fairly obvious bump in the energy dependence. Thus there seems to be some indication of the presence of the Δ; whether it can be explained in detail by the Δ-dominated TNM remains to be seen.

In summary, we have mentioned three ways in which the physics content of different classes of models vary. In principle these differences are not simply a matter of taste, but are testable. Thus information on π^- production together with sufficient calculations in models should help distinguish models including target emission from those that do not. Comparison of results for final states of different nuclear structure may indicate the importance of a single- versus a two-nucleon operator or at least suggest where the standard SNM-DWBA model needs to be supplemented. Energy dependence of cross section may test dominance of the Δ rescattering contributions (provided, of course, that the delta effects are not washed out by other aspects of the process, something which needs to be investigated more thoroughly in models).

MODELS WHICH APPROXIMATE SIMILAR PHYSICS IN DIFFERENT WAYS

Even when there is agreement on the most important ingredients of a model or on the general approach, models can differ in ways which have major numerical consequences simply by differing in the way in which the various ingredients are approximated. A trivial example of this is the use of different potentials in the standard SNM-DWBA calculations.

A more interesting example is the way in which the Δ is included in the various two-nucleon approaches. All begin with the basic diagrams in which the pion is emitted from one nucleon and rescattered via the Δ from another and thus all start with the same physics, at least in broad outline. However, the way in which this basic physics is treated is quite different, as are the details of the calculations. In the standard microscopic TNM as described by Kisslinger and Keister[6] or Dillig and Soga[16] the Δ enters via an operator—vertices and propagator—which is derived from some sort of non-relativistic reduction of the amplitudes of the diagrams of Fig. 4. Thus the result in this approach depends on the specific form of vertices and the way the off-shell pion is treated, i.e., as usually formulated, on transition potentials, and on the approximations made in the Δ propagator and in the kinematics. Differences in detail arise in such things as whether π- and ρ-exchange are included or just π-exchange, in finite versus zero range approximations, in the way s-wave rescattering is included if it is included, and in whether or not the SNM-PWBA term is added. (Note that to include the SNM-DWBA would involve some double counting.)

In the isobar-doorway model as used by Hirata[17] and as will be eventually incorporated in the calculations of Kisslinger and Keister[6] and of Dillig and Soga[16] essentially the same diagrams are assumed to dominate. However, now the intermediate Δ is treated as a Δ-nucleus state obtained by diagonalizing a Hamiltonian in a given subspace. The result thus depends on the effective interaction for producing the Δ-nucleus state and on a certain amount of phenomenology, e.g. the so-called spreading potential which is included to account for the effects of Δ propagation in the nuclear medium.

In Gibbs' model[7] the target emission part of the same set of diagrams is assumed to dominate. The diagram is evaluated, however, as a πNN vertex plus a pion-nucleon scattering amplitude. Thus here the Δ contribution originates in the experimental πN amplitude and the results depend on the particular off-shell extension made and on details of the πNN vertex and the nuclear structure.

Finally in the Ruderman or DWIA model[18] of Fig. 5 the same physics is again assumed to dominate but it is evaluated phenomenologically, by relating it to the pp → πd cross section. Thus here the Δ contribution depends on experimental information on pp → πd, on the approximations required to neglect the NN → NNπ contribution and on the off-shell to on-shell extrapolation of the pp → πd amplitude.

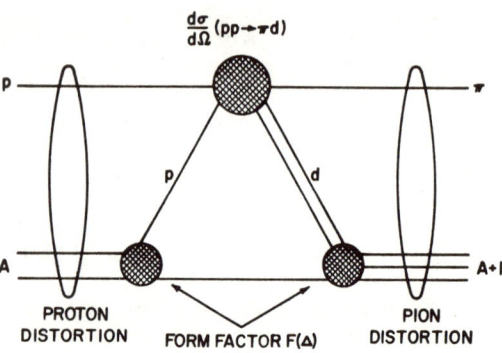

Fig. 5. Ruderman or distorted wave impulse approximation model for pion production.

Thus in all four of these cases the dominant physics, at least at the diagram level, is assumed to be the same—that of Fig. 4. The evaluation of these diagrams is done quite differently, however, and the detailed numerical results will depend on these differences. How does one choose the best approach? That seems to be an open question. There is a complicated interplay in all such models among many ingredients, so that one cannot easily isolate the consequences of somewhat technical assumptions. Until one model or the other proves to be definitely superior in fitting both cross section and analyzing power data, it is thus going to be very difficult to distinguish among models which approximate the same basic physics in different ways.

MODELS WHICH DIFFER IN EMPHASIS

Another way models can differ is in the emphasis given to certain aspects and de-emphasis to others, all within the same general physics framework. The aim of such calculations may be to explore a certain aspect to see how sensitive results are to some particular ingredient, or the emphasis may just reflect an opinion of what is important.

For example, the calculation of Cooper[19] has explored very thoroughly, all within the context of a SNM-DWBA, effects arising from a relativistic treatment of the optical model scattering and the bound state wave functions. Gibbs[7] on the one hand has particularly emphasized final states where target emission contributions should dominate and has neglected projectile emission, and on the other has apparently included the nuclear structure, via a shell model code, in a much more sophisticated way than usual. Other examples are the early analyzing power calculations, some of which included only pion distortion[20] and others only proton distortion.[21]

Such analyses, which are of course models in themselves, can be quite useful for examining sensitivities to a particular ingredient, up to a point. It is possible, however, and there are such examples in the literature, that sensitivity to one aspect may be enhanced by over-simplification or neglect of another. Thus for a process as complicated as (p,π), one must be careful until it becomes possible to handle all of the ingredients well.

PROBLEMS AND CAVEATS FOR ALL MODELS

We proceed now to a discussion of a number of particular ingredients which have not yet been done well in essentially all of the models. The idea is to delineate some of the problems and to raise questions which should be investigated more fully.

Relativistic effects

In typical (p,π) reactions momentum transfers of 500 MeV/c are obtained and incident energies range up to 800 MeV. Thus except very near threshold p/m can easily be as large as 1/2 where p is almost any momentum variable characteristic of the process. As a result one might expect that relativistic corrections of one kind or another could be large. Such corrections may originate from the small components of the wave functions, from p/m corrections to the operators, from relativistic forms for propagators, or from graphs, e.g. those involving crossed lines, which are not normally included in non-relativistic calculations. There have been a few attempts at relativistic calculations. All have been in some form of relativistic SNM and all have found sizable effects. Thus, for example, Cooper[19] found that for the momentum transfer region relevant here the "small" or p/m components of the relativistic wave functions were comparable to the "large" ones, as can be seen in Fig. 6. Brockmann and Dillig[22] found sensitivity to the "small" components (see Fig. 7), and Miller and Weber[23] found large differences between their relativistic result and a non-relativistic approximation. Clearly no one really knows how to do a fully relativistic calculation. However, equally clearly,

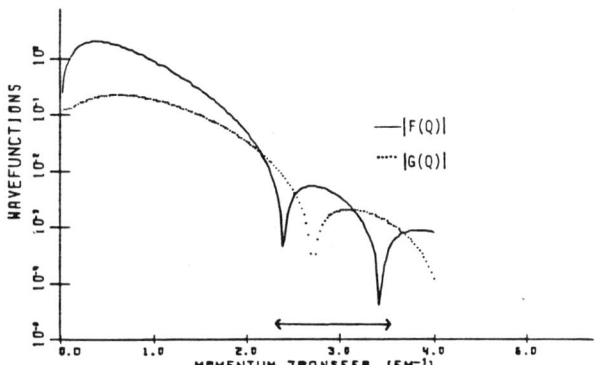

Fig. 6. Comparison, taken from Ref. 19, of the "large" [F] and "small" [G] components of the $1P_{1/2}$ bound state wave functions of ^{13}C. The arrow spans the region relevant for (p,π) reactions.

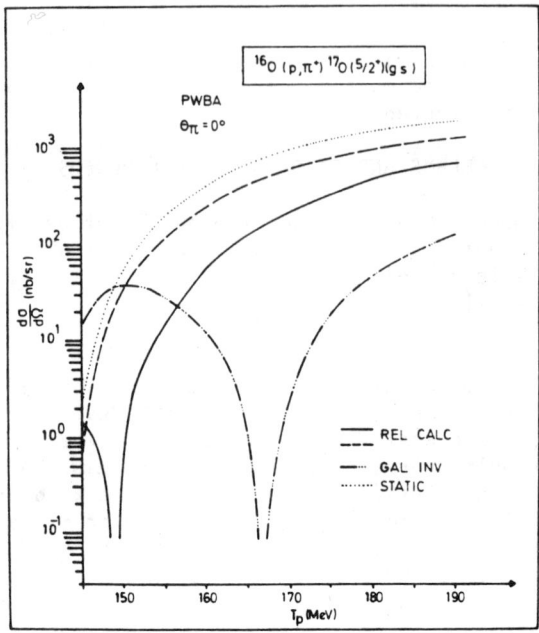

Fig. 7. Results of Ref. 22 of a relativistic plane wave Born approximation calculation showing the effect of two different ratios of large and small components in the relativistic calculations and showing how the non-relativistic reductions to Galilean invariant or static operator differ from the full calculation.

those who try seem to find non-negligible corrections from relativistic effects. Thus perhaps more thought should be put into how to make calculations more relativistically consistent, or at least how to determine which relativistic effects are important and how to include them.

Form of the interaction

A closely related, in fact almost identical, question deals with the form of the πNN interaction vertex. Several years ago there were heated discussions about $\sigma \cdot \nabla \phi_\pi$ versus Galilean invariant versus some other interaction operator as the relevant πNN vertex operator. While some of this discussion was misguided it did result in finding some situations (cf., for example, the calculations of Brockmann and Dillig,[22] Lee and Pittel[24]) where terms in the interaction such as the Galilean invariant term (whatever its coefficient) were important when used in the standard fashion in SNM-DWBA or SNM-PWBA calculations. Furthermore, attempts to show that specific forms of non-relativistic vertex operators were sensible non-relativistic reductions of a relativistic operator seemed to fail in the sense that within simple models matrix elements of the relativistic operator were very poorly approximated by the corresponding non-relativistic matrix

elements. Thus, for example, Greben and Woloshyn[25] found for a
(p,nπ) reaction that the amplitudes calculated using a relativistic
pseudoscalar or pseudovector πNN coupling were quite different from
those obtained from the static or Galilean invariant operators which
supposedly were the non-relativistic limit of the relativistic operators. An example of their results is shown in Fig. 8. It appears
then that the usual non-relativistic forms may not really be reproducing the physical picture claimed as a starting point, but instead
simply serving as a phenomenological ansatz.

These discussions of the form of the vertex operator all arose
in the context of the SNM. In more modern calculations of two-nucleon
mechanisms the effective operator is again calculated in a way which
is supposed to be a non-relativistic reduction of the relativistic pion-
rescattering-via-the-delta diagram. Invariably, however, the vertex
operators are taken as $\sigma \cdot \nabla$ for the πNN vertex and an analogous static
expression for the πNΔ vertex. In view of the difficulties with the
SNM operator it seems to be an open and important question whether
these approximations for TNM are really sufficient. Does such an
effective operator really approximate the physical situation assumed
or is it just a phenomenological starting point? How can one tell in
the absence of a fully relativistic calculation?

Tests of form of the effective interaction for TNM

It is generally acknowledged that a diagram like that of Fig. 9,
involving π- and perhaps ρ-exchange, supplemented with some s-wave
rescattering (and an impulse approximation term for pp → πd) describe
the physics of the pp → πd and NN → NNπ process. Calculations[26]
based on this physics, however, have been more or less successful depending on details of the calculation and some elements such as the
importance of the ρ are still controversial. In a nucleus, TNM calculations are sufficiently complicated, however, that to obtain an

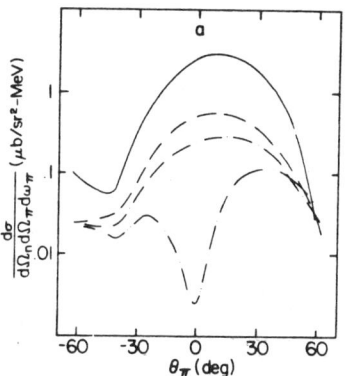
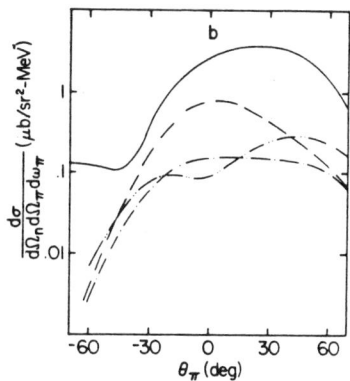

Fig. 8. Differential cross section, taken from Ref. 25, for the reaction ^{4}He(p,nπ$^+$)^{4}He for proton energy, pion energy, and neutron angle of a) T_p = 350 MeV, T_π = 80 MeV, θ_n = -10° and b) T_p = 560 MeV, T_π = 80 MeV, θ_n = -10°. The solid curve is a relativistic calculation using pseudoscalar coupling whereas the dashed curves are various non-relativistic approximations to this relativistic interaction.

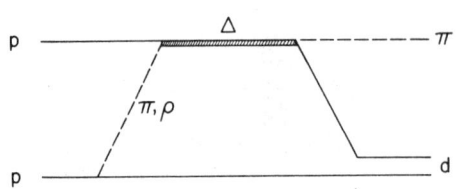

Fig. 9. Dominant contributions to pp → πd.

effective interaction from this picture drastic approximations in the effective interaction, e.g. zero range, are often made in addition to those mentioned above. Furthermore there are differences in details, e.g. differences in form factors, among various TNM calculations. Thus it is not clear in the end whether the resulting two-nucleon operator still reproduces the basic two-nucleon process NN → NNπ. Hence a test of a particular TNM may or may not be a real test of the physical picture.

Significant progress in this direction has been made, however, and some of the more modern TNM, as for example that of Dillig and Soga[16] discussed at this conference, start with much more realistic two-nucleon interactions than earlier attempts and include for example both π- and ρ-exchange, s-wave rescattering, both nucleon and Δ pole terms and appropriate form factors. Kisslinger and Keister[6] have also made important improvements, and in particular have shown the importance of a finite range calculation. Still each TNM proponent should start by showing that the effective interaction and the approximations necessary to evaluate that interaction in this particular model leave a result which still reproduces pp → dπ and NN → NNπ.

There has also been a great deal of progress in understanding the pp → πd reaction and there are now several very sophisticated models.[26] It would be useful also if each of the experts developing these models gave some thought as to how to obtain from his model some sort of simple effective interaction which could be used in Born approximation for the nuclear process, but which still reproduces the main features of the two-body process.

Double counting problems

For any calculation one must worry about over- or undercounting, but the problem is particularly severe for pion production since the "fundamental" production and rescattering vertices used to construct the operator are the same ones which contribute both to the pion rescattering and to the nucleon-nucleon force, and consequently to both proton distortion and correlations in the nuclear wave functions. The problem is aggravated by the fact that one usually does not sum a particular well-defined set of diagrams. Instead, matrix elements of some effective interaction are taken between wave functions which involve parametrized or phenomenological correlations and distortions and thus contain contributions from a large set of diagrams, but in an average sort of way.

As an example of the difficulties which can arise consider the diagrams of Fig. 10 (a) and (b) which show a contribution to the microscopic TNM description of (p,π). With the boxes drawn as in (a)

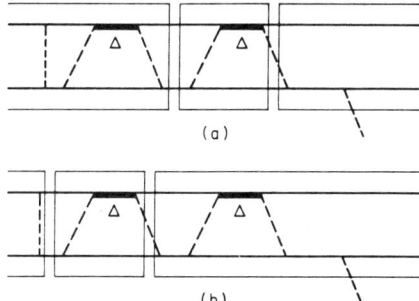

Fig. 10. An illustration of the ambiguities which can lead to double counting. In each case the box is interpreted as the basic two-nucleon operator where the scatterings to the left and right would be respectively a part of the proton distorted wave or the pion distorted wave.

the initial pion rescattering contributes to the medium-range nucleon-nucleon force and so is included in the proton distorted wave function; the box is the primary two-nucleon interaction; and the final rescattering is a non-resonant contribution to the pion distorted wave. However, the same diagram can contribute, as in (b), where now the first rescattering is the primary one and the subsequent ones would be included in the pion distorted wave. Thus in principle in the usual microscopic TNM calculations this contribution is counted twice. In practice, however, the problem may not be quite so severe as this would seem to indicate. Presumedly one wants the part of the interaction put in explicitly to account for most of the momentum transfer and most of the highly off-shell scattering. Such requirement means that the remaining rescatterings tend to be on-shell and with small momentum transfer and so more easily described via optical potential wave functions. This provides a partial way of distinguishing the "basic" interaction and so may effectively decrease the double counting in situations like Fig. 10 (a) and (b).

Observe that this double counting problem is not limited to TNM calculations but is present also in the SNM-DWBA approach. In fact these same figures can be interpreted (taking the lower line as the projectile) as contributions to a SNM-DWBA. Again ambiguities and double counting arise because of the options one has in assigning a particular πNN vertex to the proton rescattering, the basic interactions, or the pion rescattering.

Clearly all of these double counting arguments are qualitative. However, the whole double counting problem is important for a full understanding and should be looked at more carefully in the microscopic models, especially in situations where the external rescatterings are non-negligible.

Within the context of the isobar doorway model one can eliminate, or perhaps bury, this double counting problem since there one works with states which are actually obtained as solutions of scattering equations involving potentials but which effectively sum up large classes of rescattering diagrams. One pays the price, however, of the introduction of phenomenological transition potentials and a spreading potential for the Δ in the medium and either the neglect, or absorption into a background term of diagrams not involving the Δ doorway state. It remains to be seen whether the phenomenology involved here is any "worse" than that involved in say the choice of optical potential in a SNM-DWBA approach or involved in other models.

Wave functions inside the nucleus

In essentially any model the (p,π) matrix element requires the overlap of nuclear wave functions with distorted waves obtained primarily from elastic scattering information. It thus shares with any other inelastic process the problem that wave functions which are known asymptotically are needed in the interior of the nucleus. Wave functions derived from some theoretically founded potential probably are reasonably correct at nuclear distances, but one must be very careful in using wave functions derived from purely phenomenological fits using a phenomenological potential. Keister[27] looked at this problem in a SNM-DWBA using phase-equivalent transformations and found sensitivities at the order of magnitude level to the short distance behavior of the pion wave function, so this may be a quite important problem.

Analyzing powers

Analyzing power measurements promise to give information and/or distinguish among different models. Already interesting structures and energy dependences have been found experimentally.[28] We have seen how calculations so far have had difficulty fitting simultaneously analyzing powers and cross sections. A word of caution is in order, however. Even the worst of theories, i.e. those that are totally wrong, will give the correct sign of A_π in individual cases with 50% probability. Also since A_π is restricted to $-1 \leq A_\pi \leq 1$ and is zero at 0° and 180° many curves will look "qualitatively correct". So, qualitative agreement for A_π in a few individual cases is obviously not a sufficient test of any model. One must really reproduce the bumps and wiggles for several nuclei as a function of energy before claiming to have successfully fit the analyzing power.

CONCLUSIONS AND QUESTIONS

In the previous sections we have talked about some of the differences in the physical content of various models and some of the similarities with regard to difficult problems which have not yet been solved. Hopefully some nasty questions have been raised, which will provoke further work and perhaps eventually further understanding.

To end this introductory survey we list some additional more general questions which may perhaps serve as a focus for subsequent discussions.

1) What basic core of ingredients must be included in any (p,π) calculation (and in what way) for any hope of success in fitting the data available?
2) Are there major effects which have been left out of all models?
3) Are there fundamental problems with the (p,π) reaction which will prevent useful information from being obtained? That is, should we be optimistic that eventually a model can be found which will work in most cases?
4) What are the most necessary and fruitful directions for future work?
5) Are there major new experiments or classes of experiments which would be useful?

6) More specifically, would more information on (p,π^-) be important?
7) Would experiments on three-body final states, e.g. $(p,n\pi)$, be useful?

ACKNOWLEDGEMENTS

The author would like to thank M. Betz, B. Blankleider, M. Dillig, W. Gibbs, B. Keister, L. Kisslinger, J. Niskanen and G. Walker for specific comments on some of the points raised here and a number of colleagues at TRIUMF and elsewhere for more general discussions. This work was supported in part by the Natural Sciences and Engineering Research Council of Canada.

REFERENCES

1. S. Dahlgren, B. Höistad and P. Grafstrom, Phys. Lett. 35B, 219 (1971). See also the review by Höistad, Ref. 2.
2. Recent reviews include:
B. Höistad, in Advances in Nuclear Physics, eds. J.W. Negele and E.W. Vogt, vol. 11 (Plenum, New York, 1979), p. 135;
D.F. Measday and G.A. Miller, Annu. Rev. Nucl. Part. Sci. 29, 121 (1979);
H.W. Fearing, Prog. in Particle and Nuclear Physics, ed. D. Wilkinson, vol. 7 (Pergamon, Oxford, 1981), p. 113.
A bibliography of theoretical and experimental work through September 1980 can be found in H.W. Fearing, TRIUMF report TRI-80-3.
3. Z. Grossman, F. Lenz and M.P. Locher, Ann. Phys. (N.Y.) 84, 348 (1974).
4. M. Dillig and M.G. Huber, Phys. Lett. 69B, 429 (1977).
5. M. Dillig and M.G. Huber, Lettere al Nuovo Cimento 16, 293 (1976).
6. L.S. Kisslinger and B.D. Keister, these proceedings.
7. W.R. Gibbs, these proceedings.
8. L.S. Kisslinger and G.A. Miller, Nucl. Phys. A254, 493 (1975);
M. Dillig and H.G. Huber, Nuovo Cimento 16, 299 (1976);
B. Höistad et al., Phys. Lett. 73B, 123 (1978).
9. G.A. Miller, Nucl. Phys. A224, 269 (1974).
10. B. Höistad, these proceedings.
11. S. Dahlgren, P. Grafstrom, B. Höistad and A. Asberg, Nucl. Phys. A211, 243 (1973);
F. Soga, R.D. Bent, P.H. Pile, T.P. Sjoreen and M.C. Green, Phys. Rev. C 22, 1348 (1980).
12. T.P. Sjoreen, P.H. Pile, R.D. Bent, M.C. Green, J.J. Kehayias, R.E. Pollock, F. Soga, M.C. Tsangarides and J.G. Wills, Phys. Rev. C 24, 2569 (1981).
13. M. Tsangarides, Ph.D. thesis, Indiana University, IUCF internal report 79-4 (1979).
14. P. Couvert, these proceedings.
15. E. Rössle, these proceedings.
16. M. Dillig and F. Soga, these proceedings.
17. M. Hirata, Phys. Rev. Lett. 40, 704 (1978).
18. H.W. Fearing, Phys. Rev. C 11, 1210 (1975); ibid., 1493; ibid.,

        16, 313 (1977).
19. E.D. Cooper, these proceedings, and Ph.D. thesis, University of Alberta (1981).
20. S.K. Young and W.R. Gibbs, Phys. Rev. C 17, 837 (1978).
21. J.V. Noble, Nucl. Phys. A244, 526 (1975); Proc. of 4th Int. Symp. on Polarization Phenomena in Nuclear Reactions, Zürich, 1975, ed. W. Gruebler and V. Konig (Birkhäuser, Basel, 1976), p. 715.
22. R. Brockmann and M. Dillig, Phys. Rev. C 15, 361 (1977).
23. L.D. Miller and H.J. Weber, Phys. Lett. 64B, 279 (1976); Phys. Rev. C 17, 219 (1978).
24. T.S.H. Lee and S. Pittel, Nucl. Phys. A256, 509 (1976).
25. J.M. Greben and R.M. Woloshyn, Nucl. Phys. A333, 399 (1980).
26. See, for example:
M. Brack, D.C. Riska and W. Weise, Nucl. Phys. A287, 425 (1977);
J. Chai and D.O. Riska, Nucl. Phys. A338, 349 (1980);
O.V. Maxwell, W. Weise and M. Brack, Regensburg preprint (1980);
J.A. Niskanen, Proc. Fifth Int. Symp. on Polarization Phenomena in Nuclear Physics, Santa Fe, August 1980, AIPCP#69 (AIP, New York, 1981), p. 62; Nucl. Phys. A298, 417 (1978);
M. Betz and T.S.H. Lee, Phys. Rev. C 23, 375 (1981).
B. Blankleider, Ph.D. thesis, Flinders University of South Australia (1980);
B.J. VerWest, Phys. Lett. 83B, 161 (1979).
27. B.D. Keister, contributed paper at 8th Int. Conf. on High Energy Physics and Nuclear Structure, Vancouver, 1979; Nucl. Phys. A350, 365 (1980).
28. E.G. Auld, G. Jones, G.J. Lolos, E.L. Mathie, P.L. Walden and R.B. Taylor, TRIUMF preprint TRI-PP-81-27 (Phys. Rev. C Rapid Commun., in press).

SUMMARY OF THE DISCUSSION ON CONNECTIONS AMONG MODELS OF PION PRODUCTION

Harold W. Fearing
TRIUMF, 4004 Wesbrook Mall, Vancouver, B.C., Canada V6T 2A3

Following the introductory talk, given in the preceeding paper, a panel of experts, together with members of the audience, discussed the general subject of connections among the various models of the (p,π) reaction which were described at the conference and some of the difficulties with these and other models. Members of the panel included Prof. Hugh McManus of Michigan State University, Prof. James Vary of Iowa State University and Prof. George Walker of Indiana University. In this brief summary we will try to give some idea of the flavor of the presentations of the panelists and of the discussion which followed. No attempt has been made to record all of the comments verbatim. Some have been paraphrased, some omitted, and some "interpreted", though we have attempted to work in somewhere the substance of essentially all the remarks made. The author apologizes for the inevitable errors of omission or misinterpretation and the inevitable misquotations which end up in a summary of this kind.

The session began with a presentation by McManus who went "backwards in time and in technique" to describe a simple pion-rescattering-via-the-delta model of the (π^+,pp) and (π^-,pn) reactions, as for example on ^{3}He and ^{4}He where there is new data. In this model both a direct and an exchange diagram enter with one describing π^+ absorption and the other π^- absorption. The diagrams differ primarily in that one has a resonant denominator and the other does not. The interesting thing is that the ratio of the cross sections $\sigma(\pi^+,pp)/\sigma(\pi^-,pn)$, which ranges from ~100 at resonance to ~5-10 at threshold, is described fairly well by simply the ratio of the squares of these denominators. Thus McManus suggested that by analogy such simple physics, reflecting the dominance of the Δ rescattering, may also describe the important aspects of the (p,π) process. Hence all models, using these same Δ rescattering diagrams may give qualitatively the same physics and reflect the same trends in the data. Distinguishing among such models will then require numerics, i.e. require getting the absolute magnitude right, which is a much more difficult problem than simply reproducing trends.

Vary wanted to focus on those aspects of (p,π) and related reactions which might have some relevance for nuclear structure. He began by pointing out that there is a heirarchy of degrees of freedom in a nuclear system and that the relevant ones depend on the probe and on the momentum transfer. At the first level, corresponding to relatively small momentum transfers, the system can be described by the usual nuclear Hamiltonian consisting of sums of two body, three body, etc. forces between nucleons. As one probes with higher momentum transfers and samples shorter distances new degrees of freedom corresponding to Δ's and π's enter. These may be described loosely by adding a new piece $V_{\pi\Delta N}$ to the system Hamiltonian. The question then arises whether or not this new interaction can be treated perturbatively as is usually done or must one solve the full many body problem (and if so, how?). This latter would be the case if effects such as precursors or pion condensates are relevant. At still shorter distances and higher momentum transfers quark degrees of freedom presumedly enter and again one must ask whether these degrees of freedom can be treated perturbatively or does one have to solve the full many body theory in the quark variables. In any case one would then have to have a certain critical radius in the relative coordinates inside of which the dynamics of the system would be described using one set of degrees of freedom while outside of this radius a different set would be appropriate.

Walker started by addressing specifically those who have actually done (p,π) calculations. Such second generation calculations as are now in progress have become very long and very complicated. Thus it becomes an important question as to how long it will take or how practical it is to include certain detailed aspects in a calculation. It is also important to coordinate efforts among various groups. With this in mind specific questions were directed to those who spoke the previous evening. Cooper was asked what time scale would be required to add target emission and π rescattering to his model and whether this could be done in a relativistic fashion. Keister and Kisslinger and Dillig and Soga were both asked how difficult it will be to include non-locality in the Δ propagation and to modify their calculations so as to understand production very near threshold. They were also asked if they were really planning to include the SNM-PWBA contribution, omitted in most TNM, or did they feel strongly that it was negligible.

Following these presentations by the panel and a challenge by Fearing to each of the model proponents to describe what qualitative features might allow one to distinguish their model from the others, the session continued with discussion and comments from members of the audience.

Kisslinger began by emphasizing his view that the models of Keister and Kisslinger (KK) and of Dillig and Soga (DS) are essentially identical in their physics and differ only in calculational details. Miller pointed out, however, that DS have included both π exchange and ρ exchange whereas KK include only π exchange. For pp $\rightarrow$ πd the tensor force dominates so that the π and ρ contributions tend to cancel and the effect of the ρ can be treated as a form factor. In nuclei, however, there are also spin-spin contributions where the π and ρ add and so there may be some cases (as noted by Dillig) where the ρ contribution could be significant and represent new physics. Fearing commented that even when the basic physical ideas are similar calculational details can be very important and can change the qualitative results as for example in the KK work where the choice of wave functions determines in some cases whether the direct or exchange term dominates. Thus, as McManus emphasized earlier in the session, the numerics are important. It will be very interesting to understand in detail which different input and approximations are responsible for differences seen in the results.

Kisslinger also clarified (as did Gibbs later) the connection between their work and that of Gibbs. So far Gibbs has selected situations where the final states are such that one expects target emission to dominate whereas KK have looked at cases where projectile emission may be most important (although both projectile and target emission have been included in the calculation). Thus one cannot yet make a direct comparison.

Keister pointed out that π^- can be obtained within the TNM not only from target emission graphs but also from projectile emission contributions. In the latter case a π^0, emitted from the incoming proton, charge exchanges on a target neutron to give a π^-. Miller emphasized that one can get π^-, and in fact all of the physics of the projectile emission part of the TNM, also within the SNM-DWBA by an appropriate generalization of the distorting potential. For π^- one must add a charge exchange component to the pion or proton distortion and to get excitation of 2p-1h states one includes a potential coupling to these excited states as done in analyses of inelastic scattering.

With regard to relativistic corrections in TNM, Keister argued that such corrections may be less important than in SNM because of the momentum sharing which reduces the size of the individual momenta. Walker and Noble countered, however, with the observation that one gets equal sharing only with wave functions falling exponentially, e.g. Gaussians, and that with other more realistic wave functions the momentum sharing is much less. Finally in response to one of Walker's questions Keister said that while they were working on extending their calculation to include simple isobar doorway model components, to include the full Δ non-locality requires at least another three dimensional integral and probably would not get done.

Coester talked also about the relativistic problem but from a more fundamental point of view. He remarked that it was very difficult to justify the relativistic one body models which use the Dirac equation for a particle in an effective field in terms of a Dirac-Fock approximation to a quantum field theory because of difficulties in subtracting off the infinite vacuum energy arising from the filled negative states. Such objections are not relevant if one looks at these theories just as models to be used to fit data rather than as fundamental theories. A second remark emphasized that there are two aspects to relativistic effects, one having to do with the Lorentz invariance of the theory and the other with the presence or absence of antinucleon degrees of freedom. It is possible to have Lorentz invariant theories without antinucleons as done by say Betz and Lee or to have theories such as the effective field-Dirac equation theories which are not Lorentz invariant but which do contain antinucleon degrees of freedom. Finally Banerjee argued that one should avoid any sort of non-relativistic reduction, but use a relativistic theory from the start.

Noble returned to the question of high momentum components, emphasizing that we really don't know that Woods-Saxon wave functions have the right high momentum components. Furthermore in a convolution of several wave functions momentum sharing occurs only for wave functions which fall like exponentials in momentum space. If they fall like powers, one tends to get one hard event and the others as soft corrections. In the region of interest for (p,π), Woods-Saxon wave functions go like an exponential in the surface thickness. In Cooper's calculation the surface thickness may be too small, thus over-emphasizing high momentum components. Cooper later agreed that a = 0.4 as he used for ^{12}C could give too small a surface thickness but that reasonable changes in the parameter a affect the magnitude of the cross section at the factor of two level but do not change the shape.

In response to the request for specific advice to experimentalists Cooper commented that the cross section for ^{40}Ca(p,π)^{41}Ca$_{g.s.}$ was fairly well predicted up to 90° and independent of parameters in his model. Thus one should have confidence in the analyzing power predictions and such measurements would be a test of the model. Another good test would be a measurement of ^{48}Ca(p,π)^{49}Ca where predictions are that ground and first excited states should have more or less the same shape for the analyzing power and cross section, even though the structure of the states is quite different. For future work, putting in target emission will be very hard and Cooper's present plan is to use Gibbs' results for the target emission contribution to find situations where such contributions can be neglected.

Dillig pointed out that contributions coming from high momenta are not necessarily due to relativistic corrections. Experience with exchange current calculations indicates that when high momentum components come in, the usual effect is to couple in a second nucleon. Thus, important effects may come from such nucleon-nucleon correlations rather than relativistic corrections and it is therefore not clear whether a relativistic calculation using average fields is inherently "better" than a non-relativistic calculation incorporating correlations or other effects. A possible test would be to look at relativistic effects in exchange current calculations which do not involve energy transfer. If such effects are not important there, Dillig argued, then they probably would not be important when energy is transferred as then nuclear structure effects are emphasized. One also worries that theoreticians tend to underestimate difficulties with conventional effects, e.g. Coulomb or conventional distortion effects. Uncertainties in such things may hide effects due to short range or other phenomena. Dillig also agreed with others that carrying out good numerical calculations and developing a good runable code are of crucial importance. They expect to require at least another six months or more to complete and thoroughly check their calculation.

Thomas reiterated the possibility of obtaining interesting information from the related reaction (p,nπ^+) as calculated by Sherif et al. The major advantage here is that one can use an isoscalar target and leave an isoscalar final state and thus avoid the influence of the Δ. Furthermore the outgoing nucleon carries much of the momentum transfer. One can thus isolate specific effects in the production process and, for example, look at relativistic effects. It would be interesting to see if the Cooper-Sherif model could be applied here. Cooper replied later that one would need to choose a good closed-shell nucleus and be sure that elastic scattering data was available so as to pin down the distortions. Gibbs also mentioned that one could use his model to look at unbound final states and in particular that Chant was starting on such a calculation. (The nuclear states considered would be different than suggested by Thomas, however, because the T=1 nature of the pion emission operator doesn't allow isoscalar to isoscalar transitions.)

* One final point is worth mentioning which was made by Vary and a number of others in private discussions following the main session. It would be very useful to have a test case, that is a particular (or several) targets, energies, final states, etc. which everyone calculates. It might then be possible to actually make direct comparisons among models. De facto, such a case almost exists, i.e. ^{12}C at 185 MeV, since most early calculations tried to fit the original Uppsala data. This may not be the best target, however, now that there are lots of data, and certainly the energy range should be extended. Thus, I would strongly suggest that the various groups, particularly those involved in the several similar TNM calculations now in progress get together and pick a specific test case and then all calculate results for that case. (The choice should be publicized among the experimentalists also, so that there will be some data to keep things honest.) It will then be very interesting to understand, in terms of the physics involved, the reasons for any differences among model results which appear.

IV Related Reactions
Pion Induced Reactions
Complex Projectiles
Summary

Top: M.M. Sternheim, W. Benenson, W. Wharton
Middle: J.-F. Germond, J.T. Londergan, M.G. Huber
Bottom: G.E. Walker, R.D. Bent
Photos by Kent Berglund and Chuck Foster

ONE-NUCLEON KNOCKOUT REACTIONS

J.T. Londergan
Dept. of Physics, Indiana University, Bloomington, IN   47405

INTRODUCTION

In this review, we will briefly discuss the present status of reactions which are related to the subject of this workshop. Since most of the workshop has concentrated upon exclusive (π,p) reactions or their time-reversed reactions, we will confine ourselves to a review of exclusive (γ,p) or (p,d) reactions, i.e., to processes which lead to specific final states of the residual nuclei.

The reactions which we will survey are quite similar to the (π,p) reaction: they reach the same states in the final nuclei; the momentum and angular momentum transfers are similar to those in (π,p) processes. One of the most distinctive features of (π,p) reactions is high momentum transfer; the (γ,p) reaction is intrinsically a high-momentum-transfer process, and (p,d) reactions can be studied at momentum transfers comparable to those in (π,p). Finally, both (p,d) and (γ,p) reactions are thought to be "well known" compared to (π,p), although in fact the difficulties in understanding exclusive (γ,p) processes at medium energies are disturbingly similar to the problems which must be faced in a theoretical description of pion absorption.

We will concentrate here on recent results, both theoretical and experimental, particularly those where we feel that theory and experiment confront one another. We will also emphasize those areas where we can at least implicitly make connection between the pion absorption or production problem, and (p,d) or (γ,p) reactions. We make no pretense at providing a comprehensive review here, but we can refer the reader to relatively complete review articles or conference summaries.

In Section 2, we will review electromagnetic processes at medium energies. We will discuss the present situation in deuteron photodisintegration, in particular the forward cross section in that reaction. We will also discuss the validity of the impulse approximation in photonuclear reactions, and we will discuss the observed structure in the proton polarization resulting from $\gamma+d \rightarrow n+p$. We will review the situation in ^{4}He photodisintegration at low energies, and possible implications for charge-symmetry-breaking in nuclear forces. We will summarize the situation regarding (γ,p) reactions at medium energies and the role of the isobar current in exclusive (γ,p) processes.

In Section 3, we will briefly outline the situation for (d,p) and (p,d) reactions at medium energies. We will compare (where available) cross sections for (p,d), (π,p) and (γ,p) reactions. In Section 4, we will summarize our results .

0094-243X/82/79339-32 $3.00 Copyright 1982 American Institute of Physics

2. MEDIUM-ENERGY (γ,p) REACTIONS

2.1 Validity of the Impulse Approximation

Recently, there has been considerable theoretical investigation of the validity of the impulse approximation in medium-energy electromagnetic reactions.[1-4] It has become apparent that application of Siegert's Theorem[5,6] greatly simplifies calculation of photonuclear reactions, but it also tends to obscure the importance of mesonic exchange currents (MEC). We can see this by briefly reviewing the formalism for the nuclear photoeffect.

The matrix element describing the process $\gamma+A \to p+(A-1)$ can be written as

$$M_{fi}^{\lambda} = - \langle \psi_f | \int d\vec{r} [\frac{4\pi}{2k_\gamma}]^{1/2} \vec{A}^{\lambda}(\vec{r}) \cdot \hat{\vec{j}}(\vec{r}) | \psi_i \rangle \tag{1}$$

In Eq.(1), $\vec{A}^{\lambda}(\vec{r})$ is the photon electromagnetic potential,

$$\vec{A}^{\lambda}(\vec{r}) = \vec{\varepsilon}^{\lambda}(\vec{r}) e^{i\vec{k}_\gamma \cdot \vec{r}} \tag{2}$$

$\hat{\vec{j}}(\vec{r})$ is the nuclear current operator, $|\psi_f\rangle$ is the final nuclear wave function, which reduces at large distances to a continuum proton and a distinct state of the (A-1)-partical nucleus, and $|\psi_i\rangle$ is the initial A-particle target wave function. If we expand $\vec{A}^{\lambda}(\vec{r})$ in electric and magnetic multipole operators, we find that the electric multipole operator can be separated into two pieces:

$$\hat{O}_{EL} = \frac{i}{k_\gamma \sqrt{L(L+1)}} \int d\vec{r} \vec{\nabla} \cdot \hat{\vec{j}}(\vec{r}) Q_L(\vec{r}) - \frac{i}{\sqrt{L(L+1)}} \int d\vec{r} k_\gamma r j_L(k_\gamma r) Y^{(L)}(\Omega_r) \cdot \hat{\vec{j}}(\vec{r}), \tag{3}$$

$$\equiv \hat{O}_{EL}^{[a]} + \hat{O}_{EL}^{[b]},$$

where

$$Q_L(\vec{r}) \equiv (1 + r \frac{d}{dr}) j_L(k_\gamma r) Y^{(L)}(\Omega_r). \tag{4}$$

In Eq.(3), the first term $\hat{O}_{EL}^{[a]}$ dominates at low energies, and is responsible for producing the giant electric multipole resonances (we are following Arenhovel's notation here[3]). If we focus on this term, we can use the current conservation equation:

$$\vec{\nabla} \cdot \hat{\vec{j}}(\vec{r}) = -i[\hat{H}_N, \hat{\rho}(\vec{r})], \tag{5}$$

where $\hat{H}_N$ is the nuclear Hamiltonian for the A-particle system, and $\hat{\rho}(\vec{r})$ is the charge density operator. Using the fact that $|\Psi_i\rangle$ and $|\Psi_f\rangle$ are both eigenfunctions of H_N and that their eigenenergies are related by

$$E_f - E_i = k_\gamma, \qquad (6)$$

we see that in Eq. (3) we can replace matrix elements of the current operator $j(r)$ by the charge density operator $\hat{\rho}(\vec{r})$, for the operator $\hat{O}_{EL}[a]$.

We can further simplify this by expanding both $\hat{j}$ and $\hat{\rho}$ in terms of the number of participating nucleons:

$$\hat{j} = \hat{j}[1] + \hat{j}[2] + \ldots ,$$

$$\hat{\rho} = \hat{\rho}[1] + \hat{\rho}[2] + \ldots . \qquad (7)$$

In Eq.(7), $\hat{j}[1]$ is the one-body current (the current due to a moving charge or the spin current for a single nucleon), and $\hat{j}[2]$ is the two body current, one piece of which is due to virtual exchanges of charged mesons between an n-p pair (MEC). Similarly, $\hat{\rho}[1]$ and $\hat{\rho}[2]$ are one-and two-body contributions to the nuclear charge-density operators. Siegert[6] showed that the largest contribution comes from the one-body charge density. It is straightforward to show the equivalence

$$\vec{\nabla} \cdot \{\hat{j}[1] + \hat{j}[2]\} = -i[H_N, \hat{\rho}[1]]. \qquad (8)$$

We see that, using the current conservation condition and $\hat{\rho}[1]$ only is equivalent to evaluating both $\hat{j}[1]$ and the MEC term $\hat{j}[2]$. Using Eq.(8) is very efficient, but it tends to obscure the importance of the MEC term, as it is difficult to disentangle it from the one-body current term when Eq.(8) is employed.

For the electric multipole operator $\hat{O}_{EL}[a]$, the impulse approximation is recovered by taking $\hat{j}[1]$ only. To see the effect of the MEC terms for this amplitude, we can use Eq.(3) with $\hat{j}[1]$ only, and then compare with the result of Eq.(8), which includes the MEC terms. This is shown in Fig.(1). Here the solid curve is the contribution from both $\hat{j}[1]$ and $\hat{j}[2]$, while the dashed curve includes only the impulse approximation (one-body current) term, for the total cross section in the reaction $\gamma + d \to n + p$.[3]

The MEC contributions are rather large in the total cross section, even at low energies; for example, at 10 MeV the MEC increases the cross section by 25%, while at 50 MeV photon energy the cross section with MEC terms is 2.7 times the impulse approximation cross section.

The situation is even more striking for the 0° cross section. This is shown in Fig.(2).[2,3] For photon energies greater than 10

MeV, the impulse approximation is only a very small part of the photodisintegration cross section. From Figs.(1-2), we see that the MEC contributions to the γ+d → n+p cross section are large even at rather small photon energies, and that at 0° the MEC terms dominate for all energies above 10 MeV. The inadequacy of the impulse approximation may have serious consequences for electron scattering, since most electron scattering theory has been done with this approximation. The deuteron electrodisintegration has recently been re-examined by Hwang et al.[7] to test the contribution of the exchange-current contributions to this reaction.

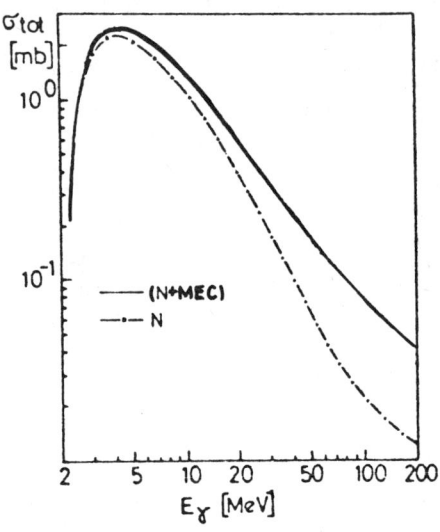

Fig. 1. Total cross section for deuteron photodisintegration. Solid curve: one-body plus two-body (MEC) currents. Dot-dashed curve: one-body currents only (impulse approximation). From Arenhovel, Ref. [3].

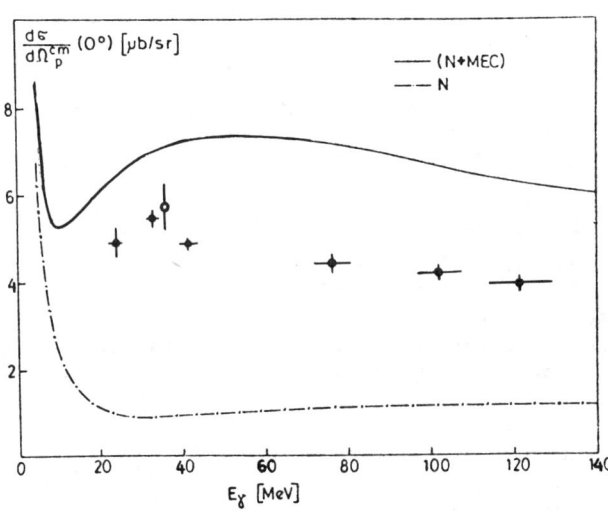

Fig. 2. Forward proton cross section [d(γ,p)n at 0°]. Notation for curves is that of Fig. 1. Data: solid circles: Hughes et al.(Mainz), Ref. [8]. Open circle: neutron radiative capture from Gilot et al. (Louvain), Ref. [17].

2.2 The reaction $\gamma+d \to n+p$ at $0°$.

The forward deuteron photodisintegration has been of great interest ever since the Mainz experiment of Hughes et al.[8] which is shown in Fig.(2). The experimental cross section was about 40% smaller than the theoretical predictions,[5] a rather unexpected result since the Partovi theory fit other experimental results to within a few percent. The theoretical results are sensitive to the D-state component of the deuteron, and much theoretical effort has been undertaken to try to explain this discrepancy[9-15]. The disagreement with experiment occurs in a region dominated by the El transitions, which should be relatively straightforward to calculate. Some work has been done on including the two-body charge density in this amplitude[12-14]; it was stated that this tended to reduce the discrepancy between theory and experiment. However, a recent paper by Jaus and Woolcock[15] claims that the inclusion of $\rho^{[2]}$ actually increases the difference between theory and experiment. These authors also state that there is a significant difference between PS and PV NNπ coupling for this reaction, and that PV coupling gives considerably better agreement with experiment[16]. At present, no theoretical calculation has obtained agreement with the data. A recent measurement by Gilot et al.[17] at Louvain of the neutron radiative capture, shown as the open circle in Fig.(2), appears to agree with the Mainz results.

2.3 Proton polarization in $\gamma+d \to n+p$.

At higher energies, the proton polarization in the reaction $\gamma+d \to n+p$ shows a rather striking structure for photon energy about 500 MeV. This was first seen by Kamae et al.[18,19] and was further studied by Ikeda et al.[20] The $90°$ proton polarization and cross section are shown in Fig.(3) as a function of photon energy. There is no observable structure in the cross section, while the polarization changes rapidly to a value of about -0.8 for photon energy about 500 MeV.

It was suggested[19] that this might be evidence for a dibaryon resonance. The argument was that such structure could be produced by a small resonant term beating against a large smoothly-varying background. Near the energy corresponding to the resonance, the change in phase of the resonant term would then produce structure in the polarization. Such an analysis remains quite speculative, although it is true that theoretical calculations which do not include a dibaryon resonance have had difficulty producing structure in the proton polarization. In Fig.(3), we show two calculations of the proton polarization in deuteron photodisintegration. The first, by Ikeda et al.[20], is based on the covariant formalism of Ogawa et al[21]. The second is a recent calculation by Anastasio and Chemtob [22]. This calculation includes the diagrams of Fig.(4), and includes the two-isobar component of the deuteron in their calculation. Such a calculation is unable to reproduce the observed structure in the polarization [23].

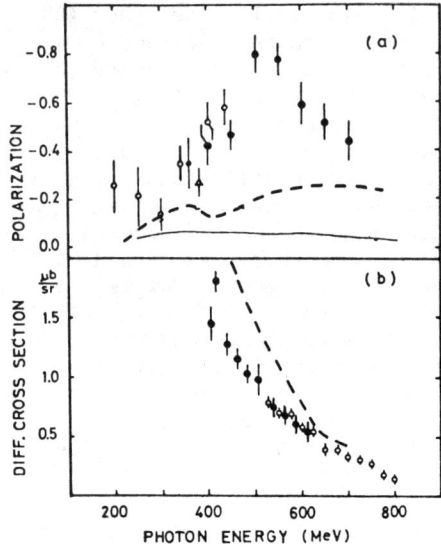

Fig. 3. Proton cross section and polarization from $d(\gamma,p)n$ at 90° c.m. vs. the photon energy. (a) Polarization. Data points are from Ref. [27] (open circles), Ref. [25] (triangle) and Ref. [18] (filled circles). (b) Cross section. Data are from Ref. [26] (open circles) and Ref. [24] (closed circles). Theoretical curves: solid curve, relativistic-covariant formulation of Ref. [21]; dashed curve: calculation of Anastasio and Chemtob, Ref. [22].

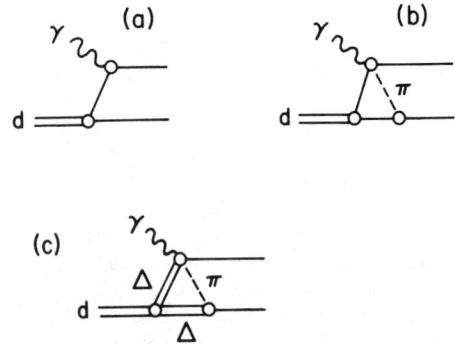

Fig. 4. Diagrams considered by Anastasio and Chemtob for $d(\gamma,p)n$, Ref. [22]. (a) Nucleon-exchange Born amplitude. (b) One-pion reabsorption term. (c) Two-isobar component of the deuteron, with pion reabsorption.

In Fig.(5) we show the angular distribution of P through this energy region[20]. The solid curve is the calculation of Ref.(21),

with nonresonant amplitudes. The dashed curves are the result of
assuming that two dibaryon resonances contribute to the
amplitudes. The spin, isospin, parity, mass and width of these
resonances are treated as free parameters.[20]

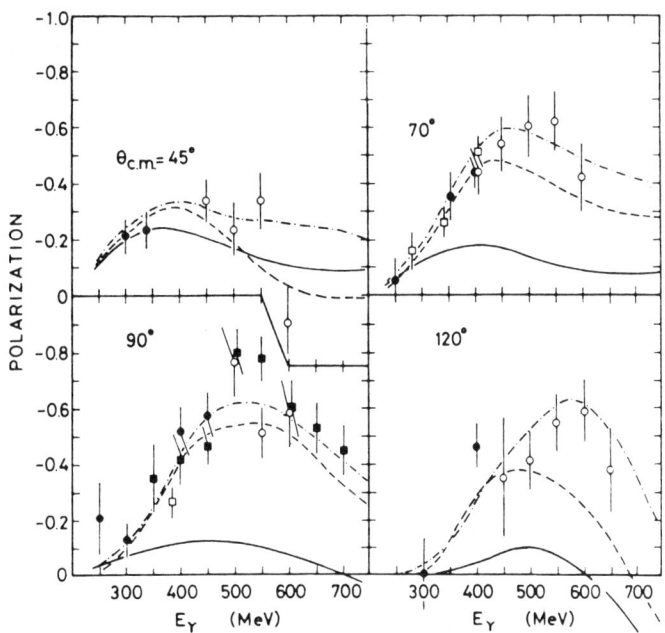

Fig. 5. Angular distribution of proton polarization from d(γ,p)n,
as a function of photon energy E_γ. Data points are from Ref. [20]
(open circles); Ref.[27] (filled circles); Ref. [25] (open squares)
and Ref. [18] (filled squares). Theoretical curves: solid curve,
covariant amplitudes of Ref. [21], with no dibaryon resonances.
Dashed curve: nonresonant amplitudes plus assumed dibaryon resonances with (J^π,T,E,Γ) assumed to be (3^-,1,2260,200) and
(3^+,0,2362,238). Dot-dashed curves: assumed dibaryon resonances
with parameters (3^-,1,2260,200) and (1^+,0,2352,342), from Ref.[20].

2.4 ^{4}He Photodisintegration and Charge Symmetry of the Nuclear Force.

The reactions γ+^{4}He → p+^{3}H and γ+^{4}He → n+^{3}He have been studied
at relatively low energies (photon energies up to 40 MeV)[28-34]. At
these energies, the reactions should be dominated by the isovector
E1 transitions. Thus, if we neglect the effect of the Coulomb
interactions in the final state (including the mass differences

between the proton and neutron final states), and if we assume
charge symmetry for the nuclear forces, then we should see equal
cross sections for p+^{3}H and n+^{3}He.

In Fig.(6), we show the angle-integrated data for the ^{4}He
cross sections to these two final states. The (γ,p) data is that
of Meyerhof et al.[28] The experimental (γ,n) cross sections are
difficult to normalize properly, and the quoted cross sections have
fluctuated considerably. However, the inverse reaction n+^{3}He →
γ+^{4}He has now been measured at TUNL by Ward et al.[33] and is in
very good agreement with the latest (γ,n) result of Berman et al.[34]

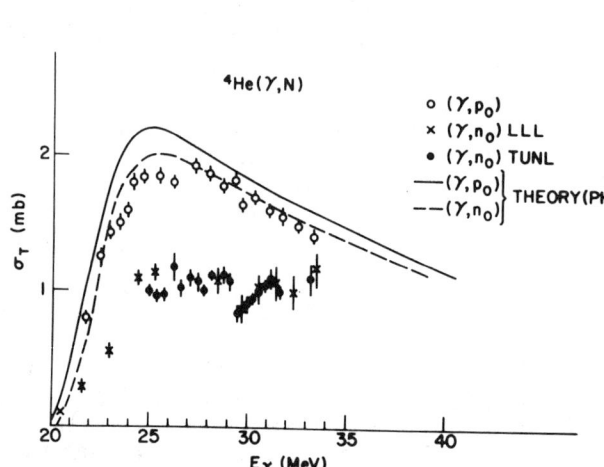

Fig. 6.
Angle-integrated cross sections for two-body photodisintegration of ^{4}He. Data: open circles: ^{4}He(γ,p)^{3}H measurements of Ref. [28]; crosses: ^{4}He(γ,n)^{3}He data from Ref. [34]; solid circles: ^{3}He(n,γ)^{4}He measurements of Ref. [33]. Theoretical curves are from a recoil-corrected continuum shell model calculation of Halderson and Philpott, Ref. [36]. Solid curve: ^{4}He(γ,p)^{3}H. Dashed curve: ^{4}He(γ,n)^{3}He.

The (γ,p) cross sections are about twice as large as the (γ,n)
cross sections for E_γ up to about 34 MeV. The ratio of (γ,p) to

(γ,n) cross sections is shown in Fig. (7); the (γ,p) results of Refs.(28) and (29) are compared with the (n,γ) data from Ref.(33). The question arises as to whether this result implies charge-symmetry violation in the nuclear force (and if so, how large a CSB-breaking force is required to fit this data).

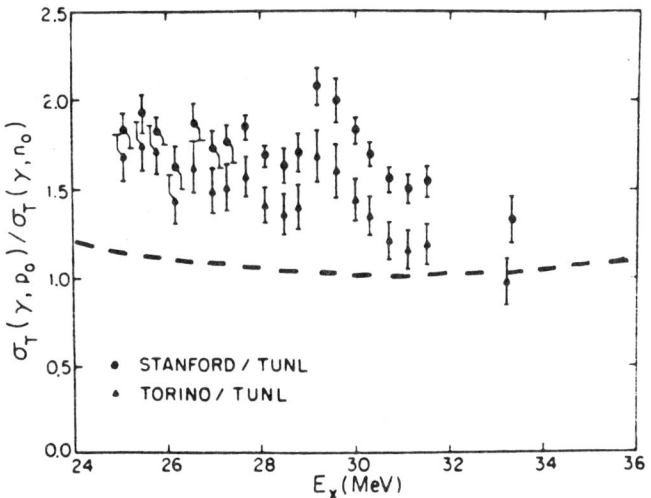

Fig. 7. The (γ,p_o) to (γ,n_o) ratio for angle-integrated photodisintegration of ^{4}He. The (γ,n_o) points are obtained from Ref. [33]. Circles: (γ,p) data from Ref. [28]. Triangles: (γ,p) data of Ref. [29]. Dashed curve: theoretical prediction of Ref. [36].

A precise theoretical estimate of this ratio requires an accurate treatment of all of the final states in the four-body system for this reaction. A preliminary result for a proper four-body treatment has been reported[35]. We show the cross sections and $(\gamma,p)/(\gamma,n)$ ratios as calculated by Halderson and Philpott[36] using a recoil-corrected continuum shell model. This calculation includes only the two-body final states and E1, E2, and M2 multipoles; they correctly include the Coulomb final-state interaction. Their calculation gives a reasonable result for the (γ,p) cross section but significantly overestimates the (γ,n) data. In Fig.(7), we show the ratio of $(\gamma,p)/(\gamma,n)$ as predicted by Halderson-Philpott vs. the Stanford[28] or Torino[29] (γ,p) data and the TUNL (γ,n) data.[33]

Gibson[35] has emphasized the necessity for reasonably complete four-body calculations in order to assess the meaning of the large excess of protons over neutrons in this decay. It would also be useful to have additional experimental data, such as $\gamma+^4\text{He}\rightarrow$ d+d near the threshold for this reaction, in order to provide stronger constraints on theoretical calculations. If this result is evidence for surprisingly large CSB nuclear forces, any theoretical prediction must simultaneously be able to reproduce other nuclear reactions in the A=4 system, such as $p+^3\text{H}\leftrightarrow n+^3\text{He}$, where there is rather accurate data available[37].

2.5 Exclusive (γ,p) Reactions for $E_\gamma < 400$ MeV.

Over the past few years there has been considerable experimental study of exclusive (γ,p), and occasionally (γ,n), reactions for photon energies 50-400 MeV. Data has been taken on nuclear targets in the 1s,1p, and 2s-1d shells[38-62], most extensively at Glasgow and Bates. Most of the data consists of (γ,p_0) reactions leading to the ground state of the residual nucleus.

Much of the original impetus for this work came from the anticipation that such reactions would enable us to "map out" single-particle momentum densities for momenta much larger than the Fermi momentum. A plane-wave impulse approximation (PWIA) for the differential cross section in a (γ,p_0) reaction gives

$$\frac{d\sigma}{d\Omega}(\gamma,p_0) \sim |\phi_{j\ell m}(\vec{q})|^2, \qquad (9)$$

where

$$\vec{q} \equiv \vec{p} - \frac{A-1}{A}\vec{k}_\gamma \qquad (10)$$

In Eq.(9), the ejected proton has quantum numbers (j,l,m); $\vec{k}_\gamma$ is the incident photon momentum, $\vec{p}$ the proton-nucleus c.m. momentum,

and A is the atomic weight of the target. By using energy
conservation, p and k_γ are related by

$$k_\gamma \sim \frac{p^2}{2\mu} + B, \qquad (11)$$

where μ is the proton-recoil nucleus reduced mass and B is the
proton binding energy. Consequently, p will be much larger than
k_γ, so that the momentum q which appears in Eq.(9) will be very
large. If the PWIA arguments were correct, then the (γ,p_0)
reaction could provide a direct measurement of the small and
poorly-known high-momentum components of single-particle wave
functions.

In Fig.(8), we plot the momentum transfer q in (p,γ_0)
reactions vs. the proton lab kinetic energy T_p. The lower curve is
appropriate for 0° cross sections, and the upper curve for 180°
reactions. This particular figure is for $p+{}^{11}B \rightarrow \gamma+{}^{12}C$, but the
curves are almost independent of the target nucleus for any heavier
target. For comparison, Fearing's review article[64] plots the same
quantity for (p,π) reactions. It is clear that the momentum
transfer in the exclusive (γ,p) reaction is large.

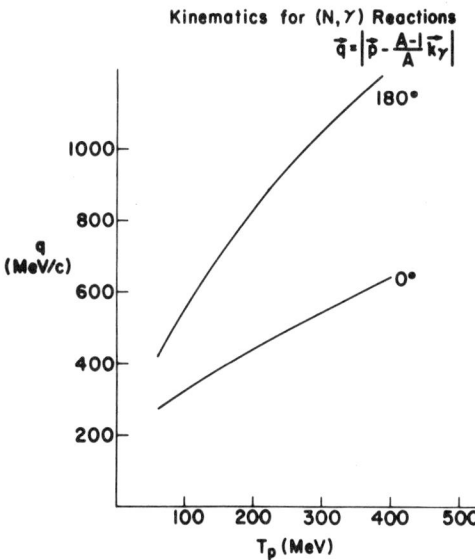

Fig. 8. Momentum
transfers q for exclusive
(p,γ) reactions vs. proton
lab kinetic energy T_p.
Curves are shown for
${}^{11}B(p,\gamma_0){}^{12}C$, but are
almost identical for
(p,γ_0) reactions on all
heavier targets. Lower
curve: 0° scattering.
Upper curve: 180°
scattering.

Analysis of (γ,p) cross sections in terms of single-particle
momentum densities seemed at first to be rather successful.
Corrections were made for proton final-state interactions, and
results seemed to agree with single-nucleon momentum wave functions
obtained from electron scattering.

One such result is shown in Fig.(9); here both (e,e'p)[65] and (γ,p$_o$) data[56] on ^{12}C have been analyzed in terms of a "modified plane wave" model[58] and the inferred momentum density $|\phi|^2$ is plotted, along with the single-particle distribution calculated from an Elton-Swift potential[66], which was obtained from elastic electron scattering.

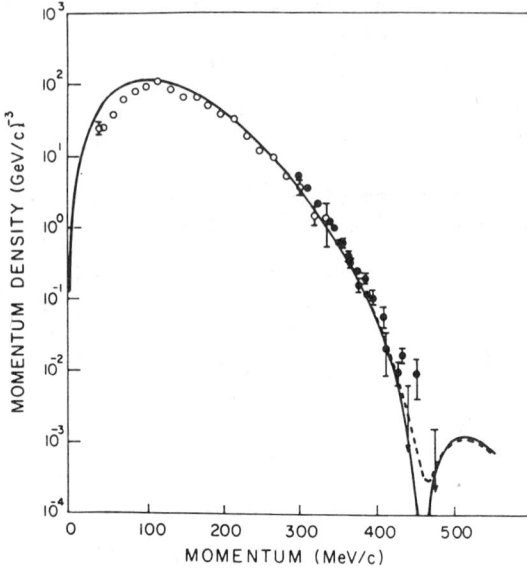

Fig. 9. The 1p-shell momentum distribution for ^{12}C deduced from (e,e'p) data (open circles, Ref. [65], and (γ,p$_o$) data (solid circles, Ref. [56]). The solid curve is the 1p$_{3/2}$ momentum distribution obtained from the potential of Elton and Swift, Ref. [66]; their potential was obtained from fitting electron scattering.

However, when (γ,p) results are compared with (γ,n) results it becomes apparent that these simple arguments are somewhat misleading. A PWIA prediction for both (γ,p) and (γ,n) cross sections gives

$$\frac{d\sigma}{d\Omega}(\gamma,p_o) \sim |\phi_{j\ell m}(\vec{q})|^2 \{\sin^2\theta + \frac{\mu_p^2}{2}\frac{k_\gamma^2}{p^2}\},$$

$$\frac{d\sigma}{d\Omega}(\gamma,n_o) \sim |\phi_{j\ell m}(q)|^2 \; \frac{\mu_n^2 k_\gamma^2}{2 \; p^2}.$$

(12)

In Eqs. (12), θ is the scattering angle between the incident γ and outgoing proton directions, μ_p and μ_n are the proton and neutron magnetic moments, respectively. Since $k_\gamma \ll p$ for these reactions, the PWIA predicts that the (γ,p$_o$) cross section should be much greater than the (γ,n$_o$) cross section (except at extreme forward or backward scattering angles).

In Fig.(10), we show the differential cross sections for $^{16}O(\gamma,p_0)^{15}N$ and $^{16}O(\gamma,n_0)^{15}O$ for $E_\gamma = 82$ MeV[57] and 79 MeV[54], respectively. The (γ,n_0) cross sections are almost as large as the (γ,p_0) cross sections at the peak, and at backward angles the (γ,n_0) cross sections are substantially larger than the (γ,p). The near equality of these two cross sections suggests important contributions from absorption of a photon by a correlated pair of nucleons. These reactions have been analyzed by Gari and Hebach [67] using phenomenological two-body currents, and also by the quasideuteron model[69,70]. The calculations of Ref.[67] give rather good fits to the existing data.

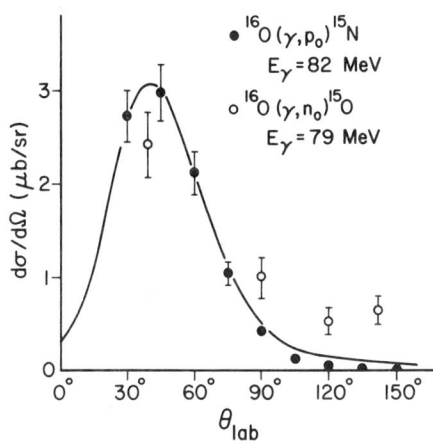

Fig. 10. Differential cross sections for photonuclear reactions on ^{16}O. Solid circles: $^{16}O(\gamma,p_0)^{15}N$ for $E_\gamma = 82$ MeV, data from Ref. [57]. Open circles: $^{16}O(\gamma,n_0)^{15}O$ for $E_\gamma = 79$ MeV, data from Ref. [54]. Solid curve: theoretical calculation of $^{16}O(\gamma,p_0)^{15}N$ from Ref. [67].

For photon energies of 100-400 MeV, we might expect Δ photo-excitation to play an important role in (γ,N) reactions. In Fig.(11) we show an isobar contribution to a (γ,p) reaction. Note that this term is analogous to an important part of the (π,p) reactions. In Fig. (12) we show the excitation functions at three angles for $^{16}O(\gamma,p_0)^{15}N$. The data are from Bates[60,62,63]. The dashed curve is a DWIA direct-reaction calculation by Leitch[62], which fails to fit the observed cross sections for momentum transfers $q \geqslant 450$ MeV/c. The solid curve is a direct-reaction plus isobar calculation of Londergan and Nixon[68] and the dot-dashed curve is by Gari and Hebach[67]. The latter include phenomenological two-body currents; in Ref. [68], the isobar photoexcitation amplitude dominates for $q \geqslant 450$ MeV/c; however, the authors of Ref. [67] state that the Δ current "is not necessary" to reproduce the observed (γ,p_0) cross sections, and they claim that other two-body currents are large for this reaction.

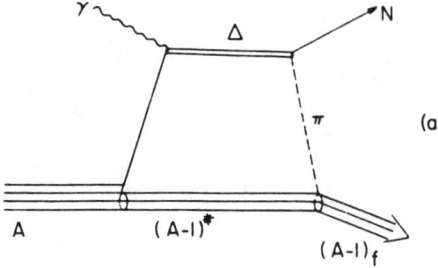

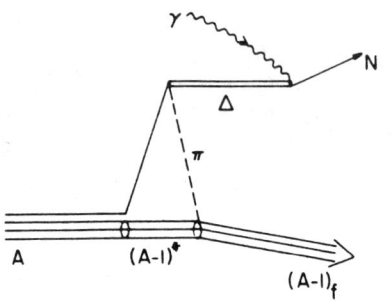

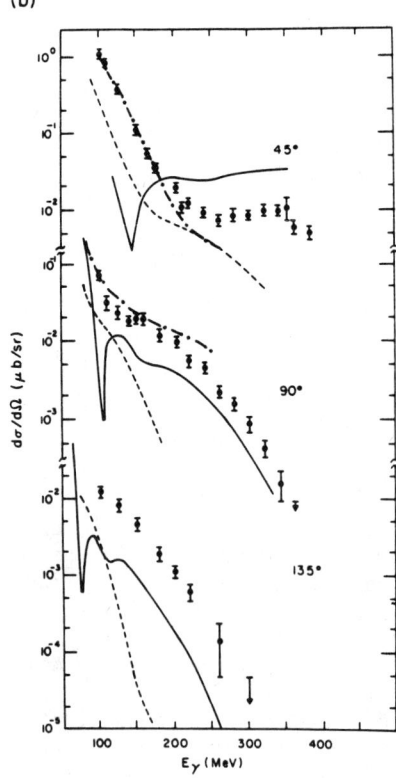

Fig. 11. Schematic representation of isobar contributions to the (γ,N) reaction. (a) Δ creation by nucleon photoexcitation; (b) the isobar is created in the nucleus and converted to a nucleon by absorbing the incident photon.

Fig. 12. Excitation function for $^{16}O(\gamma,p_0)^{15}N$ at $\theta_p = 45°$, $90°$ and $135°$ vs. the photon energy E_γ. The data is from Ref. [60]. Theoretical curves: dashed curve, direct-reaction DWIA calculation of Leitch, Ref. [62]. Solid curve, direct reaction plus isobar calculation of Ref. [68]; dot-dashed curve, phenomenological exchange currents (not including isobar) of Ref. [67].

The situation appears quite different, however, for the reaction $^{40}Ca(\gamma,p_0)^{39}K$. Here, the DWIA calculation of Leitch[62] seems to fit the available data up to q=600 MeV/c; the experimental data[47] and DWIA fit are shown in Fig.(13). For this reaction, there are at present no calculations of the exchange currents or isobar photoexcitation, but the data does not show a failure of the DWIA for any of the existing excitation functions.

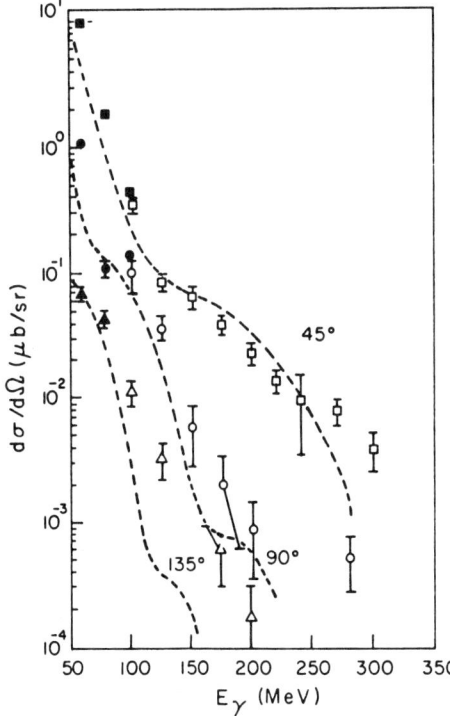

Fig. 13. Excitation function for $^{40}Ca(\gamma,p_0)^{39}K$. Data and direct-reaction DWIA calculation are from Refs. [47,62].

The present situation is rather confusing. DWIA calculations[62] suggest that the direct-reaction component of (γ,p) reactions is reasonably large: this is supported by the approximate 'scaling' of the data as shown in Fig. [9]. However, the size of the (γ,n) cross sections shows that two-body currents must be appreciable, and this is supported by theoretical calculations[67]. Finally, the importance of the Δ contribution to these reactions is not clear. Gari and Hebach[67] obtain good fits to the $^{16}O(\gamma,p_0)^{15}N$ data, but they do not include the Δ term in the large-angle scattering, so it is not clear how large a contribution this would make to their cross sections.

3. (d,p) AND (p,d) REACTIONS AT MEDIUM ENERGIES.

Reactions of the form d+A → p+B have historically served as the "proving grounds" for direct-reaction theory. Such reactions have provided us with spectroscopic factors for a great many nuclear states. The DWBA amplitude for a (d,p) reaction can be written in the form [72]

$$M_{fi}(d,p) = (N_A + 1)\int d\vec{r}_\alpha d\vec{r}_\beta \chi_f^{(-)+}(\vec{k}_\beta,\vec{r}_\beta) F \chi_i^{(+)}(\vec{k}_\alpha,\vec{r}_\alpha). \qquad (13)$$

In Eq.(13), $\vec{r}_\alpha$, $\vec{k}_\alpha$, and $|\chi_i\rangle$ are the d-A relative coordinate, momentum and distorted wave, respectively; r_β, k_β, and $|\chi_f\rangle$ are the p-B relative coordinate, momentum and distorted wave; the form factor F can be written in terms of the internal wave functions for d, A, and B, the proton spinor χ_p, and the n-p interaction v_{np} as

$$F = \langle \phi_B \chi_p | v_{np} | \phi_A \phi_d \rangle \qquad (14)$$

We will briefly review the situation for (p,d) reactions at intermediate energies by showing a few experimental results and the corresponding DWBA fits to this data. In Fig.(14), we show the

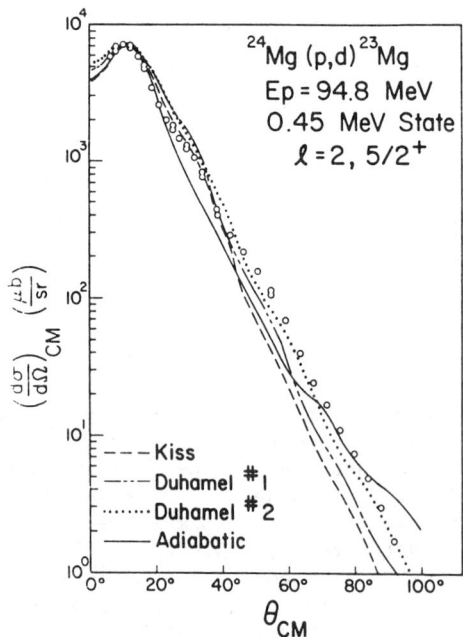

Fig. 14. Differential cross section for ^{24}Mg(p,d)^{23}Mg for E_d = 94 MeV, leading to 0.45 MeV (5/2$^+$) state in ^{23}Mg. Data is from Ref. [73]. Theoretical curves: DWBA calculations obtained using different optical potentials for the deuteron, Ref. [73].

differential cross section for ^{24}Mg(p,d)^{23}Mg, taken at IUCF for E_p= 94 MeV[73]. This cross section is to the 0.45 MeV excited state in ^{23}Mg (a J^π=5/2+, ℓ=2 reaction). The curves represent DWBA fits obtained by varying the distorting potentials for the deuteron. All of these curves give rather good fits to the data, although changes in the optical potentials produce disturbingly large changes in the spectroscopic factor. In Fig.(15) are shown the cross sections and asymmetries for the same reaction leading to the 2.36 MeV state in ^{23}Mg (J^π=1/2+, ℓ=0). Here, the DWBA fails completely to fit either the cross section or the asymmetry. This reaction, and attempts to fit it, are described in detail by Shepard et al.[74]. Qualitative agreement with the data for this case can by achieved by arbitrarily suppressing the form factor at small r; the result of this is shown as the dashed curve in Fig.(15).

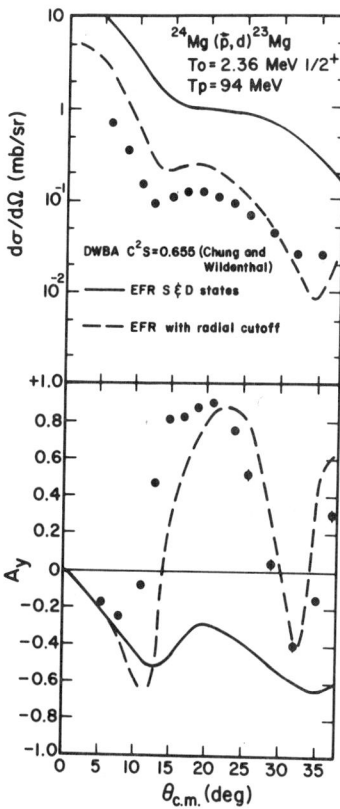

Fig. 15. Differential cross section and polarization for ^{24}Mg(p,d)^{23}Mg leading to 2.36 MeV (J^π=1/2+) state in ^{23}Mg. Data and theoretical curves from Refs. [74]. Theoretical curves: solid curve: full finite range DWBA calculation; Dashed curve: DWBA calculation with form factor arbitrarily suppressed at small r, as described in Ref. [74].

At higher energies, some (p,d) experiments have been carried out at 800 MeV at LAMPF[75,76], and there are some (d,p) measurements at 700 MeV from Saturne[77,78]. The spectra for an 800 MeV (p,d) reaction on ^{12}C are shown in Fig.(16), where they are compared with a (π^+,p) spectrum on the same target[75]. The angles are chosen so that the momentum transfer is the same in each reaction, i.e. q=2.8 fm^{-1}. The spectra look quite similar, in terms of the states which are strongly excited in each reaction.

There are two general comments which can be made regarding the (p,d) reactions at 800 MeV: (1) there are some "new" states excited at these energies which are not seen in lower-energy (p,d) studies. For example, in the ^{12}C(p,d)^{11}C reaction shown in Fig.(16), the very strongly excited group of states at E_x= 13.2 MeV (these states are also strongly seen in the (π^+,p) spectrum on the same target) is not seen at all at lower energies; (2) for transitions which are well

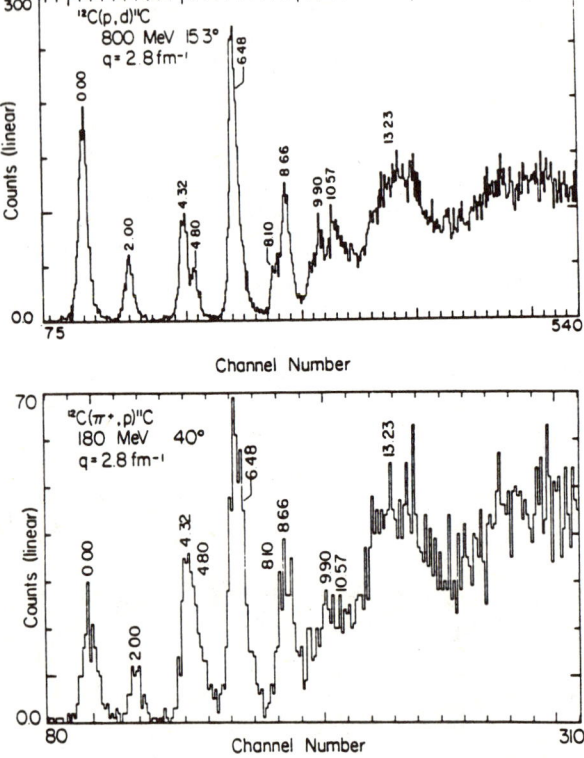

Fig. 16. Spectra for (p,d) and (π^+,p) reactions on ^{12}C at comparable momentum transfers. Data is from Ref. [75]. Upper graph: ^{12}C(p,d)^{11}C for E_p = 800 MeV. Lower graph: ^{12}C$(\pi^+,p)^{11}$C at E_π = 170 MeV.

described by the DWBA at lower energies, the DWBA seems to give a reasonable description also at 800 MeV.

To illustrate this second point, we show the cross section for $^{13}C(p,d)^{12}C$ in Fig.(17). Two exact finite range DWBA calculations[75] are also shown with the data. Both potentials give at least qualitative fits to the data. In several other (p,d) reactions at LAMPF, similar fits were obtained to the cross sections.

The Saclay group [77,78], on the other hand, conclude that the DWBA fails to reproduce their data for $^{16}O(d,p)^{17}O$. In Fig.(18), we show their (d,p) data on ^{16}O leading to the ground state and to a ($1/2^+$) excited state at Ex=0.87 MeV. The solid and dashed lines represent two different DWBA amplitudes (and have been divided by 7). The dot-dashed curve is a pion-rescattering contribution, with π-induced Δ excitation.

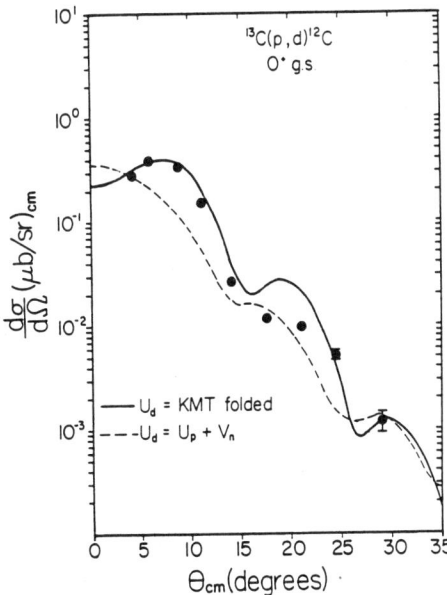

Fig. 17. The reaction $^{13}C(p,d)^{12}C_{g.s.}$ for E_p = 800 MeV. Data and calculations are from Ref. [75]. Solid and dashed curves represent DWBA fits to this reaction with two different choices of the deuteron optical potential.

Figure 18. Differential cross section for $^{16}O(d,p)^{17}O$, for E_d = 698 MeV, from Ref. [77]. Cross sections are shown to the $^{17}O_{g.s.}$ and to a $(1/2^+)$ state with E_x = 0.87 MeV. Theoretical curves: solid and dashed curves are DWBA calculations with two different deuteron optical potentials (normalized by factor 1/7); dot-dashed curve is a calculation assuming a pion rescattering reaction mechanism.

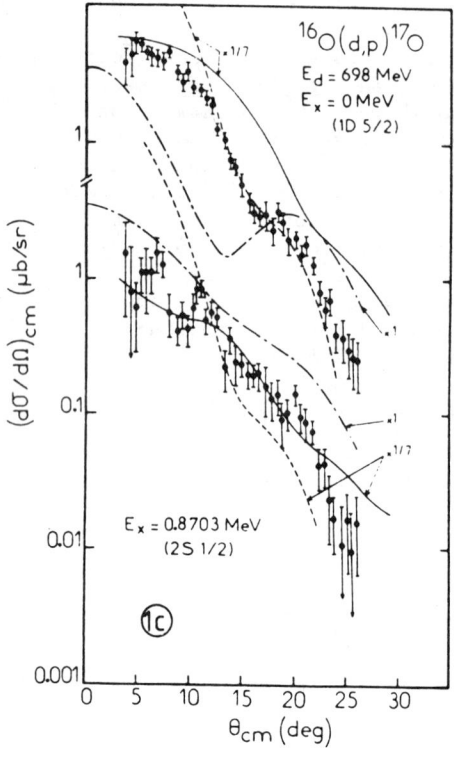

Wilkin[79] has suggested that the (d,p) reaction in this energy region, and at higher energies, might be dominated by deuteron dissociation, followed by a (p,π^+) reaction, with the π^+ being absorbed by the neutron. With some additional simplifying assumptions, Wilkin then relates the (d,p) cross section to the corresponding (p,π^+) cross section. Comparative spectra for the (p,π^+) and (d,p) reactions give qualitative support to this hypothesis. In Fig.(19), we show (d,p) and (p,π^+) cross sections on ^{6}Li taken at Saclay[77,78], at energies 698 MeV and 600 MeV, respectively. The cross sections leading to three different states in ^{7}Li, plotted against the momentum transfer, look quite similar.

Fig. 19. Comparison of ^{6}Li(d,p)^{7}Li and ^{6}Li(p,π^+)^{7}Li reactions for E_d = 698 MeV and E_p = 600 MeV. Cross sections vs. momentum transfer are shown for three different final states in ^{7}Li. Data are from Refs. [77,78].

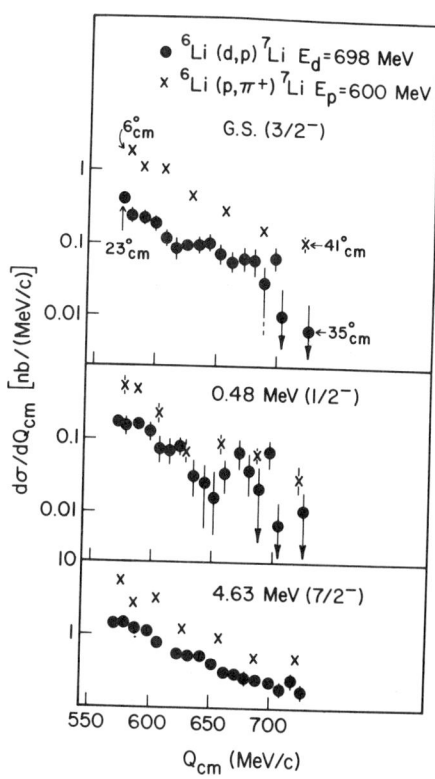

We have mentioned that those (p,d) reactions which at lower energies are well represented as "single-particle" transitions may also be well reproduced at 800 MeV by the DWBA. However, if one attempts to fit the corresponding (π^+,p) cross sections by direct-reaction theory, the results are extremely poor. In Fig.(20), we show the cross sections and attempted fits for ^{13}C(π^+,p_0)^{12}C. The theoretical calculations by Smith[75] using a Kisslinger-type optical potential completely fail to reproduce the experimental cross section. It is possible that this failure simply reflects the lack of knowledge of the pion wave function; however, at the present time a direct-reaction treatment seems to fit the (p,d) reaction well (as shown in Fig.(17)), but fails to describe the (π^+,p) reaction leading to the same final state.

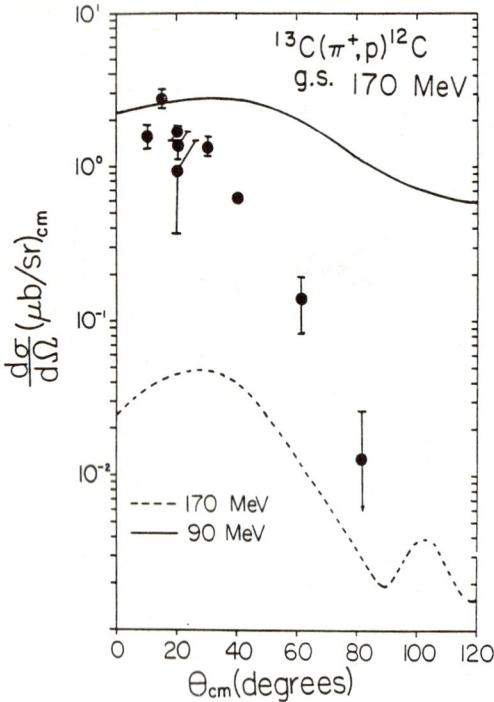

Fig. 20. Differential cross section for $^{13}C(\pi^+,p)^{12}C_{g.s.}$. Data and calculations are from Ref. [75]. Solid and dashed curves represent DWBA calculations using a Kisslinger-type optical potential for the pion.

To summarize the (p,d) and (d,p) results: (i) the small ℓ-transfer amplitudes in medium-energy (p,d) reactions are difficult to calculate, and all theoretical estimates of corrections to the DWBA seem to be too small to obtain agreement with experiment. (ii) (p,d) and (π^+,p) reactions at about 800 MeV have many qualitative features in common. They seem to preferentially excite similar groups of states. (iii) There is considerable controversy regarding the (p,d) and (d,p) reactions at 700-800 MeV. LAMPF results[75] suggest that a DWBA treatment can successfully fit many (p,d) reactions at 800 MeV; however, Saclay[77] studies of 700 MeV (d,p) reactions claim that the DWBA is incapable of describing these processes.

4. DIRECT COMPARISON OF THE (π,p) WITH (p,d) OR (γ,N) REACTIONS

There are relatively few systematic comparisons of (π,p) with (p,d) or (γ,N) exclusive reactions. At present, it is difficult to make quantitative statements or comparisons, at least in part because of the absence of any theoretical models for the (π,p) reaction which are successful in confronting a large body of data. Perhaps with the evolution of the "two-nucleon mechanism" pion production or absorption codes, we will be able to make such comparisons more directly. This should be particularly true for comparing (π,p) reactions with (γ,N) reactions in the energy region where isobar production is assumed to dominate each reaction.

We will concentrate here on just two cases where we can compare existing data. The first is a comparison of (p,π^+) and (d,p) reactions by Jacobs et al[80]. They measured $^{28}Si(p,\pi^+)^{29}Si$ for E_p= 191 MeV, and $^{28}Si(d,p)^{29}Si$ for E_d= 76 MeV. They looked at transitions leading to pairs of final states with identical spin and parity but different spectroscopic strengths. Consequently, they looked at reactions where the momentum and angular momentum transfers were both kept constant. They could then compare the excitation of these pairs of states with the (p,π^+) and (d,p) reactions.

The final states which were examined were two 3/2+ states in ^{29}Si, with E_{exc}= 1.27 MeV and 2.43 MeV; and two 5/2+ states in ^{29}Si with E_{exc}= 2.03 MeV and 3.07 MeV. The data is shown in Fig.(21), together with a DWBA fit to the (d,p) cross sections. When the (d,p) cross sections are normalized by their relative spectroscopic factors, the pairs of transitions look very similar to one another. Also, they are fit rather well by a direct-reaction calculation. The (p,π^+) cross sections, on the other hand, look very different. The transitions to the two 5/2+ states are only qualitatively similar, while the two 3/2+ transitions look totally different. As the authors remark[80], any one-nucleon mechanism for the (p,π^+) reaction would have great difficulty in reproducing the two 3/2+ transitions, and such reactions should prove a good test for any (p,π) production mechanism.

Next, we look at single-nucleon knockout reactions in ^{16}O leading to final states in ^{15}O or ^{15}N. The ground state of ^{15}O is $(1/2^-)$, and to a good first approximation can be represented as a 1p1/2 hole relative to the ^{16}O ground state. Similarly, a $(3/2^-)$ excited state of 15O at E_{exc}= 6.18 MeV is well represented as a 1p3/2 hole relative to ^{16}O. There are several one-nucleon knockout reactions leading to both the ground state and 6.18 MeV state in

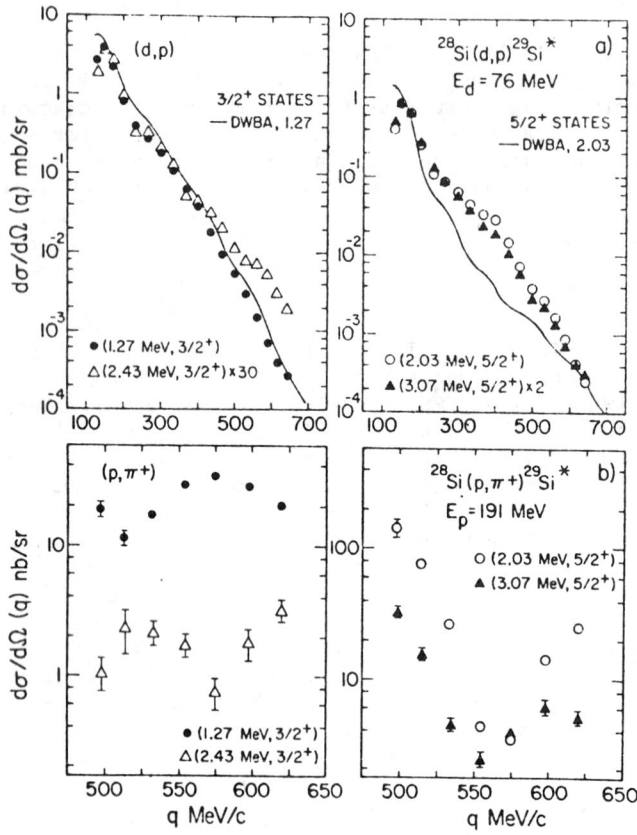

Fig. 21. Angular distributions for ^{28}Si(d,p)^{29}Si* and ^{28}Si(p,π^+)^{29}Si* to $J^\pi = 3/2^+$ (E_x = 1.27 and 2.43 MeV) and $J^\pi = 5/2^+$ (E_x = 2.03 and 3.07 MeV) states plotted vs. momentum transfer q, from Ref. [80]. For the (d,p) distributions, data for 2.43 MeV $3/2^+$ and 3.07 MeV $5/2^+$ states are multiplied by 30 and 2, respectively. The solid curves are DWBA calculations for the (d,p) reaction.

^{15}O (or to the ground state and 6.32 MeV ($3/2^-$) state in ^{15}N). These reactions are:

$^{16}O(\gamma,n)^{15}O$	Mainz E_γ= 60-180 MeV	Schoch and Goringer[71]
$^{16}O(\pi^+,p)^{15}O$	Saclay E_π= 66 MeV	Bachelier et al.[81]
$^{16}O(p,d)^{15}O$	Saclay E_p= 155 MeV	Bachelier et al.[82]
	LAMPF E_p= 800 MeV	Smith[75]
$^{16}O(^3He,\alpha)^{15}O$	Orsay E_τ= 216 MeV	Gerlic et al.[83]

We have plotted the ratio of the cross section to the ($3/2^-$) state, divided by the cross section leading to the ground state of the residual nucleus, for each of these reactions; the abscissa is the momentum transfer in each reaction; e.g., for the (π^+,p) reactions, the ratio R is given by

$$R = \frac{\frac{d\sigma}{d\Omega}(^{16}O(\pi^+,p)^{15}O^*(3/2^-, E_x=6.18 \text{ MeV}))}{\frac{d\sigma}{d\Omega}(^{16}O(\pi^+,p)^{15}O_{g.s.})} . \qquad (15)$$

The results are shown in Fig.(22). The errors are rather large; the (γ,n) data are preliminary, and the ($3/2^-$) cross sections for those reactions are subject to estimates of the cross sections leading to other excited states in this region (which cannot be resolved). However, the following points seem to emerge from this comparison:

(1) for the (p,d) reaction, $R \simeq 2$ or 3 over the entire q region;
(2) for both the (π^+,p) and (γ,n) reactions, R appears to get quite large for $q \simeq 550$ MeV/c; $R \simeq 7$-15 in this region.

If the reaction mechanism was direct knockout of a single nucleon, and if the p3/2 and p1/2 single-particle wavefunctions were identical, then we would expect R=2 (the ratio of the number of particles in the p3/2 and p1/2 shells). This appears to be true for the (p,d) results.

Preliminary data is available for $^{16}O(\gamma,p)^{15}N$, leading to the ground state and ($3/2^-, E_x$=6.32 MeV) states in ^{15}N [62,63]. The (γ,p) ratios look very similar to the (γ,n) and the (π^+,p) results. Schoch and Goringer[71] argue that the similarity of the (γ,p) and (γ,n) ratios to each other, and to the (π^+,p) ratios on ^{16}O, supports the idea that all of these reactions proceed via absorption on a pair of correlated nucleons (since it is assumed that the (π,p) reaction proceeds via the "two-nucleon mechanism").

We can use this argument to provide some historical perspective on our views of the reaction mechanism for both the (γ,N) and (π,N) processes. About ten years ago, it was proposed to make a systematic survey of (p,γ) reactions, in order to shed light on the form of the transition operator in (p,π). At that time, it seemed reasonable to assume that both (p,γ) and (p,π) would proceed via a direct, or "one-nucleon" mechanism at intermediate energies, and that the (p,γ) reaction was "well-known", in which case comparison of the two reactions would provide useful information regarding the (p,π) process. Now, it is argued that both reactions proceed via a "two-nucleon" mechanism, and in fact the (π,p) data is used to draw inferences about the (γ,N) reaction.

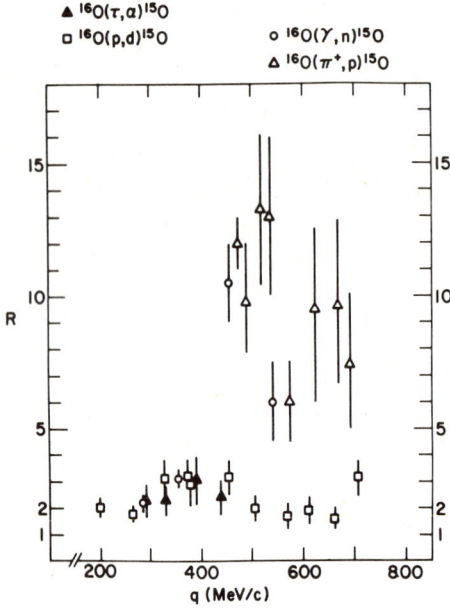

Fig. 22. Ratios of cross sections, plotted vs. momentum transfer q, leading to the ground state of ^{15}O or to the 6.18 MeV ($3/2^-$) state in ^{15}O. For example, for the (π^+,p) reaction;

$$R = \frac{\frac{d\sigma}{d\Omega}(^{16}O(\pi^+,p)^{15}O^*(J^\pi = 3/2^-, 6.18 \text{ MeV}))}{\frac{d\sigma}{d\Omega}(^{16}O(\pi^+,p)^{15}O_{g.s.})}$$

solid triangles: $^{16}O(\tau,\alpha)^{15}O$ data of Ref. [83]; open triangles: $^{16}O(\pi^+,p)^{15}O$ data of Ref. [80]; open squares: $^{16}O(p,d)^{15}O$ data of Refs. [75,82]; open circles: $^{16}O(\gamma,n)^{15}O$, data of Refs. [54,71].

Returning to the quantity R of Fig.(22), G. Miller[81,84] is able to reproduce the observed value of R for the (π,p) reactions on ^{16}O; he suggests that angular momentum matching considerations strongly favor the transition to the (3/2$^-$) state in ^{15}O, at these energies. It remains to be seen whether such an explanation would also work for the (γ,N) transitions.

SUMMARY

We have tried to present a brief overview of some topics in exclusive one-nucleon knockout or pickup reactions which could be related to (p,π) processes. For light systems, we concentrated on those aspects which suggest new phenomena or a discrepancy between theory and experiment. We mentioned the 0° deuteron photodisintegration and the failure to obtain agreement between theory and experiment; the structure in the proton polarization resulting from γ+d$\rightarrow$ n+p and the possibility of a dibaryon resonance in the N-N system; and the excess of (γ,p_0) over (γ,n_0) cross section in the ^{4}He photodisintegration.

At medium energies, exclusive (γ,N) processes share almost all of the reaction mechanism uncertainties as those in (π,N). Since both reactions are inherently high-momentum-transfer processes, comparison of (π,N) reactions with the same (γ,N) or (p,d) transitions may be quite promising. We showed a couple of examples which revealed some interesting differences in these transitions; however, much more systematic comparisons are necessary before any quantitative results can be obtained.

We express our thanks to our colleagues at Indiana University for their comments and suggestions regarding this work. Also, we thank the following people for providing us with recent data and for useful discussions about these reactions: P. Couvert, M. Dillig, D. Halderson, L. Ludeking, J. Matthews, D. Miller, G. Miller, B. Schoch, J. Shepard, and E. Stephenson.

This work was supported in part by the National Science Foundation.

REFERENCES

1. J.M. Laget, Nucl. Phys. A312, 265 (1978).
2. W-Y. P. Hwang and G.A. Miller, Phys. Rev. C22, 968 (1980).
3. H. Arenhovel, Z. Phys. A302, 25 (1981).
4. See e.g.: Mesons in Nuclei, edited by M. Rho and D. Wilkinson, Vol. II (North-Holland, Amsterdam, 1979).
5. F. Partovi, Ann. Phys. 27, 79 (1964).
6. A.J.F. Siegert, Phys. Rev. 52, 787 (1937).
7. W-Y. P. Hwang, E.M. Henley and G.A. Miller, Ann. Phys. (N.Y.), to be published.

8. R.J. Hughes, A. Zieger, H. Waffler, and B. Ziegler, Nucl. Phys. A267, 329 (1976).
9. E.L. Lomon, Phys. Lett. 68B, 419 (1977).
10. M.L. Rustgi, T.S. Sandhu, and O.P. Rustgi, Phys. Lett. 70B, 145 (1977).
11. H. Arenhovel and W. Fabian, Nucl. Phys. A282, 397 (1977).
12. E. Hadjimichael, Phys. Lett. 85B, 27 (1979); E. Hadjimichael and D.P. Saylor, Phys. Rev. Lett. 45, 1776 (1980).
13. For more references dealing with this problem, seen the review article by H. Baier, Fortschr. Phys. 27, 209 (1979).
14. M. Gari and B. Sommer, Phys. Rev. Lett. 41, 22 (1978).
15. W. Jaus and W.S. Woolcock, Nucl. Phys. A365, 477 (1981).
16. See, however, the contribution by M. Banerjee and G.E. Walker to this Workshop.
17. J.F. Gilot, A. Bol, P. Leleux, P. Lipnik, and P. Macq, Phys. Rev. Lett. 47, 304 (1981).
18. T. Kamae, I. Arai, T. Fujii, H. Ikeda, N. Kajiura, S. Kawabata, K. Nakamura, K. Ogawa, H. Takeda and Y. Watase, Phys. Rev. Lett. 38, 468 (1977); Nucl. Phys. B139, 394 (1978).
19. K. Kamae and T. Fujita, Phys. Rev. Lett. 38, 471 (1977).
20. H. Ikeda, I. Arai, H. Fujii, H. Iwasaki, N. Kajiura, T. Kamae, K. Nakamura, T. Sumiyoshi, H. Takeda, K. Ogawa, and M. Kanazawa, Phys. Rev. Lett. 42, 1321 (1979).
21. K. Ogawa, T. Kamae and K. Nakamura, Nucl. Phys. A340, 451 (1980).
22. M. Anastasio and M. Chemtob, Nucl. Phys. A364, 219 (1981).
23. See the contribution to this Workshop by C.Y. Cheung and L.S. Kisslinger who examine this problem with a 6-quark bag model.
24. P. Dougan, T. Kivikas, K. Lunger, V. Ramsay, and W. Stiefler, Z. Phys. A276, 55 (1976).
25. R. Kose, W. Paul, K. Stockhorst, and K.H. Kissler, Z. Phys. 202, 364(1967).
26. R. Ching and C. Schaerf, Phys. Rev. 141, 1320 (1966).
27. F.F. Liu, D.E. Lundquist and B.H. Wiik, Phys. Rev. 165, 1478 (1968).
28. W.E. Meyerhof, M. Suffert and W. Feldman, Nucl. Phys. A148, 211 (1970).
29. F. Balestra, E. Bollini, L. Busso, R. Garfagnini, C. Guaraldo, G. Piragino, R. Scrimaglio and A. Zanini, Nuovo Cim. 38A, 145 (1977).
30. W.R. Dodge and J.J. Murphy II, Phys. Rev. Lett. 28, 839 (1972).
31. A.N. Gorbunov, Phys. Lett. 27B, 436 (1968); JETP Lett, 8, 88 (1968).
32. J.D. Irish, R.G. Johnson, B.L. Berman, B.J. Thomas, K.G. McNeill and J.W. Jury, Can J. Phys. 53, 802 (1975).
33. L. Ward, D.R. Tilley, D.M. Skopik, N.R. Roberson and H.R. Weller, Phys. Rev. C24, 317 (1981).
34. B.L. Berman, D.D. Faul, P. Meyer and D.L. Olson, Phys. Rev. C22, 2273 (1980).

35. A. Casel and W. Sandhas, contribution to International Few-Body Conference, Eugene Oregon, 1980; for a description of these results see the review article by B.F. Gibson, Nucl. Phys. A353, 85 (1981).
36. D. Halderson and R.J. Philpott, Nucl. Phys. A359, 365 (1981); Phys. Rev. Lett. 44, 56 (1980); Phys. Rev. Lett. 42, 36 (1979).
37. For references see the review article by R.J. Philpott, Proc. of 5th Int. Symposium on Polarization Phenomena in Nuclear Physics (Santa Fe, 1980; AIP Conf. Proceedings No. 69), p. 1144.
38. P. Picozza, C. Schaerf, R. Scrimaglio, G. Goggi, A. Piazzoli, and D. Scannicchio, Nucl. Phys. A157, 190 (1970).
39. J.P. Didelez, H. Langevin-Joliot, Z. Maric, and V. Radojevic, Nucl. Phys. A143, 602 (1970).
40. A. van der Woude, M.L. Halbert, C.R. Bingham, and B.D. Belt, Phys. Rev. Lett. 26, 909 (1971).
41. N.M. O'Fallon, L.J. Koester, Jr., and J.H. Smith, Phys. Rev. C5, 1926 (1972).
42. G. Ticcioni, S.M. Gardiner, J.L. Matthews, and R.O. Owens, Phys. Lett. 46B, 369 (1973).
43. P.E. Argan, G. Audit, N. De Botton, J.L. Faure, J.-M. Laget, J. Martin, C.G. Schuhl, and G. Tamas, Nucl. Phys. A237, 447 (1975).
44. C.A. Heusch, R.V. Kline, K.T. McDonald, and C.Y. Prescott, Phys. Rev. Lett. 37, 405 (1976).
45. C.A. Heusch, R.V. Kline, K.T. McDonald, J.B. Carroll, D.H. Frederickson, M. Goitein, B. Macdonald, V. Perez-Mendez, and A.W. Stetz, Phys. Rev. Lett. 37, 409 (1976).
46. For a review of photonuclear reactions above the giant dipole resonance region, see G. Ricco, Photonuclear Reactions, edited by S. Costa and C. Schaerf, Lecture Notes in Physics, Vol. 61 (Springer, New York, 1977) p. 223.
47. For a review of (γ,p) and (γ,n) reactions at intermediate energies, see J. Matthews, Nuclear Physics with Electromagnetic Interactions, edited by H. Arenhovel and D. Drechsel (Springer, New York, 1979), p. 369.
48. J.L. Matthews, W. Bertozzi, S. Kowalski, C.P. Sargent, and W. Turchinetz, Nucl. Phys. A112, 654 (1968).
49. D.J.S. Findlay, thesis, University of Glasgow, 1975 (unpublished).
50. S.N. Gardiner, J.L. Matthews, and R.O. Owens, Phys. Lett. 46B, 186 (1973).
51. M. Sanzone, G. Ricco, S. Costa, and L. Ferrero, Nucl. Phys. A153, 225 (1969).
52. G. Manuzio, G. Ricco, M. Sanzone, and L. Ferrero, Nucl. Phys. A133, 225 (1969).
53. H.G. Miller, W. Buss, and J.A. Rawlins, Nucl. Phys. A163, 637 (1971).

54. H. Schier and B. Schoch, Nucl. Phys. A229, 93 (1974).
55. E. Mancini, G. Ricco, M. Sanzone, S. Costa, and L. Ferrero, Nuovo Cimento 15A, 705 (1973).
56. J.L. Matthews, D.J.S. Findlay, S.N. Gardiner, and R.O. Owens, Nucl. Phys. A267, 51 (1976).
57. D.J.S. Findlay and R.O. Owens, Nucl Phys. A279, 385 (1977).
58. D.J.S. Findlay and R.O. Owens, Nucl. Phys. A292, 53 (1977).
59. D.J.S. Findlay and R.O. Owens, Phys. Rev. Lett. 37, 674 (1976).
60. J.L. Matthews, W. Bertozzi, M.J. Leitch, C.A. Peridier, B.L. Roberts, C.P. Sargent, W. Turchinetz, D.J.S. Findlay, and R.O. Owens, Phys. Rev. Lett. 38, 8 (1977).
61. D.J.S. Findlay, R.O. Owens, M.J. Leitch, J.L. Matthews, C.A. Peridier, B.L. Roberts, and C.P. Sargent, Phys. Lett. 74B, 305 (1978).
62. M.J. Leitch, Ph.D. Thesis, MIT (1979, unpublished).
63. J. Matthews, private communication.
64. H.W. Fearing, in Progress in Particle and Nuclear Physics, vol. 7, ed. D.H. Wilkinson (Pergamon Press, Oxford, 1981), p. 113.
65. J. Mougey, M. Bernheim, A. Bussiere, A. Gillebert, Phan Xuan Ho, M. Priou, D. Roger, I. Sick, and G.J. Wagner, Nucl. Phys. A262, 461 (1976).
66. L.R.B. Elton and A. Swift, Nucl. Phys. A94, 52 (1967).
67. M. Gari and H. Hebach, Phys. Rep. 72, 1 (1981); H. Hebach, A. Wortberg and M. Gari, Nucl. Phys. A267, 425 (1976).
68. J.T. Londergan and G.D. Nixon, Phys. Rev. C19, 998 (1979); J.T. Londergan, G.D. Nixon and G.E. Walker, Phys. Lett. 65B, 427 (1976).
69. J.S. Levinger, Proc. Int. Conf. on Low and Intermediate Energy Nuclear Reactions (Acad. of Science, USSR, Moscow, 1967) Vol. 3, p. 411; Phys. Lett. 82B, 181 (1979).
70. B. Schoch, Phys. Rev. Lett. 41, 80 (1978).
71. B. Schoch and H. Goringer, contributed paper to Int. Conf. on Nuclear Physics, Berkeley.
72. N. Austern, "Direct Nuclear Reaction Theories," (Wiley-Interscience, New York, 1970).
73. D.W. Miller, W.P. Jones, D.W. Devins, R.E. Marrs and J. Kehayias, Phys. Rev. C20, 2008 (1979).
74. J.R. Shepard, E. Rost, and P.D. Kunz, Proc. Fifth Int. Symposium on Polarization Phenomena in Nuclear Physics; edited by G.G. Ohlsen, R.E.Brown, N. Jarmie, W.W. McNaughton and G.M. Hale (Am. Inst. of Phys. Conf. Proc. No. 69, Vol I), p. 361.
75. G.R. Smith, thesis (Los Alamos LA-8166-T, 1980, unpublished).
76. J. Shepard, private communication.
77. A. Boudard, Y. Terrien, R. Beurtey, L. Bimbot, G. Bruge, A. Chaumeaux, P. Couvert, J.M. Fontaine, M. Garcon, Y. Le Bornec, D. Legrand, L. Schecter, J.P. Tabet and M. Dillig, Phys. Rev. Lett. 46, 218 (1981).
78. P. Couvert, contribution to this Workshop, and private communication.

79. C. Wilkin, J. Phys. G6, 69 (1980).
80. W.W. Jacobs, A.G. Drentje, P.H. Pile, P.P. Singh, T.P. Sjoreen, and S.E. Vigdor, Phys. Lett. 94B, 319 (1980).
81. D. Bachelier, J.L. Boyard, T. Hennino, J.C. Jourdain, P. Radvanyi, and M. Roy-Stephan, Phys. Rev. C15, 2139 (1977).
82. D. Bachelier, M. Bernas, I. Brissaud, C. Detraz, and P. Radvanyi, Nucl. Phys. A126, 60 (1969).
83. E. Gerlic, J. Van de Wiele, H. Langevin-Joliot, J.P. Didelez, G. Duhamel and E. Rost, Phys. Lett. 52B, 39 (1974).
84. D.F. Measday and G.A. Miller, Ann. Rev. Nucl. Part. Sci. 29, 121 (1979).

DISCUSSION

Wharton (Carnegie-Mellon Univ.): I just wanted to mention that the ^{16}O to ^{15}O (g.s., $1/2^-$) transition is a very special transition. If you compare it to all other transitions of large single-particle strength in the 1p shell, its (π^+,p) cross section is found to be much weaker than all the others in that momentum region, which explains why the ratio went way up. You look at the other single-particle transitions, and that's not happening. It would be interesting to explain why that one transition behaves in such an unusual way compared to the others. [See Phys. Rev. C23, 1141 (1981)].

D. Koltun (Univ. of Rochester): Since you brought up the question of the isospin zero dibaryon and there is some interest in that, I just wanted to pose the following puzzle. One of several reasons, but a strong reason, that we have some doubts about the existence of the T=1 dibaryon, is because of the very careful and complete three-body work that has been done theoretically on those things. Do you think that it's fair to ask that one wait till one sees the comparable level of theoretical work before one decides whether there is a T = 0 dibaryon? Or has it been done? I believe it hasn't.

Londergan: First of all, I think the answer is that it hasn't. Secondly, I hope you noticed that what I presented were essentially negative inferences. I have merely shown that I know of no calculation of the relatively hard $\gamma d \to np$ cross section and polarization, which reproduces that polarization structure. It is an immense leap from that statement to say that the polarization in deuteron photodisintegration strongly confirms the existence of a dibaryon resonance. I agree with you, we should treat the appearance and disappearance of the T = 1 dibaryons as a caveat in making any strong statements about dibaryons from this experimental result.

G.E. Walker (Indiana Univ.): Do you think it would be possible, or is at all likely, that one could someday understand large energy transfer (p,γ) and (γ,p), and perhaps at the same time (γ,π), and not understand (p,π)?

Londergan: In the (γ,π) reaction, you see the dominance of isobar production rather clearly. At the appropriate energies, you will see this also in inclusive (p,π) and (p,γ) processes. For exclusive processes, the amplitudes will be considerably more complicated. There are many similarities between (p,π) reactions and (p,γ) reactions at medium energies. As a general rule, wherever you could create a pion at a given vertex, you can usually replace that with a photon. There are a few differences between pion production and photon production:
1) Since the photon interacts more weakly, photon multiple scattering can be ignored (in contrast to pion multiple scattering).
2) the photon is not an important part of the nuclear force, so some problems of double-counting and 'true absorption' which arise in pion physics can be ignored with photons.
3) with (p,γ) reactions one must insure gauge invariant amplitudes.

The first two of these differences make (p,γ) reactions easier to formulate than (p,π) reactions; the third is usually a difficulty in the (p,γ) reaction relative to (p,π).

HIGH RESOLUTION STUDY OF (π^+,pp), (π^+,pd) AND
OTHER (π^+,xx) REACTIONS ON ^{6,7}Li, ^{14}N, AND ^{16}O

W. R. Wharton
Carnegie-Mellon University, Pittsburgh, PA 15213

INTRODUCTION

We have a large amount of high quality two-particle coincidence data from pion annihilation on light nuclei at several energies between 38 and 90 MeV. All of this data is either in the reduction stage or is still the primitive on-line data collected during the monitoring of our recent experiment. The purpose of this talk is twofold: 1) to convey the measuring capability of a coincidence experiment of this type using the most modern experimental apparatus; 2) to give you a feeling for the great variety and depth of knowledge which such high quality data can provide concerning pion annihilation in nuclei. Unfortunately, during this short ten minute presentation I can show only a small selected portion of our data. Furthermore, since the data is not yet fully reduced, I can, at best, give you a sense of direction with no final conclusions.

The data was obtained by the Carnegie-Mellon nuclear physics group[†] during the last two years at the LEP channel at LAMPF, using two spectrometer systems to detect charged particles in coincidence. One spectrometer consists of a stack of eight high purity Ge-crystals which can stop protons of 200 MeV. The second spectrometer, which consists of a stack of two Si(Li) detectors followed by two high purity Ge detectors, can stop protons of 115 MeV. The first crystals in each arm can stop deuterons of about 39 MeV and tritons of 47 MeV. Energy deposits greater than 50 MeV in the first crystal are assumed to come from He ions. In front of each spectrometer are two individual wire readout proportional chambers. Each chamber has three anode planes with 1 mm wire spacing. The third anode plane, having a 45° orientation relative to the vertical and horizontal planes, enable the unsorting of more than one particle hit for each readout. The present rate handling capability of these MWPC's is in excess of 10^6 particles/sec. The front anode planes and the front of each spectrometer are usually about 10 cm and 23 cm respectively from the target. The target normal is tilted 65° to 70° relative to the beam direction resulting in a horizontal beam spot which is 10 to 15 cm long. The target thicknesses are 100 to 300 mg/cm^2 for ^{6,7}Li and 100 to 150 mg/cm^2 for NH_2 and H_2O. The liquid targets were made by stretching foils of 1 mil aluminum for the hydrazine targets and 1 mil mylar for the water targets across a stainless steal frame. The liquid target thicknesses and uniformity were determined by measuring the attenuation of a ^{90}Sr(β) source through the target. Liquid targets were made with nonuniformity less than ±0.1 mm. Absolute cross sections are obtained by normalizing to the known $D(\pi^+,p)p$ cross sections. A CD_2 target was used and our H_2O target was mixed with 5% D_2O. To use a reaction with a two-body final state to normalize a reaction with a three-body final state is possible but nontrivial. The normalization procedure must use the measured recoil-momentum distribution of the residual nucleus in the three-body final state. A Monte Carlo code has been written for the analysis.

The reactions which we studied are (π^+,xx) where xx can be any combination of two-hydrogen isotopes (p,d,t). We also measured ^{6,7}Li$(\pi^+$,He He) cross sections which is the time-reversal of the reactions measured at Orsay[1] and to be discussed at this conference. In figure 1 is shown a typical ^{6}Li$(\pi^+,pp)^4$He missing mass spectrum taken at T_π = 74 MeV, θ_1,θ_2 = -55±10, 106±10⁶. The resolution in this spectrum is 1.8 MeV FWHM. A ^{7}Li$(\pi^+,pd)^4$He mass spectrum summed over several angles at a pion energy of 38 MeV is shown in Fig. 2. We find that the ^{4}He spectra in Fig. 1 and 2 are quite different. In the ^{6}Li$(\pi^+,pp)^4$He reaction the strongest transitions are between 23 and 30 MeV excitation whereas the strongest transitions in the ^{7}Li$(\pi^+,pd)^4$He reaction is centered at 22.3±0.5 MeV. We have a reasonable explanation for this difference, but in the short time allotted me, I would rather concentrate on other features of these reactions. Furthermore, before we deal with questions involving nuclear structure we must obtain a better understanding of the pion annihilation process.

The Argonne group,[2] by detecting a single proton following pion annihilation, (π^+,p), has determined that the average number of nucleons sharing the energy and momentum of the annihilated pion increases from 3 to 5.5 as A increases from 12 to 181. However, it is difficult to determine[3,4] from this data how many of the nucleons come from the basic annihilation process and which nucleons are secondary products resulting from rescattering and final state interactions. The above ambiguity can be more easily resolved in a kinematically complete two-body coincidence experiment which can give much more information than a singles experiment.

Because we measure the momenta (both magnitude and direction) of both coincident particles we can calculate the recoil momentum. In Fig. 3 is shown the recoil momentum distribution for the ^{7}Li$(\pi^+,pp)^5$He reaction to the $3/2^-$ g.s. and $1/2^-$ 4 MeV first excited state. The pure kinematical phase space acceptance of our detection system is, for the most part, divided out

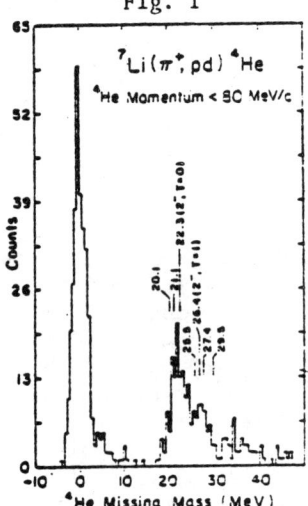

Fig. 2. The ^{7}Li$(\pi^+,pd)^4$He spectrum with recoil momentum less than 80 MeV/c.

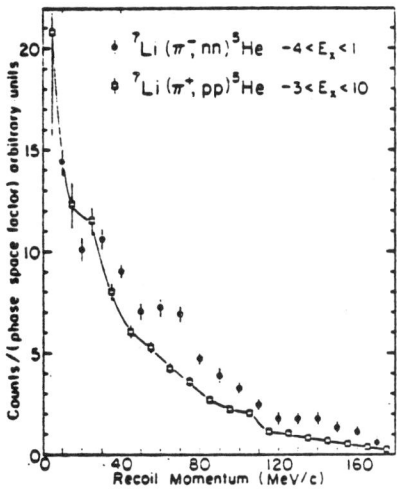

Fig. 3. Recoil momentum distributions for the two-lowest states of ^{5}He.

of the distribution. This result is still preliminary because a few minor idiosyncrasies of our detection system have not yet been included in the phase space correction. However we see that the recoil momentum distribution is reasonably consistent with earlier published ^{7}Li(π^-,nn)^{5}He data.[5] This distribution is what is expected[5] from a two-nucleon absorption process in which the ^{5}He core serves as a spectator. I show this to you because, simultaneously to the ^{7}Li(π^+,pp) measurement, we also measured the ^{7}Li(π^+,pd)^{4}He$_{g.s.}$ recoil momentum distribution and made a similar phase space correction. This is shown in Fig. 4. Also shown in Fig. 4 is the experimentally deduced momentum distribution,[6] $\phi^2(q)$, extracted from the cluster knockout reaction ^{7}Li(p,pα)T. The similarity between the two is remarkable and strongly suggests that the pion is annihilating on a "triton-cluster". What we mean by this is that all three nucleons are directly involved in the pion annihilation process. We can rule out a two-step process (π^+,$p_1 p_2$) followed by a neutron-pickup (p_2,d) in which this intermediate proton, p_2, is nearly on-shell. This two-step process would have very different kinematics from a triton-cluster annihilation. The minimum amount of momentum transfer for a 90 MeV on-shell proton to pick up a neutron is 167 MeV/c, far in the tail of our measured distribution. Other kinematic distributions such as E_p, the energy of the proton, give additional support to the above interpretation. If such a two-step process is occuring, the intermediate proton, p_2, must be far enough off-shell that the picked-up neutron is considered part of the initial annihilation process. This is what we mean by a "three nucleon cluster" annihilation.

What makes the above study so interesting is that the ^{7}Li(π^+,pd) cross section is sizeable, larger than 20% of the ^{7}Li(π^+,pp) cross section at T_π = 38 MeV, integrated over all states below 50 MeV excitation. The implication is that it is common for more than two nucleons to be involved in the pion annihilation process.

There is a warning concerning the similarity between the ^{7}Li(π^+,pd)^{4}He$_{g.s.}$ and ^{7}Li(p,pα)T momentum distributions in Fig. 4. Even if the ^{7}Li(π^+,pd)^{4}He$_{g.s.}$ reaction is a "triton cluster" annihilation with the ^{4}He as a spectator, we would not expect the recoil momentum distribution to be exactly the same as the momentum distribution between the T and ^{4}He clusters in the ^{7}Li target, $|\phi(q)|^2$. Rather, the cross section is proportional to:

$$d^3\sigma/dE_p d\Omega_p d\Omega_d \propto \phi^2(q) F(\vec{p}_p, \vec{p}_d) \qquad (1)$$

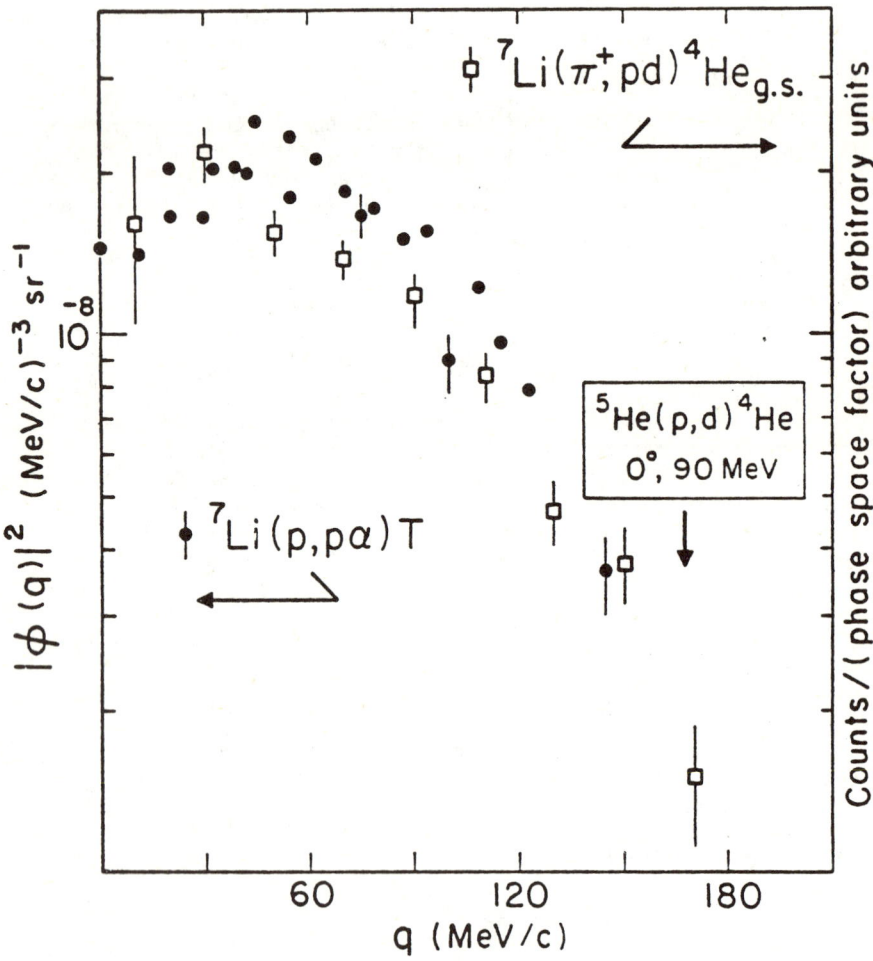

Fig. 4. The ^{7}Li(p,pα)T data (left-side scale) is from ref. 6. The scale for the ^{7}Li(π^+,pd)^{4}He$_{g.s.}$ is on the right.

where $F(\vec{p}_p, \vec{p}_d)$ is a model dependent function which depends upon short range correlations in the "triton cluster" and details of the pion annihilation mechanism. If q were independent of $\vec{p}_p$ and $\vec{p}_d$, the q distribution would be $|\phi^2(q)|$, however q is related to the final proton and deuteron momenta, $\vec{p}_p$ and $\vec{p}_d$, through energy and momentum conservation.

$$\frac{2M_N}{h^2}(T_\pi + Q) = -1/3 k_\pi^2 + (1/3 + M_N/M_\alpha)q^2 + 3/2(2/3\vec{p}_p - 1/3\vec{p}_d)^2 \\ + 2/3\vec{k}_\pi \cdot (\vec{p}_p + \vec{p}_d) \quad (2)$$

where Q is the Q-value for the reaction and M_N and M_α are the nucleon and ^{4}He masses respectively. One of the best ways to study the effect of $F(\vec{p}_p,\vec{p}_d)$ is to look for differences in the recoil momentum distributions at different recoil angles. Using our recent data, taken several weeks ago with much better statistics, we will search for these changes in the momentum distribution in both the (π^+,pp) and (π^+,pd) reactions.

Lastly, I wish to show you some of our recent data, which is of better quality than our older data. Unfortunately, I have only on-line spectra obtained during monitoring of the experiment. In figures 5 and 6 are shown part of our (π^+,pp) spectrum at T_π = 60 MeV, θ_1,θ_2 = 60°,103° on a 100 mg/cm^2 water target, 95% H$_2$O and 5% D$_2$O. Fig. 5 shows the $D(\pi^+,pp)$ peak which is shifted, by a more positive Q-value of about 20 MeV from the ^{16}O(π^+,pp) missing mass spectrum shown in Fig. 6b. The energy resolution of the $D(\pi^+,pp)$ peak is nearly 1 MeV whereas the resolution of the oxygen peaks in the same spectrum, shown in Fig. 6b, is 1.6 MeV. The reason for the worse resolution is our limited on-line analysis code and slight imperfections in our energy calibration both of which will be corrected in our off-line analysis. The on-line code cannot handle multiple hits or a no-hit in any of our wire planes. In fact four of our planes are not even used in the on-line analysis. Therefore, for most of the events, the on-line code assumes the particles go from the midpoint of the target to the midpoint of the detector. The recoil energy is calculated incorrectly resulting in a worse missing mass resolution. This only has a mild effect on the 60°,103° spectrum where the recoil energy is usually less than 1.35 MeV. However the 100°,100° spectrum in Fig. 6a has significantly worse resolution because the recoil energy is up to four times as large.

In spite of the undeveloped state of these spectra there are already interesting features coming forth. The transition to the ^{14}N 1$^+$ ground state is known to involve two units of angular momentum which can be either in the internal motion or the center of mass motion of the two removed nucleons. The transition to the 3.95 MeV 1$^+$ state is L = 0. There is a dramatic change in the ratio of these 1$^+$ states as the recoil momentum changes from P_{rec} < 200 MeV/c (Fig. 6b) to 200 < P_{rec} < 400 MeV/c (Fig. 6a). In Fig. 6b the 3.95 MeV transition is more than a factor of 10 larger than the ground state transition, whereas in Fig. 6a the two transitions are nearly equal in strength. This suggest that the angular momentum of the ground state transition is mostly in the center-of-mass motion of the two nucleons, rather than their internal motion.

In between the two 1$^+$ states is the 2.3 MeV 0$^+$,T=1 transition which is very weak. In our off-line analysis we will either see this transition or set a low upper-limit for its cross section. This transition will tell us about pion absorption on a T=1,S=0 nucleon pair. For related experimental work on this isospin dependence see ref. 7.

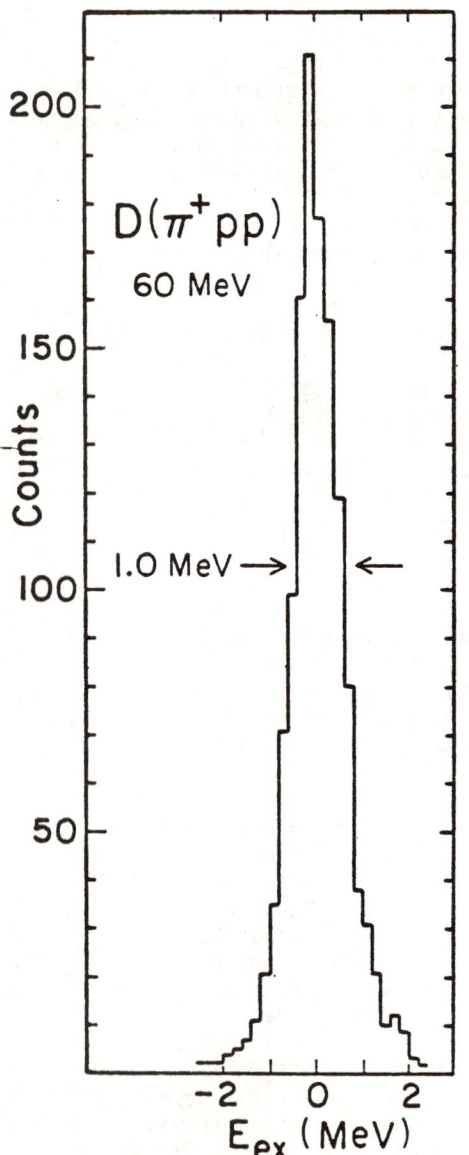

Fig. 5. The D(π^+,pp) peak is part of the spectrum shown in Fig. 6 (see text).

Lastly, I wish to acknowledge the strong nuclear theory group which has been working closely with us at Carnegie-Mellon. Kisslinger and Keister have presented at this conference their calculations of the (π^+,p) reaction. The computer code used in those calculations is presently being modified by Keister to perform (π^+,pp) calculations. I would like also to acknowledge the assistance in the data collection of Jim Amann, LAMPF, S. Dytman, MIT, K.G.R. Doss, U. of Washington and F. Takeutchi, Kyoto-Sangyo Univ.

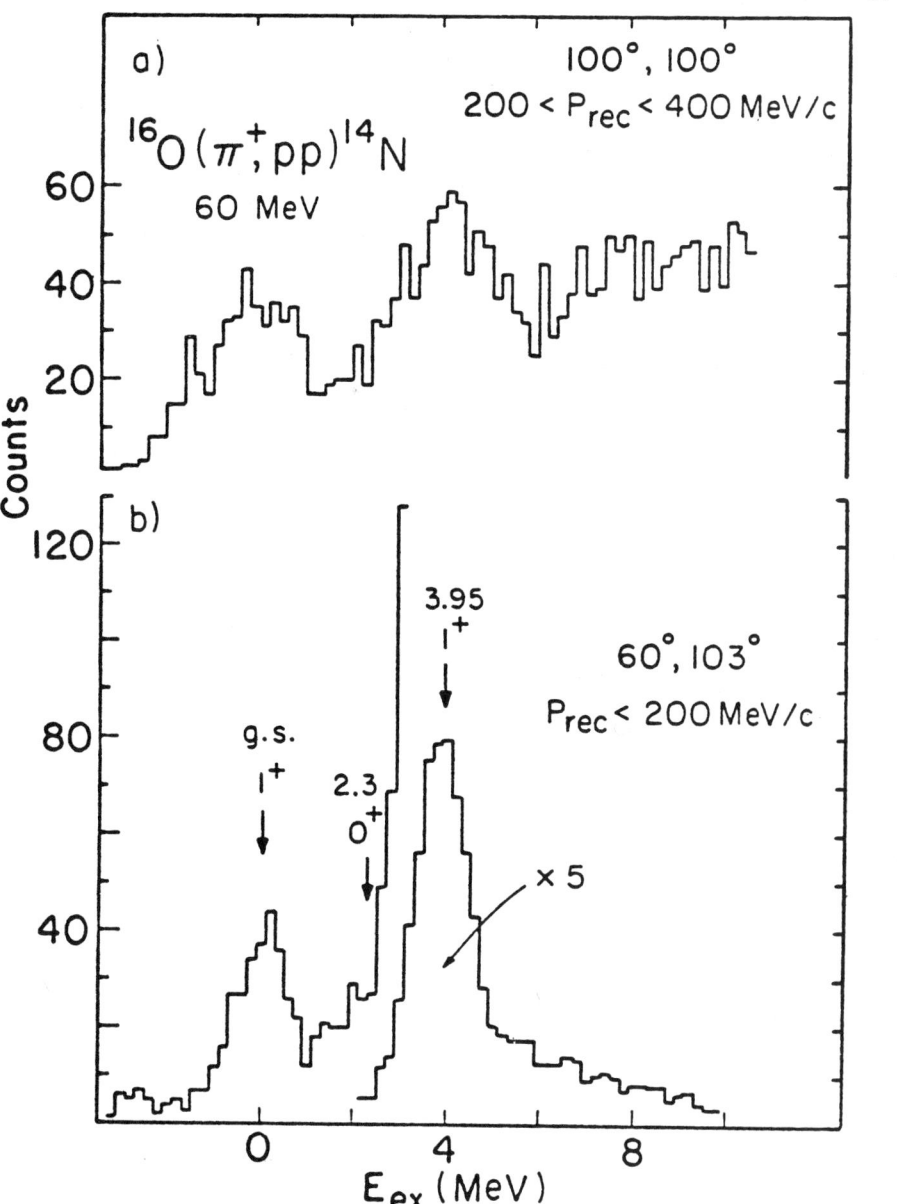

Fig. 6

REFERENCES

† CMU Medium Energy Group: P. D. Barnes, B. Bassalleck, R. A. Eisenstein, G. Franklin, R. Grace, D. R. Marlow, C. Maher, P. Pile, R. Rieder, F. Takeutchi, and W. R. Wharton.
1. Y. Le Bornec et al., Orsay preprint, June 1981.
2. R. D. McKeown et al., Phys. Rev. Lett. $\underline{44}$, 1033 (1980).
3. K. G. Doss and W. R. Wharton, Phys. Rev. C22, 1219 (1980).
4. H. C. Chiang and J. Hüfner, Nucl. Phys. A349, 466 (1980).
5. B. Bassalleck et al., Phys. Rev. C19, 1893 (1979).
6. M. Jain et al., Nucl. Phys. A153, 49 (1970).
7. D. Ashery et al., Phys. Rev. Lett. $\underline{47}$, 895 (1981).

DISCUSSION

<u>D. Koltun</u> (Univ. of Rochester): I think the question you raised about the cluster absorption is one that has always been interesting, and that kind of experiment does help us. But I think I'd like to put in a word of warning about interpretation, that I think has always been known in the business, which is that the selection of initial and final states of the nuclear targets has some effect, of course, on what you see. If you start with ^{6}Li and end up in the ground state of helium with two fast nucleons out, you're not terribly surprised if it looks like absorption on the deuteron. Similarly, if you end up in the grounds state of helium after having taking out a proton and a deuteron from ^{7}Li, you are, of course, emphasizing-just because of nuclear structure reasons-the triton aspect of it. But, it's a long way from doing that to asking about inclusive cross-sections-whether they are really dominated by that triton absorption. You are dominating the overlap, and that may dominate what you see.
<u>Wharton</u>: I just showed one particular reaction. Of course, you like to describe ^{7}Li as a triton and alpha cluster, but let me point out that in other (π,pd) reactions we see the same effects; e.g., ^{6}Li (π,pd)^{3}He. In ^{6}Li a triton cluster is in a s-state relative to the ^{3}He cluster, and we get a L=0 recoil momentum distribution that peaks at zero and comes down. Also, that excited state at 22.3 MeV in the ^{4}He spectrum has a L=0 distribution. We are analyzing our ^{16}O data. So, ^{7}Li is not a unique case. I think it's more than a simple nuclear structure effect.

<u>L.S. Kisslinger</u> (Carnegie-Mellon Univ.): One of the things that we have learned in the last couple of years is that in the so-called two-nucleon mechanism, it's really two-nucleons plus a core. That's a very important thing to keep in mind, because one of the nucleons is always in a bound state before and after the interaction for the (π,p). There is genuine momentum sharing always with at least three things. That really does take place. It's crucial to know in interpreting your results whether we really need three active nucleons plus a core. It might well be with that momentum sharing that two active nucleons and a core might in fact be enough.

<u>L. Antonuk</u> (Univ. of South Carolina): Have you taken, or do you plan to take, any data looking at a pion in the outgoing channel?
<u>Wharton</u>: We have not looked at the pion in the outgoing channel.

WHAT DO ANGULAR DISTRIBUTIONS OF SUBTHRESHOLD PIONS TELL US ABOUT HEAVY ION REACTIONS?

Walter Benenson
Cyclotron Laboratory and Physics Department
E. Lansing, Michigan 48824

ABSTRACT

An angular distribution at E/A = 138 MeV for the Ne + NaF->π+x reaction has been measured. Coulomb effects are very large at $0°$ for pion velocities near that of the projectile. Away from the Coulomb region the angular distribution is isotropic, which is in direct disagreement with the simple nucleon-nucleon first collision model. Various theoretical predictions for the magnitude of the cross section are also discussed.

The production of pions by heavy ion collisions at energies below E/A = 290 MeV has often been proposed as a method to look for coherent effects in these reactions. The low energy pions produced have a relatively long mean free path in nuclear matter and are not produced by the decay of excited remnants of the reactions. Therefore looking at pions is supposed to be a means of probing the early stages of the collision. At very low beam energies, the processes which dominate at high energies become negligible. At the beam and pion energies discussed here, the contribution of the nucleon-nucleon process is essentially zero for a sharp edged Fermi distribution. The idea of the experiment then was to look for a coherent process by comparing to simple models for the cross section and angular distribution. As we will show, the comparison is interesting but not conclusive.

Before getting to a discussion of the angular distribution taken at E/A = 138 MeV, let us first look at the previous data.[1] As expected the cross section rises very rapidly from E/A = 80 to 380 MeV. For most pion momenta the increase is about four orders of magnitude. This can be seen on Fig. 1 which is a plot of all of the $0°$ data taken in 1979 with the first version of the $180°$ spectrograph at the LBL Bevalac. A curious feature of the data was the very large π^-/π^+ ratios observed at E/A = 383 MeV. This large ratio turns out to be a universal feature of the data whenever the outgoing pions have a velocity near that of the beam as is illustrated in Fig. 2. This figure gives a large number of R measurements as a function only of the velocity relative to the projectile.

A second series of experiments was undertaken to study this ratio R with better resolution and under a wider range of conditions. These experiments will soon be reported in Phys. Rev.[2] and showed that the theory of Gyulassy and Kauffmann[3] was a very reasonable explanation of the effect. Using the formulas of ref. 3 with some modifications, one can show that about one half the charge continues on with nearly beam velocity in a manner very much like a fragmentation or peripheral process. That this persists down to E/A = 140 MeV is a surprise; in fact the present data show a huge Coulomb effect at $0°$ as can be seen in

Fig. 3. Basically, the large Coulomb effect has to occur if a projectile fragment exists, and this can be shown also semiclassically.[4]

The present experiment was undertaken to distinguish between two possible subthreshold production processes which are at the extremes of possibility. They, at least initially, gave very similar magnitude predictions at $0°$ but very dissimilar angular distributions. The first collision process as discussed by Bertsch[5] and later calculated by Hecking[6] in much more detail, is essentially a free nucleon process but with the Fermi-motion (including the tail) added in. In spite of Pauli blocking and the smearing due to the Fermi motion, this model predicts a large $0°/90°$ peaking. Thermal models, on the other hand, be they fireball or firestreak, predict complete isotropy in the center-of-mass system. This process can be thought of as black-body radiation of pions from the excited spectator region. Not everybody agrees, but I feel that thermal and first-collision production can be simply added since one is a pre- and the other a post-equilibrium process. This addition idea was used by Hecking and, as one can see in Fig. 4, the momentum distributions at $0°$ are very well represented by the theory.

The first collision process as formulated by Hecking involves a long and difficult calculation, but is not extremely sensitive to choices of parameters and the other details of the calculation. On the other hand the thermal model is very sensitive to a number of parameters. This is because black body radiation at a given wavelength has a T^5 dependence. Experience has shown that composite particle production, inclusion of isobars in the equilibrium mixture, and choice of freeze-out radius can affect the result significantly.

In Fig. 5 is given the results of the present angular distribution measurement. In Fig. 3 one sees that P_π must be >125 MeV/c to get away from the Coulomb effect. In Fig. 5 we give the results at P_π = 128 MeV/c and the various calculations available. One can see that there are three thermal models which give quite different results. The "thermal" curve is what Hecking used and comes from the paper of Kapusta[7]. The firestreak model was calculated by Westfall[8] with and without t, ^{3}He, ^{4}He production and the inclusion of isobars. The effect of switching from fireball to firestreak geometry is small, but the inclusion of composite particles is very important. Composite particle production causes the kinetic energy to be shared by fewer participating particles. This raises the temperature and enhances the yield considerably at low beam energies. The calculations of Hecking give much too much $0°/90°$ peaking, but this is a result of using too low an estimate of thermal production and of neglecting rescattering. He is now at work on this, and the agreement is much better although the cross section is still over-predicted.

What is the conclusion of this comparison? Basically we have shown that many ingredients are needed to make a correct calculation and that to search for important new effects by a comparison of data to theory is a difficult and sensitive process. The next step is clearly to make a measurement with a much heavier target and projectile. Comparison of this to A=20 results may really be a test for coherent processes.

This work was supported by the NSF under grant No. 78-22696. The project was part of a large collaborative effort with many people contributing as can be seen from refs. 1 and 2. The LBL groups were supported by the DOE office of High Energy and Nuclear Physics under contract No. W-7405-EN6-48.

REFERENCES

1. W. Benenson, G. Bertsch, G.M. Crawley, E. Kashy, J.A. Nolen, Jr., H. Bowman, J.G. Ingersoll, J.O. Rasmussen, J. Sullivan, M. Koike, M. Sasao, J. Peter, T.E. Ward, Phys. Rev. Lett. $\underline{43}$, 683 (1979), $\underline{44}$, 54 (1980).
2. J.P. Sullivan, J.A. Bistirlich, H.R. Bowman, R. Bossingham, T. Buttke, K.M. Crowe, K.A. Frankel, C.J. Martoff, J. Miller, D.L. Murphy, J.O. Rasmussen, W.A. Zajc, O. Hashimoto, M. Koike, J. Peter, W. Benenson, G.M. Crawley, E. Kashy and J.A. Nolen, Jr., LBL-11971 and Phys. Rev. in press.
3. M. Gyulassy and S.K. Kauffmann, Nucl. Phys. $\underline{A362}$, 503 (1981).
4. K.G. Libbrecht and S.E. Koonin, Phys. Rev. Lett. $\underline{43}$, 1581 (1979).
5. G. Bertsch, Phys. Rev. $\underline{C15}$, 713 (1977).
6. P. Hecking, LBL-12671 and private communication.
7. J.I. Kapusta, Phys. Rev. $\underline{C16}$, 1493 (1977).
8. G.D. Westfall, J. Gosset, P.J. Johansen, A.M. Poskanzer, W.G. Meyer, H.H. Gutbrod, A. Sandoval and R. Stock, Phys. Rev. Lett. 37, 1202 (1976).

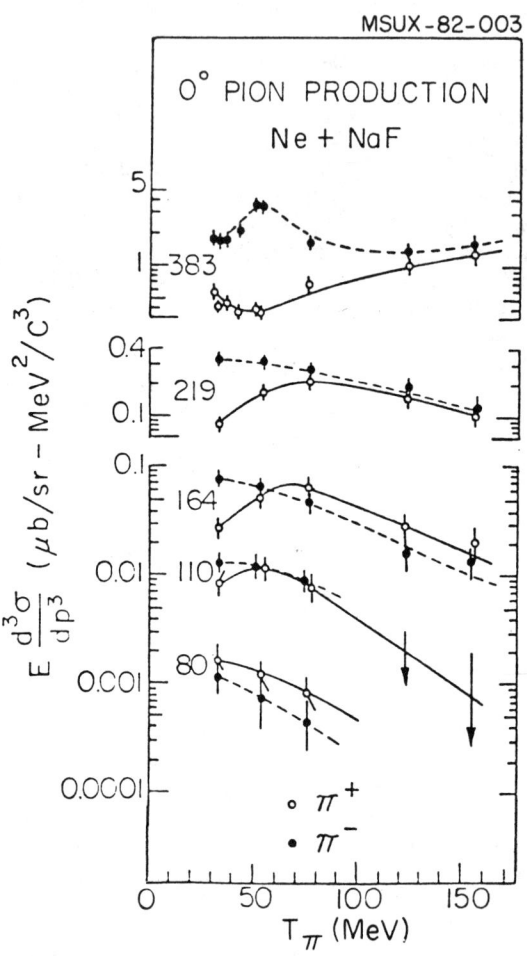

FIG. 1. Data for Ne+NaF->$\pi^{\pm}$+X at $0°$ and a variety of beam energies. The data is from ref. 1.

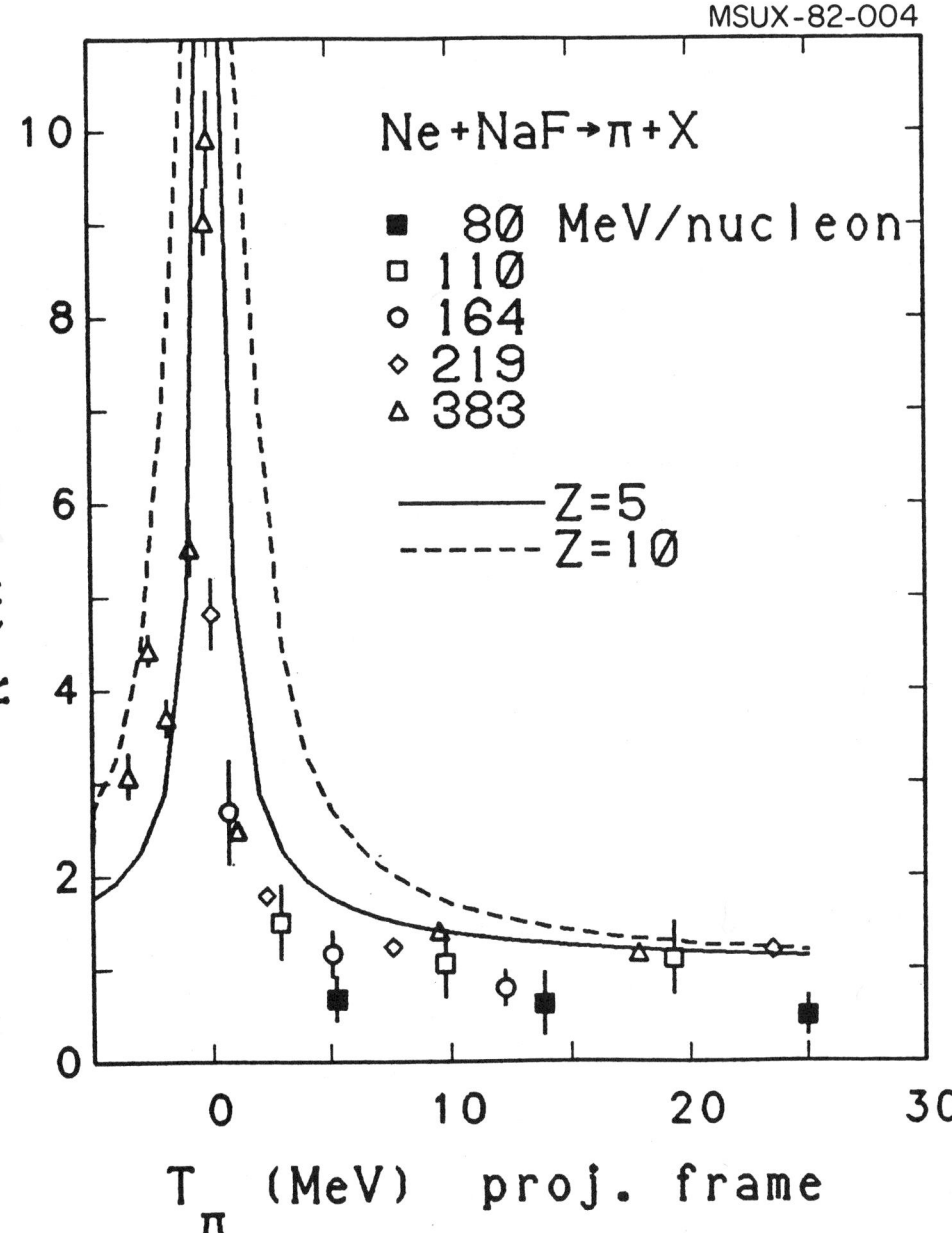

FIG. 2. Plot of R (π^-/π^+ ratio) versus kinetic energy of the pion in the projectile frame. The curves are Gamow factors for Z=5 and 10 and a source moving at the projectile velocity.

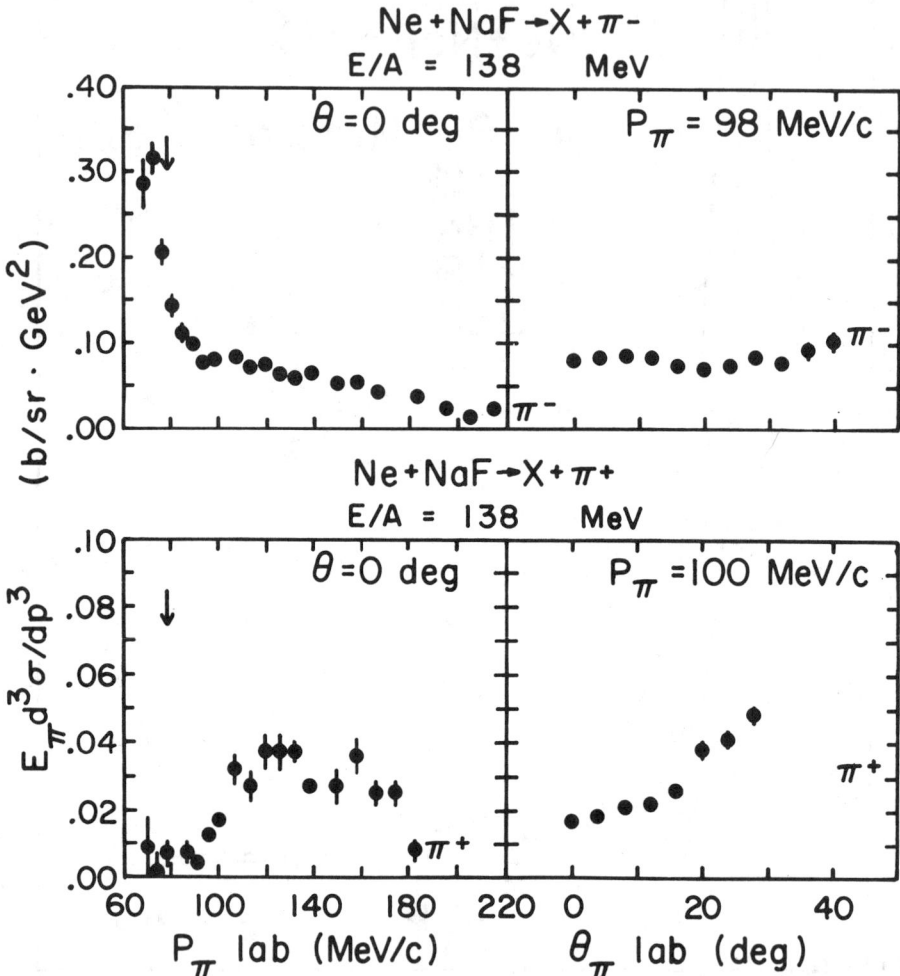

FIG. 3. The data at E/A = 138 MeV for both π^+ and π^- versus pion momentum and angle. The arrows indicate projectile velocity.

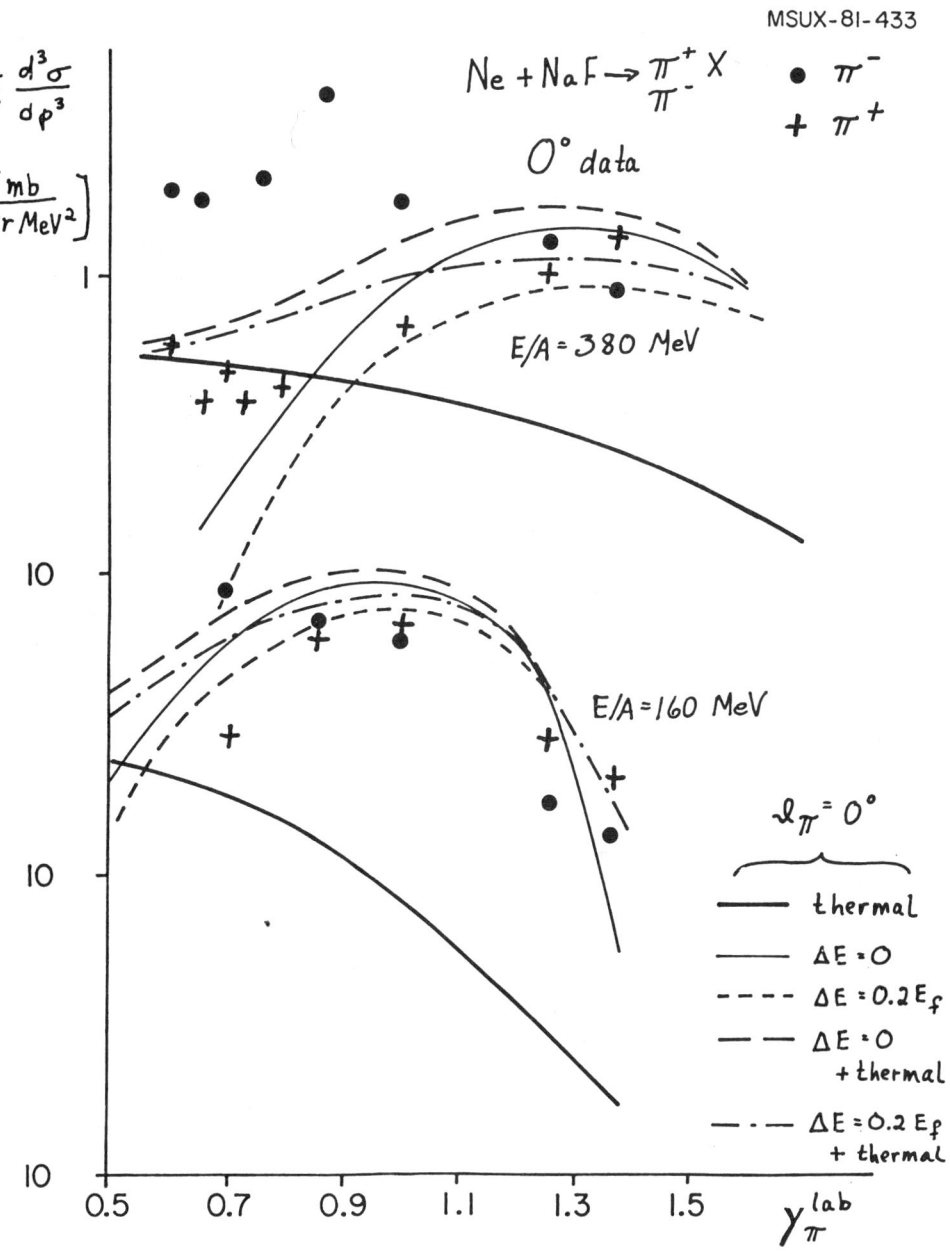

FIG. 4. Comparison of the theory of Hecking[6] to the data at E/A = 160 and 380 MeV.

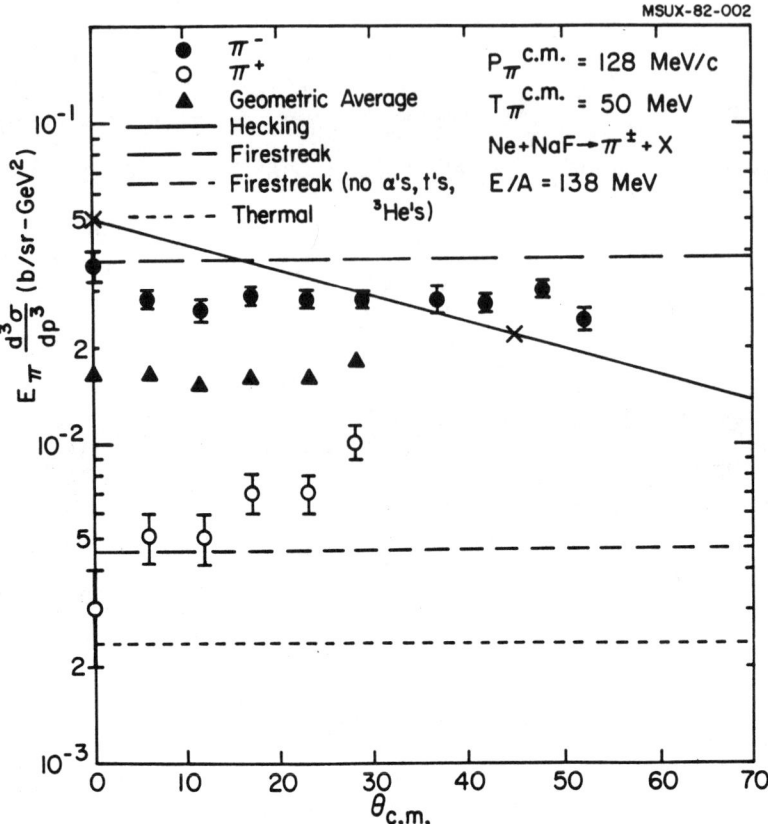

FIG. 5. Angular distributions at E/A = 138 MeV and P_π^{cm} = 128 MeV/c. The data are given for π^-, π^+ and the geometric mean of the cross sections.

PIONIC FUSION
- Coherent Pion Production in Nuclear Collisions -

M. G. Huber, K. Klingenbeck
Institute for Theoretical Physics
University of Erlangen-Nürnberg, D-8520 Erlangen

ABSTRACT

The coherent production of pions in a collision of two complex nuclei has recently been observed with a number of different target projectile combinations, mostly near the physical threshold. Such reactions are highly interesting for mainly two reasons:

- the total free energy of the entrance channel is converted into a fast pion;

- the two incident nuclei undergo a fusion process to form a bound state of the united nucleus ("ultra cold fusion").

Such a pionic fusion reaction is difficult to understand microscopically on the basis of our conventional knowledge about nuclear collisions. Its cooperative nature raises two new types of questions:

- What is the mechanism that couples the kinetic energy of the entrance channel to the pion field, avoiding the usually observed thermalisation process?

- What are the selection rules for different entrance channel fragmentations that determine the population of a specific state of a given final nucleus?

Such a reaction can also be used as a tool to investigate specific phenomena such as (i) the creation of high spin states; (ii) the coherent propagation of one and/or multiple N* excitations in nuclei; (iii) the production of new isotopes.

Some of the characteristic features of pionic fusion are discussed from a general point of view. A reaction mechanism is proposed and applied to recent data of pionic fusion of light nuclei; pertinent features of this reaction and a few possible applications are briefly discussed.

I. INTRODUCTION

Pions can be produced in a nucleus-nucleus collision provided the kinetic energy in the nuclear center of mass system is large enough:[*]

$$T_{CM} > m_\pi c^2 = 140 \text{ MeV}$$

---

[*] For the following discussion the nuclear Q-value of the reaction has been ignored.

In fact the first pions experimentally observed in a laboratory experiment[1] were produced by the bombardment of nuclei with alpha particles. The pion spectrum of the reaction

$$A_1 + A_2 \longrightarrow \pi + X \qquad (1)$$

is schematically shown in Fig. 1; it consists of two components:

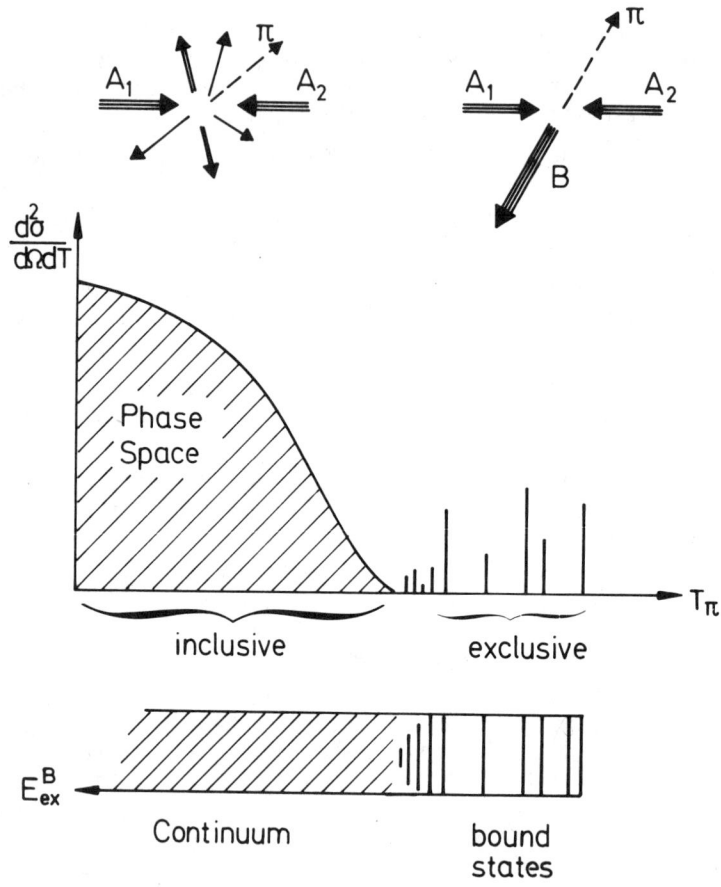

Fig. 1: A typical pion spectrum of the reaction (1).

(i) <u>exclusive spectrum</u>: the produced pions are characterized by a well defined energy; this reflects the two body character of the final channel: for the corresponding pions the residual nucleus, B, has to be bound in a specific state, $|\nu\rangle$:

$$A_1 + A_2 \longrightarrow \pi + B_\nu \qquad (2)$$

Due to the reduction of phase space the counting rates for those reactions are rather low; nevertheless, the experimental feasibility has been proven by a number of recent experiments, by Bornec et al[2,3] at Orsay, by Aslanides et al[4] at SATURNE and by Ward et al[5] at Indiana.

(ii) <u>inclusive spectrum</u>: those pions leave the nucleus in a highly excited final state with three or more particles in the continuum; the shape of the cross section is dominated by phase space factors; in fact they become more important with decreasing pion energies (for constant incident energy). For examples the reader is referred to ref. 6, 7 and 8.

The reaction (2) is highly interesting for a number of reasons; here we shall discuss only the two most important ones:

(a) the kinetic energy of the entrance channel is completely converted into the pion field; obviously, this requires a highly coherent cooperation of all the nucleons involved;

(b) the two nuclei of the entrance channel undergo a fusion reaction: they form one united nucleus in a well defined final state.

Because of those two properties we like to refer to reaction (2) as a pionic fusion process.

For a long time the reaction (2) has attracted quite some interest; we like to refer to the pioneering work of S. Wall et al[9] and to attempts to establish an upper limit for the cross section of the pionic fusion reaction[10,6].

In a conventional scheme such a reaction is considered as highly unlikely, if not impossible, simply due to the general observation that in a nucleus-nucleus collision the kinetic energy is most often thermalized: once the energy is distributed among the participating nucleons, it is indeed quite improbable to concentrate it again into one degree of freedom such as the pion field.

The pionic fusion process of eq. (2), therefore, clearly points out the need to develop a conceptual alternative to the conventional i.e. mostly statistical treatment of nuclear collisions. We have to take into account that initially the energy is not distributed among the target nucleons but rather contained in the relative motion of the two incoming nuclei. It is the central problem to find out the conditions under which this coherence among the nucleons survives the collision, and to determine the mechanisms that lead to the coherent production of the corresponding pions.

If we observe a coherently produced pion we conclude that in such a reaction the complete thermalization process must have been avoided; this indicates the need for a more or less direct coupling of the entrance channel to the pion field (see Fig. 2). Such a mechanism is difficult to visualize, since we are used to think of the pions as being produced in an elementary nucleon-nucleon interaction. We

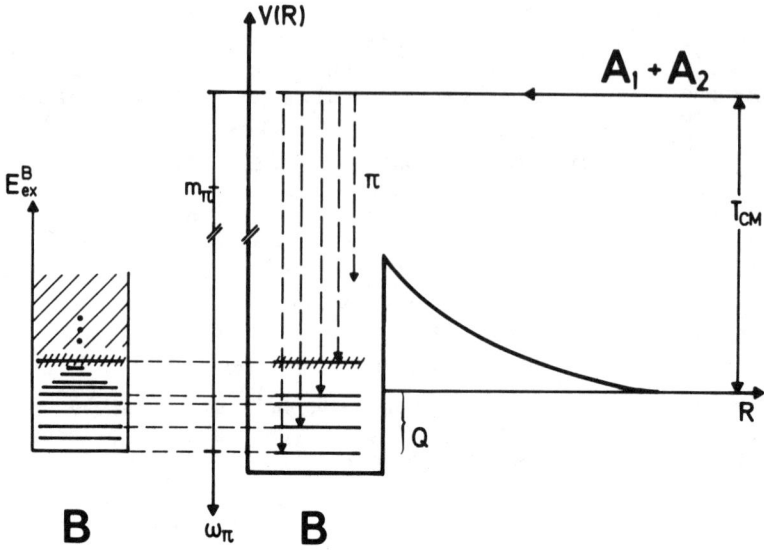

Fig. 2: Schematic picture of the pionic fusion reaction.

therefore have to find out the coupling between the relative motion of the two nuclei and the two nucleon subsystem that produces a pion which then propagates through the nucleus before it is emitted. This problem has first been formulated and qualitatively discussed by Huber and Dillig[11]. More recently, this approach has been worked out more formally by Klingenbeck et al[12] who also applied it to specific reactions[13]. In this paper we shall only briefly sketch the essentials of the model; for details the reader is referred to a forthcoming paper[14].

We like to point out the special case of eq. (2) where the projectile is a free nucleon; such reactions have been investigated quite intensively over the last few years, we refer the reader to the contributions of Hoistad[15], Nann[16], Couvert[17] and Dillig[18] to this workshop and to the summary report by P. Walden[19]. For the case of a complex projectile we have an additional degree of coherence which - at first - looks like an unwanted complication; but we shall see later, this fact opens up also a new class of phenomena that may eventually complement the available protoninduced data.

In the next chapters we develop a model for the mechanism of eq.(2) which will then be applied to a specific reaction recently investigated experimentally. In Chapters IV and V we consider several types of such cooperative reactions and discuss their use as a tool to investigate both conventional nuclear structure effects and the creation and propagation of subnucleonic excitations in a nuclear medium.

II. THE REACTION MECHANISM

As discussed previously this reaction calls for a mechanism to circumvent the thermalisation of the incoming energy among the conventional i. e. "external" degrees of freedom of the target and projectile nucleons. Within the framework of conventional Nuclear Physics it is difficult to imagine, how this can be achieved. However if one invokes the internal or quark degrees of freedom of bound nucleons one immediately ends up with a quite natural and efficient storage for energies as large as several hundreds of MeV. Such an internal quark transition of nucleons obviously offers the possibility to accommodate a large amount of energy inside of a nucleus without "heating up" the whole system. Actually from this observation, the model we propose is basically quite simple: when the two incoming nuclei collide with each other, a nucleon bound in one of the two nuclei is internally excited into an N*-resonance (more specifically: into a $\Delta(33)$); the kinetic energy of the entrance channel is thus stored in an internal or quark excitation of a bound nucleon; this configuration propagates coherently through the whole nucleus, B, until, in a last step, it decays into the final nuclear state B by emitting a pion. This reaction mechanism is schematically shown in Fig. 3. In the following it will be discussed how such a mechanism

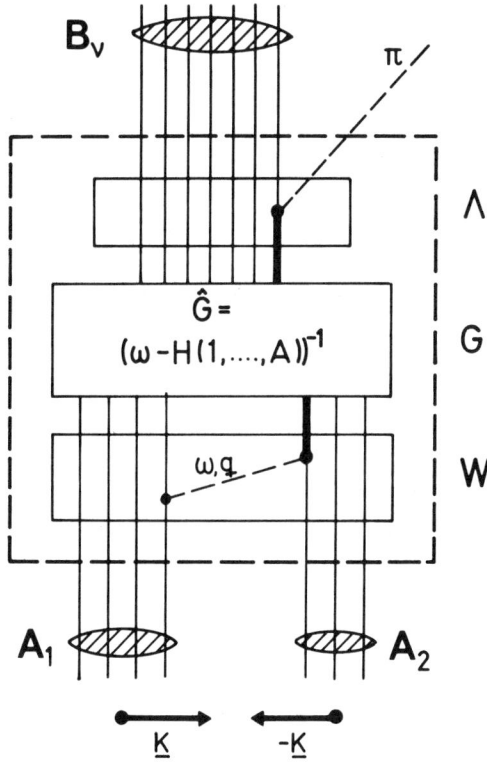

Fig. 3:
The resonant transition amplitude of pionic fusion (see eq. (10c)).

leads to a manybody operator, which couples coherently the entrance to the excit channels avoiding the thermalization procedure which is usually dominating nuclear collisions in this energy domain. We start from the differential cross section for the reaction (2):

$$\frac{d\sigma^{(\nu)}}{d\Omega}(\underline{K},\underline{k}) = \frac{1}{8\pi^2} \frac{E_1 E_2 E_B^{(\nu)}}{E_{CM}^2} \cdot \sum_f \sum_i |T_{fi}(\underline{K},\underline{k},\nu)|^2 \qquad (3)$$

with the transition amplitude:

$$T_{fi}(\underline{K},\underline{k},\nu) = \langle fin(\underline{k},\nu) | \hat{T} | in(\underline{K}) \rangle \qquad (4)$$

Here K and $\underline{k}$ denote the CM-momenta in the entrance and excit channel, respectively. In more detail the asymptotic channels are defined by:

$$| in(\underline{K}) \rangle = | A_1(j_1,t_1); A_2(j_2,t_2); \underline{K} \rangle \qquad (5a)$$

$$| fin(\underline{k},\nu) \rangle = | B_\nu(J,T); \pi; \underline{k} \rangle \qquad (5b)$$

where we have introduced the spins and isospins of the various nuclei. To derive the reaction amplitude we discuss the Nuclear Hamiltonian in an entrance channel representation:

$$H(1,2,\ldots,A) = \hat{T}_{rel} + H_1(1,\ldots,A_1) + H_2(1,\ldots,A_2) \qquad (6)$$
$$+ W_{12}(1,\ldots,A)$$

(with obvious notations).
The interaction W_{12} is given by:

$$W_{12}(1,2,\ldots,A) = \sum_{i \in A_1} \sum_{k \in A_2} w(i,k) \qquad (7)$$

where w denotes the baryon-baryon interaction of fig. 4.
It is physically motivated to devide the nuclear configuration space into a purely nucleonic part and a distinguished subspace, containing at least one excited baryon, e.g. a $\Delta(33)$. Similarly the interaction W is devided into corresponding parts (see Fig.4):

$$W_{12} = W_{12}^{nonres} + W_{12}^{res} \qquad (8a)$$

with

$$W_{12}^{nonres}(1,\ldots,A) = \sum_{i \in A_1} \sum_{k \in A_2} W_{NN \to NN}(i,k) \qquad (8b)$$

and

$$W_{12}^{res}(1,...,A) = \sum_{i\in A_1} \sum_{k\in A_2} W_{NN\to N\Delta}(i,k) \qquad (8c)$$

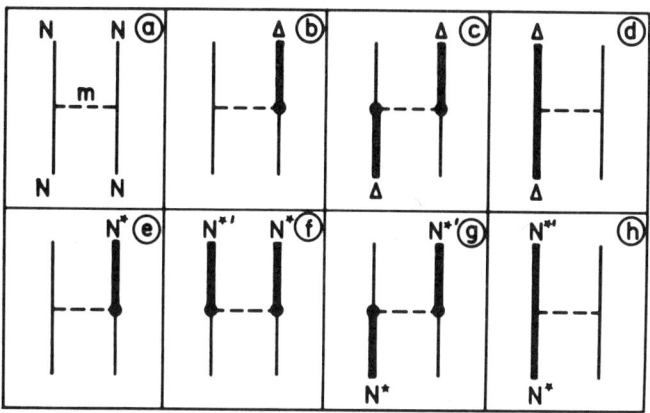

Fig. 4: The general form of the baryon-baryon interaction of eq. (7) in a meson exchange model.

Accordingly, the pion emission operator Λ of fig. 5 is split up:

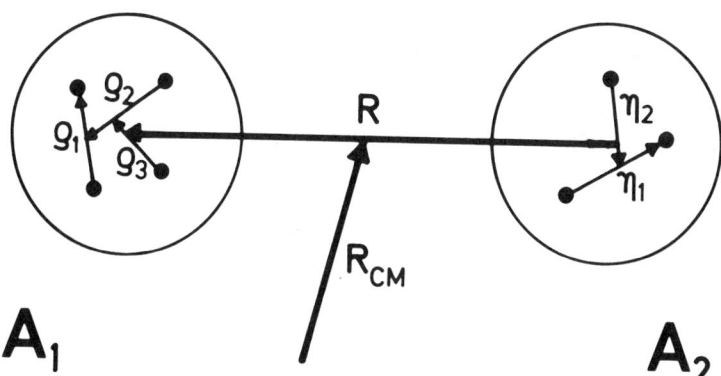

Fig. 5: Graphical representation of Jacobi Coordinates used in eq. (11)

$$\hat{\Lambda} = \hat{\Lambda}^{nonres} + \hat{\Lambda}^{res} = \sum_{j=1}^{A} \lambda_{NN\pi}(j) + \sum_{j=1}^{A} \lambda_{\Delta N\pi}(j) \qquad (9)$$

Following the methods of ref. 20, the reaction amplitude is given by:

$$\hat{T} = \hat{T}^{nonres} + \hat{T}^{res} \qquad (10a)$$

with

$$\hat{T}^{nonres} = \hat{\Lambda}^{nonres} \qquad (10b)$$

and

$$\hat{T}^{res} = \hat{\Lambda}^{res} \cdot \hat{G} \cdot W^{res} \qquad (10c)$$

For the sake of clarity we do not discuss the representation of in- and out-states of eqs. (10b, 10c). For those details the interested reader is referred to ref. 20 and a forthcoming publication[14]).

In the following we shall restrict our discussion to the resonant part of eq. (10) which contains the supposedly important and interesting $\Delta(33)$-excitation. According to eq. (10c) and to the schematic representation of fig. 3, there are three different steps:

1st step: IGNITION, W
As an effect of the operator W^{res} the kinetic energy of the entrance channel is stored in form of a quark spin excitation of one of the nucleons, either of the target or the projectile nucleus.
In Fig. 4 various such elementary process are indicated; for the purpose of this paper we want to concentrate on one possible mechanism, the (NN→ΔN) transition (see Fig. 4b); it should be pointed out, however, that the following treatment is quite general, it can also be applied to a large class of similar mechanisms (such as the ones shown in Fig. 4c,d; they will be discussed in Chapter V). According to our previous discussion it is most important to find out how such an elementary two body transition, NN→ΔN, can couple to the cm-motion of the two nuclei involved. For that we have to introduce explicitly the cm-coordinates, R_1 and R_2 of the two incoming nuclei, A_1 and A_2, respectively, and to transform the single particle coordinates accordingly. The appropriate Jacobi transform is shown in Fig. 6; we obtain the following result:

$$\underline{r}_{ik} = \underline{r}_i - \underline{r}_k = \underline{v}_i(\xi) - \underline{v}_k(\eta) + \alpha_{12} \underline{R} \qquad (11a)$$

where ξ and η denote the internal coordinates of the two nuclei, A_1 and A_2, respectively, and R the relative distance of the two centers of masses:

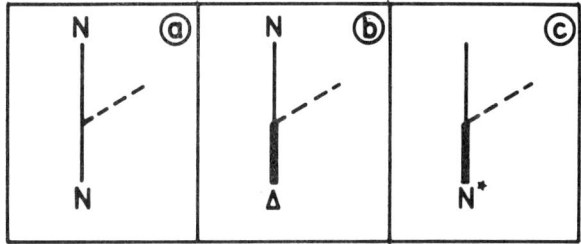

Fig. 6: Elementary pion production diagrams.

$$\alpha_{12} = \sqrt{\frac{A_1 + A_2}{A_1 \cdot A_2}} \qquad (11b)$$

For practical purposes we introduce the Fourier transform $w(q)$ of $w(i,k)$:

$$w(i,k) = \frac{1}{(2\pi)^3} \int d^3q \, w(q) \{ \hat{x}(i) \hat{y}(k) e^{-i\vec{q}\cdot\vec{r}_{ik}} + \qquad (12)$$
$$+ \hat{y}(i) \hat{x}(k) e^{i\vec{q}\cdot\vec{r}_{ik}} \}$$

where $\hat{x}$ and $\hat{y}$ denote single particle operators acting on the two nucleons involved. With the conventional representation of vertex operators we obtain for the NN → NΔ interaction of Fig. 4b:

$$\hat{x}(i) = f(q)/m_\pi \, \vec{\sigma}_i \cdot \vec{q} \, \vec{\tau}_i \cdot \hat{\phi}_\pi \qquad (13a)$$

$$\hat{y}(k) = f^*(q)/m_\pi \, \vec{S}_k^\dagger \cdot \vec{q} \, \vec{T}_k^\dagger \cdot \hat{\phi}_\pi \qquad (13b)$$

Inserting eqs. (11), (12) and (13) into eq. (8c) we arrive at the following result:

$$W = \frac{1}{(2\pi)^3} \int d^3q \, w(q) \{ X_q(s) Y_q(\eta) Z_q(R) + \qquad (14)$$
$$+ Y_q(s) X_q(\eta) Z_{-q}(R) \}$$

with

$$X_q(\xi) = \sum_{j=1}^{A_1} e^{-i q \cdot u_j(\xi)} \hat{x}(j) \qquad (15a)$$

$$Y_q(\eta) = \sum_{j=1}^{A_2} e^{i q \cdot v_j(\eta)} \cdot \hat{y}(j) \qquad (15b)$$

$$Z_q(\underline{R}) = e^{-i \alpha_{12} q \cdot \underline{R}} \qquad (15c)$$

Eq. (14) clearly demonstrates that the ignition operator W of eq. (8), affects the coordinate R, i.e. the relative motion of the two incoming nuclei. The chosen representation in momentum space clearly reveals that the momentum exchange between nucleons bound in different fragments also slows down the relative motion.

2nd step: PROPAGATION of the intermediate system, $\hat{G}$
The propagator $\hat{G} = (\omega - H)^{-1}$ now contains the generalised nuclear Hamiltonian H which treats all the baryons on the same footing irrespective of their internal state of excitation; in addition to the NN-interaction of fig. 4a also the graph of fig. 4(c-h) are accounted for in H (see ref. 21). For the case of the nuclear Δ-excitation we introduce the corresponding eigenmodes $|\mu\rangle$ of the system:

$$H |\mu\rangle = \mathcal{E}_\mu |\mu\rangle \qquad (16)$$

with the (complex) eigenenergies $\mathcal{E}_\mu$. Using the biorthogonal set of eigenfunctions we obtain:

$$\hat{G} = \sum_\mu |\mu\rangle \frac{1}{\omega - \mathcal{E}_\mu} \langle \tilde{\mu}| \qquad (17)$$

Those A* eigenmodes of eq. (17), the nuclear analogues of the elementary Δ(33) resonance, have been used in a number of investigation of pion-, photo- and electro-induced reactions on nuclei[22,23]. The results of those microscopic solutions of eq. (16) are summarized in fig. 7 in comparison with the elementary Δ-resonance:
(i) the various multipolarities ($\mu = (J^\pi, T)$) are energetically discriminated from each other by the ΔN-interaction of fig. 4;
(ii) the ΔN-dynamics shift the position of the important multipoles to lower energies by as much as 40 - 50 MeV; there is a significant broadening, particularly for the lower multipoles.

In essence there are only a very few collective resonances which dominate the spectrum of fig. 7 and which might also play an important role for the present investigation.

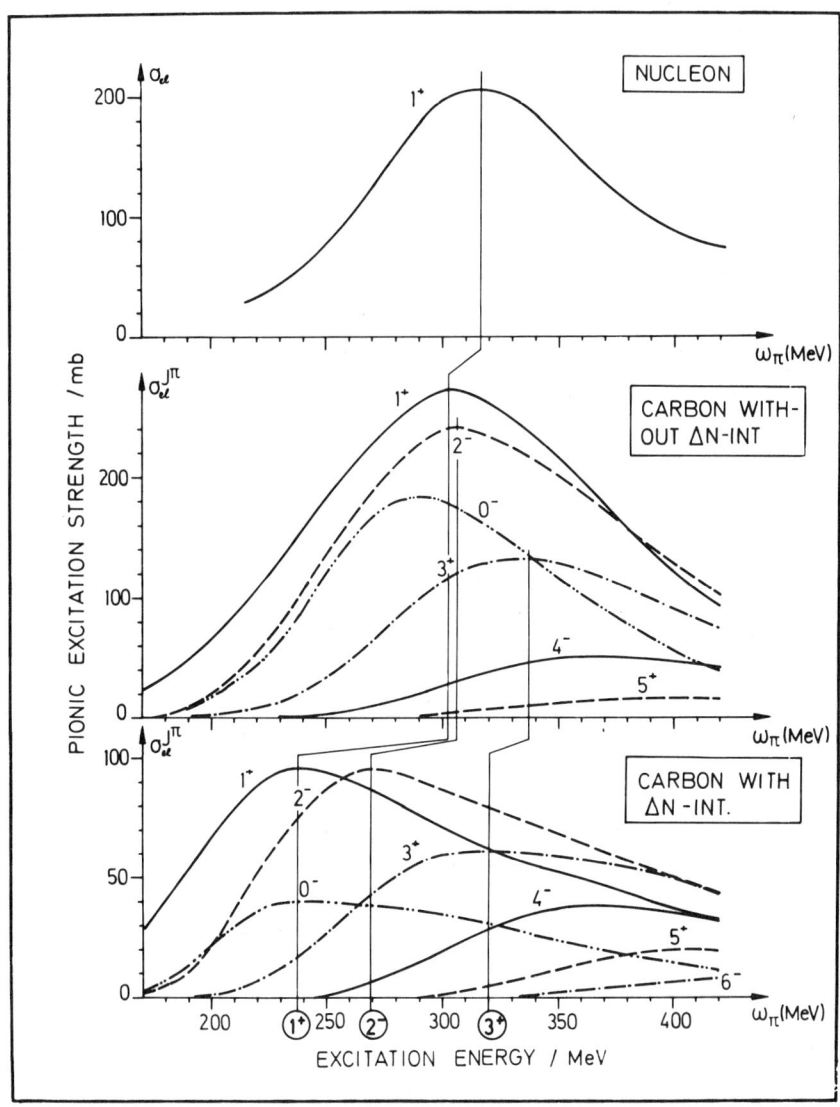

Fig. 7: Pionic excitation strength distribution of the nucleon and of the corresponding nuclear A*-multipoles for the case of ^{12}C. The effects of the ΔN interaction (see the corresponding diagrams of fig. 4) are clearly visible.

3rd step: DECAY of the nuclear N*-excitation, Λ
Here, we use a conventional pionic decay operator (see fig. 5b):

$$\Lambda^{res} = \sum_{j=1}^{A} \lambda_{\Delta N\pi}(j) = \frac{1}{(2\pi)^{3/2}} \int d^3k' \sum_{j=1}^{A} e^{-i\underline{k}'(\underline{r}_j - \underline{r}_\pi)} \hat{y}^+(j) \qquad (18)$$

This process is closely related to pion-nucleus scattering in the resonance region[20,22]. We therefore can be quite brief in this context.

In passing it should be noted that this reaction mechanism exhibits some similarities with resonant pion scattering: except for the initial, i.e. ignition phase: the propagation and the decay of the Δ in the nuclear medium is very similar to what has been studied in connection with pion induced nuclear reactions in the resonance region[20,22].

To summarize this discussion we obtain the following result for the reaction amplitude T_{fi}^{res} of eq. (10c):

$$T_{fi}^{res}(\underline{K},\underline{k},\nu) = \sum_\mu \langle fin(\underline{k},\nu) | \Lambda^{res} | \mu \rangle \cdot \qquad (19)$$

$$\cdot \frac{1}{\omega - \varepsilon_\mu} \langle \tilde{\mu} | W^{res} | in(\underline{K}) \rangle$$

We should mention the selectivity that is expressed in the structure of eq. (14): those states of the united nucleus B will be strongly populated whose cluster decomposition matches the given entrance channel fragmentation. This point will become more transparent in the next Chapter; the corresponding selection rules are quite relevant for the discussions in Chapters V and VI.
In the following we will report on an application of this model to a specific reaction.

III. APPLICATION TO A SPECIFIC REACTION

We investigate the specific pionic fusion process

$$^3He + ^3He \longrightarrow {}^6Li_\nu(J,T) + \pi^+ \qquad (20)$$

which has recently been studied experimentally at Orsay by Bornec et al[2,3]. Data were taken at threshold: T_{CM} = 141 and 135 MeV, respectively. In the actual calculation we introduce the closure approximation:

$$(\omega - \mathcal{E}_\mu)^{-1} \approx (\omega - \bar{\mathcal{E}})^{-1} \tag{21}$$

where $\bar{\mathcal{E}}$ denotes the (complex) closure energy. Then eq. (19) reduces to:

$$T_{fi}^{res,c}(\underline{K},\underline{k},\nu) = \frac{1}{\omega - \bar{\mathcal{E}}} \langle fin(\underline{k},\nu) | \Lambda W^{res} | in(\underline{K}) \rangle \tag{22}$$

Applying the treatment of the previous chapter to the operator ΛW we arrive at:

$$\Lambda \cdot W = \frac{1}{(2\pi)^{9/2}} \int d^3q \, w(q) \int d^3k' \left\{ \hat{F}_q(\xi) \hat{G}_{q,\underline{k}}(\eta) \cdot \hat{H}_{q,\underline{k}'}(\underline{R}) + \hat{G}_{q,\underline{k}'}(\xi) \hat{F}_q(\eta) \cdot \hat{H}_{-q,\underline{k}'}(\underline{R}) \right\} \tag{23}$$

The operators $\hat{F}$, $\hat{G}$ and $\hat{H}$ are closely related to $\hat{X}$, $\hat{Y}$ and $\hat{Z}$ of eqs. (15), respectively. For details see ref. 13, 14.
Introducing a cluster representation for the ^{6}Li-wavefunctions and restricting the space only to a (^{3}He x ^{3}H) fragmentation we are able to calculate the cross section for the reaction of eq. (20). The details of this calculation are reported in ref. 13. The present results, however, were obtained with slightly modified, in fact more realistic form factors. The closure energy $\bar{\mathcal{E}}$ of eq. (21) has been shifted by 50 MeV with respect to the free Δ position in order to account for the dynamical shift of the low Δ^* multipoles (see bottom part of fig. 7). The results are shown in Fig. 8. From the comparison with the experimental data it can be concluded that indeed the present model semiquantitatively accounts for the observed cross sections. In view of the microscopic approach to this complex reaction the quantitative agreement appears in fact quite surprising. We like to mention that in a pick up model analysis of Germond and Wilkin a similar agreement with the data of reaction (20) could be achieved[24,25].

With the same assumptions the energy dependence of the ground state fusion has been calculated. This is shown in Fig. 9a. We clearly see in the excitation function the role of the nuclear Δ as an intermediate resonance; its influence, however, is moderated by the kinematical constraints that become more stringent with increasing energies. The results of Fig. 9 a have been calculated with the same closure approximation used at threshold; in fact this assumption is certainly less appropriate in the proper resonance region. Still, those results might be useful as a guide; they are meant to indicate the sensitivity of the cross sections to the details of the input assumptions. From Fig. 9 there is no question that the energy dependence of the pionic fusion process contains valuable information,

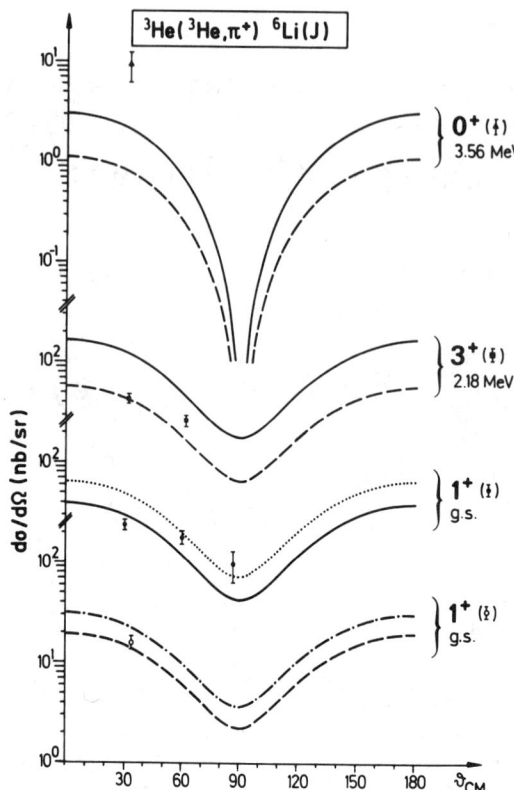

Fig. 8:
Differential cross sections for $^3He(^3He,\pi^+)^6Li$ for several states of the fusioned nucleus. Experimental data from Orsay[2,3].

mainly on the reaction mechanism.

Here, we like to emphasize the selectivity of the reaction (2) which is most clearly exhibited in eqs. (23): those states of the final nucleus B will be strongly excited whose cluster structure matches the entrance channel fragmentation. If a rearrangement between the two incoming nuclei is necessary to arrive at a given final state $|\nu\rangle$, then the corresponding reaction cross section is expected to be reduced drastically. It would be highly interesting to see whether such a selection rule actually exists. If so, this argument could be quite valuable, for example to tailor the appropriate entrance channel fragmentation for the production of a given isotope; this point will be taken up in Chapter VI.

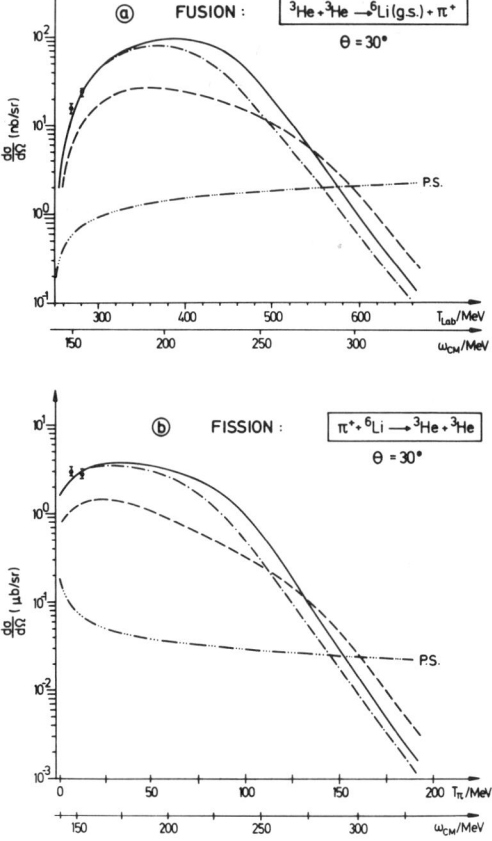

Fig. 9: (a) Excitation function for the pionic fusion reaction: ^{3}He(^{3}He,π^+)^{6}Li (g.s.). The different curves correspond to different values of the closure energy $\bar{\mathcal{E}}$ of eq. (21) within the limits indicated in fig. 7; PS denotes the corresponding phase space factor.
(b) Excitation function for pionic fission, obtained from (a) by detailed balance.

IV. PIONIC FISSION

Information about the reaction (2) can also be obtained from the inverse reaction:

$$\pi + B(g.s.) \rightarrow A_1 + A_2 \qquad (24)$$

Obviously, the ground state fission of the target nucleus into various binary fission channels can be studied (see Fig. 10) simultaneously. According to eq. (23) the pattern of the various two body decay channels depends strongly on the cluster decomposition of the target nucleus.

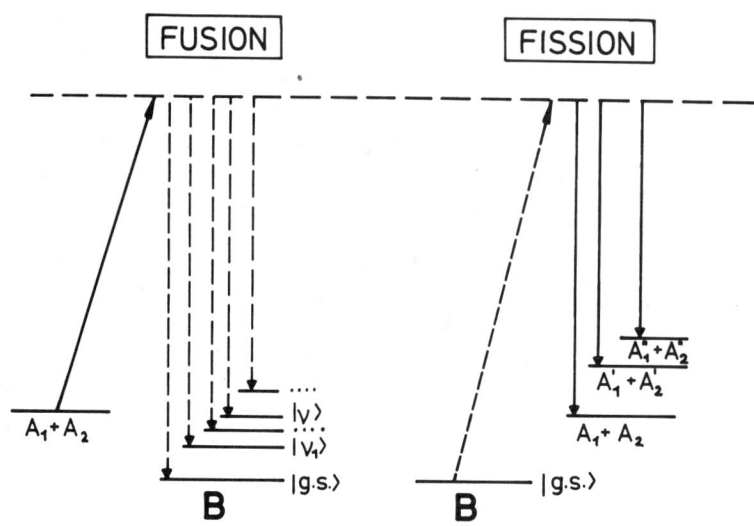

Fig. 10: A schematic comparison of pionic fusion and fission.

This is not the place for a detailed discussion of this reaction; we should point out, however, that in the vicinity of the threshold the phase space factors clearly favour the fission over the fusion process. This can be seen in Fig. 9b where the excitation function is shown for the reaction

$$\pi^+ + {}^6Li\,(g.s.) \rightarrow {}^3He + {}^3He \qquad (25)$$

It turns out that the corresponding cross section are of the order of several μb/sr. It seems that the energy dependence of the corresponding fusion/fission process could be studied in such a fission experiment quite simply, even for various (A_1/A_2) combinations. In this context we should mention that only scarce data on this fission process are available[26,27,28]; a more systematic and complete investigation, however, seems to be possible[29].

It should be pointed out, that both the fragmentation of the target nucleus and the corresponding excitation functions contain valuable information on the cluster structure of the target nucleus and on the reaction mechanism (see for example the results displayed in Fig.9b).

V. COOPERATIVE REACTIONS AS A TOOL

Pionic fusion is an example of a highly cooperative nuclear process.
It can be characterized neither by a direct nor by a statistical
reaction mechanism:
- it involves all the nucleons, there are no spectator particles;
- the process is highly coherent, the energy is contained in only
  a few degrees of freedom.

There are other possible types of such reactions; we only want to
mention the (exclusive) radiative capture of two complex nuclei: here,
the energy is taken away by a photon, the two nuclei again undergo a
coherent fusion process to form a specific bound state $|\nu\rangle$ of the
final nucleus, A:

$$A_1 + A_2 \longrightarrow A_\nu + \gamma \qquad (26)$$

As an example we like to mention the $^{12}C(\alpha,\gamma)^{16}O$ -reaction recently
studied by Sandorfi et al[30] in the region of the conventional Giant
Resonance excitations: The γ-spectrum clearly reflects the selectivity
of the reaction with respect to the population of the final nuclear
state $|\nu\rangle$ for a given projectile target combination. It looks as if
a similar mechanism (see fig. 3) like the one discussed in Chapter II
for pionic fusion would be appropriate to explain the experimental
data[31], with essentially two distinct differences:

(i) the ignition operator W leads to a conventional particle hole
    excitation (doorway configurations);

(ii) the intermediate resonances $|\mu\rangle$ of eq. (16) which act as a
     storage of the free energy are the conventional Giant Multipole
     Resonances.

In both reactions the intermediate excitation of collective nuclear
modes seems to be responsible for preventing the system to thermalize.

In the following we like to point out a few examples where pionic
fusion (and/or radiative capture) can be used to investigate specific
nuclear properties:

(i) the excitation of high spin states:
    the reactions (2) and (26) are characterized by significant
    angular momentum mismatch; therefore, we expect that - for a
    given projectile target combination - the corresponding high
    spin structures will be favoured over the low spin partners of
    the same band;

(ii) the production and investigation of new isotopes:
     clearly the reaction (2) allows to coherently produce new elements, even beyond the range of the presently available domain
     of stable isotopes; again: it will be crucial for the production
     cross section of a specific isotope B(Z,N) whether or not the
     target projectile combination matches the cluster structure of B.
     This is particular interesting for heavy ion collisions: The

energy necessary to overcome the Coulomb-barrier can be easily carried away by the emitted pion thus leaving the united nucleus even in a bound state ("ultracold fusion")*). Furthermore, the charge of the pion can be exploited to produce the same isotope with different entrance channel fragmentations or similarly, to reach different isotopes from the same entrance channel.

(iii) <u>the coherent propagation of higher N* resonances</u> or even of coherent multiple N* excitations through a nucleus: at higher incident energies also the processes shown in Fig. (4e-h) are possible; the corresponding eigenmodes $|\mu\rangle$ of eq. (16) reflect the dynamics of those excitation and at the same time determine the cross section for those reactions. It should be pointed out that some of those $(N^*\bar{N})$ particle hole configurations also couple strongly to the $(\Delta\Delta\bar{N}N)$ excitations which finally might decay by the emission of a ρ-meson or a correlated $(\pi\pi)$-pair[34,35]:

$$A_1 + A_2 \rightarrow B_\nu + \rho \qquad (27a)$$

$$A_1 + A_2 \rightarrow B_\nu + (\pi,\pi) \qquad (27b)$$

The advantage of such an exclusive reaction for the investigation of those modes of nuclear excitations is obvious; with increasing the incident energy, however, the momentum mismatch becomes more pronounced which may lead to a further reduction of the corresponding cross sections. Therefore, we expect that the applicability of such a cooperative reaction will be restricted to a few selected examples. In this context we like to refer to the $\bar{d}(p,\pi)^3$H reaction of Goldzahl et al[36] in the region of the $(\Delta\Delta\bar{N}N)$ excitation of the (A=3) system.

From the above considerations it is rather evident that pionic fusion is not only a reaction which is particularly interesting from a purely dynamical point of view but that this process can also be used as a tool to investigate different questions in the fields of conventional nuclear structure physics, of nuclear collisions and of the

---

*) Another possibility for a cold fusion reaction is to have a fast particle (preferentially a nucleon or an α particle) to carry away a large fraction of the excess energy; such a reaction has been studied for example by Gierlik et al[32]. A still another technique is the matching of the Q-value for a given reaction, a method which recently has been used by Münzenberg et al[33] to produce and identify the element Z = 107.

creation and propagation of subnuclear excitations of complex nuclei.
The extent, however, to which those reactions can efficiently be
used depends on a deeper understanding of the reaction mechanism and
- of course - on their experimental feasibility. In view of the above
mentioned perspectives it seems to be most important to further
investigate this reaction in more detail.

VI. SUMMARY AND OUTLOOK

The pionic fusion reaction discussed in the previous Chapters raises
a number of questions which touch problems usually not in the center
of interest in the field of nuclear collisions; to name only two of
the most conspicuous ones:

- What is the mechanism that prevents the thermalisation of the
  energy in a nuclear collision?
- How does the relative motion of the incident nuclei couple to the
  outgoing pion field?

At the moment we are at the beginning of an understanding of those
processes. There are still many open problems; their solution might
eventually lead us to a better comprehension of the phenomenon of
coherence among the nuclear constituents. Apart from this immediate
goal those cooperative reactions might turn out to be a useful and
even an efficient tool to investigate specific questions in the
fields of nuclear and subnuclear excitations as discussed in the
previous chapter. All those features are intimately connected with
each other; the specific nature of those exclusive processes permits,
however, a systematic investigation of this phenomenon in a number of
different ways:

- by varying the target projectile fragmentation for a given final
  nucleus, B;
- by looking for the production of different states $|v\rangle$ of the united
  nucleus;
- by varying the incident energy;
- by investigating the inverse reaction; and finally:
- by comparing various kinds of cooperative reactions with each other,
  like pionic fusion/fission and radiative capture or twobody photo/
  electrofission, respectively.

This list is by no means complete, but it indicates already that we
are dealing with a very rich phenomenon which can be investigated by
a number of quite different techniques.

Of course such a systematic investigation requires special detection
methods: due to the exclusive character of the two body final channel
the cross sections are expected to be rather low by conventional
standards (of the order of several pb/sr, with an estimated variation
of a few orders of magnitude). In order to measure such cross sections
the unique properties of the exit channel can be used by directly

measuring either the spectrum of the produced fast pions (which is free of a physical background) or the recoiling nucleus, B, which is characterized by a particularly large recoil momentum (in fact it is the largest momentum that can be achieved for a given entrance channel). Furthermore, it should be kept in mind that valuable information could be obtained in special cases from the observation of the radioactive decay products.

In view of the fundamental questions associated with the pionic fusion process an intensified effort seems to be justified, both to clarify the conceptual and theoretical problems and to provide further experimental information.

Acknowledgement

During the preparation of this talk we enjoyed close contact with Manfred Dillig and Wolfgang Knüpfer; furthermore it is a pleasure to acknowledge a number of stimulating discussions with many colleagues on the subject of this paper; we like to mention in particular N. Willis, R. Bent, L. Bimbot, Y. le Bornec, E. Hilf, H. Nann and C. Wilkin.

REFERENCES

1. E. Gardner and C. M. G. Lattes, Phys. Rev. 74 (1948), 1236A, 1558A; Science 107 (1948) 270
2. L. Bimbot, M. P. Combes, J. C. Jourdain, Y. Le Bornec, F. Reide, A. Willis and N. Willis, Contr. 9th Int. Conf. High Energy Physics and Nuclear Structure, Versailles (1981)
3. Y. Le Bornec, L. Bimbot, N. Koori, F. Reide, A. Willis, N. Willis and C. Wilkin, Phys. Rev. Lett. 47 (1981) 1870; Y. Le Bornec and N. Willis, Contr. to this workshop
4. E. Aslanides et al., to be published
5. T. Ward, private communication
6. E. Aslanides et al., Phys. Rev. Lett. 43 (1979) 1466; 45 (1980) 1738
7. W. Benenson et al., Phys. Rev. Lett. 43 (1979) 683
   L. S. Schoeder et al., Phys. Rev. Lett. 43 (1979) 1787
   I. Tanihata et al., Phys. Lett. B 87 (1979) 349
8. G. F. Bertsch, Phys. Rev. C 15 (1977) 713
   H. J. Pirner, Phys. Rev. C 22 (1980) 1962
   S. Bohrmann and J. Knoll, Nucl. Phys. A 356 (1981) 498
9. N. S. Wall, J. N. Craig and D. Erzow, Nucl. Phys. A 268 (1976) 459
10. J. Eggermann et al., Z. Phys. A 273 (1975) 273
11. M. G. Huber and M. Dillig, Proc. of the Workshop on High Resolution Heavy Ions Physics, Saclay (1978)
12. M. G. Huber, M. Dillig and K. Klingenbeck, Contr. Int. Conf. High Energy Physics and Nucl. Structure, Versailles (1981)
13. K. Klingenbeck, M. Dillig and M. G. Huber, Phys. Rev. Lett. 47 (1981) 1654
14. K. Klingenbeck, to be published
15. B. Hoistad, Contr. to this Workshop

16. H. Nann, Contr. to this Workshop
17. P. Couvert, Contr. to this Workshop
18. M. Dillig et al., Contr. to this Workshop
19. P. L. Walden, Proc. 9th Int. Conf. High Energy Physics and Nucl. Structure, Versailles (1981)
20. K. Klingenbeck and M. G. Huber, Phys. Rev. C 22 (1980) 681
21. K. Klingenbeck, Lecture Notes in Physics, Vol. 137, 102 Springer Verlag 1981
22. M. G. Huber and K. Klingenbeck, in: Studies in High Energy Phys. Vol. 2, ed. W. Bertozzi et la., Harwood Acad. Publ. (1980)
23. K. Klingenbeck, Phys. Lett. 98 B (1981) 15
24. J. F. Germond and C. Wilkin, Phys. Lett. 106 B (1981) 449
25. J. F. Germond, Contr. to this Workshop
26. J. F. Amann et al., Phys. Rev. Lett. 40 (1978) 758
27. H. C. Walter, private communication
28. H. Ullrich, private communication
29. R. R. Johnson, private communication
30. A. M. Sandorfi et al., Phys. Rev. Lett. 46 (1981) 884
31. W. Knüpfer, private communication
32. E. Gierlik et al., Z. Phys. A 295 (1980) 295
33. G. Münzenberg et al., Z. Phys. A 300 (1981) 107
34. K. Klingenbeck, Nucl. Phys. A 358 (1981) 399 c
35. M. G. Huber and K. Klingenbeck, Nucl. Phys. A 358 (1981) 243 c
36. J. Banaigs et al., Contr. 9th Int. Conf. High Energy Physics and Nuclear Structure, Versailles (1981)

DISCUSSION

D. Koltun (Univ. of Rochester): I would like to remark on one feature that you stressed, with which I completely agree. You have taken the point of view that all the physics comes from the coherence, if you like, and the reaction mechanism is rather simple. That is, if we talk in terms of interactions.
Huber: Elementary interactions.
Koltun: It's an elementary two-body interaction, and all the coherence is one coherence, which we're continually reminded is very important in nuclear physics; it is the center of mass motion. I think it's worth thinking about this in connection with the previous talk, the one about clusters and nuclei, because it's just a reminder of the distinction between reaction mechanism and the nuclear structure. Sometimes it is a little artificial, but this way of looking at -- in your case production but in the other case absorption -- also shows you can get things that look like cluster absorption, which have to do with the nuclear structure; that is, they have to do with the way the target looks and not necessarily with the absorption mechanism.
Huber: Thank you very much for this remark. I agree completely.

A.W. Thomas (TRIUMF): You talked about the reaction from the direction of two ions colliding. Would you like to comment on the relative experimental merits of doing it from the other direction?
Huber: Basically the two reactions are related to each other, but they lead to different transitions. For example, in fusion two nuclei get together. They go via some intermediate doorway state into their final states, say in the ground states. Of course, this is the same thing as shooting a pion on the ground state and you are left with the same fragmentation over there. But if you do the fusion reaction, you can get the final nucleus in excited states, and here the selection rules very clearly should show up -- some states should come strongly others less strongly -- whereas in a fission reaction you always end up in the ground state of the two outgoing partners. But again, you have an advantage: if you shoot the pion on the nucleus you may get different fragmentations out simultaneously. This is very closely related to the point Kolton made. So, the two reactions -- although they are basically very similar -- complement each other, and one should look at both.

PION-INDUCED FISSION

J.-F. Germond*
Institut de Physique de l'Université, CH-2000 Neuchâtel, Switzerland

C. Wilkin
University College London, London WC1E 6BT, UK

ABSTRACT

The $A(\pi+,^3He)B$ reaction near threshold is studied in a model where the pion is absorbed by an ^{4}He constituent of the target nucleus. The predictions of this model using harmonic oscillator cluster wave functions agree semi-quantitatively with the experimental data on the inverse reaction.

At the Berkeley conference [1] the Orsay group showed for the first time preliminary data on coherent production of pions close to threshold with a ^{3}He beam. Distinct states of ^{6}Li were observed in the ^{3}He(^{3}He,$\pi+$)^{6}Li reaction with cross sections of the order of tens of nb/sr. At this conference they [2] presented new results for other targets ^{4}He, ^{6}Li and ^{10}B. The striking feature is that whereas the ^{7}Li production rate is rather similar to that for ^{6}Li, the yields of ^{9}Be and ^{13}C are several orders of magnitude smaller.

The maximum energy of the Orsay ^{3}He beam is 283 MeV so that the pions produced have typically centre-of-mass energies of 25 MeV or less and this is not very far from the domain of pionic atoms. Indeed the scattering amplitude extracted from stopping data on ^{6}Li ($\pi-$,tt) is in very good agreement with that obtained for the charge-symmetric, time-reversed reaction ^{3}He^{3}He $\to \pi+^6$Li at Orsay [3]. There can thus be no doubting the large rates of these coherent reactions despite the fact that they require a significant transfer of momentum to three nucleons.

Not only does a theoretical model have to explain these reactions; it should also describe other rare channels of pion absorption. It should be able to account for example for the branching ratio of 1% [4] in ^{6}Li($\pi-$,td)n with a "spectator" neutron. Theories of (p,π) reactions can hardly be treated as definitive so that we should not yet seek a fully microscopic theory of the more complicated ^{3}He-induced processes. We have presented [5] a simple model which is best visualised in the direction of pion absorption so that it is convenient to indicate the reaction merely by the initial nucleus in this sense.

A tolerable nuclear model in the case of ^{6}Li is to take it as an alpha particle plus a deuteron. Since we are interested in a channel where three nucleons need to be transferred, pion absorption on the deuteron component is not favoured. On the other hand the two-body absorption $\pi+^4$He $\to$ p^{3}He is known to be very strong both

* Speaker at this Conference.

in flight [6,7] and at rest [8]. If there is a final state interaction between the produced proton and spectator deuteron we have a mechanism for driving the $\pi^+{}^6\text{Li} \to {}^3\text{He}{}^3\text{He}$. This is illustrated in general by the graph of figure 1.

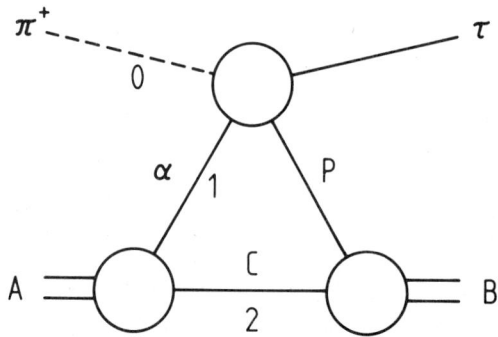

Fig. 1. Model for the $A(\pi^+,{}^3\text{He})B$ reaction.

Here we postulate a cluster model for the nucleus A consisting of an α-particle plus C, which is some multi-nucleon state. Similarly the state B, composed of C plus a proton, may be bound or unbound as in the case of the ${}^3\text{He}$-d-p final state. Due to the complexity caused by the identical ${}^3\text{He}$ in the ${}^6\text{Li}$ case let us start our discussion with the explicit example of $\pi^+{}^7\text{Li} \to {}^3\text{He}{}^4\text{He}$.

In terms of the scattering amplitude f, the unpolarised centre-of-mass differential cross section is

$$\left(\frac{d\sigma}{d\Omega^*}\right)_{\pi A \to \tau B} = \frac{p_{\tau B}}{p_{\pi A}} (2s_\tau+1)(2s_B+1) \overline{|f^2_{\pi A \to \tau B}|} \tag{1}$$

where $p_{\pi A}(p_{\tau B})$ is the initial (final) momentum, s_i are the spins and the bar denotes spin averaging. It is evident from figure 1 that we seek a linear relation between $f_{\pi^7\text{Li} \to {}^3\text{He}{}^4\text{He}}$ and $f_{\pi\alpha \to p{}^3\text{He}}$. There are two independent forms for the latter, both corresponding to the spin S=1 combination of the p and ${}^3\text{He}$.

$$f_{\pi\alpha \to p{}^3\text{He}} = (f_1 \vec{k}_\tau + f_2 \vec{k}_\alpha) \cdot \vec{S} \tag{2}$$

We here use the convention that the k's are variables in the π-α system, the p's in the π-${}^7\text{Li}$. The amplitude has an explicit momentum dependence so that the corresponding transition operator in configuration space takes the Galilean-invariant form

$$\mathcal{J}(\vec{r}_0,\vec{r}_1) = \frac{i}{m_\alpha} \vec{S} \cdot (m_p \vec{v}_0 - m_\tau \vec{v}_1) \tilde{f}_1 (\vec{r}_0-\vec{r}_1)$$
$$+ \frac{i}{(m_\alpha+\mu)} \tilde{f}_2(\vec{r}_0-\vec{r}_1) \vec{S} \cdot (m_\alpha \vec{v}_0 - \mu \vec{v}_1) \tag{3}$$

When sandwiched between plane waves this gives directly equation (2) in the centre-of-mass. However to evaluate the amplitude corresponding to fig. 1 we must take matrix elements between α-particle and ^{7}Li wave functions, ϕ_α and ϕ_{Li}, as well as the corresponding plane wave factors. After doing this we find

$$f_{\pi Li \to \tau \alpha} = K(f_1 F_1 + f_2 F_2) \tag{4}$$

where K is a kinematical factor which for low energy pions is approximately $4/\sqrt{7}$. The form factors F_i are Galilean invariant but in the centre-of-mass frame take the form

$$F_1 = \frac{1}{4} \int \left[\frac{7}{4} \vec{P}_\alpha \phi_\alpha^*(r) + 3i\vec{\nabla}\phi_\alpha^*(r) \right] \cdot \vec{S} \, e^{i\vec{Q}\cdot\vec{r}} \, \phi_{Li}(\vec{r}) \, d^3r$$

$$F_2 = \frac{1}{(m_\alpha+\mu)} \int \phi_\alpha^*(\vec{r}) e^{i\vec{Q}\cdot\vec{r}} \vec{S} \cdot [m_\alpha(1+\mu/m_{Li})\vec{P}_{Li}\phi_{Li}(\vec{r}) - \mu i \vec{\nabla}\phi_{Li}(\vec{r})] d^3r \tag{5}$$

with the momentum transfer variable

$$\vec{Q} = \frac{3}{4}\vec{P}_\alpha - \frac{3}{7}\vec{P}_{Li} \tag{6}$$

The evaluation of the form factors is simplified considerably if the ^{4}He wave function is Gaussian

$$\phi_\alpha(r) = \left(\frac{\alpha^2}{\pi}\right)^{\frac{3}{4}} e^{-\frac{1}{2}\alpha^2 r^2} \tag{7}$$

for then if we neglect terms of order μ/m_α,

$$F_1 = \frac{1}{4}\left[\frac{7}{4}\vec{P}_\alpha - 3\alpha^2 \vec{\nabla}_Q\right] \cdot \int \phi_\alpha^*(r) \, \vec{S} \, e^{i\vec{Q}\cdot\vec{r}} \, \phi_{Li}(\vec{r}) \, d^3r$$

$$F_2 = \int \phi_\alpha^*(r) \, e^{i\vec{Q}\cdot\vec{r}} \vec{S} \cdot \vec{P}_{Li}\phi_{Li}(\vec{r}) \, d^3r \tag{8}$$

If we quantise along the direction of $\vec{Q}$ only transitions from $m = \pm \frac{1}{2}$ of the ^{7}Li to $\pm \frac{1}{2}$ of the ^{3}He are allowed. For a state of ^{7}Li of orbital angular momentum L there are levels with $J = L \pm \frac{1}{2}$ and the gradient term in eqs. (8) acts rather differently in the two cases. For the non-spin-flip form factor with $J = L + \frac{1}{2}$,

$$F_1^{NSF} = -\frac{i^L}{2}\left(\frac{\pi(L+1)}{2}\right)^{\frac{1}{2}} \left[\frac{7}{4} P_\alpha^z I_L(Q) - 3\alpha^2 Q^L \frac{\partial}{\partial Q}\left(\frac{I_L(Q)}{Q^L}\right)\right] \tag{9}$$

but for $J = L - \frac{1}{2}$

$$F_1^{NSF} = -\frac{i^L}{2}\left(\frac{\pi L}{2}\right)^{\frac{1}{2}} \left[\frac{7}{4} P_\alpha^z I_L(Q) - \frac{3\alpha^2}{Q^{L+1}} \frac{\partial}{\partial Q}\left(Q^{L+1} I_L(Q)\right)\right] \tag{10}$$

with analogous formulae for the spin-flip and the F_2's.

To evaluate the radial integrals

$$I_L(Q) = \int \phi_\alpha^*(r) \, j_L(Qr) \, \phi_{Li}(r) \, r^2 dr \tag{11}$$

we need an explicit model for the levels of ^{7}Li. The ground state ($3/2^-$) and first excited state ($1/2^-$) can be given 2p harmonic oscillator wave functions but the next one ($7/2^-$) should be a 1f. Such assignments fit well the electromagnetic properties of the system and lead to I_L's in the form of polynomials times a Gaussian in Q.

For pionic atoms the f_2 amplitude should not contribute and even in the case of the Orsay production experiments its contribution should be small due to the momentum factor in eqs. (8). If we neglect completely the f_2 term we can derive immediately a relation between the π-α and π-^{7}Li cross sections with no phase ambiguities. For example in the case of the ($3/2^-$, $1/2^-$) doublet, which was not resolved in the Orsay work, we obtain

$$\left(\frac{d\sigma}{d\Omega^*}\right)_{\tau\alpha \to \pi Li(3/2^-+1/2^-)} \approx \frac{21\pi}{2} \left(\frac{p_{Li} \, p_\tau}{k_\tau k_\alpha}\right) \left[I_1\left(Q'=Q-\frac{12\alpha^2}{7p_\alpha}\right)\right]^2$$

$$\times \left(\frac{d\sigma}{d\Omega^*}\right)_{\tau p \to \pi\alpha} \tag{12}$$

where kinematic factors of order unity coming from the transformation from the π-α to π-Li centres-of-mass have been dropped. In the same spirit we interpret the formula as giving the cross section at some angle θ^* in the π-Li system as being proportional to a cross section measured at the same angle θ^* in the π-α frame. In principle there could be a small kinematic change here as well which we neglect.

Unfortunately large angle data for $p^3He \to \pi\alpha$ (viz small angle data for $^3He \, p \to \pi\alpha$) are very sparse and we are reduced to extrapolating from the forward hemisphere results. If we take

$$\left(\frac{d\sigma}{d\Omega^*}\right)_{p\tau \to \pi\alpha} = a + b \cos\theta^* + c \cos^2\theta^* \tag{13}$$

then a good fit to the Orsay [6] and IUCF [7] measurements at 200 MeV has a = 1.15, b = 0.28 and c = 0.93 µb/sr. However we really need the parameters for the same laboratory energy in the π-α and π-Li systems which involves an interpolation in energy engendering more uncertainty.

In figure 2 can be seen our model predictions for the ($3/2^-$ + $1/2^-$) and $7/2^-$ for T_{3He} = 282 MeV and the ($3/2^-$ + $1/2^-$) at 268 MeV. Given the crudity of the nuclear model the agreement is better than we have any right to expect : the model reproduces tolerably the shape of the angular distribution and its energy dependence. (T_π almost doubles between the two ^{3}He energies !) The ratio of experiment to theory is R = 0.58 for the ground state doublet but 1.33 for the 7/2⁻. Perhaps it is just a nuclear structure problem to explain these different spectroscopic factors.

Recently an IUCF group has made the first polarization measurements of the $p\uparrow {}^3He \to \pi^+ {}^4He$ reaction and found that even at 200 MeV there are sizable effects [7]. Consequently the f_2 term in eq. (2) is not completely negligible even this near threshold. Unfortunately an asymmetry measurement is sensitive to the combination $Im(f_2^* f_1)$ whereas the combinations which enter directly into our model calculations are rather $Re(f_2^* f_1)$ and $|f_2|^2$. Our preliminary calculations suggest that providing $|k_\alpha f_2| \ll |k_T f_1|$ the predictions for ^7Li should not change greatly. To illustrate this we plot also in fig. 2 the ground state prediction with the unrealistic hypothesis of taking f_1 to be zero.

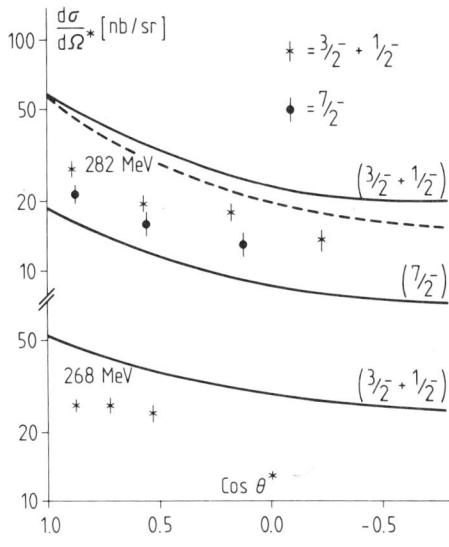

Fig. 2. $^3He\, ^4He \to \pi\, ^7Li$ cross section predictions with $f_2 = 0$ (solid curves) and $f_1 = 0$ (dashed curve) compared with the data of reference 2.

Having treated the example of ^7Li in some detail, we can discuss the results obtained for two other nuclei. The case which we considered first [5] was that of $\pi\, ^6Li \to {}^3He\, ^3He$ but this is much more complicated because of the Pauli principle. Thus we must subtract the amplitude resulting from the exchange of the two ^3He in the graph of fig. 1. Even if we keep only the f_1 amplitude we still need to evaluate the interference between $f_1(\theta^*)$ and $f_1(\pi-\theta^*)$ and as a consequence the ^6Li results depend upon the assumptions made on the angular dependence of the phase of the $\pi-\alpha$ input.

For pionic atoms the amplitudes are purely s-wave, so there is no ambiguity in the input, and we found [5]

$$|\overline{f^2_{\pi Li}}| / |\overline{f^2_{\pi\alpha}}| = 0.8 \times 10^{-2} \quad \text{(theory)}$$
$$= (0.83 \pm 0.24) \times 10^{-2} \quad \text{(experiment)} \quad (14)$$

In fig. 3 predictions of the model are presented for the first three states of ^6Li as a function of $\cos^2\theta^*$ compared to the Orsay data [3] at $T_{{}^3He} = 282$ MeV. The theory falls too low for the ground

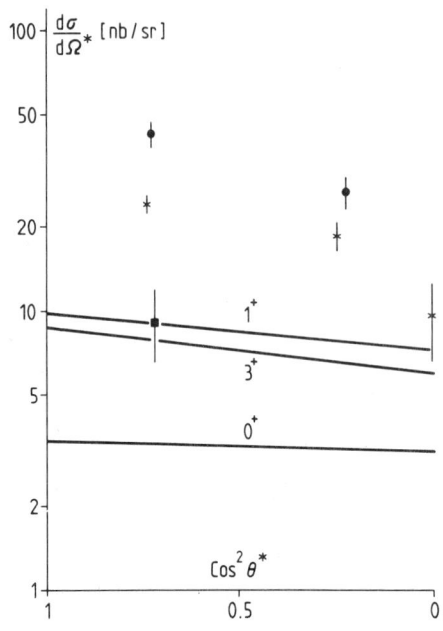

Fig. 3. ^{3}He^{3}He → π^{6}Li cross section predictions at 282 MeV compared to the data of ref. 3 for the 1$^+$ (crosses), 3$^+$ (circles) and 0$^+$ (square) levels.

state (1$^+$) by about a factor of two but we cannot yet say whether this is due to inaccuracies of harmonic oscillator wave functions at large momentum transfer or a spectroscopic factor. If the latter, we need to explain why the 3$^+$ state has a better overlap with an α-d clustering than the 1$^+$.

It is clear from eq. (2) that the f_1 and f_2 terms have opposite symmetry upon the interchange of the proton and ^{3}He so that the influence of the f_2 amplitude could be more important for ^{6}Li than ^{7}Li due to the Pauli exchange. In particular the 0$^+$ level is mainly fed by the f_1 contribution but this requires further investigation.

The data on ^{9}Be production presented at this Conference by the Orsay group ^{2} are of much poorer statistical quality since the centre-of-mass cross section for the production of the ground state is of the order of 70 pb/sr rather than the 25 nb/sr in the case of ^{7}Li. We take the conventional cluster decomposition of ^{9}Be as two α-particles plus a neutron with parameters determined from the nuclear electromagnetic properties. As in the ^{7}Li case a simple relation between the π-α and π-^{9}Be cross sections can be derived which is similar to eq. (12). However the state C in fig. 1 is an α-n pair and the algebra for evaluating the form factors is much more involved. Preliminary calculations using a local form for the transition operator in eq. (3) indicate that the ground state production at 270 MeV and 20° should be 200 pb/sr or less depending on the Ansatz used for the α-α wave function. This is not in contradiction with the Orsay results ^{2} but to be completely convincing all the different low-lying excitations will also have to be estimated.

The reduction in the ^{9}Be case compared to the lithium ones comes in part from the higher pion momentum for the same ^{3}He energy but even at threshold the matrix elements squared differ by one order of magnitude. This may be due to the additional number of nodes at the A-α-C and B-p-C vertices in fig. 1 as well as the larger nuclear size. If this is true then coherent pion production by ^{3}He projectiles may be limited in usefulness to light nuclear targets.

The low value found for the ^{13}C production ^{2} confirms this interpretation.

In the case of ^{7}Li production we see from eq. (12) that the forward angle results should be proportional to the backward angle $\pi\alpha \to p^3$He cross section. This falls fast as a function of energy ^{9} and so does the form factor $I_1(Q)$. This behaviour may be reinforced, as we enter the resonance region, by pion scattering effects which have so far been completely neglected in our calculation. The cross sections should therefore peak close to the threshold for pion production so that machines of the energy of Orsay (or Indiana) are optimal.

Is there any reason for thinking of these rare reactions in terms of pion absorption rather than production, i.e. fission rather than fusion ^{10} ? If we look at a list of rare decay channels of the pionic atom (π^6Li) the t-t branch is just one of many which a model has to explain and should not be entirely divorced from say the t-d-n one. However that work must remain for the next conference.

ACKNOWLEDGEMENTS

This work has benefited from the active help of the Orsay group, in particular Nicole Willis, Louis Bimbot and Yves Le Bornec. We are also very grateful to R.D. Bent for providing us with the IUCF ^{3}He (p,π+)^{4}He data prior to publication. This research was supported in part by the Fonds National Suisse pour la Recherche Scientifique and by the Science Research Council.

REFERENCES

1. N. Willis et al., Proc. Int. Conf. Nucl. Phys., Berkeley (1980).
2. Y. Le Bornec and N. Willis, invited paper presented at this Conference and private communication.
3. Y. Le Bornec et al., submitted to Phys. Rev. Lett.
4. H.-C. Walter, private communication.
5. J.-F. Germond and C. Wilkin, Phys. Lett. (in press).
6. N. Willis et al., J. Phys. G7, L195 (1981).
7. R. Bent, private communication.
8. M. Bloch et al., Phys. Rev. Lett. 11, 301 (1963).
9. J. Källne et al., Phys. Rev. C24 (1981) (September).
10. M. Huber et al., Contributed paper E21 to the 9th ICOHEPANS, Paris, July 1981.

DISCUSSION

W. Wharton (Carnegie-Mellon Univ.): It appears in all your calculations that you have a pion interacting with an α cluster. Do you believe that pions interacting with other kinds of clusters are much less important?
Germond: In the case of the $(\pi^+,{}^3\text{He})$ reactions yes, because we must transfer the pion momentum to at least three nucleons. But if we consider the emission of fragments lighter than ^3He, then we must also take into account the absorption on other clusters than α-clusters.

M. Sternheim (Univ. of Massachusetts): Is there any attempt to compare the data on pion induced fission to fission induced, say, by protons coming in with 140 MeV more energy than the pions —or other projectiles which lead to the same fragments ultimately?
Germond: We haven't tried that up to now.
Sternheim: The question is whether the reaction really depends so much on where the energy comes from, or is it just ....
Germond: Our model assumes that pion-induced fission is a direct reaction which depends primarily on the momentum transfer. This means that the momentum transfer is the relevant variable when comparing with reactions induced by other probes.
Sternheim: I know that on heavy nuclei there is some work comparing protons with pions, and you can't tell what the source of the energy is. You put in the energy and that's it — it doesn't depend on the source. Here, of course, there is a light nucleus, and particular states. You have possibly a more selective process.

SUMMARY OF THE WORKSHOP

G.E. Walker
Department of Physics, Indiana University, Bloomington, IN 47405

INTRODUCTION

In view of the gracious introduction I received from Professor Bent, I feel obliged to delete from my talk the off-color but extremely humorous ancedotes, (at least in the written version of this summary) lengthy diatribes on research in kaon nuclear physics and vicious and unprovoked attacks on other theorists. Unfortunately that doesn't leave much material, so I shall simply have to summarize the conference and give you my opinions concerning where we are and some things we might do next. I apologize in advance for the oversights and inaccuracies that are surely contained in this summary.

J.M. Eisenberg, in his introduction to the workshop, briefly reviewed the history of exclusive (p,π) and (π,p) reactions in nuclei with $A > 3$. He emphasized the great strides made in the quantity and quality of available experimental data in the last decade. In contrast to the experimental situation, Eisenberg tactfully described the theoretical work to date as not being an "unambiguous and unqualified" success. He suggested several questions that remained unresolved and that the theoreticians might be expected to elucidate in discussions during the conference. These included: the single-nucleon vs. two-nucleon mechanism, (strictly speaking, of course, there is no SNM) and the problem of double counting, the well-known Kisslinger-potential pion wave function catastrophe, various "relativistic" effects such as nucleon anti-nucleon contributions and/or the importance of the "lower" components in a Dirac phenomenology, lack of orthogonality between nucleon bound and continuum wavefunctions ["mercifully" these may only plague theories seriously near threshold ($T_\pi < 30$ MeV)--the main area of emphasis of the current IUCF experimental program], uncertainties regarding the high momentum components of nuclear wavefunctions and, of course, the still unknown importance of QCD, quark, bag, etc. considerations.

Eisenberg concluded his introduction by noting that one could gain some insight from an eikonal model of the (p,π) reaction perhaps applicable in the 500 MeV $< T_\pi <$ 800 MeV regime. It turns out that because of the energy transfer associated with (p,π) there is a significant longitudinal momentum transfer even at small forward angles, and thus there can be significant impact parameters deep in the nuclear interior allowing the reaction to be not surface dominated and thus relatively nondiffractive.

G. Jones reviewed the experimental situation for NN → πd and NN → NNπ, concentrating on the period after 1970. He pointed out the desirability of using orthogonal polynomials in the analysis of data instead of the earlier practice of using an expansion in powers of $\cos^2\theta$. The results shown by Jones indicate that, for a wide range of energies below ~600 MeV, the low multipoles associated with the differential cross sections and analysing powers for pp → dπ are well determined. The energy dependence of the higher partial waves as well as the complete spin dependence of the relevant amplitudes is currently under investigation and will require a continuing effort. Using the np → dπ° reaction there has been a demonstration that charge symmetry breaking effects are < 0.5%.

There are now some kinematically complete results, using polarized beams, for the reaction NN → NNπ. Since three body final states are involved, the experiments are more complex and the range of kinematic variables is much wider. Thus there needs to be more specificity by considering particular questions of interest. In this regard close cooperation with theorists at the outset in a given experiment is essential. Finally, the use of neutron beams to study the T=0 pion production mechanism was encouraged.

M. Banerjee reported on some theoretical results regarding the proper form for the πNN vertex to be used in pionic processes. It was found (independent of the choice of the underlying interaction Lagrangian) that the form $g_\pi(q^2)\gamma_5$ must be adopted. More generally the result is that, for any field theoretic multipoint function involving two nucleon lines and one or more boson lines, one should keep only that part which can survive on the mass shell of both nucleons. The remainder is not neglected but is automatically, properly, and naturally included in amplitudes involving more bosons. This result is contrary to the spirit of previously published work suggesting that experimental results can be used to motivate the adoption of a pseudo-vector πNN vertex instead of the pseudoscalar form. The result of the necessity for using the pseudo-scalar form is also applicable to exchange current treatments.

G. Miller noted that among the apparent "failures" of the "standard" nuclear model based upon two body forces and nucleon-only degrees of freedom are a) the saturation properties of nuclear matter, b) γd → np from 20 < E < 100 MeV and c) the description of the ^{3}He and ^{4}He charge density. (There are, of course, other examples associated with magnetic moments or spin degrees of freedom in nuclei). Several of these failures might be attributed to an inadequate description of the short distance part of the relative coordinate nucleon-nucleon wavefunction. Thus, one model of importance to study would be quark-bag models of hadrons-especially in regions where the bags overlap. Miller reviewed several bag models: a) the original M.I.T. bag with a bag radius of R ~ 1.0 fm. and with no pion cloud or provision for a one pion exchange potential and no PCAC relation, and b) a bag model proposed by Brown and Rho which addresses the deficiencies above but has the feature that $R \simeq .35$ fm. so the bags don't overlap at nuclear densities, and pion effects are quite important and thus cannot be treated perturbatively. Miller stressed that this last feature made reliable calculations using the model extremely difficult (impractical). The cloudy bag model (CBM) developed by Miller, Thomas, and Theberge adopts a Lagrangian essentially the same as the Brown bag. In the CBM the field equations are solved perturbatively in ϕ_π using QFT. For this case the bag radius is $R \sim .8$ FM, and pion effects while not negligible are small. For this model several properties of the nucleon and other elementary observables have been calculated and found to be in reasonable agreement with experiment. Using the idea of overlapping bags of quarks to fix up the short range behavior of the NN wavefunction, Miller and Kisslinger reported better agreement between theory and experiment for $\pi d \leftarrow pp$ than had previously been possible using a more conventional approach.

A. Thomas reviewed theories of pion production in N-N collisions. After a brief historical review of the early work, Thomas concentrated on three separate important efforts begun in the 1970s. During this decade a complete consistent set of integral equations was obtained. All orders of multiple scattering were included and such approximations as assuming fixed scatterers and adopting a completely nonrelativistic formalism were eliminated. For the reaction $pp \rightarrow \pi^+ d$, the theoretical-experimental agreement for the differential cross-section is good below 600 MeV and poor above 800 MeV. There is no general understanding of the discrepancies between theory and experiment for the various spin observables—more data will be available shortly. For the NN $\rightarrow$ NNπ sector, which is in its infancy, the theoretical-experimental agreement is good only for low momentum transfer to the produced Δ. But calculations are preliminary (the N$\Delta \rightarrow \Lambda$N transition potential is not now included). Of course a detailed study of the polarization observables is strongly motivated.

Thomas felt that the most important theoretical issue is the choice of a model at the nucleon level; should one consider models containing ρ, ω etc. as well as π, or π only with overlapping quark bags. The former is associated with hard form factors while the latter, because of the large bag size (in the MIT or CBM approaches) is associated with soft form factors. Two interesting theoretical problems would be a) to incorporate backward going pions into the few-body formalism and b) to obtain the coupled few body equations in a cloudy bag model. Finally, because the N-N exchanged pion can be on-shell near 600 MeV, pion production may not be quite as short ranged as one might have thought and thus it is important to do a good job using conventional techniques as well as considering quark bag overlap mechanisms.

A panel chaired by D. Koltun and including F. Coester, J. Niskanen, and H. Weber further discussed fundamental considerations in pionic nuclear reactions. Briefly, Koltun remarked that a sense of perspective is required--what is "fundamental" to one is a technical detail to another. Moreover, in model building, a fundamental generality may have to be balanced by the efficiency of calculation (or insight into the physics) associated with specific relevant degrees of freedom. Koltun remarked that the question of Δ's, π's, etc. vs. quark bags seemed similar to questions associated with the "model space" question met in many-body spectroscopy. Others remarked at the conference that the different approaches may continue to co-exist even after our understanding progresses, much as vibrational or rotational collective models are still used along with the nucleon shell model in nuclear structure physics.

Niskanen felt that advances in the pp = $d\pi^+$ problem would require hard work in the basics. As childhood ends in this area, he pointed to the following areas as requiring further investigation: the role of the Δ-coupling potential NN $\leftarrow$ NΔ, hard or soft form factors--the role of the ρ meson, NΔ $\rightarrow$ NΔ and ΔN, Δ "width" in presence of another nucleon, the possibility of real pions in couplings, treatment of non-resonant S and P wave rescattering, the proper inclusion of other isobars, a comparison of "rescattering" vs. the explicit Δ particle method, and of course the role of quarks.

Coester answered the question regarding the adequacy of the traditional two-body potential, nucleons only, model by pointing out that it wasn't sufficient. He also discussed the role of additional degrees of freedom. He mentioned the role of relativity depends on what people mean by including relativistic effects. (The introduction of relativistic kinematics or four momentum variables, allowance for particle creation and time orderings etc., is different from requiring a completely covariant theory). He wondered how one would build a practical phenomenology on covariant wavefunctions and what to do with a "perfect" vertex function.

Weber discussed some recent work on obtaining the N-N force from (overlapping) quark bags. He pointed out the "quark molecular approach" seems to be a failure. Weber then discussed the importance of the QCD vacuum and how its properties are essentially already properly included in a chiral bag model. Clearly developments in this area will influence the role quarks play in the short-distance N-N wave-functions in the future.

Although there are some problems in the comparison between theory and experiment for pion production in the two baryon system, the magnitude of the current problem associated with nuclear pionic processes really becomes apparent when one studies pion production from many-particle systems. A tip-off came when B. Hoistad reviewed the experimental situation in this area. First of all, he showed no theoretical curves in presenting the data. Such curves are usually associated with a demonstration of some minimal physical under-standing of the appropriate reaction mechanism! Unfortunately, as Hoistad stressed, there is just very little in the form of trends or general systematics, at least that has been identified up to now. This makes it very difficult even to find a simple mechanism for fitting the pion production data (completely forgetting about the justification for such a mechanism.) Hoistad showed that the cross section for (p,π^+) drops by ~2 orders of magnitude as one goes from He to Zr targets and that (p,π^-) reactions are suppressed ~1-2 orders of magnitude compared to (p,π^+) on the same targets. No typical shape correlation is observed for transitions believed to be "two-step" (going from supposedly closed shell nuclei to two particle-one hole states) compared to so-called single-nucleon transitions. Often the dips or minima observed in angular distri-butions occur for momentum transfers that vary with incident proton energy; such a result can signify an energy dependence in the reaction mechanism. Hoistad noted that total pion production cross sections do indicate a dependence on nuclear structure but that often there is apparently a subtle interplay between shapes and the nuclear reaction dynamics for angular distributions. At higher energies, there seems to be less nuclear structure dependence. As examples of strong energy variation, Hoistad mentioned the $^9Be(p,\pi^-)^{10}C_{g.s.}$ differential cross section. He also noted that for π^+ production $A(\theta)$ can be negative, stay flat or even go positive in the region $T_p \sim 200$ MeV. In this domain one has both structure and dramatic energy dependence. Finally, it was noted, the (p,d) and (π,p) data exhibit similar excitation functions. [The significance of this observation remains an open question.]

M.C. Green and M.A. Pickar discussed some selected recent results from IUCF. After briefly discussing the QQSP spectrometer, some conclusions regarding the ability of Coulomb effects to give significant energy dependences in the low energy ($T_p < 200$ MeV) range were noted. Recent data by Jacobs et al. were shown and compared with a simple model due to Vigdor et al. based on a two nucleon mechanism, which can relate the signs of asymmetries for special cases. Theory and experiment were in good agreement for the appropriate cases available. For the reaction $p + d \rightarrow {}^3He + \pi^\circ$ threshold production is not well described by phase space factors alone. It was also noted that even near threshold $\ell > 0$ pions play an important role.

Y. Le Bornec and N. Willis summarized recent results from Orsay using light ions with ~90 MeV/nucleon to produce pions from light targets. The results from ^{3}He(^{3}He,π^+)^{6}Li show large excitation of the low lying 3^+ state. For the reaction ^{4}He(^{3}He,π^+)^{7}Li it is not possible to resolve the lowest levels. Generally it was found that there is a strong decrease in the cross section with increasing projectile energy above ≈100 MeV/nucleon, projectile mass and target mass.

E. Rossle summarized results from SIN on (n,π) in the region $470 < T_n < 590$ MeV. He reported that a pure isobar model doesn't give the correct spectrum for np → π^+nn, perhaps indicating important non-resonant contributions. It was also reported that the ratio $R \equiv \sigma(nd \to {}^3He,\pi^-)/\sigma(nd \to t, \pi^\circ) \approx 1.76 \pm .09 \neq 2$. The deviation from two is expected from electromagnetic corrections in the associated wavefunctions.

P. Couvert from Saclay reported on the (p,π^+) reaction at higher energies. The p + d → t + π^+ reaction was studied for 700 MeV $\leq T_p \leq$ 1700 MeV at Saturne 2. A second bump was seen in the region ~3.5 GeV (CM for θ_π = 180°). It is conjectured that channels such as p + d → p + Δ + N, p + 2Δ, p + Δ (1650) → tπ^+ might be responsible for the structure seen. There was some discussion from the audience regarding the energy dependence of the resonance and speculations regarding dibaryon resonances. Couvert noted that coming attractions from Saclay include a) (SPES 1) A(p,π^+)A+1 on heavier nuclei and A(p,π^-)A+1 to study reaction mechanisms and spectroscopy; b) (SPES 4) (p,π) at higher energy $\leq$ 2.7 GeV; and c) (SPES 3) (p,pπ) (allowing perhaps Δ production or (p,ρ)!).

G.J. Lolos showed results from TRIUMF on the ^{9}Be(p,π^+)^{10}Be and ^{12}C(p,π^+)^{13}C reactions from T_p = 200-250 MeV. It was noted A(θ) was strongly dependent on energy, changing from negative to positive in going from 200 to 250 MeV in ^{12}C(p,π^+)^{13}C. Some continuum cross sections for (p,π^+) on ^{12}C for $T_p <$ 450 were also noted. I will talk more about the possible utility of such data at the end of this summary. Possibly coming in the future from TRIUMF, A(p,$\pi^\pm$)A+1 on light targets (^{10}B, ^{16}O, ^{12}C), pp → π^+ + d and NN → NNπ^+, and some pion absorption data.

H. Nann reported on an extensive three year pion production program at LAMPF using a wide range of targets from ^{1}H → ^{40}Ca. He discussed empirical trends of the data and demonstrated a lack of consistency in data to data comparisons for similar light targets.

Current models for treating pion production in complex nuclei were discussed by various proponents. Most of these models a) concentrate on selected aspects of the supposedly complicated pion production mechanism or b) are preliminary results from ambitious but uncompleted codes that eventually should allow the comparison of "realistic" theories with experiment. At the present stage of the game, a successful comparison between theory and experiment allows some joy while failure can be rationalized.

E.D. Cooper discussed the relativistic one-nucleon model using Dirac phenomenology. He pointed out that keeping the lower components without approximation in the relativistic wavefunction can be very important for the momentum transfer range appropriate for (p,π). At this stage in his investigations, using a DW "single nucleon" model, use of a pseudo-vector πNN vertex gives a qualitatively better fit to the data investigated thus far than using pseudo-scalar coupling. In light of the general results reported by Banerjee (discussed earlier), requiring PS coupling in a more general organization of pion-nucleus reactions, the meaning of Cooper's result was vigorously discussed. It will be interesting to include Cooper's one-nucleon mechanism with the two-nucleon models and compare with data. (Here one naturally will have to make some modifications to avoid some obvious double counting.) The extent to which it will be practical to expand this model to include more active nucleons, keeping the Dirac phenomenology, is a subject ripe for investigation.

B.D. Keister and L.S. Kisslinger discussed pion absorption/production and short-range phenomena. As one of three groups in attendance at this conference who are attempting to build a code for treating adequately the two-nucleon mechanism (TNM), their comments were largely based on results obtained to date in building this code. In motivating the TNM mechanism it was noted that $\pi d \to pp$ requires a pion rescattering term. K & K use a finite range for the pion propagator [not $\delta(r_1-r_2)/(k^2+m^2_\pi)$], having concluded that the zero range approximation was not useful. In their properly anti-symmetrized formalism a direct (target emission) and exchange (stripping) TNM with a virtual, intermediate Δ diagram appears. They have not yet included the non-locality of the Λ, Δ and π wavefunctions consistent with the isobar doorway model and a SNM pion plane wave piece. It is too early to see how well the theory will be able to reproduce the data. They report a troublesome, continu-ing sensitivity to changes in the πNN, $\pi N\Delta$ cutoff masses and distortions. Kisslinger pointed out that near threshold the TNM still involves a large momentum transfer for a single nucleon because there is often not an effective sharing of the momentum between the two active nucleons. Finally Kisslinger discussed a six-quark bag picture of the deuteron where the bag radius is fixed from the deuteron magnetic moment. This model seems to do a better job (at large q/short range) of reproducing the deuteron electrodisintegration data compared to standard meson exchange current calculations.

M. Dillig and F. Soga discussed a rescattering model of pion production which is similar in spirit to that reported by Keister and Kisslinger. This model also includes TNM direct and exchange terms, distorted waves, explicit pion rescattering, etc. Preliminary results were shown, but there are still many potentially important refinements to be included (such as the non-locality of the Δ and an isobar doorway consistent propagator). Dillig pointed out that it may be useful to separate pion production into three different incident proton energy regions a) $T_p > 600$ MeV, where eikonal models may be sufficient (forward scattering, strong absorption), b) $600 > T_p > 200$ MeV, where the Δ and Δ-medium effects dominate, and c) $T_p < 200$ MeV, where pion-nucleon s and p waves, pion and proton distortions, including Coulomb effects, must be carefully included. He also stressed the importance of actually carrying through the ambitious numerical calculations recently begun.

W. Gibbs concentrated his remarks on the target emission part of the (p,π) process. By investigating the ingredients of the nuclear structure input in the theory, he was able to make connection with the $Q^{JL}_{ph}(q)$ function, which can be enhanced at large q by precursor phenomena (Q is related to the driving pion field, enhancement of which leads to "precritical" phenomena). By incorporating those effects that renormalize the simple particle-hole wave functions one can effectively include effects due to "many nucleons". [These effects (if present or important) can be incorporated into any of the TNM models discussed at this conference.] Despite the fact that several potentially important effects are still not included in Gibbs' approach, he noted that for selected (p,π) experiments (where presumably the target emission piece dominates) the energy dependence of σ_T and angular distributions can be qualitatively understood. One should expand the model to include the various effects (relativity, non-locality of Δ, other diagrams, etc.) now missing.

On Saturday morning a panel, presided over by H. Fearing, and including H. McManus, J. Vary, and myself, was scheduled to discuss connections between the models reported at this conference. In fact there was a potpourri of topics briefly noted, including problems of overcounting, depending on what is included in ONM and TNM calculations, importance of studying other reactions, and the utility of being able to make connection with precursor and related many-body pionic effects. Fearing compared different models for the (p,π) reaction, noting that model calculations can differ in at least three ways: 1) including different "basic physics", 2) how the "basic physics" is approximated, and 3) the emphasis given to varying components of the calculations. Walker probed those theorists engaged in carrying out "realistic calculations" as to their timetable for obtaining "meaningful" predictions. It's going to be at least a year. Vary enlarged on the point that the relevant nuclear degrees of freedom depend on the probe and the momentum transfer. One may not be able to treat the new degrees of freedom perturbatively. McManus by describing Δ dominance in the (π^+,pp) and (π^-,pn) reactions on He, suggested all models including the Δ would give similar trends and ratios. Differences between theories would only be seen in detailed absolute magnitude calculations. [Of course Δ dominance models have not yet been shown to even reproduce trends - the meaningful calculations (see above) in the many-body environment haven't yet been completed.]

J.T. Londergan considered reactions related to pionic emission/absorption processes. He emphasized the processes (γ,p) and (p,γ) where large energy and/or three momentum transfer is delivered to the nucleus. He discussed the form and role of exchange current contributions to such processes as the total cross section for deuteron photo-distintegration, pointing out that even for $E_\gamma \sim 10$ MeV exchange current contributions (ECC) result in a 25% increase of σ_T. For $E_\gamma \sim 100$ MeV including ECC results in an increase of 260%! Londergan discussed the role of the Δ in (γ,p) reactions, noting that for light systems it gives important contributions for $q > 450$ MeV/c. (However, inclusion of the Δ does not seem to be required for fitting the $\gamma + {}^{40}Ca \to p + {}^{39}K_{gs}$ data). It is apparent that even for electromagnetic reactions the comparison between current theory and experiment is often not satisfactory at large energy and momentum transfer.

W.R. Wharton discussed high resolution studies of the (π^+,xx) reaction. He felt that the studies thus far represent perhaps only the tip of the iceberg. He conjectured that it may be common for more than two nucleons to be involved in the pion annihilation process. There are questions associated with enhancements due to clustering that will need to be resolved by choosing other targets and studying the momentum distributions of the ejected particles.

W. Benenson reported that Coulomb effects seem to be responsible for enhancement of the π^-/π^+ ratio for beam velocity pions in the reaction ${}^{20}Ne + NaF \to\ + \pi^\pm$ at E/A = 140 MeV. In the future, experiments of this kind will be done with heavier targets and projectiles. One would hope that the results will not be explainable in terms of purely Coulombic effects.

M. Huber summarized some recent work on coherent pion production in nucleus-nucleus collisions [$A_1 + A_2 \to \pi + B_{fusion}$]. He reviewed a theoretical model that allows one to go from a nucleus-nucleus relative coordinate degree of freedom to the pion degree of freedom--avoiding thermalization of the original kinetic energy. Using a cluster expansion to label the ${}^3He + {}^3He \to {}^6Li + \pi^+$ reaction as allowed while the ${}^3He + {}^6Li \to {}^9Be + \pi^+$ reaction is "forbidden", one can understand, in this model, the large rate associated with the former compared to the latter reaction. Huber suggested reactions of this type should first be studied to understand the associated reaction mechanism and then later used to study high spin states, create new isotopes, etc. The question was raised whether $A_1 + A_2 \to A + \gamma$ has a similar reaction mechanism.

Finally, J.-F. Germond discussed pion-induced fission. He noted that for $\pi^- + {}^6\text{Li}$, there is a significant contrtibution to the reaction cross section from $\to t + t$ and $\to t + d + n$. He stressed the importance of having a theory that could explain the >2 body final states as well as the two-body breakup.

Before concluding with my own suggestions for future experimental and theoretical research I would like to mention some comments I received reflecting others' views on the information presented at the conference. I promised that no names would be associated with these selected and slightly edited comments.

A proton in free space cannot emit a real pion. Thus there is no one-nucleon mechanism. What this means for practical computations is all diagrams must be calculated, and calculated well. People who do the so-called 'one-nucleon model' should not neglect target emission. People who do the 'two-nucleon model' should not neglect the one-nucleon piece. The latter must be calculated relativistically. It should be recognized, though, that the projectile emission term (on the TNM) for (p,π^+) to a single-particle state can be reliably calculated in a DWBA. In TNM calculations Niskanen's warning should be heeded --- one must include the kinetic energy of the Δ as an operator, not as a number.

Experiments should be chosen for their specific physics motivation. A good example is the work of Jacobs et al. on the 'two-nucleon' mechanism. Use of pion production to study symmetries such as time reversal and charge syummetry breaking seems very promising. Surveys are not enough.

For the (p,π) exclusive reaction, only now has theory begun to put together the main pieces we have known for some time ought to matter. Much still has to be sorted out regarding not only fits to data but also what is a consistent calculation and wht may be safely neglected.

Remark A: This is a rare situation for theorists at a conference where, having been unable to explain data, they cannot reasonably say 'more experiments are required'. Remark B: There is a need for data, partricularly in the $200 < T_p < 400$ MeV region.

There are lots of interesting absorption and production reactions other than the (p,π) exclusive case! They get disproportionately less study!

I think it is particularly encouraging to see several relatively sophisticated two-nucleon calculations in progress. It will be interesting to see how their results compare and how successful each is in reproducing the energy dependence of the reaction, which one hopes will reflect the explicit presence of the Δ.

The new data on (p,π) and related reactions is truly revolutionary. Experiment is now far ahead of theory. The Δ region offers many advantages and data should be taken in this region ($T_p = 300 - 600$ MeV), as it removes one of the greatest difficulties with present attempts of detailed calculations. More theory effort is badly needed to test even present ideas.

Pion absorption and production involves complicated momentum sharing and short range behavior. The quantitative information, such as 'ρ-meson effects', now gives us the opportunity to make new connections to quark/QCD physics.

We are getting some new theoretical information about the pion absorption operator: Banerjee et al.'s observation and Gibbs' calculations were most valuable.

It was nice to be reminded a number of times (especially by Coester) of what relativity means, for relativity (and perhaps anti-particles) are involved in pion absorption/production.

My own brief summary comments, which will repeat some points made above, follow:

EXPERIMENT

a) The recent data on total cross sections, differential cross sections and analyzing powers for $(\vec{p},\pi^{\pm})$ are generally of high quality and very impressive.

b) There will be, in the near future, new data giving more information on the spin dependence of the amplitudes. Three-body final state data ($N + N \to N + N + \pi$) will be studied further.

c) It would be useful to study (N,Nπ) on complex targets with several hundred MeV incident nucleon energies (in the TRIUMF energy range, considerably above Indiana energies; unfortunately the appropriate spectrometers for doing coincidence experiments efficiently at TRIUMF are not yet available). Such studies would be useful in further testing proposed production mechanisms in a region where the two-body input might be used in an impulse approximation.

d) Performing (p,π) experiments with associated nuclear excitation substantially above nucleon threshold might provide tests of proposed reaction mechanisms, eliminating some structure sensitivity. Here one might try to compare theory and experiment for final nuclear excitations of 30 → 60 MeV averaging over ~5 MeV bins.

e) Experiments that can be motivated on the basis of some particular selectivity are useful. [One example is the work of Jacobs et al. on the relative sign of $A(\theta)$ for particular J=1/2, 3/2 states as a test of the two nucleon mechanism.] Other possibilities include $(p,\pi^{+,0,-})$ experiments leading to final nuclei in the same isotopic multiplet. Comparison of the different cross sections for π^+, π^0, π^- production to such states can better focus on reaction mechanism questions involving "multi-step" processes because the nuclear structure should be the same for all final states belonging to the same multiplet. Identification of the appropriate state in a given spectrum can be very difficult, however, because of the contamination of unwanted isospin final states.

f) Further "surveys" for discrete final states not meeting criteria in the spirit of e) above do not appear useful at this time.

THEORY

a) Simple "single nucleon" models are very sensitive to input parameters and generally do not fit the data.

b) Large computer codes for realistically including the two-body mechanism are currently under construction by at least three groups (Keister-Kisslinger, Conte-Dillig-Soga, and Iqbal-Walker). It is very important for this partially completed work to be extended. There is much difficult work remaining and the various groups are encouraged to cooperate and keep each other informed of pitfalls and progress in their endeavors. [Of course each group has some unique focus of its own.]

c) The large momentum transfer associated with the production process accentuates the small relative distance part of the nucleon wavefunction. This means that for calculating at least the "one nucleon" piece of the process, the relativistic approach discussed by Cooper seems attractive. In addition, the sensitivity to hard/soft form-factors and the importance of bag-models for the six-quark system, etc. need to be studied further.

d) It would be interesting to incorporate some of the coupled channel formalism (used in $p + p \rightarrow d + \pi^+$, etc.) into the many nucleon environment (perhaps using a simple Fermi-gas model for Pauli blocking effects), and to calculate $(p,p\pi)$ or (p,π) in the continuum.

e) Because of the significant energy transfer involved in the pion production process, the use of static pion propagators, unfortunately, will not be adequate for treating some of the TNM diagrams.

So, in summary, the data is beautiful; but our lack of understanding of it is embarrassing. It is premature to suggest that current two-nucleon models will succeed or fail in elucidating the data, but the detailed calculations required to answer this question are long overdue. Several groups are committed to the improvement of that situation; and, although there is a critical manpower shortage, I remain optimistic that a meaningful comparison between well-motivated theory and experiment will be possible in the near future.

AUTHOR INDEX

Antonuk, L. 379
Bacher, A.D. 169,199,216,228
Benerjee, M.K. 37,240,270
Barnes, P. 62
Benenson, W. 38
Betz, M. 65
Blankleider, B. 65
Cheung, C.-Y. 60
Coester, F. 62,92,101
Conte, J. 275
Cooper, E.D. 231,318
Couvert, P. 187,216,228
Dillig, M. 36,241,275,289
Dutty, W. 171
Eisenberg, J.M. 3,45,240,271
Falk, W. 216
Fearing, H.W. 319,333
Franz, J. 171
Germond, J.-F. 411
Gibbs, W.R. 297
Green, M.C. 131
Hoistad, B. 105
Huber, M.G. 199,389
Hwang, P. 61,240,271
Hynes, M. 59,270
Iqbal, J. 273
Jones, G. 15,199,294
Keister, D.B. 265,318
Kisslinger, L.S. 13,45,92,199,240
    243,271,295,379
Klingenbeck, L.S. 389
Koltun, D. 61,92,93,101,369,379
    410
Le Bornec, Y. 155
Lehmann, L. 171
Liu, K.-F. 58,272
Lichtenberg, D.B. 58,273
Lolos, G.J. 201
Londergan, J.T. 216,270,339
Miller, G.A. 47,216,272,318
Nann, H. 219
Nicklas, G. 171
Niskanen, J.A. 36,65,101,295
Noble, J.V. 12,59,169,216,240,295
    318
Pickar, M.A. 143
Pollock, R.E. 216
Rossle, E. 36,171
Schmitt, H. 171
Sherif, H.S. 231

Silbar, R.R. 36,92,272,296
Soga, F. 275,289
Sternheim, M.M. 418
Thomas, A.W. 46,65,410
Vary, J. 59
Walker, G.E. 270,370,419
Weber, H.J. 101
Wharton, W. 369,371,418
Wilkin, C. 411
Willis, N. 155

AIP Conference Proceedings

		L.C. Number	ISBN
No.1	Feedback and Dynamic Control of Plasmas	70-141596	0-88318-100-2
No.2	Particles and Fields - 1971 (Rochester)	71-184662	0-88318-101-0
No.3	Thermal Expansion - 1971 (Corning)	72-76970	0-88318-102-9
No.4	Superconductivity in d-and f-Band Metals (Rochester, 1971)	74-18879	0-88318-103-7
No.5	Magnetism and Magnetic Materials - 1971 (2 parts) (Chicago)	59-2468	0-88318-104-5
No.6	Particle Physics (Irvine, 1971)	72-81239	0-88318-105-3
No.7	Exploring the History of Nuclear Physics	72-81883	0-88318-106-1
No.8	Experimental Meson Spectroscopy - 1972	72-88226	0-88318-107-X
No.9	Cyclotrons - 1972 (Vancouver)	72-92798	0-88318-108-8
No.10	Magnetism and Magnetic Materials - 1972	72-623469	0-88318-109-6
No.11	Transport Phenomena - 1973 (Brown University Conference)	73-80682	0-88318-110-X
No.12	Experiments on High Energy Particle Collisions - 1973 (Vanderbilt Conference)	73-81705	0-88318-111-8
No.13	π-π Scattering - 1973 (Tallahassee Conference)	73-81704	0-88318-112-6
No.14	Particles and Fields - 1973 (APS/DPF Berkeley)	73-91923	0-88318-113-4
No.15	High Energy Collisions - 1973 (Stony Brook)	73-92324	0-88318-114-2
No.16	Causality and Physical Theories (Wayne State University, 1973)	73-93420	0-88318-115-0
No.17	Thermal Expansion - 1973 (lake of the Ozarks)	73-94415	0-88318-116-9
No.18	Magnetism and Magnetic Materials - 1973 (2 parts) (Boston)	59-2468	0-88318-117-7
No.19	Physics and the Energy Problem - 1974 (APS Chicago)	73-94416	0-88318-118-5
No.20	Tetrahedrally Bonded Amorphous Semiconductors (Yorktown Heights, 1974)	74-80145	0-88318-119-3
No.21	Experimental Meson Spectroscopy - 1974 (Boston)	74-82628	0-88318-120-7
No.22	Neutrinos - 1974 (Philadelphia)	74-82413	0-88318-121-5
No.23	Particles and Fields - 1974 (APS/DPF Williamsburg)	74-27575	0-88318-122-3
No.24	Magnetism and Magnetic Materials - 1974 (20th Annual Conference, San Francisco)	75-2647	0-88318-123-1
No.25	Efficient Use of Energy (The APS Studies on the Technical Aspects of the More Efficient Use of Energy)	75-18227	0-88318-124-X

No.26	High-Energy Physics and Nuclear Structure - 1975 (Santa Fe and Los Alamos)	75-26411	0-88318-125-8
No.27	Topics in Statistical Mechanics and Biophysics: A Memorial to Julius L. Jackson (Wayne State University, 1975)	75-36309	0-88318-126-6
No.28	Physics and Our World: A Symposium in Honor of Victor F. Weisskopf (M.I.T., 1974)	76-7207	0-88318-127-4
No.29	Magnetism and Magnetic Materials - 1975 (21st Annual Conference, Philadelphia)	76-10931	0-88318-128-2
No.30	Particle Searches and Discoveries - 1976 (Vanderbilt Conference)	76-19949	0-88318-129-0
No.31	Structure and Excitations of Amorphous Solids (Williamsburg, VA., 1976)	76-22279	0-88318-130-4
No.32	Materials Technology - 1976 (APS New York Meeting)	76-27967	0-88318-131-2
No.33	Meson-Nuclear Physics - 1976 (Carnegie-Mellon Conference)	76-26811	0-88318-132-0
No.34	Magnetism and Magnetic Materials - 1976 (Joint MMM-Intermag Conference, Pittsburgh)	76-47106	0-88318-133-9
No.35	High Energy Physics with Polarized Beams and Targets (Argonne, 1976)	76-50181	0-88318-134-7
No.36	Momentum Wave Functions - 1976 (Indiana University)	77-82145	0-88318-135-5
No.37	Weak Interaction Physics - 1977 (Indiana University)	77-83344	0-88318-136-3
No.38	Workshop on New Directions in Mossbauer Spectroscopy (Argonne, 1977)	77-90635	0-88318-137-1
No.39	Physics Careers, Employment and Education (Penn State, 1977)	77-94053	0-88318-138-X
No.40	Electrical Transport and Optical Properties of Inhomogeneous Media (Ohio State University, 1977)	78-54319	0-88318-139-8
No.41	Nucleon-Nucleon Interactions - 1977 (Vancouver)	78-54249	0-88318-140-1
No.42	Higher Energy Polarized Proton Beams (Ann Arbor, 1977)	78-55682	0-88318-141-X
No.43	Particles and Fields - 1977 (APS/DPF, Argonne)	78-55683	0-88318-142-8
No.44	Future Trends in Superconductive Electronics (Charlottesville, 1978)	77-9240	0-88318-143-6
No.45	New Results in High Energy Physics - 1978 (Vanderbilt Conference)	78-67196	0-88318-144-4
No.46	Topics in Nonlinear Dynamics (La Jolla Institute)	78-057870	0-88318-145-2
No.47	Clustering Aspects of Nuclear Structure and Nuclear Reactions (Winnepeg, 1978)	78-64942	0-88318-146-0
No.48	Current Trends in the Theory of Fields (Tallahassee, 1978)	78-72948	0-88318-147-9
No.49	Cosmic Rays and Particle Physics - 1978 (Bartol Conference)	79-50489	0-88318-148-7

No.	Title		
No. 50	Laser-Solid Interactions and Laser Processing - 1978 (Boston)	79-51564	0-88318-149-5
No. 51	High Energy Physics with Polarized Beams and Polarized Targets (Argonne, 1978)	79-64565	0-88318-150-9
No. 52	Long-Distance Neutrino Detection - 1978 (C.L. Cowan Memorial Symposium)	79-52078	0-88318-151-7
No. 53	Modulated Structures - 1979 (Kailua Kona, Hawaii)	79-53846	0-88318-152-5
No. 54	Meson-Nuclear Physics - 1979 (Houston)	79-53978	0-88318-153-3
No. 55	Quantum Chromodynamics (La Jolla, 1978)	79-54969	0-88318-154-1
No. 56	Particle Acceleration Mechanisms in Astrophysics (La Jolla, 1979)	79-55844	0-88318-155-X
No. 57	Nonlinear Dynamics and the Beam-Beam Interaction (Brookhaven, 1979)	79-57341	0-88318-156-8
No. 58	Inhomogeneous Superconductors - 1979 (Berkeley Springs, W.V.)	79-57620	0-88318-157-6
No. 59	Particles and Fields - 1979 (APS/DPF Montreal)	80-66631	0-88318-158-4
No. 60	History of the ZGS (Argonne, 1979)	80-67694	0-88318-159-2
No. 61	Aspects of the Kinetics and Dynamics of Surface Reactions (La Jolla Institute, 1979)	80-68004	0-88318-160-6
No. 62	High Energy e^+e^- Interactions (Vanderbilt, 1980)	80-53377	0-88318-161-4
No. 63	Supernovae Spectra (La Jolla, 1980)	80-70019	0-88318-162-2
No. 64	Laboratory EXAFS Facilities - 1980 (Univ. of Washington)	80-70579	0-88318-163-0
No. 65	Optics in Four Dimensions - 1980 (ICO, Ensenada)	80-70771	0-88318-164-9
No. 66	Physics in the Automotive Industry - 1980 (APS/AAPT Topical Conference)	80-70987	0-88318-165-7
No. 67	Experimental Meson Spectroscopy - 1980 (Sixth International Conference, Brookhaven)	80-71123	0-88318-166-5
No. 68	High Energy Physics - 1980 (XX International Conference, Madison)	81-65032	0-88318-167-3
No. 69	Polarization Phenomena in Nuclear Physics - 1980 (Fifth International Symposium, Santa Fe)	81-65107	0-88318-168-1
No. 70	Chemistry and Physics of Coal Utilization - 1980 (APS, Morgantown)	81-65106	0-88318-169-X
No. 71	Group Theory and its Applications in Physics - 1980 (Latin American School of Physics, Mexico City)	81-66132	0-88318-170-3
No. 72	Weak Interactions as a Probe of Unification (Virginia Polytechnic Institute - 1980)	81-67184	0-88318-171-1
No. 73	Tetrahedrally Bonded Amorphous Semiconductors (Carefree, Arizona, 1981)	81-67419	0-88318-172-X
No. 74	Perturbative Quantum Chromodynamics (Tallahassee, 1981)	81-70372	0-88318-173-8

No. 75	Low Energy X-ray Diagnostics-1981 (Monterey)	81-69841	0-88318-174-6
No. 76	Nonlinear Properties of Internal Waves (La Jolla Institute, 1981)	81-71062	0-88318-175-4
No. 77	Gamma Ray Transients and Related Astrophysical Phenomena (La Jolla Institute, 1981)	81-71543	0-88318-176-2
No. 78	Shock Waves in Condensed Matter - 1981 (Menlo Park)	82-70014	0-88318-177-0
No. 79	Pion Production and Absorption in Nuclei - 1981 (Indiana University Cyclotron Facility)	82-70678	0-88318-178-9